Hardware Security

Mark Tehranipoor • Kimia Zamiri Azar •
Navid Asadizanjani • Fahim Rahman •
Hadi Mardani Kamali • Farimah Farahmandi

Hardware Security

A Look into the Future

Springer

Mark Tehranipoor
University of Florida
Gainesville, FL, USA

Navid Asadizanjani
University of Florida
Gainesville, FL, USA

Hadi Mardani Kamali
University of Florida
Gainesville, FL, USA

Kimia Zamiri Azar
University of Florida
Gainesville, FL, USA

Fahim Rahman
University of Florida
Gainesville, FL, USA

Farimah Farahmandi
University of Florida
Gainesville, FL, USA

ISBN 978-3-031-58689-7   ISBN 978-3-031-58687-3   (eBook)
https://doi.org/10.1007/978-3-031-58687-3

This Springer imprint is published by the registered company Springer Nature Switzerland AG
The registered company address is: Gewerbestrasse 11, 6330 Cham, Switzerland

If disposing of this product, please recycle the paper.

Dedicated to:
Our students and post-docs for their
dedication to the field of hardware security!

Preface

As of 2022, the global semiconductor market was valued at 580.13 billion US dollars, and it is expected to reach 1,033.5 billion US dollars by 2031. In such a demanding and dominated market, the modern semiconductor supply chain has grown tremendously throughout the past two decades. The globalization of the integrated circuit (IC) supply chain has been a significant trend in recent decades, which has reshaped the semiconductor industry by enabling companies to source components, materials, and services from various locations around the world. While the globalization of the supply chain has brought numerous benefits, such as reduced time-to-market, tapped into specialized expertise, etc., it has also allowed the inclusion of several untrustworthy entities. The growth of the semiconductor industry, as well as the involvement of more untrustworthy entities, has contributed to a rise in hardware security concerns. In addition, the complexity of design for newer generation of ICs and systems-on-chip (SoCs) has increased. It is becoming even worse as the semiconductor industry move toward heterogeneous integration (HI) of already fabricated and packaged dies (chiplets) into System-in-Package (SiP) architectures. With such high complexity, not only are microelectronic devices becoming more and more untrustworthy, but they also introduce unintentional and additional security vulnerabilities. Such circumstances make conventional testing, verification, and validation methodologies insufficient to verify trustworthiness. To overcome this challenge, more advanced security-oriented methods, spanning design, testing, verification, and validation, must be employed to identify and address these new security vulnerabilities. It is, therefore, necessary for semiconductor industry, academia, and the government to be familiar with the future trend in this field prior to developing next-generation hardware and microelectronic security solutions.

This book provides a look into the future potential problems (as well as solutions) of hardware and microelectronics security, with an emphasis on security-aware design, security verification and validation, building trusted execution environments, and physical assurance. The book highlights some critical questions that must be answered in the domain of hardware and microelectronics security in the next 5–10 years: (i) What must be done when considering security from the IP level to the system level? (ii) What would be the future of IP and IC protection against emerging threats? (iii) How security solutions could be migrated/expanded from SoC to SiP? (iv) What are the advances in power side-channel analysis with emphasis on post-quantum cryptography algorithms? (v) How can a digital twin be used to manage a semiconductor lifecycle securely? (vi) How physical assurance will look like with emerging technologies? The main aim of this book is to serve as a comprehensive roadmap for new learners and educators navigating the evolving research directions in the domain of hardware and microelectronic security. Overall, throughout 11 chapters, the book provides numerous frameworks, countermeasures, security evaluations, and roadmaps for the future of hardware security, where each chapter is briefly described below:

Chapter 1—"Quantifiable Assurance in Hardware"—provides a comprehensive demonstration of hardware design characteristics and their impact on security measurement. This chapter revisits and investigates the potential directions for building quantified and formal-oriented security measurement, particularly at the system level during the earlier phases of a design (behavioral and gate-level design abstraction).

Chapter 2—"Advances in Logic Locking"—focuses on current and future potential directions of hardware protection techniques, with an emphasis on logic locking. In this chapter, by holistically reviewing the direction and outcome of state-of-the-art logic locking techniques through the last decade, on both the attack and defense, the potential future solutions, as well as their advantages/disadvantages, are evaluated more comprehensively.

Chapter 3—"Rethinking Hardware Watermark"—first provides an extensive overview of the state-of-the-art IP watermarking countermeasures, from passive to active. This chapter then challenges the existing assumptions and characteristics of the corresponding solutions and demonstrates new path forward for potential future reliable IP watermarking approaches suitable for complex SoCs, as well as future SiPs, in which IPs/chiplets are deeply embedded.

Chapter 4—"SoC Security Verification Using Fuzz, Penetration, and AI Testing"—first demonstrates a comprehensive overview of the requirements that must be realized as the fundamentals of SoC security verification. This chapter also puts forward a realization of utilizing self-evolving techniques, such as fuzz, penetration, and AI testing, as a means of security verification in futuristic solutions.

Chapter 5—"Runtime SoC Security Validation"—emphasizes the importance of dynamicity in the hardware threat models and focuses on emerging threats, like zero-day attacks. The objective of this chapter is to establish a set of quality criteria for reconfigurable SoC security monitoring, followed by potential solutions that have been assessed separately according to these quality criteria.

Chapter 6—"Large Language Models for SoC Security"—explores the possibility of using Generative Pre-trained Transformers (GPTs) to enhance SoC security, making it more efficient and adaptable. By integrating Large Language Models (LLMs), this chapter introduces new possibilities and challenges in securing complex SoCs. By presenting and analyzing case studies and experiments, and offering practical guidelines, this chapter also discusses the prospects and challenges of using LLMs in SoC security verification.

Chapter 7—"Power Side-Channel Evaluation in Post-quantum Cryptography"—explores and evaluates potential side-channel leakages of existing post-quantum cryptography (PQC) implementations using a completely automated side-channel evaluation platform at both pre- and post-silicon levels. Using this evaluation, this chapter delineates challenges and approaches for future research directions for building more robust PQC implementations.

Chapter 8—"Digital Twin for Secure Semiconductor Lifecycle Management"—puts forward a realization of intertwined relationships of security vulnerabilities with data available from the silicon lifecycle and formulates different components of an AI-driven digital twin framework. By demonstrating how advanced modeling techniques can perform relational learning to identify such attacks, this chapter provides potential future research directions for the realization of the digital twin framework to enable secure semiconductor lifecycle management.

Chapter 9—"Secure Physical Design"—focuses on the importance of physical attributes (e.g., placement and routing) in the security of hardware designs. Then, this chapter demonstrates a secure physical design roadmap for enabling end-to-end trustworthy IC design flow. By discussing the potential use of AI/ML engines for security establishment at the layout level, this chapter enumerates the potential future directions and challenges in this domain.

Chapter 10—"Secure Heterogeneous Integration"—focuses on the security threats as the bi-product of the industry shift toward 2.5D/3D designs using heterogeneous integration (HI). This chapter investigates a three-tier security strategy to ensure secure HI, addressing supply chain security concerns across chiplet, interposer, and System-in-Package levels. This chapter also explores trust validation methods and attack mitigation for each integration level and presents a roadmap for advancing secure HI solutions.

Chapter 11—"Materials for Hardware Security"—explores the opportunities and challenges in material usage for hardware assurance in the context of HI packaging development. This chapter also discuss pre- and post-packaging hardware assurance solutions, including non-destructive inspection techniques such as infrared, X-ray imaging, and terahertz time-domain spectrum. Additionally, it investigates the integration of MEMS and NEMS for pre-packaging hardware assurance.

<table>
<tr><td>Gainesville, FL, USA</td><td>Mark Tehranipoor</td></tr>
<tr><td>December, 2023</td><td>Kimia Zamiri Azar</td></tr>
<tr><td></td><td>Navid Asadizanjani</td></tr>
<tr><td></td><td>Fahim Rahman</td></tr>
<tr><td></td><td>Hadi Mardani Kamali</td></tr>
<tr><td></td><td>Farimah Farahmandi</td></tr>
</table>

Contents

Chapter 1
Quantifiable Assurance in Hardware

1.1 Motivational Example and Corresponding Threat Models

Systems-on-Chip (SoCs) have been pervasive in current and future electronic products. Almost every computing system (e.g., mobile phones, payment gateways, IoT devices, medical equipment, automotive systems, avionic devices, etc.) is built around the SoC. A modern SoC contains a wide range of sensitive information, commonly referred to as security assets (e.g., keys, biometrics, personal info, etc.). The assets are necessary to build security mechanisms and need to be protected from a diverse set of attack models, such as IP piracy [1], power side-channel analysis [2], fault injection, malicious hardware attack [3], supply chain attack [4], etc.

Often, hardware designs cannot be patched to fix security vulnerabilities after the design is fabricated and deployed in the field. Any changes to the hardware design after fabrication would require redoing the entire design, which costs money and time in addition to losing market share. Hence, it is crucial to fix the vulnerabilities at the earlier design phases, such as RTL or gate level. Additionally, a quantifiable assurance, such as quantitative estimation and measurement of the security, is much needed to evaluate the security of the whole design.

While extensive research has been done in the testing domain [5–13], security verification is still on the rise, and the modern Electronic Design Automation (EDA) tools still lack the notion of automatic security evaluation and optimization [14–16]. In most cases, security is considered an afterthought. To address this, researchers have developed different detection [17–20] and defense [21–28] mechanisms to build secure designs. At the same time, several methodologies have been developed by the hardware security community to assess the defense mechanisms quantitatively. However, all these approaches can be only applied to intellectual property (IP) blocks, which are the primary building blocks of the platform. Note that the terms "Systems-on-Chip (SoCs)" and "Platform" will be used interchangeably in the rest of this chapter. The security of an IP does not necessarily remain the

Fig. 1.1 Hierarchical approach for platform-level (system) security estimation

same after it is integrated into the platform. During the integration, additional parameters are introduced that directly impact the security of an IP at the platform level. These parameters either degrade or upgrade the security of the IP after being placed in the platform. It leads to the development of new platform-level security estimation and measurement methodologies. While the security estimation methods should accurately estimate the silicon-level security of the platform at the earlier design stages, the measurement approach should accurately measure the platform-level security by exhaustive simulation or emulation. Another important aspect is that threats are usually considered individually during mitigation. Mitigating vulnerabilities for one threat can potentially make the design more vulnerable to another threat. Thus, security optimization is needed to obtain the best possible collective security considering the diverse threat model. A high-level overview of the platform-level security estimation approach is shown in Fig. 1.1. It follows the bottom-up approach starting with the IP-level security estimation following the platform integration and the platform-level security estimation.

This chapter tries to demonstrate how to build a comprehensive approach to estimate and measure the platform-level security at the earlier phases of a design. It first discusses the IP-level security metrics and different IP-level parameters that contribute to the security at the IP level for five different threats: (1) IP piracy, (2) power side-channel analysis, (3) fault injection, (4) malicious hardware, and (5) supply chain. Then, the transition from IP to platform will be discussed, and the definition of different parameters during this transition will be depicted. Then, an approach for the platform-level security estimation and measurement will be discussed, followed by its evaluation over two case studies for the estimation and measurement of resiliency against IP Piracy and the Power Side-Channel (PSC) analysis attacks at the platform level.

Now a motivating example will be discussed showing why there is a need to develop SoC-level security estimation and measurement techniques. For this example, we consider the power side-channel (PSC) analysis as the threat model of the interest. Figure 1.2 shows the transition from IP to the platform. Let us have a look at a cryptographic core, IP_2, shown in Fig. 1.2, assuming that IP_2's security against the PSC vulnerability has already been evaluated as a standalone module.

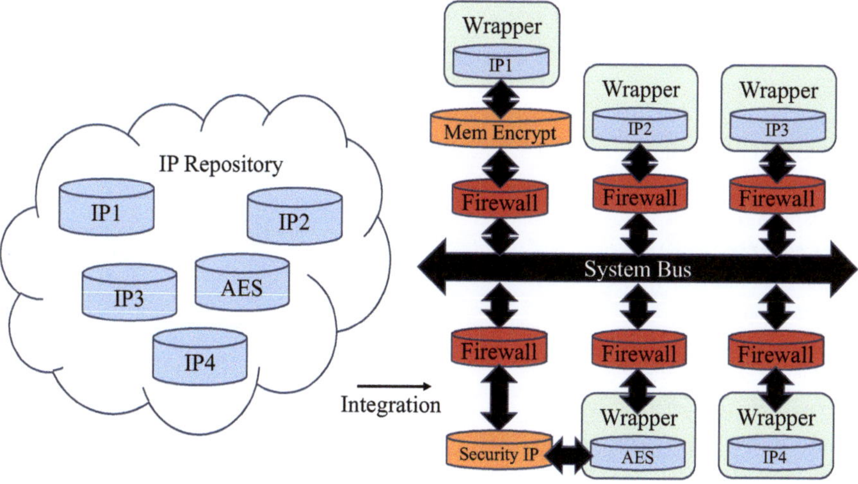

Fig. 1.2 System integration flow: transition from IPs to the platform (system)

Now, we want to check if the IP_2's robustness against the PSC attacks remains the same after the IP is integrated into a system. If not, does it decrease or increase, and by how much? To understand how SoC parameters affect the PSC robustness on IP_2, let us consider PDN. The first step to perform power side-channel leakage analysis on the crypto core, IP_2, is to collect the power traces of the IP. However, depending on how PDN is implemented at the SoC level, different power rails are shared among different IPs. Therefore, it is impossible to collect power traces only from IP_2 at the platform level. On the other hand, one assumption while assessing the IP-level PSC robustness is that the attacker has the privilege to measure the power trace of the IP itself. However, this assumption cannot be applied to the platform, since the attacker can only collect the power trace of the entire platform, which is the accumulated power trace of all active IPs along with IP_2. This might make the collected power traces noisier than the power traces collected from standalone IP_2, potentially making the attackers' job more challenging. In other words, the security of IP_2 against PSC attacks might be stronger when it is integrated into the SoC unlike what is measured at the IP level. Thus, the platform-level security of an IP is subject to change depending on different parameters introduced during the transition from IP to the platform. So, considering IP-level parameters demonstrated in Table 1.1, a list of such parameters affecting the platform-level security for five different threats is included in Table 1.2. An accurate estimation greatly affects the design choice. For example, in the case of PSC assessment, if the estimated SoC-level security of IP2 is high enough, we may not shield the IP through additional countermeasures, which significantly saves design effort, area, power, and cost. This example implies that the transition from IP to platform affects the security at the platform level by introducing different parameters, leading to the need for the platform-level security estimation and measurement methodology.

Table 1.1 IP-level parameters for hardware security measurement and estimation

Parameters	Threat				
	IP Piracy	PSC	Fault Inject	HW Trojan	Supply chain
Locking key size	✓	–	–	–	–
Locking key distribution	✓	–	–	–	–
Output corruption	✓	–	–	–	–
Locking mechanism	✓	–	–	✓	–
Key error rate (KER)	✓	–	–	–	–
Input error rate (IER)	✓	–	–	–	–
Mode of operation	–	✓	–	–	–
Cryptographic key size	–	✓	–	–	–
# clock cycles per crypto. opr.	–	✓	–	–	–
Data-driven switching activity	–	✓	–	–	–
Isolated switching activity	–	✓	–	–	–
Gate type/sizing	–	–	✓	–	–
Nearby cells	–	✓	✓	–	–
Static probability	–	–	–	✓	–
Signal rate	–	–	–	✓	–
Toggle rate	–	–	–	✓	–
Fan-out	–	–	–	✓	–
Immediate fan-in	–	–	–	✓	–
Lowest controllability of inputs	–	–	–	✓	–

Now we explain the threat models for the five different attacks. We describe how these attacks are performed at the IP level. This lays a solid foundation for presenting security estimation and measurement methods at the IP and platform levels.

IP Piracy This attack refers to making an illegal copy or cloning of the IP. Usually, an attacker makes little or no modification to the original IP and sells it to a chip designer claiming its ownership. Logic locking has been a popular technique to protect IP piracy [29]. In this approach, the design is locked by inserting a set of key gates at the gate-level netlist. Only with the correct value of the key inputs, the IP is expected to function correctly. For all other wrong key values, a corrupted output will be produced so that the attacker cannot retrieve the functionality of the IP [30–32]. In recent years, Boolean Satisfiability Checking-based attack (SAT attack), and its variants, has been a very efficient technique to retrieve the correct key from a locked gate-level netlist [33–36]. It has been shown that the SAT-based attack can retrieve the correct key in a few hours for a considerably large keyspace. This attack utilizes the modern SAT solver and iteratively solves a SAT formula, Conjunctive Normal Format (CNF) of the design, and prunes out the wrong keys till the correct key is retrieved. The attack model is summarized as follows:

Table 1.2 IP and platform-level design parameters for IP Protection, Fault Injection, Power Side Channel, Malicious Hardware, and Supply Chain

Threat	IP-level parameters	Platform-level parameters
IP Piracy	Locking key size	Platform-level testing architecture
	Locking key distribution	Bypassing Capability
	Output corruption	Wrapper Locking
	Locking mechanism	Accessibility to the internal flip-flops
	Key error rate (KER)	Compression Ratio
	Input error rate (IER)	Scan Obfuscation
Power Side Channel	Mode of operation	On-chip voltage regulator
	Cryptographic key size	Pipeline depth
	Number of clock cycles for one cryptographic operation	CPU Scheduling
	Data-driven switching activity	IP-level parallelism
	Isolated switching activity	Shared power distribution network (PDN)
	–	Dynamic voltage and frequency scaling (DVFS)
	–	Clock jitter
	–	Decoupling capacitance
	–	Communication bandwidth
	–	Clock-gating
Fault Injection	Gate type/sizing	Power distribution network
	Nearby cells	Decoupling capacitance
Malicious Hardware	Static probability	IP wrapper/Sandboxing
	Signal Rate	IP firewall
	Toggle Rate	Bus monitor
	Fan-out	Security policy
	Immediate Fan-in	Probability of the malicious IP's activation during critical operation
	Lowest Controllability of Inputs	Isolation of trusted and untrusted subsystems
Supply Chain	–	Electronic chip ID (ECID)
	–	Physically unclonable function (PUF)
	–	Path delay
	–	Aging
	–	Wear-out
	–	Asset management infrastructure (AMI)

- The attacker has access to the locked IP's gate-level netlist.
- The attacker also has a functional copy of IC, called *Oracle*, in which she can apply inputs and observe the correct outputs.
- The attacker has access to the scan chain of the IC.

For sequential designs, it is important to have access to the scan chain to be able to perform the SAT attack [37]. The access to the scan chain and the ability to perform the scan shift operation provide full controllability of the circuit's internal states and reduce the complexity of SAT attacks, just like applying it on a combinational design [38]. The SAT-solving tools produce *distinguishing input patterns (DIPs)* [33] and prune out the wrong keyspace, resulting in finding the correct key at the end. DIP is an input pattern, x_d, for which there are at least two different key values, k_1 and k_2, that generate two different outputs, o_1 and o_2. A SAT-solving tool first finds out the DIP, x_d, which is then applied to the functional IC to obtain the correct output o_d. This correct input–output pair (x_d, o_d) is then compared to the input–output pair of the locked design. If they do not match, then the key values used to find the DIP are pruned out. Note that all the wrong keys cannot be pruned out with a single DIP. This process continues until no further DIP is found. Consequently, all the wrong key values will be pruned out, leaving only the correct key.

Power Side-Channel (PSC) Leakage   The threat model for the power side-channel attack assumes that there are distinct assets (keys) in the critical IPs that are subject to the risk of being revealed by the power consumption analysis. Power side-channel attacks utilize the traces of power consumption of an IP to extract the secret key from a cryptographic implementation such as advanced encryption standard (AES) [39]. In the case of the AES algorithm, the secret key is used to generate a number of round keys. The plaintext goes through a series of round operations, and each round consists of a byte substitution, row shifting, mix column, and add round key operations. The power consumption of a device is determined by the transistor/switching activity of the chip. However, it turns out that the power consumed at different phases of the cryptographic operation is related to the secret key's value. For example, the s-box computation in the first round allows power traces to be statistically distinguished based on the value of the least significant bit of the s-box output [40].

For platform-level security estimation, we consider crypto IPs where the attacker's goal is to leak the key through differential power analysis (DPA) or correlation power analysis (CPA) attacks [2, 41]. The attacker is assumed to have complete access to the power consumption measurement of the target chip. An adversary begins by collecting power traces during cryptographic operations on the chip for a large set of known plaintexts in order to leak the secret key. The adversary then employs a divide-and-conquer strategy to piece together the secret key byte by byte. The adversary takes into account all 256 hypothetical values for each byte and divides the traces into two sets based on the value of the first s-box output's least significant bit (0 or 1). Then two sets are compared to see if there is a statistical difference; the correct hypothetical key causes the two sets to differ significantly [40].

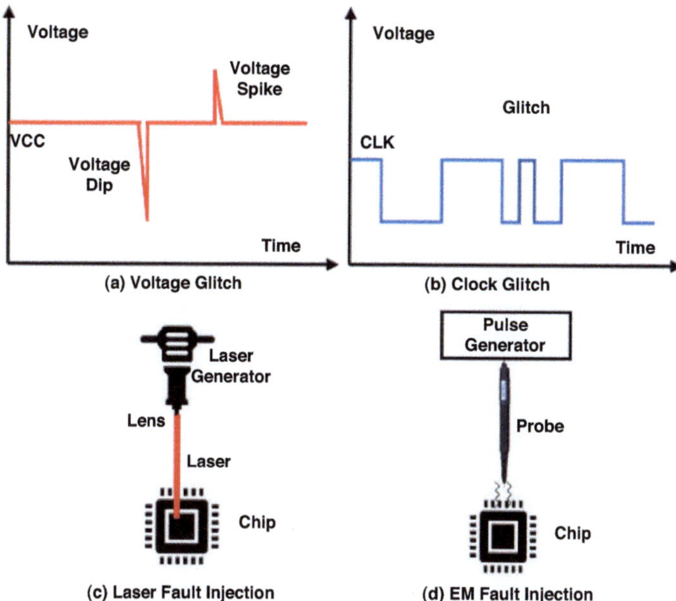

Fig. 1.3 Different fault injection techniques (voltage, clock, laser, and EM)

Fault Injection In hardware, fault injection attacks are among the most common non/semi-invasive side-channel attacks [42, 43]. An attacker creates transient or permanent faults during the device's normal operation and observes the erroneous outputs. It allows an attacker to drastically reduce the number of experiments needed to guess the secret, a.k.a. Differential Fault Analysis (DFA) [44].

Prominent fault injection techniques include voltage and clock glitching, EM/radiation injection, and laser/optical injection [43]. Voltage glitching involves a momentary power drop or spike in the supply voltage during the operation, as shown in Fig. 1.3a. These voltage spikes or drops affect the critical paths, thus causing latching failures of the signals [45]. A voltage glitch can be caused by either disturbing the main power supply causing a global effect or running a power-hungry circuit like ring oscillators (ROs) to cause a localized voltage drop. In recent years, it has been demonstrated that remote fault injection attacks are also possible by disturbing the supply voltage remotely by using either software-based registers to control the supply voltage or ROs to cause voltage drop in remote FPGAs [46, 47]. Clock glitching involves adding glitches in the clock supply or disturbing normal behavior, as shown in Fig. 1.3b. These clock glitches can cause setup and hold-time violations [48]. Since voltage starvation can also impact the clock and circuit timing, voltage and clock glitching attacks are combined to cause timing faults.

Optical/laser fault injection attacks use different wavelength laser/optics to inject transient faults from either the front- or backside of the chip, as shown in Fig. 1.3c. In order to attack from the front-side higher wavelength laser (1300 μm) is used, whereas from the backside near-infrared laser (1064 μm) is used due to its lower absorption coefficient. The injected laser generates electron–hole pairs in the active region, which drift apart under the electric field's influence, resulting in transitory currents. These transient currents cause the transistors to conduct and thus cause the charging/discharging of the capacitive loads. Laser/optical fault injection generally requires depackaging of the chip to expose the die and thus qualifies as a semi-invasive attack. However, with the laser/optics, an attacker can target the precise locations and time of the laser injection, thus making it a powerful attack [49]. Similarly, EM fault injection uses electric or magnetic field flux to influence the device's normal functioning, as shown in Fig. 1.3d. Electromagnetic field causes voltage and current fluctuations inside the device, leading to the faults [50, 51].

Malicious Hardware Malicious hardware is a significant security concern in modern SoC platforms. This attack usually originates by inserting malicious hardware or performing malicious modifications in the original design [52–54]. Malicious hardware can be implanted by third-party IP vendors and untrusted design, fabrication, and test facilities. It can potentially violate either of the following:

- *Confidentiality:* It can leak a sensitive information to untrusted observable points.
- *Integrity:* It can illegally alter/modify a security-critical data.
- *Availability:* It can affect the availability of the device.

Supply Chain Modern SoC design supply chain comprises several steps, such as defining design specification, RTL implementations, IP integration, verification, synthesis, design-for-test (DFT) and design-for-debug (DFD) structure insertion, physical layout, fabrication, testing, verification, packaging, and distribution [17, 55–62]. The globalization of these steps makes the IC design increasingly vulnerable to different supply chain attacks. As the design moves through the supply chain, the risk of the design being exposed to different supply chain attacks also increases. Common supply chain attacks that put the hardware design flow at risk are recycling, remarking, and cloning attacks.

- **Recycling:** Recycling refers to claiming a used electronic component as a new product. In this type of attack, the attackers extract the chip from a used system. Then with little or no modification, they sell the chip as a new product. The recycled chip usually shows a degraded performance and a shorter lifetime [53, 55–65, 65–68, 68–72].
- **Remarking:** Each electronic chip is marked with some information so that it can be uniquely identified. Examples of such information are part identifying number (PIN), lot identification code or date code, device manufacturer's identification, country of manufacture, electrostatic discharge (ESD) sensitivity identifier, certification mark, etc. [53, 73]. This marking information plays an important role. For example, a space-graded chip can withstand extreme conditions such as a wide range of temperatures and radiation that commercial-graded chips are not capable of tackling. Hence, the space-graded chips cost more. The attacker can

remark the commercial-grade chip and sell it in the market at a higher price. In addition, using this remarked chip in spacecraft will surely lead to a disastrous failure.

- **Cloning:** Attackers can clone a design due to the malicious intent of IP piracy. For example, a dishonest platform integrator can clone the IP and sell it to another platform integrator.

1.2 IP-Level Security Metrics and Design Parameters

This section defines IP-level security metrics for five different threat models. We also discuss the design parameters that contribute to the security metric at the IP level.

1.2.1 Metrics for IP Piracy and Reverse Engineering

This subsection briefly describes the existing metrics to evaluate security against IP Piracy. As discussed in Sect. 1.1, logic locking has been a prominent solution to prevent IP piracy by locking the design with a key. However, attackers leverage modern SAT solvers to retrieve the key in a very reasonable time frame [74]. Hence, as a measure of the robustness against IP Piracy, we aim to discuss different metrics that represent the SAT attack resiliency. The most commonly used metrics to evaluate the resistance against IP Piracy are listed below.

Output Corruptibility [75] Output corruptibility is defined as the probability that the two outputs from a locked IP with the wrong key and the functional unlocked chip are not equal for the same input pattern [76]. If C_e and C_o represent the locked and functional IP, respectively, and the input pattern space and wrong keyspace are represented by I and K, respectively, the output corruptibility C_r is

$C_r(C_e, C_o) \equiv P_r[C_e(i, k) \neq C_o(i)], \text{ where } i \in I, k \in K.$

Corruptibility is a very efficient measure of hiding the design's functionality. However, higher corruptibility may help SAT-solving tools to a quicker convergence.

Number of SAT Iterations [77] This metric represents how many iterations the SAT attacker needs to reveal the key. Usually, the higher the number, the better the resiliency is. However, the higher iteration number does not necessarily ensure that the key retrieving time is also higher. It depends on different initial conditions and how the SAT solver walks through the DIP search space. For two different designs, one having a higher number of iterations does not ensure it has higher resiliency than the other one. Also, the time for each iteration is not necessarily equal. However, usually, a very high number of iterations are expected to take longer in terms of time.

CPU Time [78] CPU time indicates the time needed by the SAT attacking engine to extract the correct key. In this case, timeout is the expectation for an ideal SAT resilient design. The higher the time is, the better the robustness of the design.

1.2.2   IP-Level Parameters for IP Piracy Security Metrics

The IP-level parameters refer to the different design parameters of an IP at RTL/gate level that contribute to quantifying the resiliency against a particular threat. This subsection identifies a set of such parameters that play an important role while evaluating IP Piracy robustness against the SAT attack. Table 1.1 includes different important IP-level parameters for all the five threats.

Locking Key Size Key size is one of the essential IP-level parameters that contribute to IP Piracy resiliency. The ideal SAT attack resilient locking mechanism expects to put the attacker in a situation where she cannot retrieve the key without brute force. In practice, the locking mechanism is implemented in a way so that the number of iterations to retrieve the key becomes as close as the number of iterations during brute force. Under this assumption, the increasing key size should increase the SAT resiliency exponentially.

Locking Key Distribution The key distribution represents the fact of whether one locking key is shared among more than one IP or not. In the case of a shared key locking mechanism, retrieving a single key helps to retrieve the functionality of more than one IP [79, 80].

Output Corruption The output corruption represents the deviation of the output of a locked IP from the correct output. As the main idea of locking an IP is to hide the functionality of the design, the deviated output with the wrong key serves this purpose and prevents the attacker from retrieving the functionality of the IP. Usually, the output corruption is measured in terms of the Hamming distance between the correct and the wrong output. Ideally, a 50% hamming distance is considered the highest deviation. Now, from the SAT attackers' point of view, output corruption plays a reverse role than the locking point of view. The highly deviated output usually helps the SAT solver prune out more DIPs in a single iteration. The ideal case for the SAT resilient locking mechanism is that the attacker should be able to prune out only one DIP for one wrong key which eventually leads the attacker to perform a brute force attack [81].

Locking Mechanism The locking mechanism refers to the fact that how the locking is actually implemented in the design. To be specific, it represents how the key gates are placed inside the design, how many fan-outs are affected by the key gate, and eventually, what the output corruption is. The researchers have shown that locking with randomly inserted key gates is susceptible to sensitizing the key bit to the output that leads to retrieving the correct key [82].

Key Error Rate (KER) KER [83] of a key represents the fraction of the input minterms corrupted by the key [83]. For an input length of n bits, if X_K is the set of the corrupted input minterms by the key K, then the KER for key K is defined as

$$KER = \frac{|X_k|}{2^n} \tag{1.1}$$

The higher value of KER tends to rise the overall output corruptibility of the design.

Input Error Rate (IER) IER [83] of an input minterm is represented as the ratio of the number of wrong keys that corrupt the input minterm to the total number of wrong keys [83]. For a given input minterm X, if it is corrupted by the set of wrong keys K_X and the K_{WK} denotes the set of all wrong keys, then IER is

$$IER = \frac{|K_X|}{|K_{WK}|} \tag{1.2}$$

IER contributes to the functional corruptibility of the design, hence higher IER helping in hiding the correct function of the design.

1.2.3   *Metrics for IP-level Power Side-Channel (PSC) Attacks*

This subsection discusses some existing security metrics and IP-level design parameters for power side-channel analysis at the IP level.

Signal-to-Noise Ratio (SNR) SNR is the ratio of the variance of actual power consumption to the additive noise [84]. If P_{signal} and P_{noise} denote the power consumption of the target gates and the additive noise, respectively, the SNR is

$$SNR = \frac{Var(P_{signal})}{Var(P_{noise})} \tag{1.3}$$

where var represents the variance of a function. The signal-to-noise ratio is a measure of the difficulty level to retrieve the correct key by analyzing the power traces.

Measurement to Disclose (MTD) Measurement to disclose (MTD) is the required number of power traces to reveal the correct key successfully [85]. MTD depends on both SNR and the correlation coefficient ρ_0 between the power model and the signal in the power consumption. Mathematically, MTD is defined as

$$MTD \propto \frac{1}{SNR * \rho_0^2} \tag{1.4}$$

Test Vector Leakage Assessment (TVLA) TVLA [86] utilizes Welch's t-test to evaluate the side-channel vulnerability. TVLA works on two sets of power traces. One is generated from the fixed key and fixed plain text and another from the same fixed key and random plain text. Then, hypothesis testing is performed on the two sets of traces, assuming a null hypothesis that the two sets of traces are identical. The accepted null hypothesis represents that the collected traces do not leak information about the key. A rejected hypothesis indicates that the traces can be exploited to retrieve sensitive information. TVLA is defined as

$$TVLA = \frac{\mu_r - \mu_f}{\sqrt{\frac{\sigma_r^2}{n_r} + \frac{\sigma_f^2}{n_f}}} \tag{1.5}$$

where μ_f and σ_f represent the mean and standard deviation of the set of traces with fixed key and fixed plain texts. μ_r and σ_r are the mean and standard deviation of the set of traces with the same fixed key but random plain texts. n_r and n_f are the numbers of traces in the set of traces with random and fixed plaintext, respectively.

Kullback–Leibler (KL) Divergence KL divergence [87] is used to measure the statistical distance between two different probability distribution functions. The notion of the statistical distance is leveraged in power side-channel analysis to identify the vulnerable design. Suppose the distribution of the power consumption of a cryptographic algorithm is different for different keys [88]. In that case, it indicates that the power consumption can be exploited to reveal information about the key.

If $f_{T|k_i}(t)$ and $f_{T|k_j}(t)$ are the two probability density functions of the power consumption for the given keys k_i and k_j, respectively, the KL divergence is

$$D_{KL}(k_i||k_j) = \int f_{T|k_i}(t) \log \frac{f_{T|k_i}(t)}{f_{T|k_j}(t)} dt \tag{1.6}$$

A larger value of KL divergence implies that the power consumption for different keys differs significantly and that the design can be easily exploited by power side-channel analysis. On the other hand, the lower magnitude of the KL divergence indicates greater robustness against the power side-channel attacks.

Success Rate (SR) Success Rate refers to the ratio of the successful attack to retrieve the correct key to the total number of attacks attempted [89]. It is defined as

$$SR = \frac{Number\ of\ successful\ attacks}{Total\ number\ of\ attacks} \tag{1.7}$$

Side-Channel Vulnerability (SCV) The side-channel vulnerability (SCV) metric is functionally comparable to the widely used SNR. However, SCV metric can be utilized in formal methods based on information flow tracking (IFT) to assess PSC vulnerability with a few simulated traces at the pre-silicon design stage [90], as opposed to thousands of silicon traces required in SNR metric. It is defined as

$$SCV = \frac{P_{signal}}{P_{noise}} = \frac{P_{T.hi} - P_{T.hj}}{P_{noise}} \tag{1.8}$$

where $P_{T.hi}$ and $P_{T.hj}$ denote the average power consumption of the target function when the Hamming Weight (HW) of the output is $hi = HW(Ti)$ and $hj = HW(Tj)$ for ith and jth input patterns, respectively. For the PSC assessment, the difference between $P_{T.hi}$ and $P_{T.hj}$ is estimated as signal power.

1.2.4   IP-Level Parameters for PSC Security Metrics

We explored and extracted the IP-level design parameters that contribute to the IP-level PSC metric. We briefly discuss those parameters as follows.

Mode of Operation   The total power consumed during a cryptographic operation can be greatly influenced by the mode of operation. For example, the AES block cipher algorithm can operate in multiple modes such as Electronic Code Book (ECB), Cipher Block Chaining (CBC), Cipher Feedback (CFB), Counter (CTR) mode, etc. PSC resiliency for different modes of operation will vary [91] as the power consumption and noise level will be different due to the additional logic, use of initialization vector or counter, nonce, etc.

Cryptographic Key Size   Key size of the cryptographic algorithm is another important factor because it determines the number of round operation in AES algorithm. The rounds of operation for 128, 192, and 256 bit keys are 10, 12, and 16, respectively. The parallel activities will vary depending on the number of rounds, resulting in varying resiliency against side-channel attacks.

Number of Clock Cycles per Cryptographic Operation   When evaluating resiliency against side-channel attacks, the number of clock cycles required to perform an AES operation in the implementation is crucial. A loop unrolling architecture may be used in the AES implementation, in which one or more rounds of operations are performed in the same clock cycle. In the simplest case, only one round of the algorithm is implemented as a combinational processing element, and the results of the previous clock cycle are stored in data registers. Multiple data blocks are processed simultaneously in a clock cycle in a pipelined architecture, and the implementation usually requires multiple sets of registers and processing elements. Furthermore, the key expansion can be done at the same time as the round operations or during the first clock cycle. As a result, the number of cycles needed for the algorithm will have an impact on the parallel operation and noise in the power trace.

Data-Dependent Switching Activities   Data-dependent switching activities are the transistor activities directly related to the key. The operations in the round that involve the use of key and outputs generated influence the data-dependent activities. For example, the first add round key and substitute operations involve the use of the key value that contribute directly to the power consumption. Moreover, the value of the key and the plaintext also affect the data-dependent activities.

Data-Independent Switching Activities Data-independent switching activities are the transistor activities that are not correlated to the input key and add noise to the power trace. For example, the first round in AES operation adds enough confusion and diffusion to the plaintext that the correlation between the key and power will reduce significantly in the later rounds. Moreover, additional logic and circuitry for different implementation and architecture of AES algorithm will provide different amount of data-independent switching. Furthermore, the traces include the power consumption from parallel activities in other IPs present in a system. Depending on the design and application-specific parameters of the system, the additional activity added to the noise will vary. If no other IP is allowed to operate during an AES operation, it will result in the worst-case scenario from a security perspective.

1.2.5 Metrics for IP-Level Fault Injection Attacks

This section discusses parameters that are crucial in fault injection attacks and can impact the overall feasibility of the fault injection attacks, as discussed below.

Spatial Controllability It allows an attacker to target a specific net/gate in the design. Considering the large design size, not all components (nets/gates/registers) are crucial for a successful attack. An attacker tries to inject faults in specific components, if violated, helps his case to exploit and cause integrity and confidentiality violations. Therefore, the higher the spatial controllability of the fault injection attack, the higher is the design's susceptibility. Several parameters can contribute to spatial controllability, i.e., fault method (clock, voltage, laser, etc.), design's timing information (path delays, clock, etc.), and library information (cell/gate types). For example, laser and optical fault methods provide more spatial controllability to an attacker to target specific locations on the chip. In contrast, clock and voltage glitching methods can violate multiple paths in the design, thus injecting multiple faults in the design. Similarly, delay distribution of the paths can dominate which registers will be impacted by the clock and voltage glitching.

Temporal Controllability It allows an attacker to control the fault injection time during the design's execution. For example, to effectively perform differential fault analysis on an AES with minimal faults, an attacker would like to inject fault in the eighth round of the execution [92, 93]. However, attacking after the eighth round would require more faults to retrieve the entire key and before the eighth round would make the differential fault analysis complex, rendering key retrieval futile. Therefore, if an attacker can control the triggering of the clock, voltage, or laser injection, it directly impacts the attacker's capability to control the faults and thus impacts the design's susceptibility against fault injections.

Fault Type and Duration Type of faults, i.e., permanent, semi-permanent, and transient faults, can also impact the fault injection attacks. For example, a laser injects transient faults in the device, but higher laser power can break the silicon, causing permanent faults in the device. Where transient faults cause bit-flips at the gate's output, permanent faults can lead to stuck-at-0/1 at the gate's output. Therefore, different types of faults require different analyses from the attacker to leak information [94]. Similarly, fault duration can also impact the design's susceptibility. For example, if a clock glitch is too small or within the slack relaxation of the timing path, the fault may not get latched to impact the design's functionality. Similarly, the laser's duration could affect the number of clock cycles during which the transient effect of laser current would manifest in the design.

Fault Propagation It is required to ensure if the injected faults can propagate to the observable points (e.g., ciphertext in AES encryption). Faults injected by clock/voltage glitching, laser injection, etc., if not latched to the register, do not go through logical flow, having no impact on the design's execution. Timing information of the paths, laser stimulation period, fault duration, system's clock, and other factors can impact if the faults will be successfully latched or not.

Fault Method Fault injection method can also play a major role while evaluating/developing fault injection security metrics. For example, clock and voltage glitching-based fault injection methods are global in nature. Meaning: it is hard to control the fault location, and a single glitch can cause a fault in multiple paths across the design. In contrast, localized fault injection methods such as laser and optical fault injection are local in nature, and an adversary can target specific fault nodes to inject fault. Also, the physical parameters involved with different methods are very different. For example, laser injection requires the backside of chip to be exposed, whereas clock glitching requires access to the system's clock.

The above discussed parameters are crucial to define the fault metrics to determine the design's susceptibility. At the IP level, faults can occur at the datapath or the control path. Datapath comprises the data flow through the design from register to register via combinational gates. In contrast, a control path consists of control registers controlling the datapath flow, e.g., finite state machines (FSMs).

Security metric to evaluate fault injections in the datapaths is challenging, and no security metrics exist to the best of our knowledge. To perform the evaluation, one has to consider the underlying design (e.g., type of crypto) and the threat model. For example, it has been shown that faults injected in the eighth, ninth, and tenth rounds of AES can assist in leaking keys, whereas faults in earlier rounds are more difficult to exploit [44]. The hypothesis and the mathematical models developed to exploit faults differ across crypto algorithms.

However, faults on the control paths, like FSM, can allow an attacker to bypass the design states. For example, in AES, an attacker can bypass round operations to directly reach the done state, thus, rendering the security from encryption futile. Thus, to measure the overall vulnerability of FSM to fault injection attacks, the vulnerability factor of fault injection (VF_{FI}) has been proposed in [45] as

$$VF_{FI} = PVT(\%), ASF \tag{1.9}$$

$$PVT(\%) = \frac{\sum VT}{\sum T}, ASF = \frac{\sum SF}{\sum VT} \tag{1.10}$$

The metric is composed of two parameters, i.e., PVT(%) and ASF. PVT(%) is the ratio of the number of vulnerable transitions ($\sum VT$) to the total number of transitions ($\sum T$), whereas ASF is the average susceptibility factor ($\sum SF$), defined as

$$SF_T = \frac{\min(PV) - \max(PO)}{avg(P_{FS})} \tag{1.11}$$

where $min(PV)$ is the minimum value of delays in path violated, and max(PO) is the maximum value of delays in paths not violated. Accordingly, the higher the PVT(%) and ASF value, the more susceptible the FSM is to fault injection attacks.

1.2.6 Metrics for IP-Level Malicious Hardware

This section enumerates the existing metric to evaluate the malicious hardware in a design. We discuss some of these metrics in the rest of this subsection.

Controllability and Observability Controllability [95] of a component within a design refers to the ability to control the inputs of the component from the design's primary inputs. On the other hand, observability [95] is the ability to observe the output of a component from the primary outputs of the design. The primary inputs of a design are assumed as perfectly controllable, and the primary outputs are perfectly observable. The measurement of the controllability and observability can be normalized between 0 and 1.

For controllability, let us consider a component within a design that has input variables x_1 to x_n and output variables z_1 to z_n. If the controllability is represented by CY, then the value of CY for each output of the component can be calculated by

$$CY(z_j) = CTF \times \frac{1}{n} \sum_{i=1}^{n} CY(x_i) \tag{1.12}$$

Here, CTF is the controllability transfer function of the component, defined as

$$CTF \cong \frac{1}{m} \left(\sum_{j=1}^{m} 1 - \frac{|N_j(0) - N_j(1)|}{2} \right) \tag{1.13}$$

where $N_j(0)$ and $N_j(1)$ are the number of input patterns for which z_j has values 0 and 1, respectively.

For observability, OY for each input of a component in the design is

$$OY(x_i) = OTF \times \frac{1}{m} \sum_{j=1}^{m} OY(z_j) \tag{1.14}$$

where OTF is the observability transfer function which is the indicator that a fault in the input of a component will propagate to the outputs. If NS_i is the number of input patterns for which x_i changes the output, then the OTF is

$$OTF \cong \frac{1}{n} \sum_{i=1}^{n} \frac{NS_i}{2^n} \tag{1.15}$$

Statement Hardness It is the measure of the vulnerability analysis of malicious hardware insertion at the behavioral level [96]. To be specific, it represents the difficulty of executing a statement in the RTL source code. In a design, the statements with a lower value of statement hardness are vulnerable to malicious hardware insertion attacks. An attacker is most likely interested in targeting this area to insert malicious hardware, hoping that it will be activated rarely with a certain trigger condition.

Hard to Detect It is proposed in [97] to quantify the areas in the gate-level netlist that are susceptible to malicious hardware insertion. These areas are the common targets of the attacker, as inserting malicious hardware in this area will reduce the probability of being detected during the validation and verification steps [98].

Code Coverage Code coverage [99] is defined as the percentage of the design source code executed during the functional verification. A higher code coverage represents the lower probability of malicious hardware insertion as it indicates the lower suspicious area in the design. However, it depends on the quality of the testbench used during the functional verification. In [97, 100], the authors utilize the code coverage analysis and propose a technique called *Unused Circuit Identification (UCI)*. It is used to find out the line of codes in the RTL design which are not executed during the functional verification. These lines of code are the potential target of an attacker to insert the malicious hardware.

Observation Hardness (OH) Observation hardness [101] is defined as the percentage of a primary input propagating to an intermediate node. To determine the OH value, a stuck-at-0 or stuck-at-1 fault is injected into a primary input. This fault can be fully detected, potentially detected, or undetected at an intermediate node, representing the intermediate node's dependency on the primary input. The observation hardness of an internal node can be defined as

$$OH = \frac{Total\ No.\ of\ Detected\ Faults}{Total\ number\ of\ faults\ injected} \tag{1.16}$$

1.2.7   IP-Level Parameters for Malicious Hardware Metrics

This subsection lists IP-level parameters for the maliciousness of an IP.

Static Probability Static probability of a signal in the design refers to the fraction of time the signal value is expected to be driven at a logic high (1'b1). Ideally, the probability of being at logic high and low should not be skewed. A signal having a significantly higher value of the probability of being logic high indicates that the signal is rarely driven by a logic low, which can be the potential trigger condition of malicious hardware [102].

Signal Rate The signal rate of a wire in the design refers to the number of transitions from logic high to logic low or logic low to logic high at that wire per second [102]. A wire having lower signal rates can be a potential candidate for the trigger of a malicious circuit in a design.

Toggle Rate For a signal, it is defined as the rate at which the signal switches from its previous value [102]. Attackers usually utilize a signal having a very low toggle rate to implement the trigger condition of the malicious circuit.

Fan-Out Fan-out is an essential feature for malicious hardware insertion attacks. Fan-out of a net is the number of other logic elements the net propagates to. Usually, malicious hardware is inserted to cause a significant impact on the design. In the case of a denial-of-service attack, the attacker would most likely target a payload that impacts a larger part of the design. Thus a net having higher fan-out may be a good choice of an attacker [102].

Immediate Fan-In A large number of immediate fan-in usually indicates that a large function with low entropy has been implemented with rare activation conditions. Hence, the immediate fan-in becomes an important IP-level parameter of the malicious hardware attack [102].

Lowest Controllability of Inputs Input signals of a group of logic within the design that have a lower impact on the design outputs are considered suspicious nets [103]. Usually, this type of nets has very low controllability from the primary inputs of the design. The difficulty in controlling these nets makes them a potential candidate for malicious hardware insertion.

1.2.8   Metrics for IP-Level Supply Chain Attacks

As discussed in Sect. 1.1, we consider recycling, remarking, and cloning as part of the supply chain attack. This section briefly presents some existing metrics used in assessing these threats.

Metrics for Cloning Over the years, strong PUFs [104] have been very effective in detecting cloned chips. The confidence of detecting cloned chips depends on the quality of the PUF, which leads to the PUF quality metrics being a great measure to assess whether a chip is cloned or not with a certain confidence. Major PUF quality metrics include *uniqueness, randomness,* and *reproducibility.*

Regarding uniqueness, it is the measure of distinctive challenge–response pair (CRP) between chips. Ideally, a PUF referring to one chip should produce a unique CRP that other chips cannot produce, which makes PUF so effective in authenticating a chip while detecting cloned one. Inter-chip hamming distance (inter-HD) is a common measure to mathematically calculate the uniqueness over multiple PUFs [105]. For n number of PUFs, if k is the bit-length of a PUF response and R_i and R_J are from PUF_i and PUF_j, respectively, then inter-HD is calculated as

$$HD_{inter} = \frac{2}{n(n-1)} \sum_{i=1}^{n-1} \sum_{j=i+1}^{n} \frac{HD(R_i, R_j)}{k} \times 100\% \tag{1.17}$$

An inter-HD value of 50% is considered to be the ideal *uniqueness* as it represents the maximum difference between two PUF responses.

For randomness, it indicates the unpredictability of a PUF's response. A PUF can only be used to identify cloned chips when an attacker cannot predict its response. Ideally, a PUF's response is free from all correlations and hence cannot be properly modeled. A PUF with good diffusive property shows good randomness in its response. The diffusive property refers to the fact that a slight change in the input challenge causes a significant variation in the output response.

For reproducibility, it measures a PUF's ability to produce the same challenge–response pair in different environmental conditions and over time. Quantitatively, the reproducibility is calculated as the intra-PUF hamming distance as follows [105]:

$$HD_{intra} = \frac{1}{m} \sum_{y=1}^{m} \frac{HD\left(R_i, R'_{i,y}\right)}{k} \times 100\% \tag{1.18}$$

Here, m is the number of samples for a PUF, k is the length of the response, and $HD(R_i, R'_{i,y})$ represents the hamming distance between the response R_i and the response of y_{th} sample $R'_{i,y}$).

An ideal PUF is expected to always produce the same response to a particular challenge under different operating conditions with 0% intra-PUF HD.

Metric for Recycling and Remarking Guin et al. [66] developed a metric called Counterfeit Defect Coverage (CDC) to evaluate the counterfeit detection techniques.

Counterfeit Defect Coverage (CDC): CDC represents the confidence level of detecting a chip as counterfeit after performing a set of tests. It is defined as

$$CDC = \frac{\sum_{j=1}^{n}(X_{Rj} \times DF_j)}{\sum_{j=1}^{m} m} \times 100\% \tag{1.19}$$

where DF_j is the defect frequency of the defect j, which indicates how frequently the defect appears. XR_j is the confidence level of detecting defect j by test method R.

These three attacks suit well in platform-level discussion, and hence, there is no exhaustive analysis on IP-level parameters so far. We continue our discussion on these three attacks while discussing the platform-level security for different threats.

1.3 Transition from IP to Platform

In this section, we steer our discussion from IP to platform. So far, we have discussed different IP-level security metrics and design parameters contributing to those security metrics for all five threats. This section focuses on how the design parameters evolve while moving from IP to the platform. In other words, we show different design parameters introduced during the transition from IP to the platform, potentially impacting security metrics at the platform level.

While moving from IP to platform, an IP gets surrounded by its neighboring IPs. The glue logic is implemented to connect the IPs, which is not considered while developing metrics at the IP level. A platform-level testing architecture is introduced that might affect the security at the platform level. To get an exhaustive list of such additional parameters, we investigate the transition from IP to platform and identify several parameters introduced during the integration, which impact the security of the entire platform. We also categorize those parameters based on their impact on a particular threat. We call these additional parameters platform-level parameters. Table 1.2 summarizes both the IP- and platform-level parameters for all five threats. Parameters in the rightmost column are the platform-level parameters that show up after the integration. These parameters are not considered during the assessment of security at the IP level. However, they have a significant impact on security while assessing from the platform point of view. This section briefly discusses all these additional parameters and their impact on IP piracy, PSC analysis, fault injection, malicious hardware, and supply chain attacks at the platform level.

1.3.1 Platform-Level Parameters for IP Piracy

To understand the platform-level parameters for IP Piracy, let us consider Fig. 1.4. It shows different platform-level parameters that affect the SAT resiliency at the

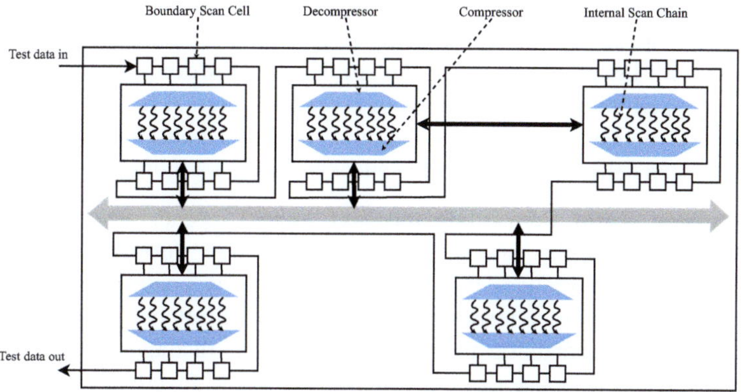

Fig. 1.4 Platform-level parameters affecting resiliency against IP piracy

platform level. For example, each IP is wrapped around a wrapper. Compression and decompression circuits have been added to accelerate testing. The wrappers can vary with some of the features such as bypassing capability, access to the internal scan chain, etc. Due to the platform-level parameters, the way IPs can be attacked while they are standalone significantly differs, while they are placed in the platform. As discussed in Sect. 1.1, SAT attack on a sequential design greatly depends on the access to the internal scan chain. As shown in Fig. 1.4, the scan access to the target IP inside the platform is not the same as the standalone IP. Now, the additional boundary scan chain comes with the wrapper. A brief description of these SoC-level parameters affecting the SAT attack resiliency against IP Piracy is discussed below.

Type of Platform-Level Testing Architecture Platform-level SAT resiliency greatly depends on the testing structure of the platform, as the SAT attack leverages this structure to retrieve the correct key. At the IP level, the retrieval of the key from a locked sequential IP by SAT attack is facilitated by the scan chain of that IP. Consequently, at the platform level, the attack's success greatly depends on the testing structure of the platform. Different types of testing structures bring a different level of impact on the IP Piracy resiliency. An example of the platform-level testing structure includes flattened, direct I/O, concatenated scan chain, etc.

Bypassing Capability It refers to the capability of bypassing some IPs in a platform and targeting the specific IP to perform SAT attack. Having bypassing capability in the platform-level testing structure provides the attacker with the same attacking capability as the IP. However, preventing bypassing capability, although reducing the testing capability, increases the SAT attack's difficulty.

Wrapper Locking As an industry standard, all the IPs are wrapped around a wrapper during the platform integration. A locked wrapper boosts the SAT attack resiliency one step ahead as the attacker needs to break the wrapper lock to be able to perform SAT attack at the platform level.

Accessibility to the Internal Flip-Flops As a critical parameter that has a significant impact on platform-level IP Piracy, the bypassing capability provides attackers with the ability to target the specific IP. However, suppose the wrapper is not equipped with the instructions to access the internal flip-flops. In that case, the attackers will be left with only the input–output response of that IP through its boundary scan cells, which makes the attackers' job a lot more complicated.

Compression Ratio Performing the SAT attack depends on the controllability/observability of the internal state of the design through the scan chain. (De)compression introduces the difficulty of precisely controlling internal states and observing the exact response of the internal flip-flops. Note that the SAT attack's efficacy depends on the ability to apply the distinguished input patterns and compare the response of both the locked and the functional copy of the IC. The decompressor prevents the attacker from applying the inputs of her choice to the internal flip-flops through the scan chain. On the other hand, the compactor prevents the attacker from comparing the original responses as the shifted output is compacted. Ideally, the attackers need to reverse the compacted output to retrieve the original response before compaction. Usually, the existing compaction circuits used in industries act as a one-way function. It means retrieving the original response from the compacted response introduces some difficulty to the SAT attackers by reducing observability.

Scan Obfuscation Scan obfuscation is a great technique to hide the access of the scan chain from the adversaries [106–110]. With the help of the scan obfuscation, the scanned-in and scanned-out responses are scrambled, and only with the help of the correct key, the original values of the scan chain are retrieved. Hence, it greatly reduces the controllability and the observability of the internal flip-flops in a design. As we already discussed, the SAT attackers' success greatly depends on the capability of controlling and observing the scan chain circuitry; the scan obfuscation indeed introduces resiliency for the SAT attack at the platform level.

1.3.2 Platform-Level Parameters for Power Side-Channel Analysis

The platform-level PSC analysis is greatly affected by several parameters that are introduced after the platform integration, such as:

Pipeline Depth For the sake of simplicity, while we discuss the Power Side-Channel Analysis attack, we consider leaking key from a cryptographic engine by measuring and analyzing the power trace. In the platform-level view, while the cryptographic IP is placed in the platform, it is neighbored by different IPs and different communication buses. The Central Processing Unit (CPU) is one of the essential IPs in the platform. Thus, different features associated with the CPU now become the platform-level parameters. The pipeline depth of the CPU is one of the platform-level parameters that indicates the instruction-level parallelism of that CPU. Usually, more parallelism means more activity occurs at the same time. Thus, a higher pipeline depth is likely to increase other activity at the same time when the cryptographic IP operates, eventually inducing noise in the power trace of the platform. The increased noise will mask the power consumption of the crypto operations and improve resiliency against power side-channel vulnerability. However, this is under the assumption that the instruction-level parallelism will likely increase different IPs' activity.

CPU Scheduling CPU Scheduling indicates the quality of the CPU in performing the parallel operations. The higher CPU utilization indicates a higher probability of the simultaneous activity along with the cryptographic operation, leading to higher noise insertion in the measured power.

IP-Level Parallelism It has been discussed above that the higher pipeline depth potentially increases the instruction-level parallelism. Although the deeper pipeline increases the probability of increasing simultaneous operation, it does not necessarily guarantee that more IPs will operate simultaneously. However, the IP-level parallelism indicates more simultaneous activity from different IPs, which most likely contribute to the noise insertion in the power trace of the platform, leading to the potential increase in the robustness against the Power Side-Channel Analysis attack. One important aspect should be noted that the noise we have been discussing so far is not coming from the total power consumption of the other IPs. Rather the differential power created by the other IPs contributes to the noise. Switching activity from additional IPs and communication buses are the most important factors to contribute to this differential power.

Shared Power Distribution Network (PDN) Shared PDN refers to the fact where different components share the same power rail of the power distribution network (PDN) with the target cryptographic IP. Shared PDN information provides us the notion of an estimated noise that might be inserted in the main rail from where the attackers measure the power. A higher number of IPs sharing the same power rail with the target IP will increase the likelihood of more noise insertion.

On-Chip Voltage Regulator Modern integrated circuits (ICs) include on-chip voltage regulators to achieve lower noise, better transient response time, and high power efficiency. Such on-chip voltage regulators can be utilized to make the design more resilient against power analysis attacks [111]. When such techniques are used to scramble or enhance the entropy of the device's power profile, the platform's robustness for PSC vulnerability improves.

Dynamic Voltage and Frequency Scaling (DVFS) Additional noise is used in random dynamic voltage and frequency scaling approaches to randomize power consumption and reduce data-dependent correlation. Such techniques can be employed as a platform-level countermeasure against DPA attack [112].

Glitch Glitching is a common phenomenon in the COMOS circuit. On a high level, glitching refers to the multiple switching of a gate within a single clock cycle.

Circuits with higher-order glitches tend to inject more noise in the power grid than the glitch-free design [113]. Thus the neighboring IPs having higher-order glitches may help to increase the resiliency of a cryptographic IP at the platform level.

Clock Jitter An essential step while performing the differential power analysis attack is synchronizing the collected power traces through alignment by the clock edges. However, the clock generator may introduce the clock jitter at the platform level, potentially making the analysis more difficult [114].

Communication Bandwidth The bandwidth of the communication bus will impact the probable parallel activities in other IPs. A higher bandwidth allows more IPs to be active during crypto operations. The additional switching activities are not correlated to the crypto operations and increase noise in the power trace. So, a higher communication bus bandwidth improves resistance against side-channel attacks.

Clock-Gating It is for reducing the dynamic power consumption of the clock signal that drives large load capacitance. Power dissipation is minimized in the clock-gated circuits as the registers are shut off by the control logic during the idle period. Clock-gating techniques can be utilized to improve the power side-channel resiliency of a platform. For example, latch-based random clock-gating [115] can obfuscate the power trace with a time-shifting approach. Thus, the presence of clock-gating logic can affect the power side-channel vulnerability metric of a platform.

1.3.3 Platform-Level Parameters for Fault Injection

The following is platform-level parameters for IP resistance against fault injection:

- **Nearby IPs:** Nearby IPs to the critical IPs can also have a major impact on the fault resistivity of the cell. For example, a critical IP surrounded by very high switching IPs can diminish the decaps effect, thus increasing the noise margins. This can amplify the effect of voltage and clock glitching attacks.
- **Interconnect:** Interconnects can also have an impact on fault tolerance. A higher density of metal interconnects can make it harder to perform laser fault injection from the front-side. Similarly, interconnect length and width can impact the delays, thus impacting timing faults.
- **Power Distribution Networks:** Power and ground rails form the major mesh in the chip, forming many horizontal and vertical loops. Therefore, they are the most susceptible to EM injections. EM injection can cause voltage drops/shoots at both power and ground rails. In addition, both rails can experience drops/shoots at a different rate with respect to the EM pulse power due to asymmetrical couplings [50]. This voltage swing (i.e., $V_dd - Gnd$) propagates toward the pads while being attenuated along their path. The propagation of the voltage swings results in sampling faults during EM injection.

- **Decaps:** Power grid is the major source of noise in the circuit. Reduced power supply voltages have helped in reducing these noise margins. However, power-hungry circuits (such as ROs) can increase these noise variations, causing delay variations and voltage drops. Decaps help to reduce these noise margins in the supply voltage. Therefore, the distance of the critical registers from decaps and power grid, supply voltage, the width of power lines, etc. can have a major impact on the susceptibility of the critical register to voltage fault injection.
- **Error Correction and Redundant Circuitry:** Error correction circuits such as bus parities can also impact the design's resiliency against fault injection. For example, parity bits can help resolve single or few bit faults in the data bus depending on the error correction circuit. Similarly, redundant (time or spatial) circuitries in the design can help verify if faults were injected in the design [116]. However, such measures incur heavy area and latency overhead on the design.

1.3.4 Platform-Level Parameters for Malicious Hardware

In this section, we discuss the platform-level parameters that affect the malicious hardware attack at the platform level [117, 118].

IP Wrapper IP wrapper is usually used at the platform level, which acts as a checkpoint for the IP not to perform any unauthorized action. Being monitored and controlled by the wrapper, an IP, even including malicious hardware, cannot easily perform the malicious activity. Hence, the IP wrapper impacts the platform-level maliciousness to some degree.

IP Firewall Like the IP wrapper, an IP firewall monitors and controls the activity of an IP and ensures the expected behavior. The firewall containing a good firewall policy can block an IP's activity from performing malicious action, potentially protecting against the malicious hardware attack at the platform level.

Bus Monitor A bus monitor is a component that continuously monitors the activity in the BUS connecting different IPs. Monitoring the BUS activity can significantly reduce the malicious data snooping and message flooding with the intent of the denial-of-service attack.

Bus Protocol Bus protocol is the set of rules that each component in a platform should follow while using the bus. The goal of the bus protocol is to ensure the correct and secure use of the bus. A well-developed bus protocol has the potential to reduce the risk of malicious activity during communication through the bus.

Security Policy Security policies are a set of policies/rules that are enforced by a policy enforcer in a platform. It aims to maintain secure functionality and avoid any potentially unauthorized activity. An exhaustive list of security policies can significantly reduce the malicious activity inside the platform. However, developing the security policy list is a non-trivial task.

Probability of the Malicious IP's Activation During Critical Operations Not all the IPs are meant to interact with each IP in a platform. Based on the design requirements, each IP has a probabilistic value to interact with a particular IP. The interaction of an IP having a security asset with a malicious IP is more of a concern than the interaction with a non-malicious IP. Hence, the interaction probability becomes an essential parameter in quantifying platform-level security.

Isolation of Trusted and Untrusted Subsystems Isolating the untrusted IPs from the security-critical IPs significantly reduces the probability of the occurrence of any malicious activity.

Memory Management Unit (MMU) Policy Memory is one of the most important components of a platform. MMU policy ensures that no unauthorized application can access the protected memory region, reducing the risk of any malicious activity in the secure memory.

1.3.5 Platform-Level Parameters for Supply Chain

The platform-level supply chain attack resiliency is significantly affected by several parameters introduced at the chip level. Some of these parameters are discussed in the rest of this section.

Electronic Chip ID (ECID) ECID is a unique number that is used to tag the chip platform throughout the supply chain. It is used in detecting the remarked chip by tracking its ECID.

Physically Unclonable Function (PUF) It utilizes the inherent process variation introduced in every chip during the manufacturing process [104]. This process variation is unpredictable and uncontrollable, making a unique fingerprint of the individual chip used in chip identification and authentication [119, 120]. PUF is very efficient in detecting the cloning of a platform chip.

Path Delay In [121], the authors utilize the path delay as the fingerprint to detect the recycled chip. As the recycled chip has been in use for a particular duration of time, it is usual that the performance of the chip will be degraded. Due to the negative/positive bias temperature instability (NBTI/PBTI) and hot carrier injection (HCI), the path delay of a used chip becomes sufficiently larger than it is used to detect whether the chip is recycled or not. The higher the path delay, the higher probability that the chip is recycled.

Aging and Wear-Out Temperature instability (BTI), hot carrier injection (HCI), electromigration (EM), and time-dependent dielectric breakdown (TDDB) are the key parameters for the aging degradation and wear-out of a platform chip [4, 66, 122–124]. An aged and worn-out chip usually represents a significantly degraded performance compared to the new chip.

Asset Management Infrastructure The asset management infrastructure (AMI) is a cloud-based infrastructure that facilitates testing a platform after it is fabricated. It can establish secure communication with the chip on the manufacturing floor. It is also responsible for unlocking and authenticating the chip following the secure protocol, consequently preventing the overproduction of the chip.

All the parameters discussed above contribute to platform-level security to some degree. These parameters either help or hurt the base security (IP-level security) while quantifying the security at the platform level. Section 1.5 shows how these parameters are included when the security measurement and estimation is performed for two different threats: IP piracy and power side-channel analysis.

1.4 Security Measurement and Estimation: Definition

In previous sections, we started with an IP-level discussion. Then we discussed the transition from IP to the platform and discussed how the parameters evolve during this stage. In this section, we introduce two novel terms called security measurement and estimation. In particular, we discuss both the measurement and estimation from the platform point of view.

Security Measurement It refers to the exhaustive security verification of the design through simulation, emulation, formal verification, etc. [125]. In other words, security measurement includes performing actual attacks and analyzing the resiliency of the design against a particular threat. For example, measuring the PSC robustness of a platform requires performing the power side-channel analysis attack on the platform by collecting and analyzing its power traces. If we consider MTD as the security measure for power side-channel analysis, then the number of power traces required to reveal the cryptographic key from the platform would indicate the measured security of the platform. Similarly, measuring security for IP piracy would require performing SAT attack on the platform and measuring the time to reveal the locking key [126]. To measure the maliciousness of platform, an exhaustive security property verification may be a good choice to measure the confidentiality violation through information leakage. This approach can also be used to measure the data integrity violation. PUF, ECID, and age detecting sensors can be used to measure the probability of a chip being recycled, remarked, or cloned.

Security Estimation With no dependency on the simulation or emulation, it leverages the modeling of the design and different parameters to estimate the security. As a result, the estimation is a lot faster than the measurement. Security estimation is a very efficient technique that can enhance the process of building a secure design. For example, with the help of the estimation, the designer can quickly estimate the security of the entire chip earlier in the design phase. Based on this estimated security, the designer can easily make the necessary modifications to meet the other PPA (Power, Performance, and Area) metrics while not compromising the security. As the core component of the estimation, first, a model of the impact factor of the SoC-level parameters is developed. Then this model can predict/estimate the security of an unknown SoC while not needing the actual value of the parameters. Unlike measurement, estimation can be very useful for a larger design. Ideally, the estimation should not depend on the design size. It brings a great benefit to making the designer capable of choosing among different components and a different SoC configuration that provides the best result in terms of security. However, the estimated security lags behind the measured security in terms of accuracy.

1.5 Showcasing Security Measurement and Estimation

This section explores estimation and measurement approaches for IP piracy and power side-channel analysis attack, as two case studies. We explain each of the steps of both the estimation and measurement with some example design.

1.5.1 Security Measurement/Estimation for IP Piracy

1.5.1.1 Platform-Level SAT Resiliency Measurement Flow

This section provides the step-by-step process for the measurement of platform-level SAT resiliency of an IP. As mentioned in Sect. 1.3, the transition from the IP to platform introduces different additional parameters that significantly impact SAT resiliency. Thus, to measure the SAT resiliency at the platform level, we first mimic the platform-level environment by adding the platform-level parameters with the target IP. The first step of performing SAT attack on a design involves converting the design from the gate-level netlist to the conjunctive normal format (CNF). Sometimes this conversion of the design is referred to as SAT modeling. To measure the platform-level SAT resiliency, we first perform the SAT modeling of the design along with the additional platform-level parameters. Then we perform the SAT attack to measure the time the SAT attacking tools take to extract the key at the platform level. For the sake of simplicity in the demonstration, we are considering one platform-level parameter, such as decompression/compaction. The goal is to measure the SAT attacking time for breaking the lock of an IP, while the compression/decompression circuits are associated with that. The measurement is comprised of two steps: platform-level SAT modeling of the design and performing SAT attack. The following subsections present these two steps in detail.

SAT Modeling for SoC-Level Attack The existing SAT attacking tools are not capable of performing attack directly on the HDL code. The design needs to be converted into the Conjunctive Normal Form (CNF) before feeding to the SAT attacking tool. The CNF conversion flow for both the combination and sequential design is shown in Fig. 1.5.

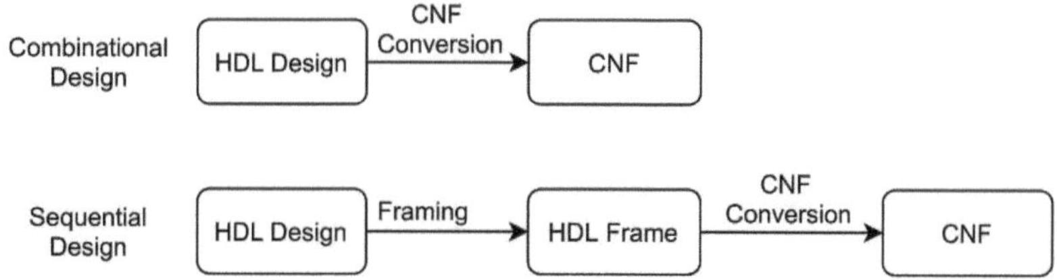

Fig. 1.5 Modeling of the HDL design for the SAT attack

Modeling of Combinational Designs: The netlist of the combinational designs can readily be converted to the CNF. There are open-source synthesis tools such as ABC [127] that can covert the HDL design of a combination circuit into the CNF.

Modeling of Sequential Designs: To understand the platform-level modeling with the platform-level parameters, we should first have a solid understanding of IP-level modeling. This section describes the modeling of the sequential design for performing SAT attacks at both IP and platform levels.

- *IP-Level SAT Modeling:* The conversion of the sequential designs into CNF format requires an additional step named "framing." The existing SAT attacking tools have a limitation of performing SAT attack on sequential designs. To address this issue, the sequential design needs to go through the framing process. Framing converts the sequential design into a one-cycle equivalent design. In other words, framing converts the sequential design to an equivalent combinational design. For example, we want to perform SAT attack on an AES design that takes 11 clock cycles to complete one cryptographic operation. By framing the AES implementation, we can break the design into 11 pieces of a one-cycle design. To get the exact behavior of the sequential design, the 11 frames can be stitched together. However, to perform the SAT attack for measuring SAT resiliency, one frame is sufficient. It is because the SAT attack proceeds by comparing the output of the locked and the unlocked design and pruning the wrong keys away from the search space. For the comparison of the outputs, only one frame is adequate. However, both the locked and unlocked designs should be framed. One question that arises here is, how can one frame the physical chip that is used as an oracle. Here, the scan access to each flip-flop does the same job as framing. Shifting the value through the scan chain, running the design for one clock cycle in functional mode, and shifting out all the flip-flops value give the same output as one frame would provide. An example of framing of the sequential design is shown in Fig. 1.6. The sequential design is the combination of the sequential (flip-flops) and combinational parts. During the framing, all flip-flops are removed from the design, and the corresponding Q pins and D pins are added as primary inputs and outputs, respectively, while the connection between the D and Q pins with the combinational parts is maintained. This frame is then converted into the CNF format.

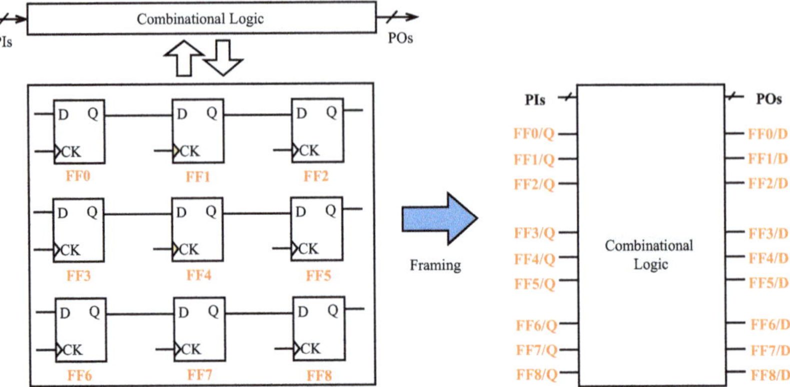

Fig. 1.6 Framing of sequential design (the impact of state registers)

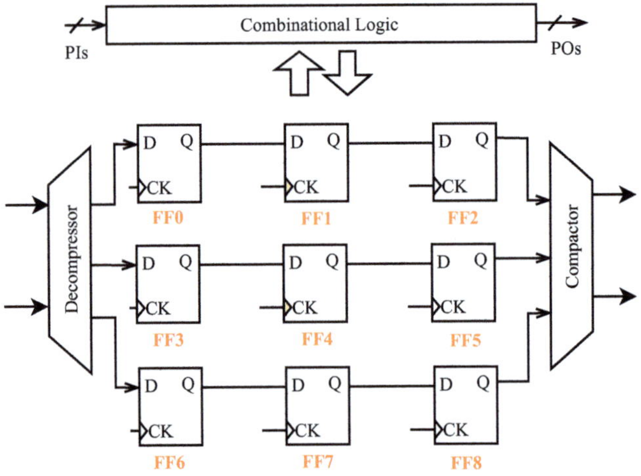

Fig. 1.7 Sequential design with the decompressor and compressor

- *Platform-Level SAT Modeling:* Modeling of the design for the platform-level SAT attack needs to take the platform-level parameters into consideration. In this example of measurement flow, we consider only the impact of the compression ratio (CR) as this is the most important factor in measuring platform-level SAT resiliency. Figure 1.7 shows a typical sequential design with a decompressor and compactor circuit. Consider the design in Fig. 1.7, which has three scan chains with three flip-flops in each chain. While applying the desired input pattern, it needs to go through the decompressor, and after three clock cycles, the pattern is shifted into all nine flip-flops. Similarly, while capturing the values out of the flip-flops, they pass through the compactor circuits, and it takes three clock cycles to shift out all the flip-flops' values. To model the design with the decompressor and the compactor, we need to mimic the exact same behavior in the framed design. Note that the framed design is a one-cycle design. So to mimic the shifting in and shifting out of the data in three cycles, three copies of the decompressor and three copies of the compactors should be instantiated instead of one copy. For example, let us consider the SAT modeling with the compactor circuit. While shifting out the values, FF2, FF5, and FF8 are applied to the compactor in the first clock cycle. Similarly, the group of FF1, FF4, and FF7 and the group of FF0, FF3, and FF6 are applied to the compactor at the second and third clock cycle, respectively. This phenomenon reflects the scenario of resource sharing. In other words, one compactor provides the compression to three groups of flip-flops at different times. Now, if we are interested in shifting out the values of the three groups in one clock cycle, one possible solution would be to connect a separate compactor to each of these groups. In other words, if we connect three compactors to these three groups, we can shift out all the values in one cycle. We leveraged this while modeling the design with the decompressor and the compactor circuit. Figure 1.8 shows our approach to model a framed design with the compactor circuit. In a framed design, all the flip-flops' D and Q pins are introduced as primary outputs and primary inputs, respectively. This makes the connection of the compactor even easier. We instantiate three copies of the same compactor and assign each of those into each group of flip-flops. We can see that the one compactor is connected to the group of FF2, FF5, and FF8. Similarly, two other copies of the compactor are connected to the group of FF1, FF4, and FF7 and FF0, FF3, and FF6. This connection can be made both in Verilog and in bench format. In our case, we first converted the design without the decompressor and the compressor into the bench format (CNF). Then we connected the bench format of the decompressor and the compactor with the design bench.

Performing SAT Attack The next step of the platform-level SAT resiliency measurement is performing the actual attack. For this purpose, we used an open-source SAT tool named Pramod [128]. We followed the modeling approach discussed above for both the locked and oracle designs with the compression and decompression circuit. Then both of these models are fed to the spammed tool to measure the number of iteration and the CPU time to extract the key.

1.5.1.2 Platform-Level SAT Resiliency Estimation Flow

We developed a data-driven estimation methodology for estimating platform-level SAT attacking time, while the IP-level SAT attacking time and the value of the SoC-level parameters (in this case, compression ratio (CR)) are given. Toward this goal, we first built a large dataset by measuring the platform-level SAT attacking time for different IPs with different combinations of key length and compression ratios. Analyzing the results, we found an interesting relationship between the compression ratio and SAT attacking time. We observe that nearly all the platform-level SAT attacking time for all the IPs follows a general trend with the increasing value of

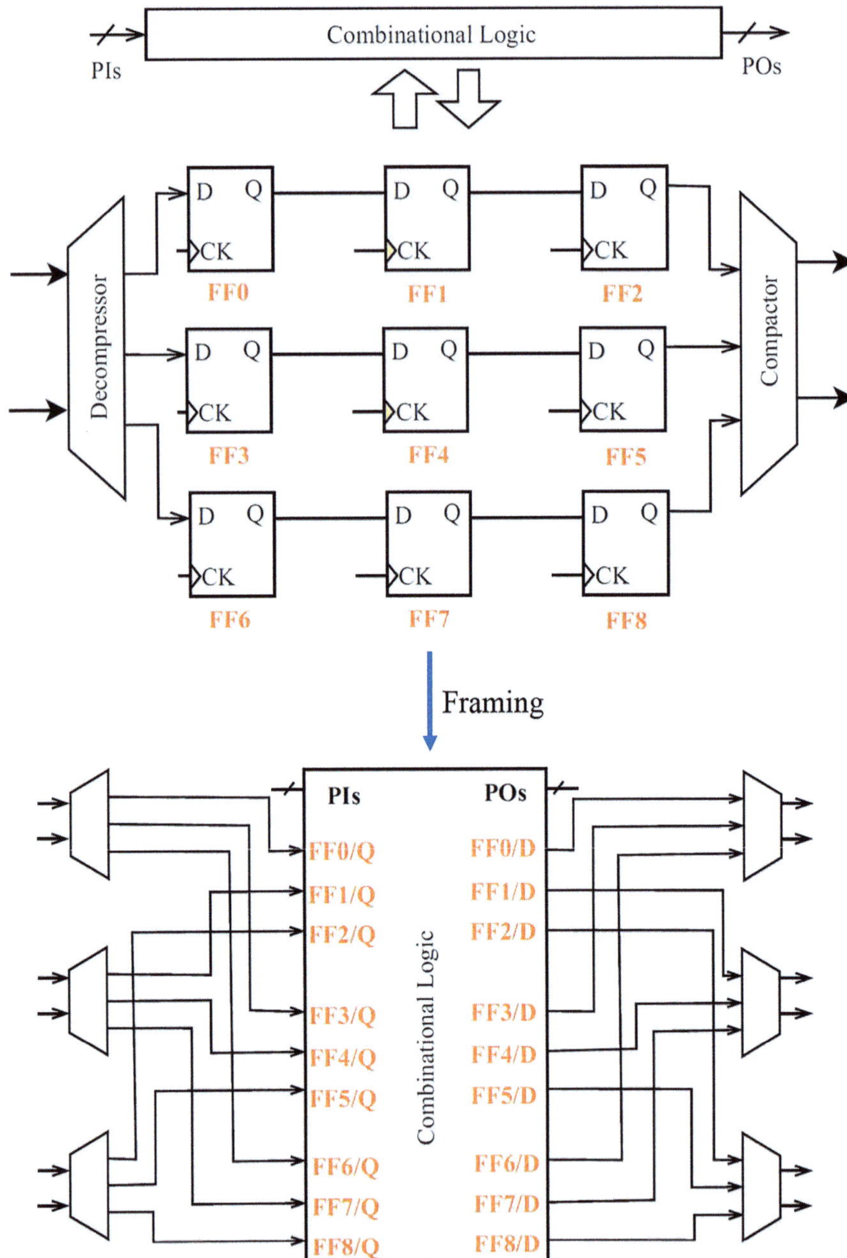

Fig. 1.8 Modeling sequential design with compactor

Table 1.3 Summary of benchmark circuits and details of implementation

Category	Benchmark	# of gates	Compression ratio	# of keys	# of experiments
Small	c499	212	1,2,4,8,16	4	20
	c880	404	1,2,4,8	4	16
	c1355	574	1,2,4,8,16	3	15
	c1908	925	1,2,4,8	4	16
Medium	k2	1908	1,2,4,8,16	3	15
	c3540	1754	1,2,4,8	3	12
	c5315	2427	1,2,4,8,16	3	15
Large	seq	3697	1,2,4,8,16	3	15
	c7552	3695	1,2,4,8,16	2	10
	apex4	5628	1,2,4,	4	2
Total					146

compression ratio. We leveraged this behavior to build our estimation model, which is later used to predict the SAT time for an unknown IP at the platform level.

Data Collection and Modeling We modeled and performed SAT attack for ten different benchmarks with different values of the compression ratio. We divided the benchmarks into three categories: small, medium, and large. We categorized the benchmarks in terms of the number of gates. The benchmarks that have a number of gates less than 1000 are defined as the small benchmark. Benchmarks containing 1000–2500 gates are considered a medium, and the benchmarks having more than 2500 gates are defined as the large benchmark. Each benchmark is locked with a random locking mechanism with four different key lengths. Each version of the design is modeled with the decompressor and compactor of the compression ratio of up to four different values (1, 2, 4, 8, 16). With all these combinations, we performed a total of 146 experiments, shown in Table 1.3.

From each of the experiments, we extracted the following features and their values: key length, number of gates, number of primary inputs, number of primary outputs, number of scan equivalent I/O (FF IO), compression ratio, CPU time, number of total inputs, number of total outputs, and number of iterations. Analyzing the results, we notice a dominant pattern that exhibits the characteristics of most of the designs. Figure 1.9 shows the dominant pattern of four different benchmarks: c499_enc50, k2_enc10, c5315_enc25, and c7552_enc10. Basically, this figure shows how the SAT attacking time (in the Y-axis) changes with the increase of the compression ratio (shown in the X-axis). The SAT attacking time at compression ratio 1 is equivalent to the IP-level SAT attacking time. The other compression ratio values mimic the platform-level scenario having the decompressor/compactor with the corresponding compression ratio. As an example plot for k2_enc10 benchmark, SAT attacking time at compression ratio 1 is around 1 second. This indicates the IP-level scenario. In other words, if the k2_enc10 was attacked as a standalone IP, the time to retrieve the key would have been around 1 second. However, let us consider this IP is placed in a platform where the decompressor and compressor circuit with

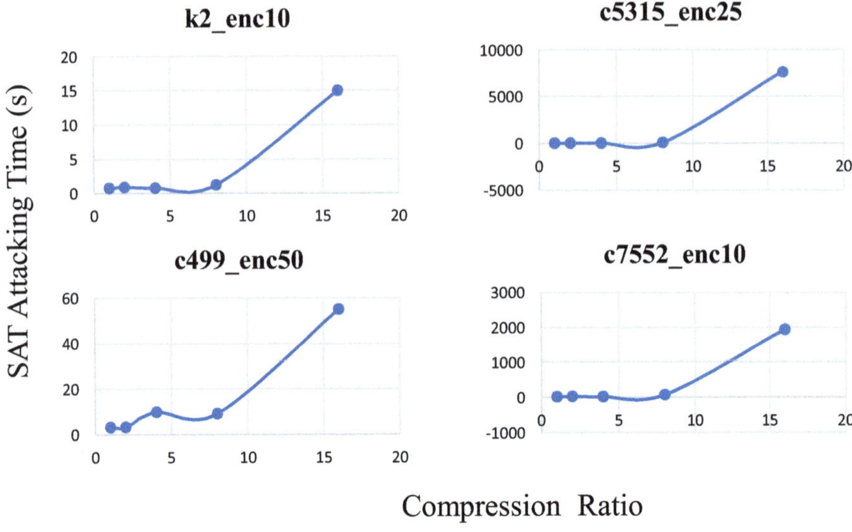

Fig. 1.9 Dominant pattern of CPU time vs. CR for four different benchmarks

compression ratio 16 is implemented. In the plot, we can see that the SAT attacking time now increases to 15 seconds (the last point in the plot). This indicates that the IP-level SAT attacking time does not remain the same at the platform level. In this case, higher compression ratio helps to increase the SAT resiliency at the platform level. Similarly, the impact of other platform-level parameters should also be investigated. However, this study is out of the scope of this chapter.

The next step is how to build a model with the collected data to estimate the SAT resiliency of an unknown IP at the platform level. The dominant patterns shown in Fig. 1.9 seem to follow the same nonlinearity. However, when they are plotted on the same scale, their rising rate seems to be different. Figure 1.10 shows the plot of the dominant patterns on the same scale. From this figure, it is obvious that one single model cannot be used to estimate/predict the SAT attacking time for an unknown IP at the platform level. To address this issue, we carefully selected 20 experimental data out of the total 146 experiments to use in the security estimation modeling. While choosing those 20 experimental data, we focus on bringing the possible diversity. These 20 experimental data are the training dataset of the estimation model based on which the SAT resiliency of unknown IPs at platform level is estimated.

The estimation process comprises two steps: model selection and estimation. As discussed above, the estimation model is built with 20 submodels. While estimating the SAT resiliency for an unknown design, first, it should be identified which submodel (one of the 20 submodels) best suits the unknown design. Toward this goal, we leverage some IP-level metadata of a locked IP. The metadata consists of (1) key length, (2) number of gates, (3) number of primary inputs, (4) number of primary outputs, and (5) number of flip-flops or equivalent ports.

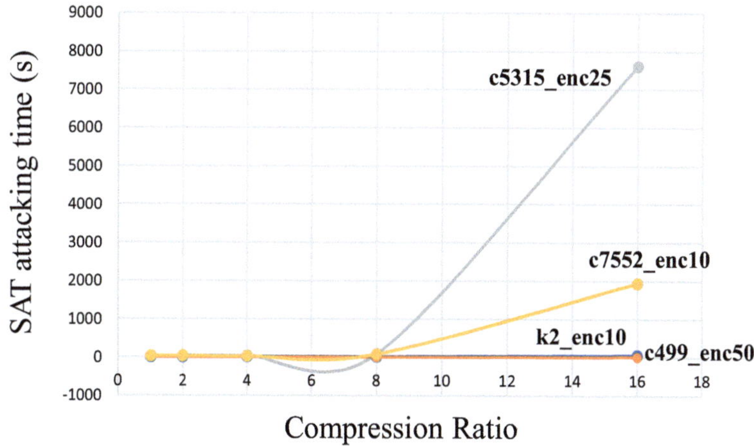

Fig. 1.10 Dominant pattern of CPU time vs. compression ratio for four different benchmark circuits (at the same scale)

We utilize the cosine similarity metric for the best match of the metadata of the unknown design with the submodel. The cosine similarity metric is usually used to find out the angular distance between two vectors. In our case, the metadata is treated as the vector of the items. Then the cosine similarity metric is calculated between the metadata of unknown design and all the 20 submodels. The submodel with the highest similarity metric is selected as the estimation model.

We applied the curve fitting technique to all experimental data to build the submodels. For example, Fig. 1.11 shows the curve fitting approach of one dataset for c499_enc50 benchmark in Table 1.4 with second and third order polynomials. We chose the second order polynomial over the third order for modeling the relationship between SAT attacking time and the compression ratio to avoid any possible overfitting.

1.5.2   Security Measurement for Power Side-Channel Analysis

The PSC measurement and estimation approach focuses on evaluating the PSC vulnerability at the earliest pre-silicon design stage, i.e., register-transfer level (RTL). PSC vulnerability measurement is performed based on the fine-grained simulation of the RTL design. This provides a more accurate approximation of power consumption in each clock cycle of the encryption process. On the other hand, PSC vulnerability estimation depends on a very rough estimate of power consumption based on a few IP attributes rather than an actual simulation of the switching activities. This gives a quick estimate of PSC robustness, but it does not reflect the precise resiliency that the measurement provides. It is worth noting that

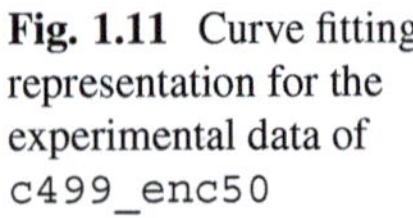

Fig. 1.11 Curve fitting representation for the experimental data of `c499_enc50`

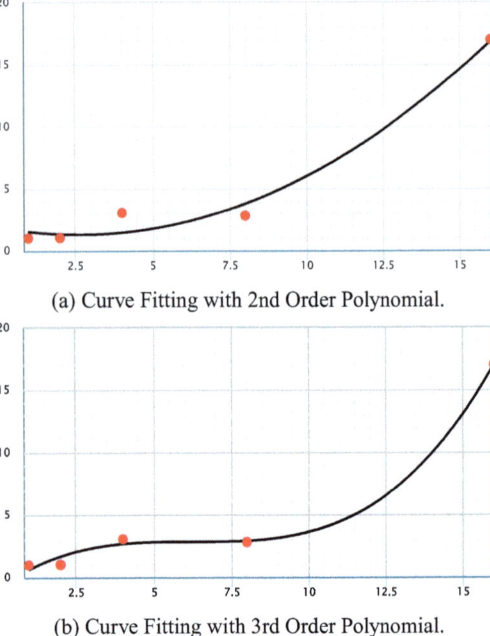

(a) Curve Fitting with 2nd Order Polynomial.

(b) Curve Fitting with 3rd Order Polynomial.

Table 1.4 SAT attack time vs. compression ratios 1–16 for `c499_enc50`

Compression Ratio	1 [normalized]	2	4	8	16
SAT Attack Time (s)	1	1.028257	3.0492296	2.8186724	16.9930236

we limited the scope of the demonstration to see how the most important platform-level factor, i.e., IP-level parallelism, influences power side-channel vulnerability.

1.5.2.1 PSC Vulnerability Measurement Flow

The framework includes two main parts: RTL Switching Activity Interchange Format (SAIF) file generation and identification of vulnerable designs and blocks based on the vulnerability metrics. Then the vulnerable blocks in the design that are leaking information the most are identified for further processing.

Subsystem Definition In order to analyze the influence of extra noises injected into the power trace by concurrently active non-crypto IPs, such as CPUs or peripherals, we consider a small subsystem for preliminary measurements. Figure 1.2 depicts such a subsystem built from standalone IPs, which includes a sample AES core and a few other IP blocks that execute random logic operations unrelated to cryptography. The AES encryption key is the asset that needs to be protected against side-channel attacks. RT level functional simulation is performed to generate power profile/switching activity (toggle count) of design that is used as power traces.

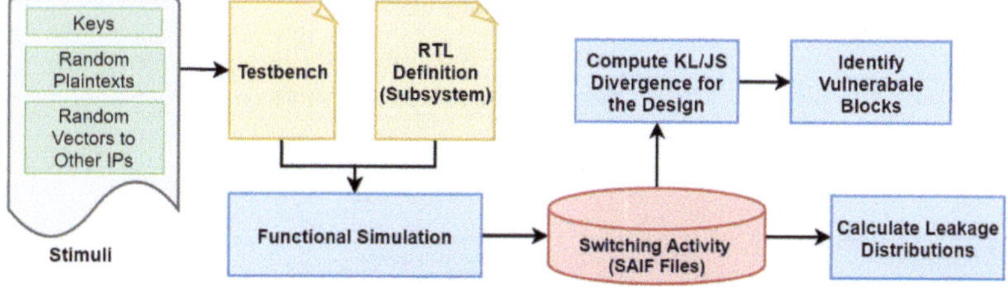

Fig. 1.12 Power side-channel (PSC) vulnerability measurement workflow

We also assume that there is a scheduler in place that controls the simultaneous operation of the additional IPs. For our measurement purposes, we use ISCAS'89 benchmark circuits that differ in terms of the number of inputs/outputs/gates/flip-flops, etc.

Workflow Figure 1.12 illustrates the workflow of the PSC vulnerability measurement to provide a security score for an RTL design of a system/subsystem. The measurement approach starts with the functional simulation of the crypto core and additional IPs. AES core is feed with a set of random plaintexts for two keys from a set of predefined keys or random keys. The additional IP blocks perform random logic operations and add noise to the power trace. The keys are chosen such that each key consists of the same subkey, e.g., $key0 = 0 \times 1515\ldots15$. For $key1$ and $key2$, the hamming distance between two different subkeys is maximum. For $key1$ and $key2$ with a number of random plaintexts, two sets of SAIF files are generated that contains the switching activities/toggle counts, considered as the power trace in the RTL of abstraction. Toggle count for each key per input plaintext is stored into separate SAIF files (Switching Activity Interchange Format). Then the SAIF files are parsed through to generate switching activity distribution profiles for the subsystem and individual IPs for $key1$ and $key2$. Finally, JS divergence metric is calculated between the switching activity to assess RTL-based PSC vulnerability of the subsystem.

Key Pair Selection For the AES-LUT implementation, the encryption operation takes 11 cycles to finish. We collect switching activity traces for 1000 random plaintexts with two different keys for each clock cycle. The best key pair among all possible key pairs is expected to provide the maximum KL divergence for vulnerability evaluation in the worst-case scenario. However, it is impossible and impractical to find the best key pair since the keyspace is huge, i.e., $\binom{2^{128}}{2}$. So, the key pairs are selected empirically such that each key consists of the same subkey, and the hamming distance between two different subkeys is maximum. Table 1.5 shows such a key pair used for AES crypto operations.

Evaluation Metric JS divergence metric estimates statistical distance between two different probability distributions. For instance, if power leakage probability

Table 1.5 Sample key pair with maximum hamming distance

Key_1	0x0000_0000_0000_0000
Key_2	0xFFFF_FFFF_FFFF_FFFF

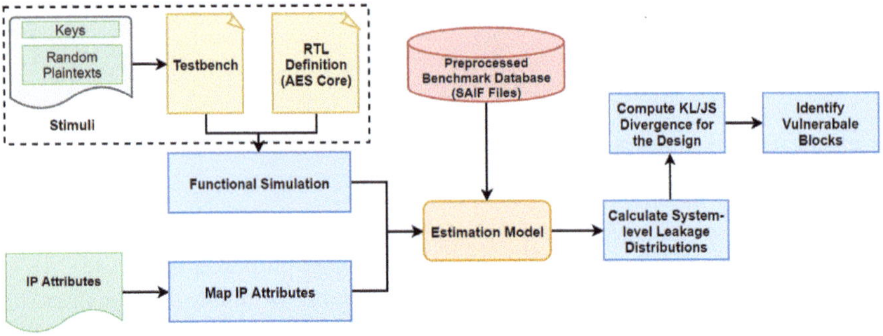

Keys

Random Plaintexts

Stimuli

Testbench

RTL Definition (AES Core)

Preprocessed Benchmark Database (SAIF Files)

Compute KL/JS Divergence for the Design

Identify Vulnerabale Blocks

Functional Simulation

Estimation Model

Calculate System-level Leakage Distributions

IP Attributes

Map IP Attributes

Fig. 1.13 Power side-channel (PSC) vulnerability estimation workflow

distributions based on two different keys are distinguishable, JS divergence between these two distributions is high, which provides indication on how vulnerable the implementation is. The higher the difference between two power leakage distributions is, the higher impact the key has on the power consumption of the design/blocks, the more susceptible the design/blocks are to power analysis attack. For instance, if power leakage probability distributions based on two different keys are distinguishable, KL divergence between these two distributions is high, which indicates how vulnerable the implementation is. The JS divergence value for a design is translated to a security score between 1 and 5 based on predefined threshold value. The higher security score value refers to a more secure design.

1.5.2.2   PSC Vulnerability Estimation

Figure 1.13 illustrates the workflow of the PSC vulnerability estimation to provide a security score for an RTL design of a system/subsystem. Extensive simulation is not required for the estimation, rather each IP is mapped to an IP from a set of preprocessed IP database whose switching activity profiles are already generated. The mapping is performed based on the IP attributes such as the number of inputs/outputs, D-flip-flops, inverters and number of gates (AND, NAND, OR, and NOR), etc. Then, the switching activity for the system is calculated from the AES core and mapped IPs. The rest follows the measurement process to calculate the JS divergence and provide an estimated security score.

Preprocessed Benchmark Circuits   Unlike measurement, the estimation approach does not include the exhaustive simulation/emulation process. In order to map the IPs of the system to preprocessed IPs, we created a database with a number of IPs

Table 1.6 Preprocessed benchmark circuits used for sample subsystems

Bench name	# of inputs	# of outputs	D-FF	Inverters	Gates	AND	NAND	OR	NOR
s298	3	6	14	44	75	31	9	16	19
s344	9	11	15	59	101	44	18	9	30
s386	7	7	6	41	118	83	0	35	0
s400	3	6	21	58	106	11	36	25	34
s420	18	1	16	78	14	49	19	18	34
s444	3	6	21	62	119	13	58	14	34
s510	19	7	6	32	179	34	61	29	55
s526	3	6	21	52	141	56	22	28	35
s641	35	24	19	272	107	90	4	13	0
s713	35	23	19	254	139	94	28	17	0
s820	18	19	5	33	256	76	54	60	66
s832	18	19	5	25	262	78	54	64	66
s838	34	1	32	158	288	105	57	56	70
s953	16	23	29	84	311	49	114	36	112
s1196	14	14	18	141	388	118	119	101	50
s1238	14	14	18	80	428	134	125	112	57
s1423	17	5	74	167	490	197	64	137	92
s1488	8	19	6	103	550	350	0	200	0
s5378	35	49	179	1775	1004	0	0	239	765
s9234	36	39	211	3570	2027	955	528	431	113
s13207	62	152	638	5378	2573	1114	849	512	98
s15850	77	150	534	6324	3448	1619	968	710	151
s38417	28	106	1636	13470	8709	4154	2050	226	2279
s38584	38	304	1426	7805	11448	5516	2126	2621	1185

from the ISCAS'89 benchmark suites. The properties of the benchmarks circuits are shown in Table 1.6. The benchmark circuits are simulated with random input vectors to mimic the activity in a real system-on-chip, where the additional noises are uncorrelated to the crypto operations.

1.5.2.3 Results

Now, in order to measure the system-level vulnerability of a design, we utilize the PSC vulnerability estimation framework to generate the switching activity of each IP with varying degrees of parallel activity. For a small subsystem with an AES core and few other IP blocks performing random logic operations, PSC vulnerability is affected by the additional noises in the power trace introduced by the platform-level parameters that translate to additional toggle count in the RTL. In our preliminary measurement of a small subsystem, we present the effects of these additional switching activities.

Table 1.7 An exemplary illustration of switching activity profile of a subsystem and the corresponding individual IPs for a given key

Subsystem (uut)	AES core	s832	s953	s1488	s5378
22379	11580	242	1314	291	8213
22532	11622	250	1317	311	8282
22750	11496	323	1313	370	8498
22749	11586	325	1315	294	8483
22397	11388	255	1311	294	8464
22555	11461	261	1313	367	8421
22752	11614	243	1311	369	8459
22579	11569	249	1311	369	8459
22528	11470	260	1313	313	8395
22512	11428	253	1311	309	8497

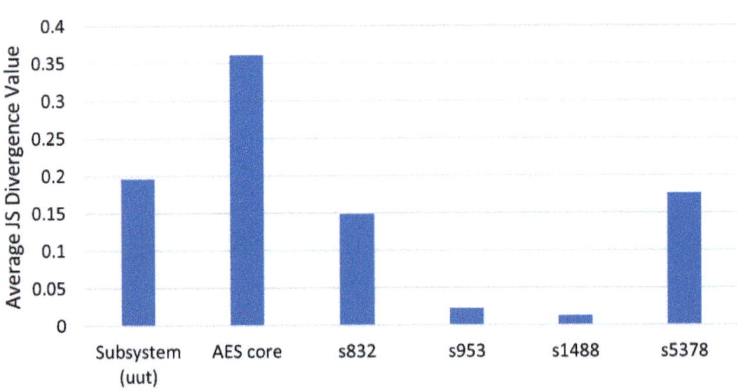

Fig. 1.14 PSC vulnerability measurement results for a sample subsystem

Measurement Results With our framework, we initially perform the functional simulation of a system with the AES-LUT and four benchmark circuits with the stimulus as mentioned above. From the generated SAIF files, the distribution of switching activity is generated for individual IPs, shown in Table 1.7. Here each of the rows corresponds to the toggle count during an encryption operation on a random plaintext for a given key. Column A shows the samples of toggles count for the subsystem, and columns B–F show the toggle counts of AES core and four benchmarks. The additional switching activities increase the noise in the distribution. The switching activity (toggle rate) of the sample subsystem and individual IPs are calculated for each clock cycle and each plaintext with key_1 and key_2. The two sets of data are used to compute the Kullback–Leibler or Jensen–Shannon divergence to find similarity between the two distributions. The calculated JS divergence matrix for each clock cycle from key1 and key2 is shown in Fig. 1.14. From the KL divergence matrix, we observe that the AES module has relatively higher correlation of power traces with keys compared to benchmark IPs. It is

Table 1.8 Estimated JS divergence matrix for multiple AES implementations (with different technologies) with varying number of IPs

AES cores	Estimated JS divergence value for subsystems			
	AES	AES+2IPs	AES+4IPs	AES+6IPs
AES-LUT	0.3125	0.2353	0.2012	0.1174
AES-GF	0.2962	0.2877	0.2700	0.2457
AES-CTR	0.1881	0.1897	0.1861	0.1302

expected as the benchmark IPs are based on random inputs, they show very low level of JS divergence. Now, if we look at the JS divergence of the sample subsystem, we see that the additional noises reduce the correlation significantly.

Estimation Results For estimation, individual AES implementations are preprocessed to generate the switching profiles. Next, we use the IP mapping with the input metadata of the IPs and form the composite distribution of switching activities based on the additional noise level. Here, in Table 1.8, the average JS divergence value of three different AES implementation is presented with a different number of additional IPs. For the demonstration purpose, we estimate the power trace for a few sample subsystems with different degrees of additional noises. As we move to the right, more random activities are added to the AES switching activities which decrease the JS divergence value as it successfully hides the correlation of power traces.

Limitations and Scopes We presented the proof of concept for PSC score measurement and estimation for small subsystems. We also demonstrated how additional noises from IP-level parallel activities impact the vulnerability score in a system. However, the estimation model provides very crude approximation of noise as the input attributes of IPs are limited. With additional information such as the IPs that share the power rail, the timing overlap of the IPs, etc., a robust estimation can be modeled to estimate the additional noise level from the metadata of the IPs provided as user input. Then, the composite switching activity profiles for a platform can be generated by adding an appropriate number of benchmark IPs and tuning the weights based on the input platform-level parameter. In order to validate the estimation score, extensive simulation measurements of the platform will be required.

1.6 The Notion of Security Optimization

There have been a lot of research for power, area, and performance optimization of a design [129–133]. However, the security optimization has been explored till date. This section describes a new concept called *Security Optimization*. Power, performance, and area (PPA) optimization has already been an essential part of the design flow. Existing industry-level EDA (electronic design automation) tools contain the feature to optimize these three attributes. However, there is no such concept of security optimization to date, and thus, the tools are not equipped with the security optimization methodology. The fundamental difference between security and the other three attributes: power, performance, and area is that these three attributes are one-dimensional while security is multidimensional. For example, in the perspective of this chapter, the term security is composed of five different threats: IP Piracy, Power Side-Channel Analysis, Fault Injection, Malicious Hardware, and Supply Chain. Researchers have always been considering building countermeasures and detecting and fixing vulnerabilities for individual threats.

To date, there is no comprehensive solution that focuses on improving security as a whole, given the list of threats to be considered. We argue that the security against an individual threat is not entirely orthogonal to others. Security improvement against one threat may weaken the security against others, thus making security optimization a non-trivial task. For example, let us consider the fact described in Sect. 1.5.1.2, different DFT features (i.e., decompressor/compressor) greatly benefit security against IP piracy. However, on the other hand, DFT features might facilitate malicious hardware attacks by leaking secret information. Thus, it is important to develop a methodology to get optimum combined security against the threat model.

Toward this approach, the first and most important task is to identify which parameters' contribution to the security expands across the threat space. For example, DFT is a parameter that affects security against both IP piracy and malicious hardware attack. The second step is to identify the impact factor of the parameters on security enhancement or degradation against individual threats. As shown in Sect. 1.5.1.2, the DFT features (de)compressor enhances the robustness against IP piracy by a specific amount. Similarly, the quantification of security degradation against malicious hardware by information leakage through DFT should also be calculated. Thus, to get optimum security, threats should not be considered individually; instead, a collective approach should be taken to achieve security against all threats without benefiting or hampering the security against any particular threat.

1.7 Challenges in Security Measurement and Estimation

This section presents challenges during estimation, measurement, and optimization of security at the platform level. We list out the challenges in three categories: (1) Challenges in Platform-Level Security Estimation and Measurement, (2) Challenges in Achieving Accurate Estimation, and (3) Challenges in Security Optimization.

1.7.1 Challenges for Quantifiable Platform-Level Security

The major challenges in estimating and measuring the platform-level security are:

1. Identifying the parameters contributing to the platform-level security

2. Modeling the impact of the platform-level parameters on security estimation
3. Platform-level parameters introduction and establishment at the design time while performing the actual attack at the measurement (verification) time

During the transition from IPs to platform, several new parameters are introduced at different stages and abstraction levels of the design flow. However, not all these parameters impact the system's security. Identifying the parameters that impact security against different threats is a challenging task, as there is no systematic approach to accomplish it. Modeling the impact of different platform-level parameters on security is also a non-trivial task. A list of challenges for different threats are:

In IP piracy, (1) the RTL platform does not provide information about the test structure introduced in later design stages. The robustness of a platform against IP piracy through SAT attack greatly depends on the test structure of that platform. Hence, estimating the security based on parameters that do not exist becomes quite challenging. (2) As discussed in Sect. 1.5, the proposed estimation model for IP piracy is built based on a data-driven model, in which a set of data for different designs and the values of the platform-level parameter (in this case, compression ratio) are used to predict/estimate the security of an unknown design. However, selecting a suitable set of such designs and the values of the platform-level parameters is difficult. (3) While estimating an unknown platform's security against IP piracy, the first step is to match the metadata of that design with all the designs used in modeling. A bad choice of metadata can potentially steer the estimation method toward a poor security estimation outcome. However, developing a representative set of metadata is a very challenging task. Also, with the developed metadata, the simple cosine similarity-based matching algorithm does not accurately select the suitable model, which leads to the need for future machine learning (ML)-based approaches [134].

In PSC analysis, (1) parallel activities on the system bus and active IPs that share the power rail with the target IP are the most critical elements in platform-level PSC vulnerability assessment. Without extensive human effort from verification engineers, identifying and controlling IP activities in complex SoC designs becomes extremely challenging. Moreover, HW/SW co-verification for a platform will be required to compute the simulated power traces required for PSC vulnerability estimation and measurement. (2) Because RTL designs do not include any physical information about the blocks, the simulated power model may fail to capture factors such as shared power distribution network, voltage regulator, clock jitters, etc. Analysis at a lower abstraction level (e.g., gate-level and layout-level) may improve the accuracy of the power models. However, each abstraction level increases the analysis time by tenfold, making it nearly impossible to perform gate-level or layout-level analysis for a full-blown SoC design. (3) The data-driven model proposed for the fast estimation of PSC vulnerability may not be able to predict accurate switching activity of the IPs. It will be significantly reliant on the IP properties chosen to match the switching activities of the existing IP repository. ML-based algorithms could be useful for mapping metadata in order to estimate switching activity of the IPs. However, the training phase will have to incorporate power models for a large number of IPs with varying workloads, temporal overlapping, etc. This becomes a challenging task due to the sheer volume of the training data and time required to train the model.

In fault injection, (1) modeling of physical parameters involved with fault injection methods, such as laser, clock, voltage, etc., is the most challenging part when it comes to assessing fault injection susceptibility of a pre-silicon design/SoC, and (2) considering the large size of SoCs, the possible number of fault locations grows exponentially in the design/SoC. Therefore, exhaustive simulation of all fault nodes and analyzing their impact on the design's security is a challenging part.

In malicious hardware, (1) a major platform-level parameter contributing to the platform-level security against malicious hardware is the set of security policies. It is very unlikely that an exhaustive set of security policies will be available at the earlier design stages, which significantly hinders the estimation capability. Even with a set of security policies, quantifying the contribution to platform-level malicious operation is a daunting task. (2) The interaction probability between IPs significantly impacts the security of the entire platform. An IP with a security asset having a high interaction probability with a malicious IP can lead to a greater risk of malicious operation than the interaction with a non-malicious IP. However, getting information about this parameter at an earlier platform stage is a challenging task.

In the supply chain, as discussed in the previous section, the widely used counterfeit IC detection mechanism utilizes the security primitive PUF. As PUF is implemented using the process variation of the chip, it is practically impossible to measure the security against cloning attacks at the RT level, as there is no notion of process variation in this stage of the design.

1.7.2 Challenges in Achieving Accurate Estimation

The security estimation should provide a highly accurate estimate of the platform-level security, closely resembling the security measured in a full-blown SoC in silicon. The challenge is that the estimation is performed at the earlier stages (e.g., RTL and gate level), and we aim to estimate the security at the SoC in silicon. Consider estimating the platform-level security against PSC attacks at the RT level. It is quite challenging to accurately model the impact of the platform-level parameters, such as PDN, Decap, and DVFS, on the security at the RT level. These parameters are not available at the RT level and become available later in the design flow. The only information about the power traces that can be achieved at this stage is the switching activity, which potentially leads to a poor estimation. Thus, estimating the security at the RT level closely representative of the platform-level security at silicon is a very challenging task. This is mainly because most platform-level information is unavailable at the RTL or gate level.

1.7.3 Challenges in Security Optimization

Security optimization is another challenge as improving security against one threat may impact the robustness against the other threat models. Unlike power, performance, and area, security is not a single term; instead, it comprises a set of diverse threat models. For example, if a particular chip is claimed to be x% secure, it does not provide specific information. Does it mean that the chip is x% secure against all possible attacks? It is not possible to claim that a chip is secure against all threats. This is because the threats are not confined to a particular number. There are already a large number of threats available, and many more threats are yet to come. Thus, as a first step, the threat models should be clearly defined before estimating and measuring the security of a design. The security against power side-channel attacks may be orthogonal to the security against other threats such as IP protection through logic locking. Hence, techniques for measuring security against different threats should be developed separately, but the impact of one on another threat model must be measured or estimated within the SoC as well.

References

1. M. Yasin, B. Mazumdar, J.J.V. Rajendran, O. Sinanoglu, SARLock: SAT attack resistant logic locking. in: *2016 IEEE International Symposium on Hardware Oriented Security and Trust (HOST)* (IEEE, New York, 2016), pp. 236–241
2. P. Kocher, J. Jaffe, B. Jun, Differential power analysis, in *Annual International Cryptology Conference* (Springer, New York, 1999), pp. 388–397
3. M. Tehranipoor, F. Koushanfar, A survey of hardware Trojan taxonomy and detection. IEEE Des. Test Comput. **27**(1), 10–25 (2010)
4. U. Guin, X. Zhang, D. Forte, M. Tehranipoor, Low-cost on-chip structures for combating die and IC recycling, in *2014 51st ACM/EDAC/IEEE Design Automation Conference (DAC)* (IEEE, New York, 2014), pp. 1–6
5. N. Ahmed, M. Tehranipoor, V. Jayaram, A novel framework for faster-than-at-speed delay test considering IR-drop effects, in *Proceedings of the 2006 IEEE/ACM International Conference on Computer-Aided Design* (2006), pp. 198–203
6. N. Ahmed, C.P. Ravikumar, M. Tehranipoor, J. Plusquellic, At-speed transition fault testing with low speed scan enable, in *23rd IEEE VLSI Test Symposium (VTS'05)* (IEEE, New York, 2005), pp. 42–47
7. M. Tehranipoor, K. Peng, K. Chakrabarty, *Test and Diagnosis for Small-delay Defects* (Springer, Berlin, 2011)
8. M. Yilmaz, K. Chakrabarty, M. Tehranipoor, Test-pattern selection for screening small-delay defects in very-deep submicrometer integrated circuits. IEEE Trans. Comput. Aided Des. Integr. Circuits Syst. **29**(5), 760–773 (2010)
9. J. Lee, S. Narayan, M. Kapralos, M. Tehranipoor, Layout-aware, IR-drop tolerant transition fault pattern generation, in *Proceedings of the Conference on Design, Automation and Test in Europe* (2008), pp. 1172–1177
10. N. Ahmed, M. Tehranipoor, V. Jayaram, Supply voltage noise aware ATPG for transition delay faults, in *25th IEEE VLSI Test Symposium (VTS'07)* (IEEE, New York, 2007), pp. 179–186

11. M.H. Tehranipour, N. Ahmed, M. Nourani, Testing SoC interconnects for signal integrity using boundary scan, in *Proceedings of the 21st VLSI Test Symposium, 2003* (IEEE, New York, 2003), pp. 158–163

12. N. Ahmed, M.H. Tehranipour, M. Nourani, Low power pattern generation for BIST architecture, in *2004 IEEE International Symposium on Circuits and Systems (IEEE Cat. No. 04CH37512)*, vol. 2 (IEEE, New York, 2004), pp. II–689

13. J. Ma, J. Lee, M. Tehranipoor, Layout-aware pattern generation for maximizing supply noise effects on critical paths, in *2009 27th IEEE VLSI Test Symposium* (IEEE, New York, 2009), pp. 221–226

14. K.Z. Azar, M.M. Hossain, A. Vafaei, H. Al Shaikh, N.N. Mondol, F. Rahman, M. Tehranipoor, F. Farahmandi, Fuzz, penetration, and ai testing for soc security verification: challenges and solutions, in *Cryptology ePrint Archive* (2022)

15. H. Al-Shaikh, A. Vafaei, M.M.M. Rahman, K.Z. Azar, F. Rahman, F. Farahmandi, M. Tehranipoor, Sharpen: Soc security verification by hardware penetration test, in *Proceedings of the 28th Asia and South Pacific Design Automation Conference* (2023), pp. 579–584

16. N.N. Mondol, A. Vafaei, K.Z. Azar, F. Farahmandi, M. Tehranipoor, RL-TPG: Automated Pre-Silicon Security Verification through Reinforcement Learning-Based Test Pattern Generation, in *Design, Automation and Test in Europe (DATE)* (IEEE, New York, 2024), pp. 1–6

17. H. Salmani, M. Tehranipoor, J. Plusquellic, A layout-aware approach for improving localized switching to detect hardware Trojans in integrated circuits, in *2010 IEEE International Workshop on Information Forensics and Security* (IEEE, New York, 2010), pp. 1–6

18. C. Lamech, R.M. Rad, M. Tehranipoor, J. Plusquellic, An experimental analysis of power and delay signal-to-noise requirements for detecting Trojans and methods for achieving the required detection sensitivities. IEEE Trans. Inf. Forensics Secur. **6**(3), 1170–1179 (2011)

19. K. Ahi, N. Asadizanjani, S. Shahbazmohamadi, M. Tehranipoor, M. Anwar, Terahertz characterization of electronic components and comparison of terahertz imaging with x-ray imaging techniques, in *Terahertz Physics, Devices, and Systems IX: Advanced Applications in Industry and Defense*, vol. 9483 (International Society for Optics and Photonics, New York, 2015), 94830K

20. J. Villasenor, M. Tehranipoor, Chop shop electronics. IEEE Spectr. **50**(10), 41–45 (2013)

21. N. Karimian, Z. Guo, M. Tehranipoor, D. Forte, Highly reliable key generation from electrocardiogram (ECG). IEEE Trans. Biomed. Eng. **64**(6), 1400–1411 (2016)

22. M.T. Rahman, D. Forte, Q. Shi, G.K. Contreras, M. Tehranipoor, CSST: Preventing distribution of unlicensed and rejected ICs by untrusted foundry and assembly, in *2014 IEEE International Symposium on Defect and Fault Tolerance in VLSI and Nanotechnology Systems (DFT)* (IEEE, New York, 2014), pp. 46–51

23. M.S. Rahman, A. Nahiyan, S. Amir, F. Rahman, F. Farahmandi, D. Forte, M. Tehranipoor, Dynamically obfuscated scan chain to resist oracle-guided attacks on logic locked design, in *Cryptology ePrint Archive* (2019)

24. H.M. Kamali, K.Z. Azar, S. Roshanisefat, A. Vakil, H. Homayoun, A. Sasan, ExTru: a lightweight, fast, and secure expirable trust for the Internet of things, in *2020 IEEE 14th Dallas Circuits and Systems Conference (DCAS)* (IEEE, New York, 2020), pp. 1–6

25. A. Nahiyan, F. Farahmandi, P. Mishra, D. Forte, M. Tehranipoor, Security-aware FSM design flow for identifying and mitigating vulnerabilities to fault attacks. IEEE Trans. Comput. Aided Des. Integr. Circuits Syst. **38**(6), 1003–1016 (2018)

26. X. Zhou, B. Ahmed, J.H. Aylor, P. Asare, H. Alemzadeh, Data-driven Design of Context-aware Monitors for Hazard Prediction in Artificial Pancreas Systems. arXiv preprint arXiv:2104.02545 (2021)

27. M.S.U.I. Sami, F. Rahman, F. Farahmandi, A. Cron, M. Borza, M. Tehranipoor, Invited: End-to-End Secure SoC Lifecycle Management. in *2021 58th ACM/IEEE Design Automation Conference (DAC)* (2021), pp. 1295–1298. https://doi.org/10.1109/DAC18074.2021.9586106

28. M.S.U.I. Sami, F. Rahman, A. Cron, D. Donchin, M. Borza, F. Farahmandi, M. Tehranipoor, POCA: First Power-on Chip Authentication in Untrusted Foundry and Assembly, in *2021 IEEE International Symposium on Hardware Oriented Security and Trust (HOST)* (2021)

29. K.Z. Azar, H.M. Kamali, F. Farahmandi, M. Tehranipoor, *Understanding Logic Locking* (2023)
30. J. Lee, M. Tehranipoor, C. Patel, J. Plusquellic, Securing scan design using lock and key technique, in *20th IEEE International Symposium on Defect and Fault Tolerance in VLSI Systems (DFT'05)* (IEEE, New York, 2005), pp. 51–62
31. J. Lee, M. Tebranipoor, J. Plusquellic, A low-cost solution for protecting IPs against scan-based side-channel attacks, in *24th IEEE VLSI Test Symposium* (IEEE, New York, 2006), 6–pp
32. J. Lee, M. Tehranipoor, C. Patel, J. Plusquellic, Securing designs against scan-based side-channel attacks. IEEE Trans. Dependable Secure Comput. **4**(4), 325–336 (2007)
33. P. Subramanyan, S. Ray, S. Malik, Evaluating the security of logic encryption algorithms, in *2015 IEEE International Symposium on Hardware Oriented Security and Trust (HOST)* (IEEE, New York, 2015), pp. 137–143
34. K.Z. Azar, H.M. Kamali, H. Homayoun, A. Sasan, SMT attack: next generation attack on obfuscated circuits with capabilities and performance beyond the SAT attacks, in *IACR Transactions on Cryptographic Hardware and Embedded Systems (TCHES)* (2019), pp. 97–122
35. S. Roshanisefat, H.M. Kamali, H. Homayoun, A. Sasan, RANE: an open-source formal de-obfuscation attack for reverse engineering of logic encrypted circuits, in *Great Lakes Symposium on VLSI (GLSVLSI)* (2021), pp. 221–228
36. K.Z. Azar, H.M. Kamali, F. Farahmandi, M. Tehranipoor, Warm Up before Circuit De-obfuscation? An Exploration through Bounded-Model- Checkers, in *International Symposium on Hardware Oriented Security and Trust (HOST)* (2022), pp. 1–4
37. K.Z. Azar, F. Farahmand, H.M. Kamali, S. Roshanisefat, H. Homayoun, W. Diehl, K. Gaj, A. Sasan, COMA: communication and obfuscation management architecture, in: *Int'l Symposium on Research in Attacks, Intrusions and Defenses (RAID)* (2019), pp. 181–195
38. M. El Massad, S. Garg, M. Tripunitara, Reverse engineering camouflaged sequential circuits without scan access, in *2017 IEEE/ACM International Conference on Computer-Aided Design (ICCAD)* (IEEE, New York, 2017), pp. 33–40
39. J. Daemen, V. Rijmen, *The design of Rijndael*, vol. 2 (Springer, Berlin, 2002)
40. P. Kocher, J. Jaffe, B. Jun, P. Rohatgi, Introduction to differential power analysis. J. Cryptogr. Eng. **1**(1), 5–27 (2011)
41. E. Brier, C. Clavier, F. Olivier, Correlation power analysis with a leakage model, in *International Workshop on Cryptographic Hardware and Embedded Systems* (Springer, Berlin, 2004), pp. 16–29
42. M.-C. Hsueh, T.K. Tsai, R.K. Iyer, Fault injection techniques and tools. Computer **30**(4), 75–82 (1997)
43. H. Ziade, R.A. Ayoubi, R. Velazco, et al., A survey on fault injection techniques. Int. Arab J. Inf. Technol **1**(2), 171–186 (2004)
44. P. Dusart, G. Letourneux, O. Vivolo, Differential fault analysis on AES, in *International Conference on Applied Cryptography and Network Security* (Springer, Berlin, 2003), pp. 293–306
45. A. Nahiyan, K. Xiao, K. Yang, Y. Jin, D. Forte, M. Tehranipoor, AVFSM: a framework for identifying and mitigating vulnerabilities in FSMs, in *2016 53nd ACM/EDAC/IEEE Design Automation Conference (DAC)* (IEEE, New York, 2016), pp. 1–6
46. K. Murdock, D. Oswald, F.D. Garcia, J.V. Bulck, D. Gruss, F. Piessens, Plundervolt: software-based fault injection attacks against Intel SGX, in *2020 IEEE Symposium on Security and Privacy (SP)* (IEEE, New York, 2020), pp. 1466–1482
47. J. Krautter, D.R.E. Gnad, M.B. Tahoori, FPGAhammer: remote voltage fault attacks on shared FPGAs, suitable for DFA on AES, in *IACR Transactions on Cryptographic Hardware and Embedded Systems* (2018), pp. 44–68
48. M. Agoyan, J.-M. Dutertre, D. Naccache, B. Robisson, A. Tria, When clocks fail: on critical paths and clock faults, in *International Conference on Smart Card Research and Advanced Applications* (Springer, Berlin, 2010), pp. 182–193

49. J.G.J. Van Woudenberg, M.F. Witteman, F. Menarini, Practical optical fault injection on secure microcontrollers, in *2011 Workshop on Fault Diagnosis and Tolerance in Cryptography* (IEEE, New York, 2011), pp. 91–99
50. M. Dumont, M. Lisart, P. Maurine, Electromagnetic fault injection: how faults occur, in *2019 Workshop on Fault Diagnosis and Tolerance in Cryptography (FDTC)* (IEEE, New York, 2019), pp. 9–16
51. M.R. Muttaki, M. Tehranipoor, F. Farahmandi, FTC: a universal fault injection attack detection sensor, in *IEEE International Symposium on Hardware-Oriented Security and Trust (HOST)* (IEEE, New York, 2022)
52. M. Tehranipoor, H. Salmani, X. Zhang, M. Wang, R. Karri, J. Rajendran, K. Rosenfeld, Trustworthy hardware: Trojan detection and design-for-trust challenges. Computer **44**(7), 66–74 (2010)
53. S. Bhunia, M. Tehranipoor, *Hardware Security: A Hands-on Learning Approach* (Morgan Kaufmann, Burlington, 2018)
54. P. Mishra, S. Bhunia, M. Tehranipoor, *Hardware IP Security and Trust* (Springer, Berlin, 2017)
55. N. Asadizanjani, M.T. Rahman, M. Tehranipoor, Counterfeit detection and avoidance with physical inspection, in *Physical Assurance* (Springer, Berlin, 2021), pp. 21–47
56. A. Stern, U. Botero, F. Rahman, D. Forte, M. Tehranipoor, EMFORCED: EM-based fingerprinting framework for remarked and cloned counterfeit IC detection using machine learning classification. IEEE Trans. Very Large Scale Integr. VLSI Syst. **28**(2), 363–375 (2019)
57. X. Wang, Y. Han, M. Tehranipoor, System-level counterfeit detection using on-chip ring oscillator array. IEEE Trans. Very Large Scale Integr. VLSI Syst. **27**(12), 2884–2896 (2019)
58. K. Yang, D. Forte, M.M. Tehranipoor, CDTA: A comprehensive solution for counterfeit detection, traceability, and authentication in the IoT supply chain. ACM Trans. Des. Autom. Electron. Syst. (TODAES) **22**(3), 1–31 (2017)
59. N. Asadizanjani, M. Tehranipoor, D. Forte, Counterfeit electronics detection using image processing and machine learning. J. Phys. Conf. Ser. **787**(1), 012023 (2017). IOP Publishing
60. N. Asadizanjani, S. Gattigowda, M. Tehranipoor, D. Forte, N. Dunn, A database for counterfeit electronics and automatic defect detection based on image processing and machine learning, in *ISTFA 2016* (ASM International, New York, 2016), pp. 580–587
61. K. Xiao, D. Forte, M.M. Tehranipoor, Circuit timing signature (CTS) for detection of counterfeit integrated circuits, in *Secure System Design and Trustable Computing* (Springer, Berlin, 2016), pp. 211–239
62. K. Yang, D. Forte, M. Tehranipoor, An RFID-based technology for electronic component and system counterfeit detection and traceability, in *2015 IEEE International Symposium on Technologies for Homeland Security (HST)* (IEEE, New York, 2015), pp. 1–6
63. M.M. Tehranipoor, U. Guin, D. Forte, Counterfeit test coverage: an assessment of current counterfeit detection methods, in *Counterfeit Integrated Circuits* (Springer, Berlin, 2015), pp. 109–131
64. S. Shahbazmohamadi, D. Forte, M. Tehranipoor, Advanced physical inspection methods for counterfeit IC detection, in *ISTFA 2014: Conference Proceedings from the 40th International Symposium for Testing and Failure Analysis* (ASM International, New York, 2014), p. 55
65. U. Guin, K. Huang, D. DiMase, J.M. Carulli, M. Tehranipoor, Y. Makris, Counterfeit integrated circuits: a rising threat in the global semiconductor supply chain. Proc. IEEE **102**(8), 1207–1228 (2014)
66. U. Guin, D. DiMase, M. Tehranipoor, A comprehensive framework for counterfeit defect coverage analysis and detection assessment. J. Electron. Test. **30**(1), 25–40 (2014)
67. M. Tehranipoor, D. Forte, Tutorial T4: All you need to know about hardware Trojans and Counterfeit ICs, in *2014 27th International Conference on VLSI Design and 2014 13th International Conference on Embedded Systems* (IEEE, New York, 2014), pp. 9–10

68. M. Tehranipoor, H. Salmani, X. Zhang, Counterfeit ICs: Detection and Prevention of Recycled ICs Using On-Chip Sensors, in *Integrated Circuit Authentication* (Springer, New York, 2014), pp. 179–205
69. U. Guin, D. Forte, M. Tehranipoor, Anti-counterfeit techniques: from design to resign, in *2013 14th International Workshop on Microprocessor Test and Verification* (IEEE, New York, 2013), pp. 89–94
70. U. Guin, M. Tehranipoor, On selection of counterfeit IC detection methods, in *IEEE North Atlantic Test Workshop (NATW)* (2013)
71. U. Guin, M. Tehranipoor, *Counterfeit Detection Technology Assessment* (2013)
72. U. Guin, D. Forte, M. Tehranipoor, Design of accurate low-cost on-chip structures for protecting integrated circuits against recycling. IEEE Trans. Very Large Scale Integr. VLSI Syst. **24**(4), 1233–1246 (2015)
73. S. Ray, E. Peeters, M.M. Tehranipoor, S. Bhunia, System-on-chip platform security assurance: architecture and validation. Proc. IEEE **106**(1), 21–37 (2017)
74. H.M. Kamali, K.Z. Azar, F. Farahmandi, M. Tehranipoor, Advances in logic locking: past, present, and prospects, in *Cryptology ePrint Archive* (2022)
75. K. Zamiri Azar, H. Mardani Kamali, H. Homayoun, A. Sasan, Threats on logic locking: A decade later, in *Proceedings of the 2019 on Great Lakes Symposium on VLSI* (2019), pp. 471–476
76. H.M. Kamali, K.Z. Azar, K. Gaj, H. Homayoun, A. Sasan, LUT-Lock: A Novel LUT-based Logic Obfuscation for FPGA-bitstream and ASIC-hardware Protection, in *IEEE Computer Society Annual Symposium on VLSI (ISVLSI)* (2018), pp. 405–410
77. Y. Xie, A. Srivastava, Mitigating SAT attack on logic locking, in *International Conference on Cryptographic Hardware and Embedded Systems* (Springer, 2016), pp. 127–146
78. P. Morawiecki, M. Srebrny, A SAT-based preimage analysis of reduced KECCAK hash functions. Inf. Process. Lett. **113**(10–11), 392–397 (2013)
79. H.M. Kamali, K.Z. Azar, H. Homayoun, A. Sasan, Full-Lock: hard distributions of SAT instances for obfuscating circuits using fully configurable logic and routing blocks, in *Proceedings of Design Automation Conference (DAC)* (2019), p. 89
80. H.M. Kamali, K.Z. Azar, H. Homayoun, A. Sasan, InterLock: an intercorrelated logic and routing locking, in *IEEE/ACM International Conference on Computer-Aided Design (ICCAD)* (2020), pp. 1–9
81. S. Roshanisefat, H.M. Kamali, K.Z. Azar, S.M.P. Dinakarrao, N. Karimi, H. Homayoun, A. Sasan, DFSSD: deep faults and shallow state duality, a provably strong obfuscation solution for circuits with restricted access to scan chain, in *VLSI Test Symposium (VTS)* (2020), pp. 1–6
82. J. Rajendran, Y. Pino, O. Sinanoglu, R. Karri, Security analysis of logic obfuscation, in *Proceedings of the 49th Annual Design Automation Conference* (2012), pp. 83–89
83. Y. Liu, M. Zuzak, Y. Xie, A. Chakraborty, A. Srivastava, Robust and attack resilient logic locking with a high application-level impact. ACM J. Emerg. Technol. Comput. Syst. (JETC) **17**(3), 1–22 (2021)
84. Y. Yano, K. Iokibe, Y. Toyota, T. Teshima, Signal-to-noise ratio measurements of side-channel traces for establishing low-cost countermeasure design, in *2017 Asia-Pacific International Symposium on Electromagnetic Compatibility (APEMC)* (IEEE, New York, 2017), pp. 93–95
85. S. Mangard, Hardware countermeasures against DPA–a statistical analysis of their effectiveness, in *Cryptographers' Track at the RSA Conference* (Springer, Berlin, 2004), pp. 222–235
86. B.J. Gilbert Goodwill, J. Jaffe, P. Rohatgi, et al., A testing methodology for side-channel resistance validation, in *NIST Non-invasive Attack Testing Workshop*, vol. 7 (2011), pp. 115–136
87. S. Kullback, R. ALeibler, On information and sufficiency. Ann. Math. Stat. **22**(1), 79–86 (1951)
88. M. He, J. Park, A. Nahiyan, A. Vassilev, Y. Jin, M. Tehranipoor, RTL-PSC: automated power side-channel leakage assessment at register-transfer level, in *2019 IEEE 37th VLSI Test Symposium (VTS)* (IEEE, New York, 2019), pp. 1–6

89. F.-X. Standaert, T.G. Malkin, M. Yung, A unified framework for the analysis of side-channel key recovery attacks, in *Annual International Conference on the Theory and Applications of Cryptographic Techniques* (Springer, Berlin, 2009), pp. 443–461
90. A. Nahiyan, J. Park, M. He, Y. Iskander, F. Farahmandi, D. Forte, M. Tehranipoor, Script: a cad framework for power side-channel vulnerability assessment using information flow tracking and pattern generation. ACM Trans. Des. Autom. Electron. Syst. (TODAES) **25**(3), 1–27 (2020)
91. D. Jayasinghe, R. Ragel, J.A. Ambrose, A. Ignjatovic, S. Parameswaran, Advanced modes in AES: are they safe from power analysis based side channel attacks?, in *2014 IEEE 32nd International Conference on Computer Design (ICCD)* (2014), pp. 173–180. https://doi.org/10.1109/ICCD.2014.6974678
92. M. Tunstall, D. Mukhopadhyay, S. Ali, Differential fault analysis of the advanced encryption standard using a single fault, in *IFIP International Workshop on Information Security Theory and Practices* (Springer, Berlin, 2011), pp. 224–233
93. N.F. Ghalaty, B. Yuce, M. Taha, P. Schaumont, Differential fault intensity analysis, in *2014 Workshop on Fault Diagnosis and Tolerance in Cryptography* (IEEE, New York, 2014), pp. 49–58
94. F. Zhang, X. Lou, X. Zhao, S. Bhasin, W. He, R. Ding, S. Qureshi, K. Ren, Persistent fault analysis on block ciphers, in *IACR Transactions on Cryptographic Hardware and Embedded Systems* (2018), pp. 150–172
95. K.R. Kantipudi, Controllability and observability, in *ELEC7250-001 VLSI Testing (Spring 2005), Instructor: Professor Vishwani D. Agrawal* (2010).
96. H. Salmani, M. Tehranipoo, Analyzing circuit vulnerability to hardware Trojan insertion at the behavioral level, in *2013 IEEE International Symposium on Defect and Fault Tolerance in VLSI and Nanotechnology Systems (DFTS)* (IEEE, New York, 2013), pp. 190–195
97. M. Tehranipoor, H. Salmani, X. Zhang, Integrated circuit authentication. Switzerland: Springer, Cham. doi **10**, 978–983 (2014)
98. M.M. Hossain, K.Z. Azar, F. Farahmandi, M. Tehranipoor, TaintFuzzer: SoC security verification using taint inference-enabled fuzzing, in *International Conference On Computer Aided Design (ICCAD)* (IEEE, New York, 2023), pp. 1–9
99. A. Nahiyan, M. Tehranipoor, Code coverage analysis for IP trust verification, in *Hardware IP Security and Trust* (Springer, Berlin, 2017), pp. 53–72
100. N. Farzana, M.M. Hossain, K.Z. Azar, F. Farahmandi, M. Tehranipoor, FormalFuzzer: formal verification assisted fuzz testing for SoC vulnerability detection, in *Asia and South Pacific Design Automation Conference (ASP-DAC)* (IEEE, New York, 2024), pp. 1–6
101. N. Farzana, A. Ayalasomayajula, F. Rahman, F. Farahmandi, M. Tehranipoor, SAIF: automated asset identification for security verification at the register transfer level, in *2021 IEEE 39th VLSI Test Symposium (VTS)* (IEEE, New York, 2021), pp. 1–7
102. T. Hoque, J. Cruz, P. Chakraborty, S. Bhunia, Hardware IP trust validation: learn (the untrustworthy), and verify, in *2018 IEEE International Test Conference (ITC)* (IEEE, New York, 2018), pp. 1–10
103. A. Waksman, M. Suozzo, S. Sethumadhavan, FANCI: identification of stealthy malicious logic using Boolean functional analysis, in *Proceedings of the 2013 ACM SIGSAC Conference on Computer & Communications Security* (2013), pp. 697–708
104. C. Herder, M.-D. Yu, F. Koushanfar, S. Devadas, Physical unclonable functions and applications: a tutorial. Proc. IEEE **102**(8), 1126–1141 (2014)
105. A. Maiti, V. Gunreddy, P. Schaumont, A systematic method to evaluate and compare the performance of physical unclonable functions, in *Embedded Systems Design with FPGAs* (Springer, Berlin, 2013), pp. 245–267
106. M.S. Rahman, A. Nahiyan, F. Rahman, S. Fazzari, K. Plaks, F. Farahmandi, D. Forte, M. Tehranipoor, Security assessment of dynamically obfuscated scan chain against oracle-guided attacks. ACM Trans. Des. Autom. Electron. Syst. (TODAES) **26**(4), 1–27 (2021)

107. X. Wang, D. Zhang, M. He, D. Su, M. Tehranipoor, Secure scan and test using obfuscation throughout supply chain. IEEE Trans. Comput. Aided Des. Integr. Circuits Syst. **37**(9), 1867–1880 (2017)
108. K.Z. Azar, H.M. Kamali, H. Homayoun, A. Sasan, From cryptography to logic locking: a survey on the architecture evolution of secure scan chains. IEEE Access **9**, 73133–73151 (2021)
109. H.M. Kamali, K.Z. Azar, H. Homayoun, A. Sasan, On designing secure and robust scan chain for protecting obfuscated logic, in *Great Lakes Symposium on VLSI (GLSVLSI)* (Virtual Event, China, 2020), pp. 217–222
110. H.M. Kamali, K.Z. Azar, H. Homayoun, A. Sasan, SCRAMBLE: the state, connectivity and routing augmentation model for building logic encryption, in *IEEE Computer Society Annual Symposium on VLSI (ISVLSI)* (2020), pp. 153–159
111. W. Yu, *Exploiting On-Chip Voltage Regulators as a Countermeasure Against Power Analysis Attacks* (University of South Florida, South Florida, 2017)
112. K. Baddam, M. Zwolinski, Evaluation of dynamic voltage and frequency scaling as a differential power analysis countermeasure, in *20th International Conference on VLSI Design Held Jointly with 6th International Conference on Embedded Systems (VLSID'07)* (IEEE, New York, 2007), pp. 854–862
113. S. Mangard, T. Popp, B.M. Gammel, Side-channel leakage of masked CMOS gates, in *Cryptographers' Track at the RSA Conference* (Springer, Berlin, 2005), pp. 351–365
114. https://www.invia.fr/Pages/Tutorials/clock-randomizationagainst-side-channel-attacks.aspx
115. K. Tanimura, N.D. Dutt, LRCG: latch-based random clock-gating for preventing power analysis side-channel attacks, in *Proceedings of the Eighth IEEE/ACM/IFIP International Conference on Hardware/Software Codesign and System Synthesis* (2012), pp. 453–462
116. S. Patranabis, A. Chakraborty, P.H. Nguyen, D. Mukhopadhyay, A biased fault attack on the time redundancy countermeasure for AES, in *International Workshop on Constructive Side-Channel Analysis and Secure Design* (Springer, Berlin, 2015), pp. 189–203
117. M.M.M. Rahman, S. Tarek, K.Z. Azar, F. Farahmandi, EnSAFe: enabling sustainable SoC security auditing using eFPGA-based Accelerators, in *2023 IEEE International Symposium on Defect and Fault Tolerance in VLSI and Nanotechnology Systems (DFT)* (IEEE, New York, 2023), pp. 1–6
118. M.M.M. Rahman, S. Tarek, K.Z. Azar, M. Tehranipoor, F. Farahmandi, Efficient SoC security monitoring: quality attributes and potential solutions, in *IEEE Design & Test* (2023)
119. Y. Alkabani, F. Koushanfar, Active hardware metering for intellectual property protection and security, in *USENIX Security Symposium* (2007), pp. 291–306
120. X. Wang, M. Tehranipoor, Novel physical unclonable function with process and environmental variations, in *2010 Design, Automation & Test in Europe Conference & Exhibition (DATE 2010)* (IEEE, New York, 2010), pp. 1065–1070
121. X. Zhang, K. Xiao, M. Tehranipoor, Path-delay fingerprinting for identification of recovered ICs, in *2012 IEEE International Symposium on Defect and Fault Tolerance in VLSI and Nanotechnology Systems (DFT)* (IEEE, New York, 2012), pp. 13–18.
122. X. Zhang, M. Tehranipoor, Design of on-chip lightweight sensors for effective detection of recycled ICs. IEEE Trans. Very Large Scale Integr. VLSI Syst. **22**(5), 1016–1029 (2013)
123. H. Dogan, D. Forte, M.M. Tehranipoor, Aging analysis for recycled FPGA detection, in *2014 IEEE International Symposium on Defect and Fault Tolerance in VLSI and Nanotechnology Systems (DFT)* (IEEE, New York, 2014), pp. 171–176
124. S. Wang, J. Chen, M. Tehranipoor, Representative critical reliability paths for low-cost and accurate on-chip aging evaluation, in *Proceedings of the International Conference on Computer-Aided Design* (2012), pp. 736–741
125. M.M. Hossain, A. Vafaei, K.Z. Azar, F. Rahman, F. Farahmandi, M. Tehranipoor, SoCFuzzer: SoC vulnerability detection using cost function enabled fuzz testing, in *2023 Design, Automation & Test in Europe Conference & Exhibition (DATE)* (IEEE, New York, 2023), pp. 1–6

126. M.S. Rahman, K.Z. Azar, F. Farahmandi, H.M. Kamali, Metrics-to-methods: decisive reverse engineering metrics for resilient logic locking, in *Proceedings of the Great Lakes Symposium on VLSI 2023* (2023), pp. 685–690
127. https://github.com/berkeley-abc/abc
128. https://bitbucket.org/spramod/host15-logic-encryption/pull-requests/
129. W. Chuang, S.S. Sapatnekar, I.N. Hajj, Timing and area optimization for standard-cell VLSI circuit design. IEEE Trans. Comput. Aided Des. Integr. Circuits Syst. **14**(3), 308–320 (1995)
130. O. Schliebusch, A. Chattopadhyay, E.M. Witte, D. Kammler, G. Ascheid, R. Leupers, H. Meyr, Optimization techniques for ADL-driven RTL processor synthesis, in *16th IEEE International Workshop on Rapid System Prototyping (RSP'05)* (IEEE, New York, 2005), pp. 165–171
131. B. Ahmed, S.K. Saha, J. Liu, Modified bus invert encoding to reduce capacitive crosstalk, power and inductive noise, in *2015 2nd International Conference on Electrical Information and Communication Technologies (EICT)* (IEEE, New York, 2015), pp. 95–100
132. S.K. Saha, B. Ahmed, J. Liu, A survey on interconnect encoding for reducing power consumption, delay, and crosstalk, in *2015 2nd International Conference on Electrical Information and Communication Technologies (EICT)* (IEEE, New York, 2015), pp. 7–12
133. M. Tehranipoor, M. Nourani, N. Ahmed, Low transition LFSR for BIST-based applications, in *14th Asian Test Symposium (ATS'05)* (IEEE, New York, 2005), pp. 138–143
134. K.Z. Azar, H.M. Kamali, H. Homayoun, A. Sasan, NNgSAT: neural network guided SAT attack on logic locked complex structures, in *IEEE/ACM International Conference On Computer Aided Design (ICCAD)* (2020), pp. 1–9

Chapter 2
Advances in Logic Locking

2.1 Introduction to Logic Locking

Modern integrated circuit (IC) supply chain has evolved dramatically in the last two decades for multiple reasons, such as ever-increasing cost/complexity of IC manufacturing, huge recurring cost of IC fab maintenance and troubleshooting, aggressive time-to-market, the expedition of the IC supply chain flow, involvement of multiple third-party Intellectual Property (IP) vendors, engagement of cutting-edge technologies, being a primary forerunner in the semiconductor market, etc. [1]. Over time, the economy of scale has pushed for ever-increasing adoption of the IC supply chain's horizontal model. In the IC supply chain's horizontal model, separate entities fulfill various stages of design, fabrication, testing, packaging, and integration of ICs, forming a globally distributed chain.

Plunged in the globalization ocean of the IC supply chain, given the complexity of originally implementing major components of a chip, the design team will bring and acquire multiple 3rd party IPs from numerous IP owners to reduce time-to-market. Additionally, given the overall cost of fabrication, wafer sort, dicing, packaging, package test, and getting access to state-of-the-art technologies, performing most, if not all, these steps in the offshore facility may be the pre-ferred approach for design houses. Outsourcing and the involvement of numerous stakeholders in various stages of the supply chain dramatically reduce the cost and time-to-market of the chip.

Retaining of being competitive is even getting worse especially in the current postpandemic market, where chip demand is exceeding foundry fab capacity, causing tremendous shortages in the market, which even results in increasing the share prices of major contract chipmakers, including TSMC, UMC, and SMIC [2]. In such a market with this unprecedented demand, we will face a more panic IC design, implementation, manufacturing, and testing by original equipment manufacturers (OEMs), which is done precariously to steal the market/contracts. So, with fewer precautions taken by the OEMs to meet the market demand, and by

© The Author(s), under exclusive license to Springer Nature Switzerland AG 2024
M. Tehranipoor et al., *Hardware Security*,
https://doi.org/10.1007/978-3-031-58687-3_2

getting more globalization, OEMs and/or IP vendors face a drastic reduction of the control and monitoring capability over the supply chain.

Despite the benefit achieved by the globalization of the IC supply chain, after the involvement of multiple entities within the IC supply chain with no reciprocal trust and lack of reliable monitoring, the control of original manufacturers and IP owners/vendors over the supply chain will be reduced drastically, resulting in numerous hardware security threats, including but not limited to IP piracy, IC overproduction, and counterfeiting [3, 4].

To address threats associated with the horizontal IC supply chains, a variety of design-for-trust countermeasures have been investigated in the literature. Watermarking, IC metering, IC camouflaging, and hardware obfuscation are examples of passive to active design-for-trust countermeasures [5–8]. In comparison to other design-for-trust countermeasures, logic locking as a proactive technique for IP protection has garnered remarkable attention in recent years, culminating in a plethora of research over the last two decades on designing a variety of robust solutions at different levels of abstraction [9].

Logic locking enables the IP/IC designers to provide limited postfabrication programmability to the fabricated designs, thereby concealing the underlying functionality behind an ocean of options. The resultant locked circuit functionality is governed by the logic locking secret, known as the key, which is known only to authorized/trusted entities, such as IP owners or OEM, and by loading the correct key value, the design house can unlock the circuit. The concept of locking a circuit against malicious or unauthorized activities was first introduced in *Lock and Key* technique [8, 10, 11]. In this technique, a key-based mechanism has been introduced that adds programmability (randomness) to the design-for-testability (DFT) subcircuits (subchains) when being accessed by an unauthorized user. Over time, this programmability has been extended to other parts of the design, like combinational parts [12, 13], sequential (non-DFT) parts [14, 15], and even parametric/behavioral (non-Boolean) aspects of the design [16]. With high applicability and adaptability of logic locking, a wide range of logic locking techniques currently being explored in both the academic context and semiconductor industries, such as Mentor Graphic's TrustChain platform, enabled by logic locking and the newest Defense Advanced Research Projects Agency (DARPA) project on Automatic Implementation of Secure Silicon (AISS) [17, 18].

Logic locking countermeasures appeared as a promising protection mechanism against IP piracy and IC overproduction. However, the development of attacks on logic locking by white hatter researchers, known as logic deobfuscation attacks, is also essential for the evolution of the logic locking paradigm. This helps the researchers to distinguish between weak and robust countermeasures, highlighting the weaknesses of existing countermeasures and illuminating the need for and giving direction to future research in this area. Hence, for more than a decade, numerous studies investigated logic locking in both defense and attack perspectives. It results in the evolution of both the attacks and countermeasures to become increasingly sophisticated. Furthermore, over time, the emergence of cutting-edge technologies, e.g., failure analysis (FA) equipment [19–21], application of state-of-the-art approaches like the usage of machine learning [22, 23], and deeper infiltration by the adversaries even into trusted facilities [24], shows that the logic locking countermeasures are not yet as mature as what promised from the theoretical point of view.

In this chapter, we first holistically review the direction of logic locking through the last decade, in both the attack and defense sides. We will define the parameters, characteristics, and assumptions, either directly or indirectly affecting the outcome of logic locking or the deobfuscation attacks. Then, we will categorize all defenses and attacks and will evaluate each category based on all predefined characteristics and parameters, separately. By providing a very comprehensive comparison between different categories, in both the attack and defense sides, this chapter helps to delineate the future of logic locking and new possible directions.

2.2 Background of Models and Assumptions in Logic Locking

Figure 2.1 demonstrates the main steps of modern IC design flow, in which by starting from the design specification, multiple parties will be involved horizontally in the process of IC manufacturing [25], and by outsourcing that can be engaged at different stages, the OEMs have the least reliable form of control to their belongings (the circuitry or the IP), which results in introducing these contracted and mainly offshore entities as the untrusted parties. As for malicious/untrusted entities, from each IC manufacturing stage, per Fig. 2.2, given any representation of the IC, including integrated design, physically synthesized netlist, layout, packaged IC, or IC under test, the functionality can be reverse-engineered. The toughest yet possible form of circuit reconstruction is usually referred to as physical reverse engineering, which can be done by malicious end-users. Successful reverse engineering, in any form, allows the malicious (untrusted) entities to steal the IP/design and/or illegally overproduce or reuse it.

Given such circumstances, in which the IP vendors or design teams have to protect their belongings [IP and the design (integrated)] against any form of reverse engineering demonstrated in Fig. 2.2, we are witnessing different broad categories of design-for-trust techniques, which mainly are (i) IC camouflaging [7, 26–29], (ii) split manufacturing [30–33], and (iii) logic locking [11–13, 34, 35]. Considering the possibility of having reverse engineering at different layers (as demonstrated in Fig. 2.2), both IC camouflaging and split manufacturing are applicable to a subset of these threats. For instance, IC camouflaging uses high-structural similar logic gates (or other physical structures such as dummy vias) with different functions. However, it is only effective against postmanufacturing attempt(s) of reverse engineering

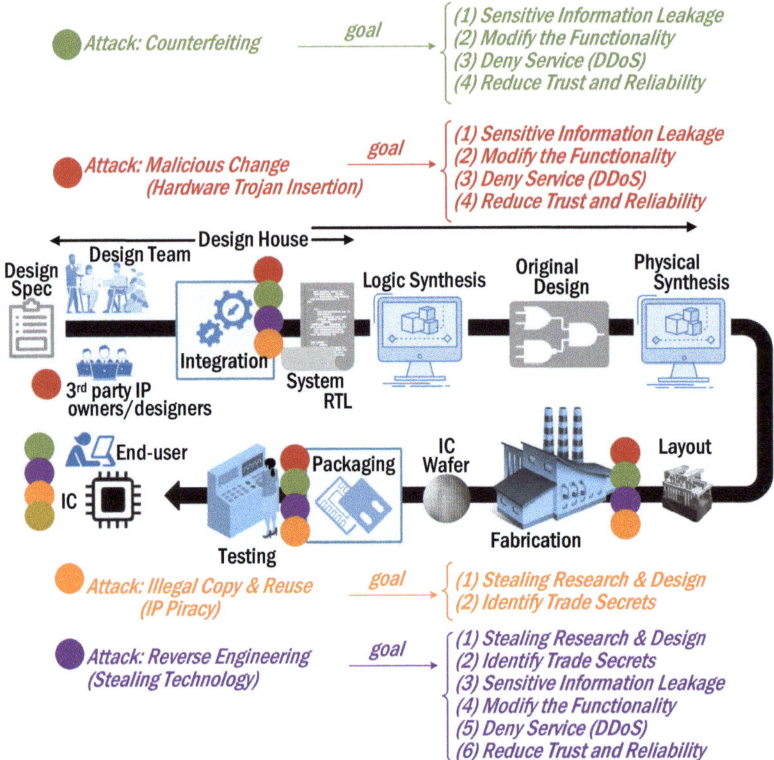

Fig. 2.1 Conventional (globalized) IC design flow and its trustworthiness issue

(by the malicious end-user), and it provides no limitations against a foundry's attempt at reverse engineering, as a foundry has access to all masking layers and is not trapped by structural ambiguity for being able to logically extract a netlist. Split manufacturing, on the other hand, is the integration of the transistors and lower metal layers (a.k.a. Front End Of Line (FEOL) layers) fabricated in cutting-edge technology nodes by an untrusted (and mostly offshore) high-end foundry, with higher metal layers [a.k.a. Back End Of Line (BEOL) layers] fabricated at the design house's trusted low-end foundry. This countermeasure alleviates the security risks at the untrusted foundry. However, this methodology still cannot protect the design against malicious end-users and malicious insiders that threaten the confidentiality of the proprietary technology. On the contrary, logic locking—if implemented meticulously—can be a proactive hardware-for-trust technique that can protect against all previously mentioned threats through the IC supply chain.

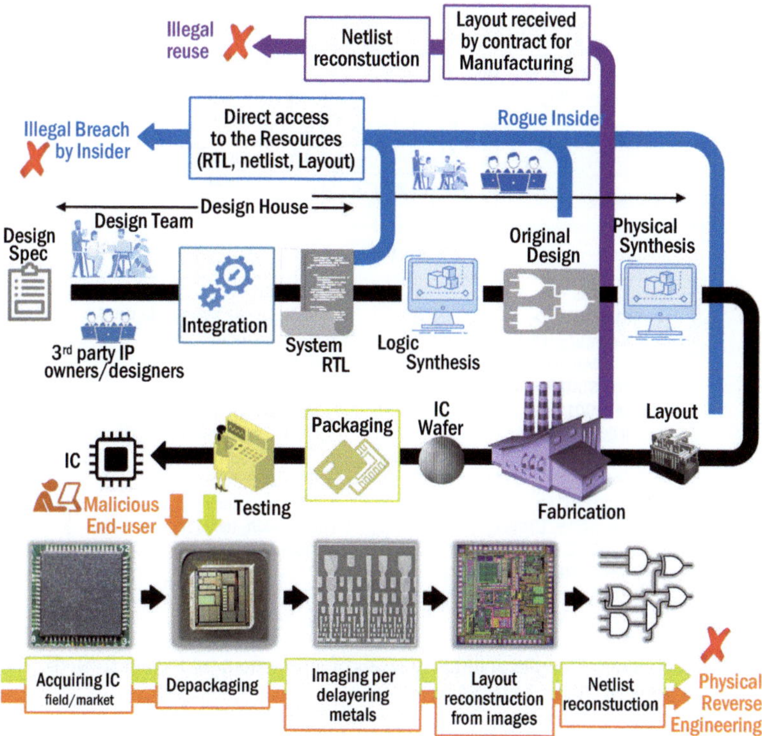

Fig. 2.2 Different threat models at different stages of IC design flow

2.2.1 Basic Definitions of Logic Locking

Logic locking[1] is the capability of adding postfabrication programmability that
could be added using some extra gates, known as key-programmable gates (key
gates) that are driven from the secret of logic locking, i.e., the key. Logic locking
techniques could be implemented at different levels of abstraction. Figure 2.3
demonstrates a simple example of logic locking in different levels of abstraction.
For instance, at layout level as shown in Fig. 2.3a, the metal–insulator–metal
(MIM) structure, which connects two adjacent metal layers, has been engaged
as key-based programmable unit for routing-based locking [36]. Table 2.1 shows
general specification of logic locking at different layers of abstraction. In general,
moving from layout level to RTL or HLS level mitigates the implementation effort.
However, at a lower level of abstraction, finding a logic locking countermeasure at

---

[1] Although the term *logic locking* indicates a specific (gate level) variation of *hardware obfuscation*, both terms have been widely used interchangeably in the literature. In this chapter, we also used *logic locking* as the synonymous word for *hardware obfuscation*.

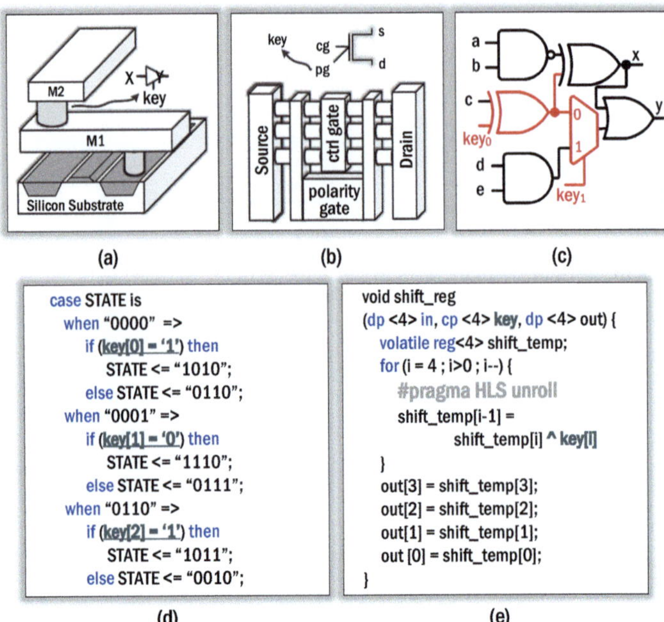

Fig. 2.3 Logic locking examples at different levels of abstraction: (**a**) Layout level, (**b**) Transistor level, (**c**) Gate level, (**d**) RTL, (**e**) HLS level

Table 2.1 Logic locking specification at different abstraction layers

Circuit	Granularity	Overhead	Implementation Effort
Layout-level	Bitwise, wiring	Near to zero	High
Transistor-level	Bitwise, switching, wiring	Low	High
Gate-level	Bitwise, logical	Variant	Medium
RTL-level	Bitwise, operational, behavioral	Mid-high	Low
High-level (HLS)	Bitwise, operational, behavioral	Mid-high	Low

lower overhead is easier to be achieved. Furthermore, moving to a higher level of abstraction (like RTL and HLS) can also provide some form of protection against a subset of insider threats. Although at the moment, more than 90% of the existing logic locking techniques are introduced and implemented at the gate level, mostly done as a postsynthesis stage on the synthesized gate-level netlist in the supply chain, we will demonstrate RTL/HLS-based logic locking is one of the pivotal and widely used current trends in logic locking.

As *bitwise* form of logic locking is achievable at any abstraction layer, based on the type of key gates, either used for logic locking or obtained after synthesis/compilation, we can also categorize logic locking into three main groups: (1) *XOR*-based, (2) *MUX*-based, and (3) *LUT*-based. As their names imply, they are

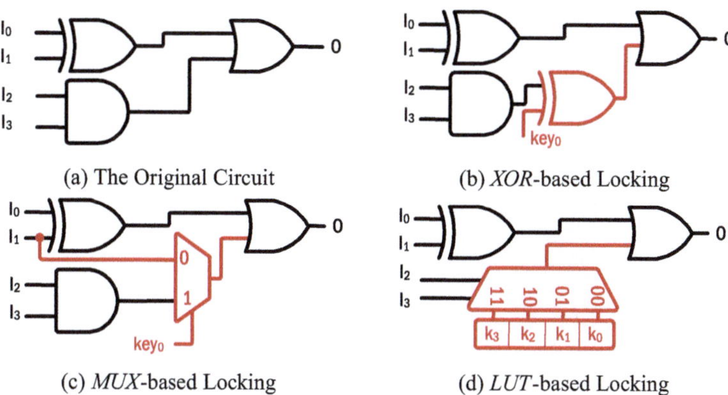

(a) The Original Circuit   (b) *XOR*-based Locking

(c) *MUX*-based Locking   (d) *LUT*-based Locking

Fig. 2.4 Basic gates used for logic locking, (**a**) original, (**b**) correct key ($k_0 = 0$), (**c**) correct key ($k_0 = 1$), (**d**) correct key ($k_{0:3} = 0001$)

using **eXclusive-OR**,[2] **MUltipleXer**s, **Look-Up-Table**s for obfuscation, respectively. Figure 2.4 depicts a simple example of each of these models, demonstrated at gate-level abstraction. It is worth mentioning that except layout-level and transistor-level locking that can be applied for concealment of wiring or switching, for both RTL and HLS level that can target operational or behavioral, the locking part in the resultant netlist generated after synthesis has been implemented using one (or a combination) of these three key gates. During the last decade, different logic locking techniques commonly have engaged these gates with different structures/functions for locking purposes. Based on some properties of these key gates, such as location, structure, count, intercorrelation, etc., the countermeasures provide various levels of robustness against the existing deobfuscation attacks.

A crucial property of logic locking techniques is the **output corruptibility**. Output corruptibility is a very efficient measure of hiding the design's functionality, while the logic locking is in place. Corruptibility means that when an incorrect key is applied to the locked circuit, (1) for how many output pins, and more importantly (2) for how many of the input patterns, the primary output (PO) will be corrupted once the key is incorrect. Usually, the output corruption is measured in terms of the hamming distance (HD) between the correct and wrong output. Ideally, a 50% hamming distance is considered the highest deviation. Based on the location, structure, count, intercorrelation, etc., of the key-based *XORs/MUXs/LUTs* that are engaged for locking purposes, the corruptibility will change. Corruptibility directly affects the resiliency of the countermeasures against the existing attacks.

---

[2] For simplicity, we describe it as XOR-based. It can be trivially extended to be XNOR-based as well.

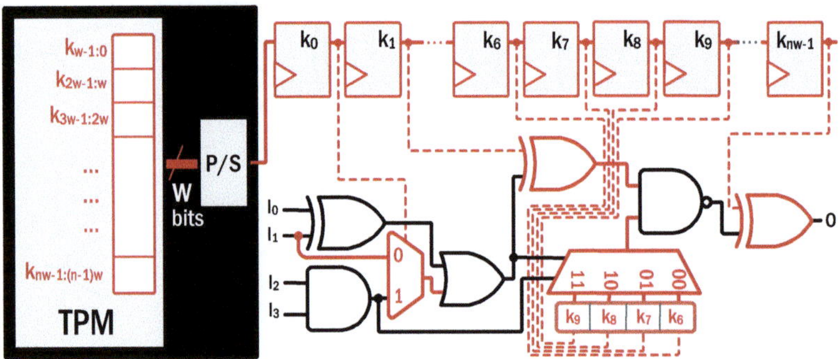

Fig. 2.5 Logic locking kcy initialization from TPM

For instance, if the corruptibility is low, it allows the adversary to look for a specific way for only those POs affected or those specific input patterns that produce output corruption. For a well-designed logic locking countermeasure, the corruptibility must be high to avoid such vulnerabilities.

As demonstrated in Fig. 2.4, the secret of logic locking, referred to as the *key*, must be provided to recover the correct functionality of the locked circuit. The **key initialization** of logic locking must be accomplished at a trusted facility and will be stored in TPM after the fabrication. Hence, the key management infrastructure around logic locking determines how the key will be initiated [37–39]. At power UP of a locked IC, as a part of the boot process, the content of TPM must be read and loaded into temporary registers connected to the locked circuit.[3] Figure 2.5 shows a simple example of key initialization structure when logic locking is in place. Building the competent and secure infrastructure for key initiation[4] has been studied tremendously in the literature [37–43], whose details can be found in [44]. This part of the design consists of (1) TPM that consists of the logic locking key, (2) TPM wrapper which serializes the logic locking key via parallel-in to serial-out (P/S) module, and (3) temporary registers that store the logic locking key, while the IC is power ON, and (4) all infrastructure or customized cells required for securing the logic around key registers.

---

[3] This is when the logic locking key size is large enough (in the range of 1k–2k key bits). For smaller key sizes, range of <200, although not recommended for incurred routing land overhead challenges, the key could be simultaneously loaded from the TPM with no shift process but still needs to be registered for testability purposes.

[4] In these infrastructures, regardless of the logic locking technique, different factors must be considered, e.g., the mechanism of key loading, integrating of key registers with DFT structure, removing the key leakage possibility through the scan chain, retaining the test coverage at high rates, etc.

2.2.2 Models/Assumptions in Attacks on Logic Locking

Based on the threat models and assumptions evaluated in the deobfuscation attacks on logic locking, the attacks could be categorized into different subgroups. From a malicious end-user point of view, some of the attacks require access to one *additional* activated version of the fabricated circuit (at least two in total: one for reverse engineering and obtaining the locked netlist, and one as the reference known as oracle). This group of attacks could be referred to as **oracle-guided** attacks. On the other hand, those attacks with no need for having access to the oracle are called **oracle-less** attacks. During the last decade, most of the attacks are members of oracle-guided attacks. However, in many real cases, the adversary cannot obtain one additional activated chip, and the fulfillment of this requirement is hard to be achieved. So, the adversary has to rely on only oracle-less attack models.

Many of the deobfuscation attacks are **invasive**. They require access to the netlist of the chip. Acquiring the netlist of the chip could be accomplished differently at various stages of the IC supply chain, as demonstrated in Fig. 2.2. For instance, the malicious end-user as the adversary can obtain the fabricated IC from the field/market and then reconstructs the netlist through physical reverse engineering (orange path), in which the main steps of physical reverse engineering are depackaging, delayering, imaging, image (of metal layers) processing, and reconstructing the netlist. In this case, during the physical reverse engineering, since the key is stored in TPM, it will be wiped out in the depackaging stage, and the obtained netlist would be locked.

Regarding the obtained (reverse-engineered) netlist, the same happens in all other cases, and the extracted netlist would be the locked version with no (correct) key. Another example is when the adversary might be at the foundry, and once they receive the GDSII of the chip from the design house to be fabricated, the GDSII is provided without the correct key (locked). In this case, although no delayering or physical infiltration is required, GDSII is required to be accessed and evaluated for netlist extraction, and thus we consider it a *weak* invasive model. Even while the adversary is a rogue insider in the design house, except verification engineer[5] who needs the key for verification purposes, there is no necessity of sharing the key for stages like integration, synthesis, floorplanning, etc., showing that the obtained design is available with no key.

Unlike invasive attacks, there exists a very limited number of **semi-invasive** and **noninvasive** deobfuscation attacks. In these attacks, the adversary relies on optical probing, such as electro-optical probing (EOP) and electro-optical frequency management (EOFM). Such attacks focus on pinpointing and probing the logic gates and flip-flops of the circuits containing the secrets. So, regardless of the logic locking technique used in the circuit, this group of attacks, which will be discussed

---

[5] We assume that the verification team is always trusted.

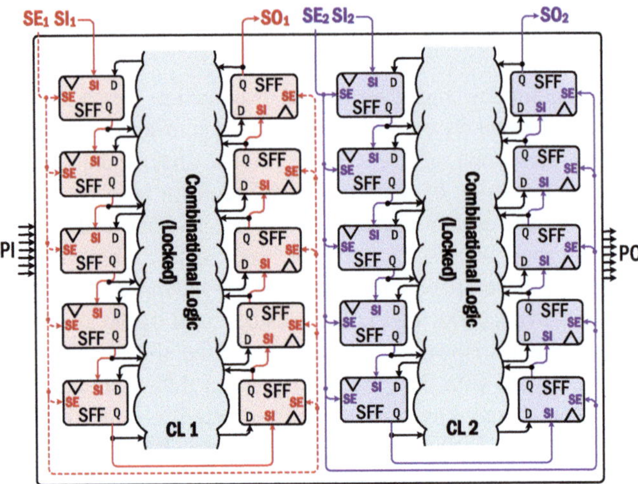

Fig. 2.6 Design-for-testability (scan chain) architecture in ICs

through the chapter, would be able to be a real threat for revealing the security assets like logic locking key.

The availability of design-for-testability (DFT) structure, i.e., scan chain architecture, for testability/debug purposes in ICs opens a big door for the attackers to assess and break logic locking techniques. Hence, many of the attacks on logic locking assume that the **scan chain is OPEN**. Figure 2.6 shows a simplified scan architecture with two chains. Assuming that the scan chain is *OPEN*, SE, SI, and SO pins would be available. So, the adversary can reach (control and observe) each combinational part, e.g., CL_1 and CL_2 in Fig. 2.6, whose FFs are part of the scan chain. The scan chain access allows the adversary to divide the deobfuscation problem into a bunch of much smaller subproblems (each CL) and assess them independently. However, it is very common for an IC to limit/restrict access to the scan chain for security purposes. But, even while **the access to the scan chain is NOT OPEN** (e.g., SO pins are burned), some other deobfuscation attacks have studied and demonstrated the possibility of retrieving the correct key/functionality of the locked circuit via primary inputs/outputs (PI/PO).

Furthermore, the adversary located at the foundry might have the capability of priming and manipulating the GDSII to insert hardware Trojan for different purposes, such as key leakage through PO after the activation. The capability of inserting stealthy Trojans around the logic locking circuitry will be able to (1) disable the test access or (2) retrieve the logic locking key through PO. Hence, the capability of adding hardware Trojans when logic locking is in place, and being nondetectable by Trojan detection techniques could be a part of adversary capabilities that required meticulous consideration by the designers.

2.3   Logic Locking: Countermeasures

Starting 2008, numerous logic locking techniques have been introduced in the literature each trying to introduce a new countermeasure against the existing deobfuscation attacks. Figure 2.7 demonstrates a top view of existing notable logic locking techniques introduced so far. Based on the previously discussed basics and definitions of logic locking, we categorized them into different groups: (1) primitive,

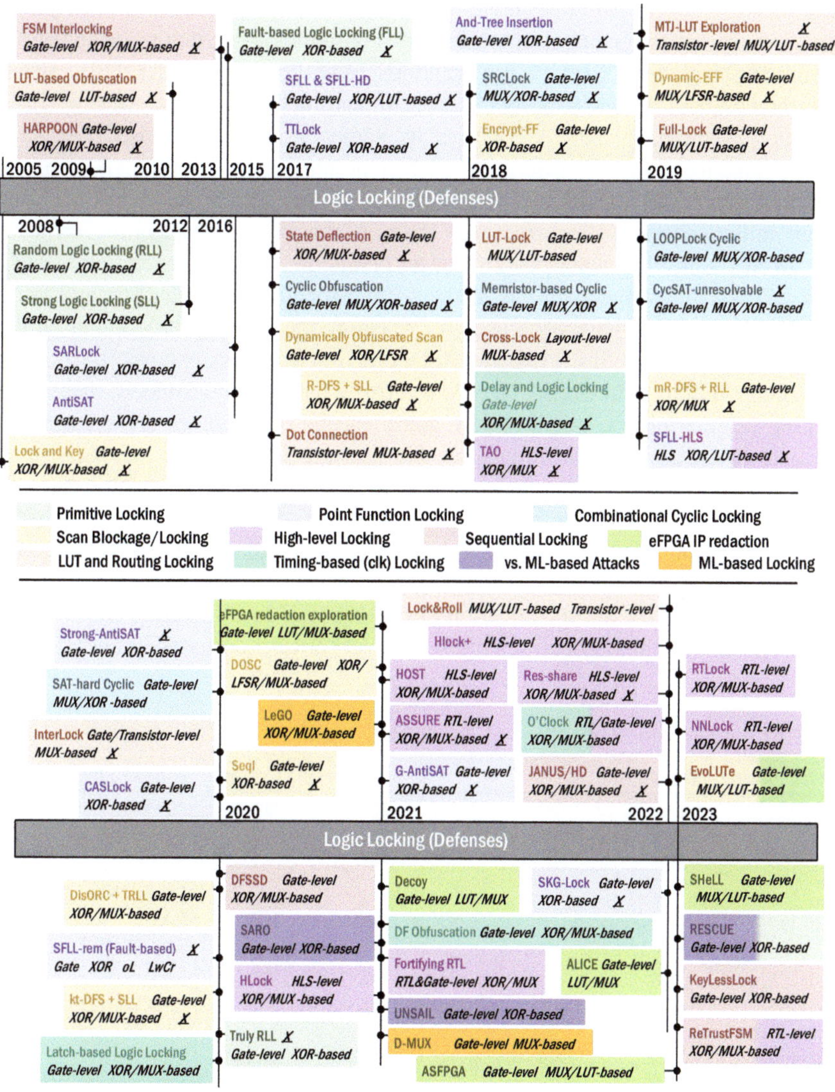

Fig. 2.7 Logic locking techniques

(2) point function, (3) cyclic, (4) LUT/routing, (5) scan-based locking/blocking, (6) sequential/FSM, (7) timing-based, (8) eFPGA-based, and (9) high level (RTL/HLS). The main specification of any member of each countermeasure is illustrated in Fig. 2.7. In this section, regardless of the existing successful attacks, we will briefly review the main specification of each logic locking category.

2.3.1 Primitive Logic Locking

The first group of countermeasures on combinational circuits is primitive techniques, including EPIC *a.k.a.* random logic locking (RLL) [12], strong (interference-based) logic locking (SLL) [13], and fault-based logic locking (FLL) [45]. For example, in EPIC (RLL), as its name implies, XOR-based key gates will be inserted at some arbitrarily chosen points in the circuit. All primitive techniques are XOR-based and implemented at the gate level. Since a locked circuit initiated with an incorrect key corrupts the PO by propagating errors at POs, in SLL and FLL, some features of automatic test pattern generation (ATPG) tools and testability specification, such as controllability/observability, and faults propagation/masking, have been used for selecting the location of XOR-based key gates. For instance, in SLL [13], specifications like key gates exclusion, isolation, cascading (running), mutability, and convergence have been examined, thereby by forming an interference graph of key gates, the best candidates are selected for key gate insertion, helping to enhance the strength of the logic locking against testing-based attacks (in comparison with RLL). However, using these features results in a notable reduction of corruptibility in SLL and FLL compared to that of RLL. In addition, in [43], a truly RLL has been defined that relies on XOR/XNOR insertion around inverters or buffers that is for randomly locking of signal polarities. Table 2.2 summarizes the specification of this breed of logic locking, showing that all are attacked by the *Boolean satisfiability* (SAT) attack, discussed in Sect. 2.4.1.2.

Table 2.2 Specification of primitive logic locking techniques

Logic locking	Mechanism	Overhead	Corrupt	Attacked by
RLL [12]	Insertion of XOR key gates at *random* places	■□□□□	■■■	ATPG: Sect. 2.4.1.1[a], SAT: Sect. 2.4.1.2
FLL [13]	Insertion of XOR key gates at *lower* testable points	■■□□□	■■□	SAT: Sect. 2.4.1.2
SLL [45]	Interference-based key gate insertion (non-mutuable, non-isolated, etc.)	■□□□□	■■□	SAT: Sect. 2.4.1.2
TRLL[b] [43]	Insertion of XOR key gates at buffers and inverters	■□□□□	■■■	SAT: Sect. 2.4.1.2

[a] Referring to the Section of the Book covering these attacks
[b] Here TRLL is evaluated as an individual locking technique

2.3.2 *Point Function Logic Locking*

The main aim of point function techniques is to minimize the number of available input patterns showing that a specific key is incorrect.[6] This breed was the first attempt against the Boolean satisfiability (SAT) attack, which can prune the keyspace by ruling out the incorrect keys in a fast-convergence approach [46, 104]. In the literature, this group of logic locking techniques is known as provably logic locking techniques. A logic locking technique is *provably* secure once they are algorithmically resilient (cannot be broken) against any type of I/O query-based attacks. SARLock and Anti-SAT are the very first logic locking techniques in this category [48, 49]. As demonstrated in Fig. 2.8a, the main structure of point function techniques relies on a *flipping* (corrupting) circuitry that flips (corrupts) the limited PO(s) only for a very limited number of input patterns (e.g., 1) per each incorrect key. Also, a *masking/restore* (correcting) circuitry has been engaged in point function techniques to reflip the impact of flipping circuitry, guaranteeing the correct functionality when the correct key is applied.

Point function techniques could be applied on the function-modified (stripped) version of the circuit, known as stripped function logic locking [52]. In such techniques, the original part is modified, and in at least one minterm, the (affected) POs are *ALWAYS* flipped/corrupted. This is done using *cube stripper* module as demonstrated in Fig. 2.8b. The *restore* unit builds the flipping and masking circuitry. In the stripped function techniques, for each incorrect key, there exists a very limited number of input patterns (e.g., 1) *plus* one extra input pattern (caused by the stripped function) that corrupts the POs (double point corruption). In fact, approaches with no stripping accomplish the flipping at one point (one input pattern per an incorrect key), while others with enabled stripping do the flipping at two points.

Point function techniques could be categorized as XOR-based techniques. In SFLL-flex [52], LUTs are also engaged for minterm generations of corruption/-

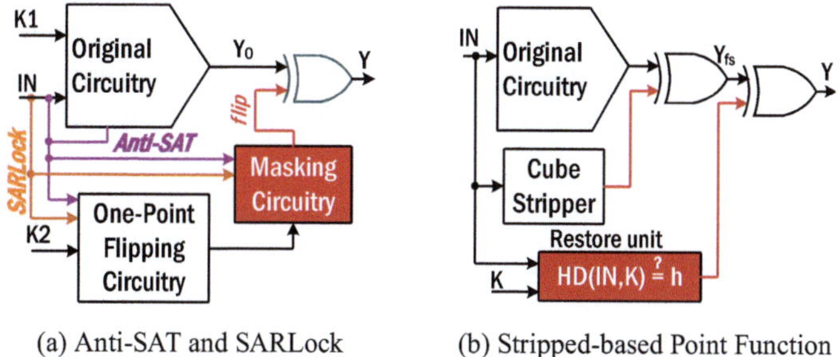

(a) Anti-SAT and SARLock   (b) Stripped-based Point Function

Fig. 2.8 The structure of point function techniques

---

[6] The best case is *ONE* input pattern per each incorrect key.

correcting subcircuits. Except for SFLL-HLS [57] that is implemented at the high level (HLS), all techniques in this category are implemented at the gate level (postsynthesis). Since the point function subcircuitry will be added for a limited (e.g., 1) number of POs and corrupts the PO for a very limited number of input patterns, the corruptibility of this breed of obfuscation is very low. However, there also exist more recent point function techniques that try to overcome the low corruptibility of such techniques with the introduction of different flipping (corrupting) circuits [54–56]. Table 2.3 summarizes the main specification of point function techniques. As shown, two new studies, i.e., sparse prime implication and Valkyrie (EDA-based), break all the variants in this group, showing the big structural issue behind this breed of logic locking.

2.3.2.1 Compound Logic Locking

The point function logic locking limits the corruption per each incorrect key to (i) a very little set of input patterns and (ii) observable at a very little set of POs.[7] Hence, this breed suffers from a very low output corruption that significantly undermines its strength. Hence, a new paradigm was first introduced after point function techniques, in which the composition of different logic locking has been discussed, known as **compound** logic locking techniques. In compound logic locking, two or more different logic locking techniques can be used and applied simultaneously to a circuit *if and only if* none of them weakens the other ones. For instance, since the locking parts of the point function techniques are completely decoupled from the original part, as demonstrated in Fig. 2.8, and since the primitive logic locking techniques provide high corruptibility, as demonstrated in Fig. 2.9, these two can easily be combined to mitigate the low-corruptibility issue of point function techniques, and also getting the benefit of high resiliency from point function techniques.

The concept of compound logic locking can be used for any feasible combination. For instance, in [58], a bilateral logic encryption has been introduced in which a low-corruption logic locking technique and a routing-based approach have been integrated to show the effectiveness of compound logic locking against a wider range of attacks. Table 2.4 shows general definition of compound techniques.

2.3.3 Cyclic-Based Logic Locking

As its name implies and as shown in Fig. 2.10, cyclic logic locking will add key gates that control the possibility of adding/removing combinational cycles into the circuit.

---

[7] The best case (in terms of enhancing the security) is that *ONE* input pattern can only rule out *ONE* incorrect key, and the corruption per each incorrect key can be witnessed at only *ONE* PO.

Table 2.3 Specification of point-function logic locking techniques

Logic locking	Mechanism	Overhead	Corrupt	Attacked[a] by
SARLock [48]	Adding flipping circuit to corrupt only ONE input pattern per each incorrect key + masking circuit for correct key	■□□□□	■□□	Removal: Sect. 2.4.1.3, Approximate: Sect. 2.4.1.2, EDA-based: Sect. 2.4.1.3, Bypass: Sect. 2.4.1.2
AntiSAT [49]	Merging of ANDed two toggled functions (g & $\bar{g}$) as the flipping+masking circuitry together	■□□□□	■□□	SPS/Removal: Sect. 2.4.1.3, Approximate: Sect. 2.4.1.2, EDA-based: Sect. 2.4.1.3, Bypass: Sect. 2.4.1.2,
And-Tree [50]	hard-coded AND trees as flipping circuitry + a generic masking circuitry	■□□□□	■□□	CASUnlock: Sect. 2.4.1.3, EDA-based: Sect. 2.4.1.3
TTLock [51]	SARLock + Stripping original circuit for one minterm	■■□□□	■□□	EDA-based: Sect. 2.4.1.3, GNNUnlock: Sect. 2.4.3.3, FALL: Sect. 2.4.1.3
SFLL-HD [52]	SARLock + Stripping original circuit for $d = {}^{k}C_h$ minterms (h: HD and k: key size)	■■□□□	*variant* (w.r.t. h)	EDA-based: Sect. 2.4.1.3, GNNUnlock: Sect. 2.4.3.3, FALL: Sect. 2.4.1.3
SFLL-flex [52]	Adding the flexibility of protecting user-defined input patterns in a point-function manner + LUT-based restore circuitry	*variant* (w.r.t. user-defined list)	*variant* (w.r.t. user-defined list)	EDA-based: Sect. 2.4.1.3
SFLL-rem [53]	removing logic for creating the corruption based on fault insertion + a generic restore circuitry	■□□□□	*variant* (w.r.t. s-a fault location)	EDA-based: Sect. 2.4.1.3
G-AntiSAT [54]	Merging of ANDed two toggled functions (f & g) as the flipping+masking circuitry together	■■□□□	*variant* (w.r.t. specs of f & g)	EDA-based: Sect. 2.4.1.3
S-AntiSAT [55]	App-level input pattern protection + generic restore circuitry	*variant* (w.r.t. input patterns list)	*variant* (w.r.t. input patterns list)	EDA-based: Sect. 2.4.1.3
CAS-Lock [56]	Variant AND-OR tree as variant corruptible circuitry	■■□□□	*variant* (w.r.t. OR gates)	EDA-based: Sect. 2.4.1.3, CASUnlock: Sect. 2.4.1.3

[a] Per each logic locking technique, there might be more successful attacks than the ones listed here, and we listed the most notable ones directly applicable to

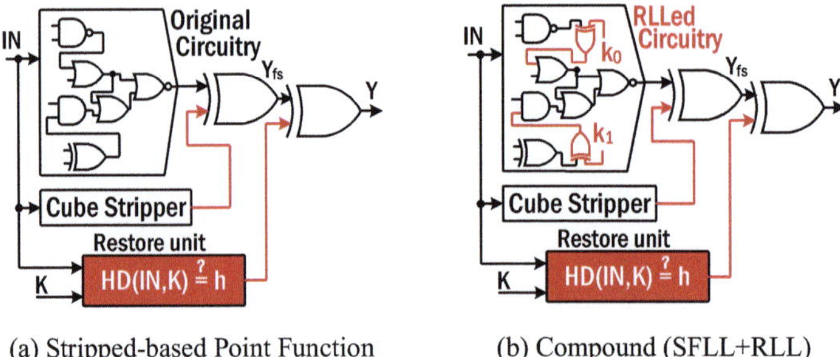

(a) Stripped-based Point Function   (b) Compound (SFLL+RLL)

Fig. 2.9 The structure of compound logic locking techniques

Table 2.4 Specification of compound logic locking techniques[a]

Logic locking	Mechanism	Overhead	Corrupt	Attacked by
X+Y	Combination of primitive + point function	■■□□□	■■■	Approximate: Sect. 2.4.1.2, compound attacks: Sect. 2.4.1.2

[a]This table is only for primitive + point function techniques
X: Any primitive logic locking technique
Y: Any point-function logic locking technique

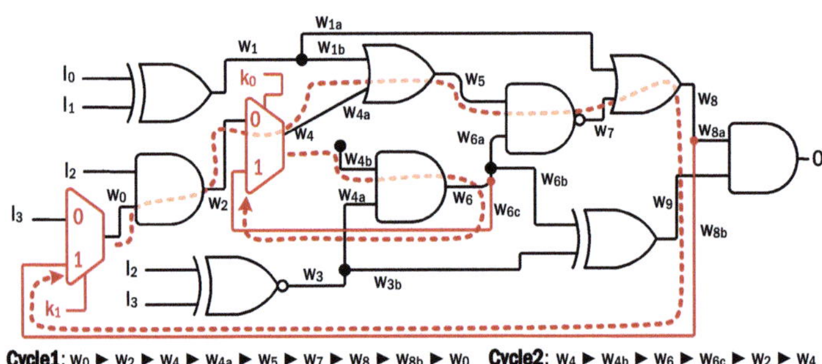

Cycle1: $W_0 \blacktriangleright W_2 \blacktriangleright W_4 \blacktriangleright W_{4a} \blacktriangleright W_5 \blacktriangleright W_7 \blacktriangleright W_8 \blacktriangleright W_{8b} \blacktriangleright W_0$ **Cycle2**: $W_4 \blacktriangleright W_{4b} \blacktriangleright W_6 \blacktriangleright W_{6c} \blacktriangleright W_2 \blacktriangleright W_4$

Fig. 2.10 An example of cyclic obfuscation using 2-to-1 MUXes

Having combinational cycles will add difficulties for the CAD tools (like synthesis and timing analysis) to deal with such circuits. Many CAD tools do not allow the designers to have combinational cycles in the circuit. However, the designer would be able to handle the combinational cyclic paths during the physical design in a manual manner, like adding constraints for false paths. Hence, combinational cycles are used commonly as a means of logic locking recently. In cyclic-based logic locking, different approaches are considered through different studies:

(i) *Adding False Cycles*: In this case, similar to examples demonstrated in Fig. 2.10, some combinational cycles have been added into the design that should not be remained once the circuit is unlocked [59]. Having such cycles

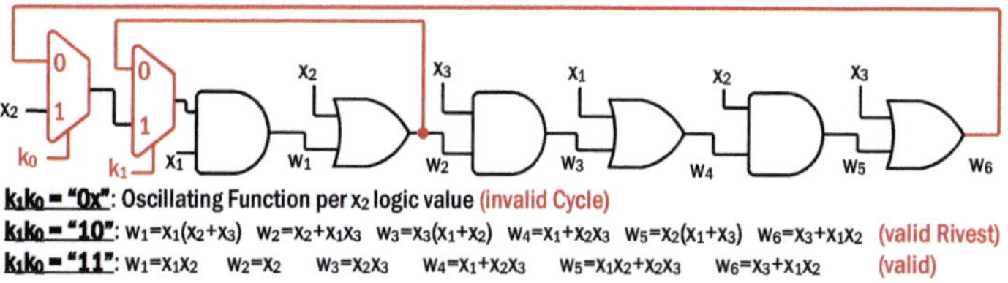

Fig. 2.11 Rivest subcircuit as decoy-based cyclic logic locking

in the design creates uncertainty, glitches, and malfunction, while the correct key is not provided.

(ii) *Adding Misguiding Combinational Cycles*: Since the design must have no combinational cycles by itself, some approaches engage a set of valid misguiding combinational cycles, such as Rivest circuits [60] as shown in Fig. 2.11, to violate such basic assumption [61, 62], in which cycles are part of the original functionality. However, template-based logic subcircuits will be used for building this kind of cycle in a design that makes them vulnerable to structural analysis.

(iii) *Exponential increase of cycles*: To increase the complexity of cyclic-based logic locking, some studies have evaluated the possibility and techniques that can be applied to exponentially increase the number of cycles *w.r.t.* the number of feedback [61, 63]. Exponentially increasing the number of cycles will exponentially enhance the complexity of cyclic analysis. However, the existing approaches will also raise the area overhead significantly.

We are also witnessing some other studies that are using other techniques for increasing the complexity of cyclic-based logic locking, such as the implementation of cyclic locking using memristor-CMOS logic circuits, and inserting cycle pairs (duplicating subcircuits), which are useful for hiding and misguiding the adversary for any kind of analysis on added/inserted cycles [64–66].

In general, since rerouting is required to generate the combinational cycles, all cyclic logic locking techniques use key-based MUX gates at different levels of abstraction. Also, in some cases, key-based XOR gates are used to build the model. All existing cyclic logic locking techniques are implemented at the gate level or transistor level. Since rerouting increases the correlation of wiring in different logic cones of the circuit, the corruptibility of this group of logic locking would be high. Table 2.5 shows the main specification of this breed of logic locking. Although some of these techniques are not broken so far, they are not considered as a mainstream in logic locking since it results in challenging industry adoption specifically in implementation and physical-based EDA tools.

Table 2.5 Specification of cyclic logic locking techniques

Logic locking	Mechanism	Overhead	Corrupt	Attacked by
Cyclic [59]	adding false combinational cycles into the circuit	■■□□□	■■■	CycSAT: Sect. 2.4.1.2
SRCLock [61]	exponentially increasing false combinational cycles w.r.t. the number of feedbacks	■■■□□	■■■	CP&SAT: Sect. 2.4.1.2
memristor-based cyclic [64]	unresolvable combinational cycles + memristor cells for camouflaging cycles	■■□□□	■■■	iCySAT: Sect. 2.4.1.2
CycSAT-Unresolvable [62]	unresolvable combinational cycles	■■□□□	■■■	iCySAT: Sect. 2.4.1.2, BeSAT: Sect. 2.4.1.2
SAT-hard cyclic [63]	exponentially increasing false combinational cycles w.r.t. the number of feedbacks + rivest cycles + non-occuring cycles	■■■□□	■■■	NONE
LOOPLock [65]	creating pairs of cycles that increasing the complexity of cycle analysis	■■□□□	■■■	NONE

2.3.4   LUT/Routing-Based Logic Locking

Some logic locking techniques derive the benefit from the full configurability of look-up tables (LUTs). Relying on the fact that a u-input LUT can build all 2^{2^u} possible functions, as demonstrated in Fig. 2.4, in existing LUT-based logic locking techniques, some actual logic gates of the original design are replaced with LUTs (same-size or larger LUTs), and the initialization (configuration) values of the LUTs are considered as the secret (key) and would be initiated after the fabrication. In existing LUT-based logic locking, the following have been investigated as the crucial factors [67–71]:

(i) *Size of LUTs*: A point-to-point LUT-based replacement has been implemented in the existing LUT-based techniques, in which each selected gate will be replaced by a *same-size* (LUT2x1, LUT2x2 in Fig. 2.12) or *larger* (LUT3x1, LUT4x1, and LUT4x2 in Fig. 2.12) LUTs. Enlarging LUTs (increasing the number of inputs) will exponentially increase the complexity of the logic locked circuit; however, it also increases the area/power overhead exponentially. In enlarged LUTs, the extra (unused) inputs can be used as the key to expand the complexity space (e.g., $k_{0:2}$ in Fig. 2.12).

(ii) *Number of LUTs*: The number of gates selected to be replaced with same-size/larger LUTs directly affects the complexity of the logic locked circuit, either functionally or structurally. However, similar to the side effect of LUT size, using more LUTs will significantly increase the overhead of the logic locking approach.

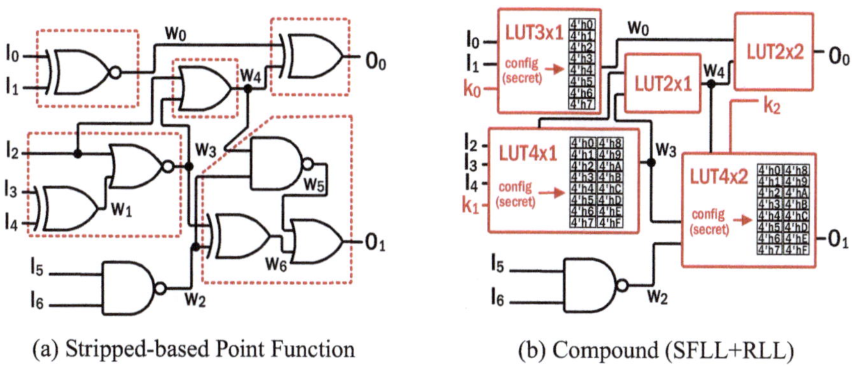

<table>
<tr><td>(a) Stripped-based Point Function</td><td>(b) Compound (SFLL+RLL)</td></tr>
</table>

Fig. 2.12 LUT-based logic locking technique: an example

(iii) *Replacement Strategy*: The location of replacement in LUT-based logic locking plays a crucial role in making the resultant circuit more complex in the domain of logic locking. The work LUT-lock [68] has investigated how a good replacement strategy can render a higher grade of complexity, while a heuristic approach has been used.

In general, although LUT-based logic locking can provide more reliable resilience against existing attacks, it extremely suffers from incurred overhead, which limits the application of this breed of logic locking. The existing LUT-based techniques are all implemented at the gate level or transistor level, and based on the placement strategies used for LUT insertion, the corruptibility of these techniques is also high.

Similar to LUT-based logic locking, which can be built using a MUX-based (tree of MUXes) structure, numerous studies have exploited the concept of MUXes for another breed of logic locking, known as routing-based locking. In routing-based locking, false/decoy/invalid paths could be added using rerouting modules in different levels of abstraction. For instance, at the gate level of abstraction, MUXes could be used for rerouting, and at layout level, metal–insulator–metal (MIM) could be used for rerouting between metal layers as a means of logic locking [72].

Dealing with the complexity of the routing modules has been evaluated for a long time as a part of physical P&R in both FPGAs/ASICs [73–75]. Hence, this complexity is borrowed here as a means of locking in these techniques. But the concept of routing-based locking was firstly derived from the usage of wiring/connection concealment in split manufacturing techniques [76–78]. In these techniques, different forms of rerouting, such as perturbation, lifting nets, etc., have been engaged to hide the connections between BEOL and FEOL. Hence, a similar approach has been used in a circuit for locking purposes. For instance, Fig. 2.13 shows two different routing-based techniques, i.e., cross-lock [79] and full lock [80]. As demonstrated, in routing-based locking, some wires of the design will be selected, and the routing module, whose configuration is the locking key, will conceal the connection between sources and sinks.

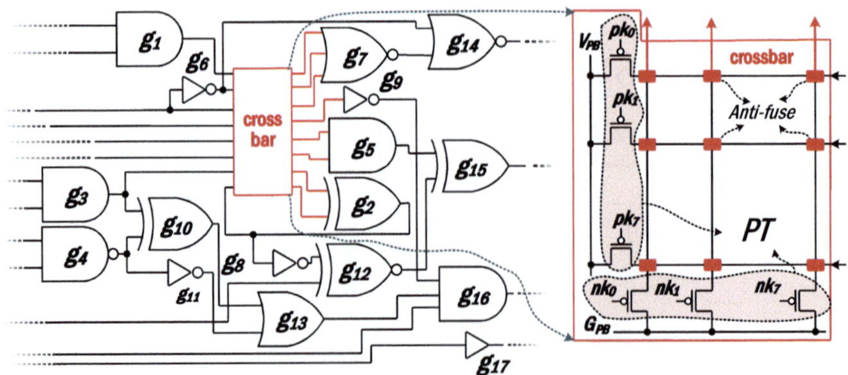

(a) Crossbar-based Routing-based Locking

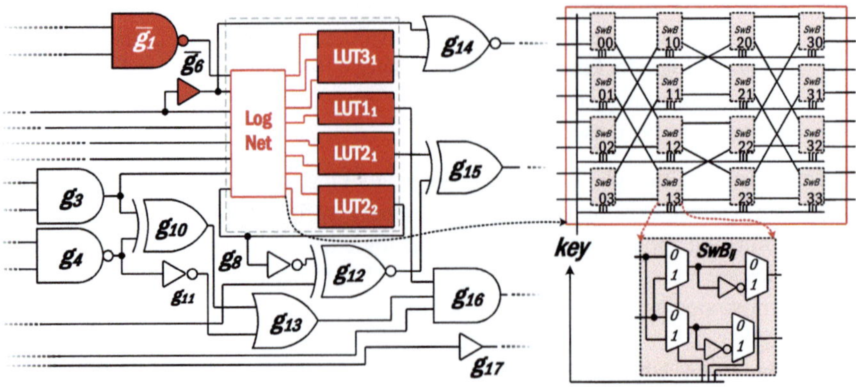

(b) Logarithmic-based Routing-based Locking

Fig. 2.13 Routing-based locking technique

Unlike point function techniques that decrease the pruning power of the existing I/O query-based attacks (e.g., the SAT attack) by exponentially increasing the required number of I/O pairs, routing-based techniques show how a routing module can extremely increase the complexity of design per each stimulus [80]. Routing-based locking techniques have investigated different factors that significantly affect the outcome of this breed of locking solutions as listed here:

(i) *Abstraction Layer of Implementation*: Routing modules can be implemented at different layers of abstraction. Moving toward lower levels (e.g., layout) significantly increases the implementation efforts and challenges; however, if implemented correctly at lower levels, it incurs considerably lower overhead. Techniques like cross-lock and interconnect-based locking are implemented at the layout level [79, 81]. Full lock implements routing module at the gate level [80]. An extension on full lock is also implemented at gate level [82], and interlock is also implemented at both gate level and transistor level [83].

(iii) *Topology of Routing Modules*: Once a routing module is in place, the topology of interconnections is a determining characteristic. Hence, in routing-based locking techniques, different topologies have been used, such as crossbar in [79, 81], logarithmic networks in [80, 82–84], and irregular [85]. The type of topology can directly affect the resiliency of the countermeasure against different attacks. For instance, logarithmic networks with deeper layers can extremely increase the complexity of the model in any graph-based and routing-based analysis. However, the topology can negatively affect the overhead, particularly delay of timing paths selected for routing locking.

(iv) *Cycle Involvement in Routing-Based Locking*: Based on the wire selection in routing-based locking techniques, cycles might be created once an incorrect key will be applied. For instance, as shown in Fig. 2.13, the output of g2/LUT2$_2$ is back and connected to the input of routing module. Although the creation of cycles will increase the complexity of locking techniques, to avoid any design/implementation flow challenges dealing with cycles, in more recent techniques [82, 83], since actual timing paths of the design are embedded into routing modules, there exists no possibility of cycle creation.

 (v) *Logic Embedding in Routing Module*: To increase the complexity of routing modules, recent works [80, 82, 83] have investigated the embedding of logic gates into (switch) layers of routing module. For instance, switch boxes of LogNet in Fig. 2.13b provide configurable inversion. Hence, Gates like g1 and g5 is toggled (stripped) and the LogNet will recover the correct logic values. Hence, in these techniques, with the full configuration capability, stripping the functionality of the original circuit would be an extra available option.

Table 2.6 summarizes the specification of routing-based locking countermeasures. In comparison with other breeds, routing-based locking will incur higher overhead. However, more recent studies, such as coarse-grained eFPGA-based IP redaction (Sect. 2.3.9), show that this form of locking that inherits full reconfigurability for locking purposes still has significant potential as a means of logic locking. As demonstrated, all existing routing-based locking solutions are broken, mostly by very recent ML-based attacks described in Sect. 2.4.3.3.

2.3.5 Scan Chain Logic Locking/Blocking

Providing the access to the internal parts of the circuit for test/debug purposes is almost inevitable in modern/complex ICs. As discussed previously in Sect. 2.2.2, design-for-testability (DFT)-based synthesis in ASIC design flow provides this capability for the designers by adding scan (register) chain structures into the circuit. The DFT-based scan chain architecture has been widely used in most modern ICs. Even in cryptographic circuits, which have very sensitive information, such as

Table 2.6 Specification of routing-based locking techniques

Logic locking	Mechanism	Overhead	Corrupt	Attacked by
Cross-lock [79]	Insertion of key-based layout-level cross-bar with combinational cycles	■■□□□	■■■	CP&SAT, NNgSAT: Sect. 2.4.1.2
Dot connection [81]	Insertion of key-based layout-level cross-bar with combinational cycles	■■□□□	■■■	CP&SAT, NNgSAT: Sect. 2.4.1.2
Full-lock [68]	Insertion of gate-level key-based logarithmic routing modules with inversion stripping	■■■■■	■■■	CP&SAT, NNgSAT: Sect. 2.4.1.2
Modeling routing [68]	Insertion of gate-level key-based logarithmic routing modules with logic embedded	■■■■■	■■■	Untangle: Sect. 2.4.3.3
InterLock [68]	Insertion of gate or transistor-level key-based logarithmic routing modules with logic embedded + stripping	■■■□□	■■■	Untangle: Sect. 2.4.3.3

encryption key, to have a high fault coverage, the test/debug step requires access to the scan chain to control and observe the internal states of the design under test (DUT). The full controllability and observability requirement in DFT-based (scan-based) testing, however, might pose security threats to ICs with security assets, such as locked circuits that keep their own secret, i.e., the unlocking key. Hence, DFT access allows the adversary to split and divide the bigger problem into a set of independent smaller problems by splitting the whole circuit into c smaller and only combinational logic parts, whose access to their register elements are available via scan chains (e.g., $CL1$ and $CL2$ of Fig. 2.6 through $\{SI_1, SE_1, SO_1\}$ and $\{SI_2, SE_2, SO_2\}$, respectively). We further demonstrate that scan chain access availability is one of the main assumptions of the SAT attack, discussed in Sect. 2.4.1.2.

Since restricting the scan chain access can enlarge the problem space domain from small combinational subcircuits to a whole large sequential circuit, a breed of logic locking techniques focus on different methodologies that only and primarily target the locking of the scan chain architecture, known as scan-based logic locking techniques [86–91]. By using these techniques, the scan chain pins, i.e., scan-enable (SE), scan-in (SI), and particularly scan-out (SO), would be limited/restricted for any unauthorized access, and these approaches do not allow the adversary to get the benefit of these pins, and they lose the chance of direct/independent controlling/observing the combinational parts of the circuit. Similar to scan-based logic locking techniques, some approaches *BLOCK* the access to the scan chain pins, particularly *SO* [37–39, 41, 43], called scan blockage techniques. The scan blockage will happen based on a sequence of specific operations in the scan chain structure. For example, assuming that (part of) the key is loaded into the scan chain after activation, switching *SE* to 1 (enabling shift mode) might be perilous. Hence, shift operation would be limited after the activation. Figure 2.14 shows the top view of both scan locking and scan blockage techniques. Compared to the scan-based

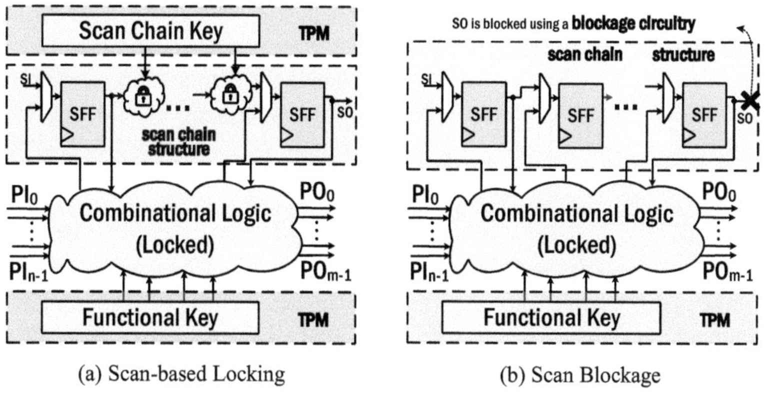

(a) Scan-based Locking   (b) Scan Blockage

Fig. 2.14 Scan chain locking/blocking technique

logic locking techniques, the scan blockage techniques incur less area overhead. However, they have some limitations during the test phase, such as limiting the functional test and increasing the test time and complexity.

The following describes the main specifications and principles of scan chain locking/blockage techniques:

(i) *Combination with Functional Logic Locking*: Scan chain locking/blockage is orthogonal to other functional logic locking techniques. In fact, the scan-based logic locking techniques only lock the scan structure. So, even while an incorrect key is initiated in the circuit, it only affects the scan chain functionality and has no impact on the functionality of the circuit. Hence, these techniques need to be combined with one of the other discussed combinational logic locking techniques, and the existing techniques mostly use one of the primitive logic locking techniques. Some techniques use RLL [38], some other techniques use SLL [39, 86, 92], and one scan blockage technique is combined with truly-RLL (TRLL) [43]. Hence, the corruptibility of these approaches is dependent on the engaged functional logic locking techniques.

(ii) *Implementation Abstraction Layer*: Since scan chain locking/blockage techniques must be applied to the scan chain structure, they only could be done after design-for-testability (DFT) synthesis. So, these techniques could be implemented at the gate level, the transistor level, or the layout level. All the existing techniques are implemented at the gate level.

(iii) *Test Coverage/Overhead*: Scan chain locking/blockage can negatively affect the testability metrics, e.g., test coverage, resulting in the reduction of reliability of the circuits. Hence, one crucial criterion in this breed is to keep the test coverage as close as possible to that of the original circuit before scan locking/blocking. Additionally, in some of the blockage techniques, extra test pins have been added to support high testability coverage and efficient resiliency [37–39, 41]. However, adding extra pins can extremely increase the die size of the manufactured chip.

(iv) *Locking Operation Method*: In scan chain locking, the locking part can operate statically or dynamically. In the dynamic approach, components like LFSR and/or PRNG have been engaged that change the configuration at runtime [86, 88, 91, 92]. However, in techniques with statically scan chain locking [87, 89], the configuration is always fixed. The dynamicity—if implemented correctly—is always a huge bar for the adversary, specifically, once they rely on I/O query-based attacks, and dynamicity invalidates all learned information acquired on previous acts and enforces the adversary to restart the process.

(v) *The Security of Scan Chain Architecture*: Scan chain locking/blocking brings some modification into the structure of scan chain(s). This modification particularly comes from scan blockage techniques. For instance, with a more robust scan chain architecture, the designer might integrate all regular FFs and key-dedicated FFs into a common scan chain. However, this can undermine the security of the logic locking key, regardless of locking techniques, if a leakage possibility could be found through the modified scan chain structure or cells. So, the security of the scan chain must be evaluated and guaranteed, while scan chain locking/blockage is applied to the circuit.

Table 2.7 summarizes the main specification of existing scan-based locking/blocking techniques. At the moment, a combination of one scan-based logic locking or scan blockage with functional logic locking techniques shows high resiliency and received significant attention. Scan-based protection extremely increases the size and complexity of formulation for any attacking model on logic locking. We further discuss that this type of locking is an integral part of reliable and security-guaranteed logic locking techniques against all existing threats.

2.3.6 FSM/Sequential Logic Locking

One basic assumption in all previously discussed logic locking techniques' threat model is that the availability of the scan chain is NOT restricted. However, in a notable portion of real ICs/SoCs, this availability is blocked for critical security reasons. Assuming that the scan chain access is already limited/restricted, some logic locking techniques focus on locking the whole circuit, which could be considered as sequential logic locking techniques. Most of them focus on the locking of the state (FSM) of the circuit. In existing FSM locking techniques [14, 15, 93–98], the original FSM has been targeted and altered in different ways, as demonstrated in Fig. 2.15: (i) adding few extra sets (modes) of states to the original state transition graph (STG), such as locking/authentication mode states, (ii) adding traps such as black hole, (iii) altering the deepest states of the circuit that makes the timing analysis longer and more complex, (iv) adding shadow states which act like decoy states, and (v) making the FSM combinational fan-in-cone subcircuitry key-dependent.

Table 2.7 Specification of scan-based locking/blocking techniques

Logic locking	Type	Mechanism	Overhead	Attacked by
DOS [86, 92]	Locking by PRNG	Dynamic LFSR-based Shuffling and Toggling with Shadow Scan Insertion	■■□□□	ScanSAT: Sect. 2.4.2.3
Encrypt-FF [87]	Locking by XOR	XOR-based key gate insertion within the scan chain	■□□□□	ScanSAT: Sect. 2.4.2.3
Robust design for security (R-DFS) [41]	Blockage	A custom secure DFF for key storage + SO blockage circuitry for post-activation	■■□□□	shift&leak: Sect. 2.4.2.3
dynamic-EFF [88]	Locking by PRNG	Adding PRNG and malfunctioning by PRNG for incorrect keys	■■□□□	DynUnlock: Sect. 2.4.2.3
Extended R-DFS [38]	Blockage	A custom secure DFF for key storage + SO blockage circuitry for post-activation	■■□□□	shift&leak: Sect. 2.4.2.3
seql [89]	Locking by XOR	XOR-based key gate insertion between functional and scan chain paths	■□□□□	ScanSAT: Sect. 2.4.2.3
kt-DFS [39]	Blockage	A custom secure DFF for key storage + SO blockage circuitry for post-activation	■■□□□	NONE
DisORC [43]	Blockage	SO Blockage circuitry with full shift disable after activation + disabling shift + oracle dishonesty	■■□□□	NONE
DOSC [91]	Locking by PRNG + counter	Dynamic LFSR-based Shuffling and Toggling with Shadow Scan Insertion	■■□□□	NONE

In the primitive FSM-based logic locking techniques, there exists no dedicated port/pin/wiring for key values, and the traversal sequence of these extra added states (sequence of input patterns), like traversal of locking/authentication modes, is the locking/authentication key, and a correct traversal allows the user to reach and traverse the original part of the FSM. Hence, unlike all other logic locking techniques, key inputs are implicitly added, and we can consider this form of logic locking as keyless logic locking. Also, the output generated by the correct traversal of authentication states serves as a watermark. In addition to these groups, a set of studies evaluate FSM locking without adding any extra state. However, the complexity and overhead (area) of this approach are higher compared to other schemes [95]. More recent FSM-based logic locking studies also evaluate the combination of state traversal and key-based logic locking [96–98]. Table 2.8 summarizes the main specifications of existing FSM-based logic locking techniques.

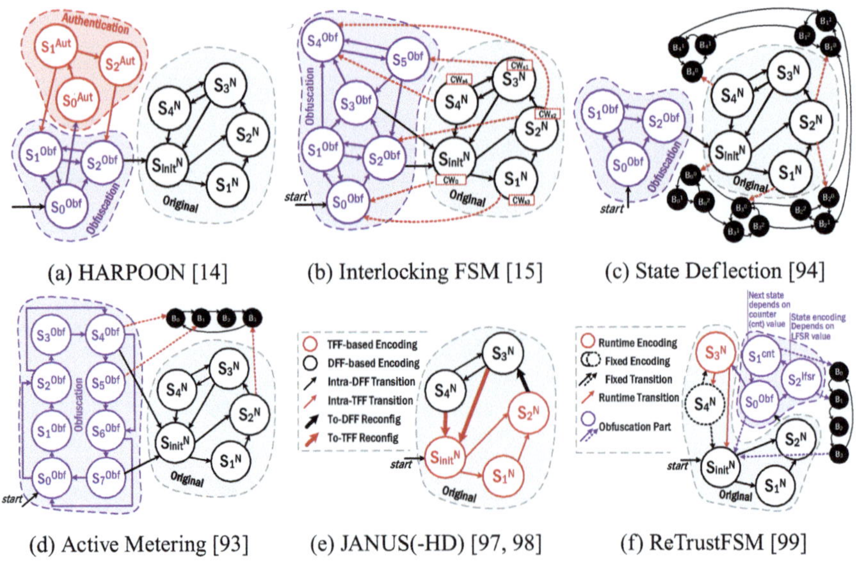

Fig. 2.15 Existing FSM obfuscation solutions

2.3.7 *Behavioral Timing-Based Locking*

Unlike all previous techniques that focus on the functionality of the design, some logic locking techniques went one step further and lock the behavioral properties of the circuit, such as timing. For example, DLL [16] introduces a custom tunable delay and logic gate, demonstrated in Fig. 2.16a which could have different delays based on the value of the key. Since most of the CAD tools are dealing with Boolean nature, it would be hard for the adversary to deal with this form of ambiguity. Some more recent techniques use multicycle paths, key-controlled clock handling, and latch-based structure, in which the timing of the circuit would be changed in an asynchronous/clock-gated manner [47, 101–103, 105]. For instance, in latch-based and asynchronous-based techniques [47, 103], the time of storage of data in the FFs are asynchronized and controlled with the key. More particularly in data flow obfuscation [47], as demonstrated in Fig. 2.16c, *key-based c-element* of asynchronous circuits has been used to control the timing/flow of data within the design. So, without having the correct key, the flow of data movement will be changed within the circuit, which consequently corrupts the functionality and may result in appearing some halt in the circuit.

The timing-based and latch-based technique shows promising results against different adversary' actions. However, due to the lack of full EDA (electronic design automation) tool support for asynchronous designs, replacing the flip-flops with latches and implementing asynchronous latch-based designs raise burdensome challenges in the IC design process, and it makes the usage of asynchronicity almost

Table 2.8 Specification of FSM-based logic locking techniques

Logic locking	Mechanism	Overhead	Corrupt	Attacked by
HARPOON [14]	Adding extra authentication and obfuscation states before the original initial state	*variant* (w.r.t. size of added states)	High at new extra states	RANE:2.4.2.2, Fun-SAT: 2.4.2.2
FSM interlocking [15]	Adding traps per each original state for incorrect key or sequence	■■□□□	■■■	Fun-SAT: 2.4.2.2
Active metering [93]	Adding extra obfuscation states + black holes for specific transitions	*variant* (w.r.t. size of added states)	High at new extra states	Fun-SAT: 2.4.2.2
Dynamic deflection [94]	Inserting code-word transitions based on input pattern + extra states before initial state	*variant* (w.r.t. size of added states)	High at new extra states	Fun-SAT: 2.4.2.2
FSM reconfig [100]	reconfigurable logic for FSM circuitry (Only on FPGA)	■■■□□	■■■	NONE
DFSSD [96]	Adding faults at deep states using counter + point function	■□□□□	■□□	NONE[a]
JANUS/HD [97, 98]	Concealment of state transition circuitry using a configurable control unit	■■□□□	■■■	NONE[b]
ReTrustFSM [99]	Counter- & LFSR-based logic locking + combining both key-less and key-based locking	■□□□□	■■■	NONE

[a] It could be vulnerable to structural-based removal attack, if no camouflaging is in place for some locking gates

[b] It may be vulnerable to the sequential SAT attack due to its simplicity. Not tested by the main paper

impractical for complex SoCs. A very recent study, called O'clock, relies on widely used clock-gating techniques and targets the clock-gating enabling circuitry for obfuscation purposes with full EDA-based support. One of the main features of these timing-based techniques is that they control and manipulate the time of data capturing at storage elements, i.e., at FFs. Hence, the adversary cannot track and follow the exact and correct timing of data capturing, and there is no additional benefit for the design team to restrict/block the scan chain. So, unlike the scan locking or blockage, they can keep scan chain available (imperative to perform in-field debug and test but exploited by SAT attack [46]) to the untrusted foundry and end-users while resisting a wider range of I/O query-based attacks, like SAT-based attacks. Table 2.9 summarizes the main specifications of existing timing-based logic locking techniques. As demonstrated, except for delay locking that is broken using a theory-based I/O query-based attack, called SMT attack [106], all others are not broken, showing the robustness of this breed of logic locking.

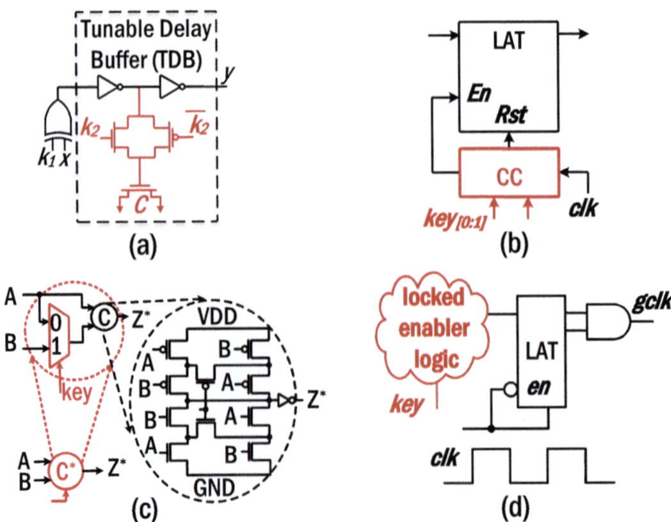

Fig. 2.16 Behavioral timing-based locking: (**a**) Delay-based [16], (**b**) Latch-based [103], (**c**) Asynchronous locking [47], (**d**) Clock-gated locking [105]

Table 2.9 Specification of timing-based logic locking techniques

Logic locking	Mechanism	Overhead	Corrupt	Attacked by
Delay locking [16]	Insertion of tunable delay buffers that control both function and delay of key gates	■■□□□	■■■	SMT: Sect. 2.4.1.2
Latch-based [103]	adding key-based latches + decoy-based latch insertion	■■□□□	■■■	NONE
Data-flow [47]	controlling the data flow using asynchronicity + decoy latches + decoy logic	■■□□□	■■■	NONE
O'Clock [105]	locking the clock gating circuitry + transition-based stripping functionality	■□□□□	■■■	NONE

2.3.7.1 Beyond-CMOS and Mixed-Signal Logic Locking

In comparison to CMOS technology, emerging technologies, such as spintronics, memristors, FinFet, CNTFETs, and NWFETs, which are also compatible to be integrated with CMOS technology, promise and provide unique properties that can be engaged for security purposes, features like variability/randomness, runtime reconfigurability, or polymorphic behavior. The utilization of these features can help to obtain resilience against reverse engineering, to build unique PUF/TRN generation units, to protect IPs, and to build masking against side-channel leakage [107]. Hence, a set of existing logic locking (and more on camouflaging) techniques have used such technologies. Many of these approaches utilize these technologies to achieve (i) *reconfigurability/dynamicity* that helps to invalidate continuous analysis on the locked circuit (requires frequent restart), (ii) *resiliency against reverse engineering* that helps against diverse acts by the adversary for revealing the secrets, such as probing or retrieving the unlocking key, or retrieving the netlist by the untrusted foundry. The utilization of emerging technologies for IP protection through logic locking can be summarized as follows:

(i) *Spintronic-Based*: With a nonvolatile switching mechanism and other related concepts like spin-transfer torque, spin electronics technology can provide both computation and storage/memory capabilities (STT) [108–110]. All-spin logic (ASL) for camouflaging [111], spintronics-based reconfigurable LUTs [69, 112–114], fully programmable polymorphism based on giant spin-Hall effect (GSHE) [115, 116] are some approaches that leverage this technology for locking+camouflaging at lower overhead compared to CMOS-based counterparts.

(ii) *Memristor-Based*: Memristor-based (memory-resistor) cells are basically able to retain their internal resistive state w.r.t. the voltage/current applied, which can be used for building Boolean logic [117, 118]. In [64, 119], the concept of reconfigurable (polymorphic) cyclic logic locking has been proposed in which the memristor cell(s) has been used to protect against adversaries in the foundries and test facilities.

(iii) *FET-Based*: Tunnel field-effect transistors, carbon nanotube field-effect transistors (CNTFET), and Nanowire FETs (NWFETs) are other leading emerging technology candidates to replace CMOS FinFET and DRAM technologies, which are compatible with CMOS technology [120–124]. Similarly, these technologies could be used for implementing polymorphic gates (PLG) as a means of locking+camouflaging [125–128]. In [127], silicon nanowire (SiNW) FETs are utilized to implement different PLGs for making the locking part less traceable. The authors of [125] propose silicon NWFETs for camouflaging+locking, by using controlled ambipolarity of NWFETs, helping to build primitives like NAND, NOR, XOR, and XNOR functions.

Apart from beyond-CMOS technologies, some approaches also investigated the applicability and application of logic locking in analog and mixed-signal (AMS) circuits, which is a large subclass of analog ICs, including data converters, phase-locked loops (PLLs), radio frequency (RF) transceivers, etc. [129–135]. Many of these approaches consider the locking of digital parts that directly or indirectly affects the analog side [129, 130, 135]. However, to expand the locking over both analog and digital sections, other approaches target modules like analog-to-digital or digital-to-analog (ADC/DAC) converters [131]. Another study uses chaotic computing and proposes a mixed-signal locking by adopting chaogate [136] to get two important advantages of this mechanism, i.e., (1) Ability to build all 2^{2^2} functions of a 2-input Boolean gate using a single chaotic element (like LUT but at lower overhead), and (2) Dynamic function update capability [134]. At the layout level, few studies also investigate the sizing and hiding as a camouflaging approach by means of fake contacts in the active geometry of the layout components [133].

2.3.8 High-Level Logic Locking

Although a high portion of existing logic locking techniques is implemented at the
gate level (or even transistor/layout level), they may be incapable of targeting all
semantic information (defined and described in higher level, e.g., RTL or HLS) since
logic synthesis and optimizations will convert/simplify/twist much of it into the
netlist. Hence, to (1) be able to directly target any algorithmic and semantic of the
design for locking purposes, and (2) to get the benefit of synthesis and conversion
done by the optimizations for twisting logic locking part into the original part(s), a
bunch of recent studies utilizes new methodologies for locking at a higher level of
abstractions [57, 137–145].

Another advantage of high-level logic locking techniques is that they can
potentially protect the design against a wider range of untrusted entities. As shown
in Fig. 2.2, with the notion of insider threats (e.g., a rogue employee at the design
house), high-level locking allows the design team to have the IP locked and
protected from far earlier stages of IC design flow. As demonstrated in Fig. 2.17,
high-level locking can be categorized into the following groups. Also, Table 2.10
covers the main specification of existing countermeasures in this breed of logic
locking:

(i) *High-Level Locking Before Synthesis/Transition*: Some of the approaches
    apply the logic locking at the highest level (e.g., C/C++) code before syn-
    thesizing the design through the HLS steps [141]. In such cases, the HLL code
    provided to the HLS tool is already locked, and conversion/absorption/transfor-
    mation on locked semantic will happen through (i) the HLS intermediate steps,
    including allocation, scheduling, and binding, as well as (ii) RTL synthesis.
(ii) *High-Level Locking+Synthesis (HLS Extension)*: Some studies implement
    and integrate locking with the intermediate steps of the HLS engine [57,
    137, 140, 145]. Such approaches analyze intermediate representations (IRs)
    generated through the HLS flow and will extend these IRs by applying

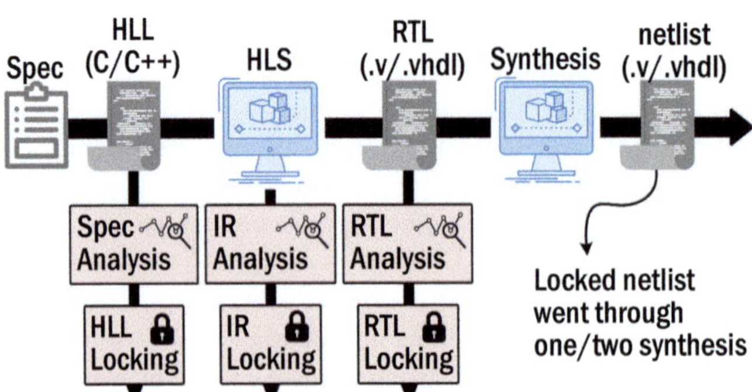

Fig. 2.17 High-level logic locking technique

Table 2.10 Specification of high-level locking techniques

Logic locking	Type	Mechanism	Overhead	Attacked by
TAO [137]	within HLS	Targeting loops, constants, branches, and manipulation of HLS scheduling	*variant* (w.r.t.) features	SMT: Sect. 2.4.1.2
SFLL-HLS [57]	within HLS	point function locking through HLS flow	■□□□□	SPI: Sect. 2.4.1.3
ASSURE [139]	at RTL	Targeting branches, constants, and arithmetic operations	■■■□□	SMT: Sect. 2.4.1.2
HOST [143]	at RTL	creating RTL with numerous abstraction behavior + building FSM-based SMT-hard instance	■■□□□	NONE
RTLock [138]	at RTL	ILP optimization for locking at RTL + building multiple locking techniques	■□□□□	NONE
HLock [141]	at HLL	Targeting function calls, control flow, arithmetic operation, constants + ILP optimization	■□□□□	NONE
HLock+ [146]	at HLL	Targeting function calls, control flow, arithmetic operation, constants + ILP optimization + point function based compound logic locking	■□□□□	NONE
Fortifying RTL [142]	at RTL & Gate-level	Targeting branches, constants, and arithmetic operations + gate-level scan blockage	■□□□□	NONE

the locking part. In these approaches, the HLL code is not locked, but the RTL generated by the HLS tool is locked, and the locked RTL will face conversion/absorption/transformation through RTL synthesis.

(iii) *Register-Transfer Locking (Post-HLS or Direct RTL)*: Techniques in this subgroup apply logic locking at RTL, either on the output of the HLS engine, i.e., the RTL generated by the HLS tool [139, 143] or RTL designed/developed directly by the designers [144]. This group analyzes the specification of RTL (like abstract syntax tree) or other graph-based representation to find the best candidates for selection and locking, and then the locked RTL will go through the RTL synthesis.

(iv) *Compound Locking (High Level + Gate Level)*: We also witness a compound form of logic locking in which RTL locking and gate level have been engaged as a single countermeasure. In [142], RTL locking is combined with a scan-based logic locking to thwart threats induced by both untrusted foundry and malicious users.

Moving from gate level to higher levels allows the designers to target higher-order elements, such as semantic information: (1) *Constants*: Sensitive hard-coded information through the computation (e.g., filter coefficients, encryption IVs, etc.).

(2) *Arithmetic Operation*: Critically determining functional arithmetic operation (e.g., multiplications, shifting, adding, subtraction, etc.). (3) *Conditional Branches*: Branches defining the execution flow (based on the control flow graph). (4) *Function Calls*: Important function calls that build the main hierarchy of the design. (5) *Memory Access*: Address, read, and write for the memory blocks with sensitive information. Based on these elements targeted for locking, all high-level logic locking techniques can provide high corruptibility. Table 2.10 summarizes the specification of existing logic locking at higher level.

2.3.9 eFPGA-Based IP-Level Locking

Some recent studies have investigated a coarse-grained form of logic locking, in which redaction by the usage of embedded FPGA (eFPGA) has been done at SoC level [147–151]. In this case, after selecting specific (targeted) module(s), they will be replaced with fully reconfigurable soft embedded eFPGA or readymade eFPGA-hard macro. So, the eFPGA, with configurable logic blocks (CLBs) containing look-up tables (LUTs), flip-flops, and routing logic, can be fabricated and programmed to realize the desired functionality. In this case, the secret of the design would be the bitstream that determines the functionality of eFGPA, and the adversary must recover the complete bitstream to implement the correct functionality in each eFPGA.

The eFPGA-based IP redaction can be considered as a superset of LUT/routing-based logic locking, as described in Sect. 2.3.4. Because of the symmetric structure and uniform CLB/LUT distribution, structural attacks are difficult to be applied. From the functionality point of view, fully configurability and growth of bitstream size result in a significant enhancement of resiliency against I/O query-based attacks. A recent exploration has investigated eFPGA parameters for IP redaction, in which it is shown that similar to routing-based locking, coarse-grained eFPGA IP redaction is an SAT-hard instance that cannot be broken using any of the existing I/O query-based attacks. However, compared to LUT/routing-based logic locking, their incurred overhead is even getting worse [150].

2.4 Logic Locking: Attacks

This section provides a more comprehensive evaluation of the existing successful attacks introduced so far on logic locking. Based on the models and assumptions previously discussed in Sect. 2.2.2, all attacks on logic locking could be categorized into different groups as demonstrated in Fig. 2.18 [152]. The categorization of attacks on logic locking is heavily dependent on the availability of the target chip in activated/unlocked mode (oracle). The availability of oracle will help the adversary to build up more algorithmic attacks. Most of the *oracle-guided* attacks

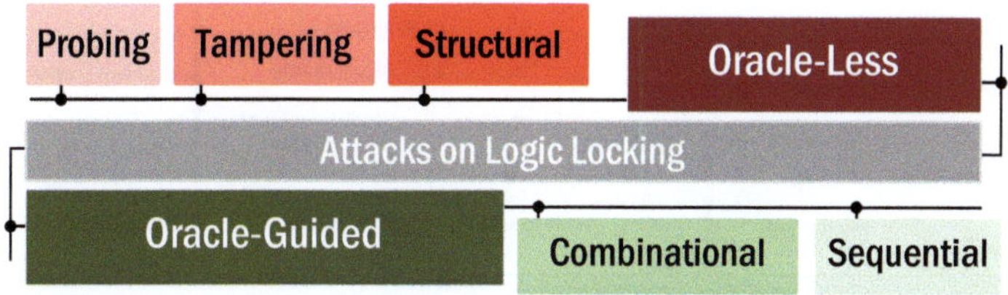

Fig. 2.18 The general categorization of the attacks on logic locking

could be also known as algorithmic attacks, in each, a systematic flow has been proposed that results in the exposure of either logic locking key or the correct functionality of the locked circuit. Almost all attacks in this category could be considered as *(weak) invasive*, and they require to have access to the netlist of the locked circuit. Based on the applicability, they could be categorized into attacks on *combinational* circuits (when scan access is available) or *sequential* circuits (when scan access is NOT available). However, *oracle-less* attacks, on the other hand, mostly rely on CAD tools such as synthesis tools, or physical attributes of the circuit such as side-channel information to accomplish the attack flow. Based on the structure/mechanism, there exist three main subcategories of *oracle-less* attacks, which are *structural*, *tampering*, and *probing* attacks. Figure 2.19 illustrates almost all existing attacks on logic locking and their categorizations. In the following of this chapter, we thoroughly evaluate the members of each subgroup of this hierarchy, and we will cover the following crucial information per each subgroup:

(1) A uniform assessment on each attack algorithm based on their proposed mechanism
(2) The main purpose of each attack, the capability, as well as applicability of each attack on the existing countermeasures
(3) The challenges and limitations of each attack in terms of implementation effort/feasibility, applicability, design time/complexity, overhead, etc.
(4) Existing or potential countermeasures that could break each attack

2.4.1 Oracle-Guided (OG) on Combinational Circuits

The name of each category used in Fig. 2.18 represents the attack model used for that category. Here the oracle-guided (OG) attacks on combinational circuits show that the deobfuscation attacks of this category have been introduced based on the following assumptions:

It is oracle-guided: The attacker requires to have access to one additional activated/unlocked version of the chip (oracle).

Fig. 2.19 Attacks on logic locking techniques

It is on combinational circuits: Since almost 100% of real application ICs are
   sequential, it implies that having access to the DFT structure, i.e., scan chain
   pins, is available to provide the access to each combinational part (CL) of the
   circuit.

It is invasive:[8] The attacker requires to have access to the netlist of the locked circuit
   (locked GDSII at the foundry or reverse-engineered of chip acquired from the
   field/market).

---

[8] Except for one preliminary deobfuscation attack, called *hill climbing*, all other oracle-guided
deobfuscation attacks on combinational circuits require access to the netlist.

2.4.1.1 OG Combinational ATPG-Based Attacks

Once the logic locking key is incorrect, function corruption will happen, and since the propagation of corrupted signal(s) resembles fault propagation flow, the possibility of using testability and fault analysis attributes was firstly investigated as a means of attack on logic locking. These deobfuscation attacks exploit almost the same techniques/algorithms that are widely used for automatic test pattern generation (ATPG), such as exhaustive testing, randomness used with algorithmic methods for testing and debug, symbolic difference check, and path sensitization methods [153].

Sensitization Attack

Similar to the logic-level path sensitization in ATPGs with three main steps, i.e., (1) fault sensitization, (2) fault propagation, and (3) line justification, the sensitization attack [13] consists of the same steps. In the sensitization attack, each key bit at any arbitrary gate will be treated as a stuck-at fault. Then, it will be propagated to the primary/scan output (PO/SO) using (1) fault sensitization and (2) fault propagation. After determining the input pattern (state) that propagates the value of the key to the PO/SO, the attacker applies the same input pattern (state) to the oracle, and since the correct key value is already loaded in the oracle, the correct key value will be propagated to the PO/SO. Then, the attacker could observe and record this output as the value of the sensitized key.

The main purpose of the sensitization attack is to break RLL [12] as a member of primitive logic locking solutions described at Sect. 2.3.1. The attack results on RLL show that the sensitization attack could determine individual key values of the RLL-locked circuit in a time linear with respect to the size of the key. It should be noted that, in the sensitization attack, the propagation of a key bit to the PO/SO is heavily dependent on the location of the key. Hence, they classify key gates based on their location and discuss corresponding attack strategies for each case. The summary of strategies and techniques used in the sensitization attack is reflected in Table 2.11.

Table 2.11 Classification of key gates in sensitization attack [13]

Term	Description	Strategy used by attacker
Runs of KGs	Back-to-Back KGs	Replacing by a Single KG
Isolated KGs	No path between KGs	Finding unique pattern per KG (Golden pattern (GP))
Dominating KGs	$k1$ is on every path between $k0$ and POs	Muting $k0$, sensitizing $k1$
Concurrently mutable convergent KGs	Convergent at a third gate, both can be propagated to POs	Muting $k0/k1$, sensitizing $k1/k0$
Sequentially mutable convergent KGs	Convergent at a third gate, one can be propagated to POs	Determining $k1$ by GP, update the Netlist, target $k0$
Non-mutable convergent KGs	Convergent at a third gate, none can be propagated to POs	Brute force attack

There exist some limitations/challenges in the sensitization attack: (1) Similar to path sensitization in ATPGs, it is only applicable for acyclic combinational circuits. In the presence of feedback, it faces an infinite loop. (2) The efficiency of this attack would be low if the key gates are located at nonsynthesizable points. Thus, in SLL, as another member of primitive logic locking solutions [13], nonsynthesizable points have been widely exploited, to introduce a countermeasure against this attack.

Random-Based Hill-Climbing Attack

In [154], another ATPG-based attack has been introduced that finds specific test patterns to apply them to the locked circuit, and by observing the responses, it can lead the preliminary guessed (random) key values to the correct ones. Unlike the sensitization attack [13] and all other OG attacks on combinational logic locking, the hill-climbing attack does not require netlist access and could be considered as a *noninvasive* attack. It uses a randomized local key-searching algorithm to search the key that can satisfy a subset of correct input/output patterns. It first selects a random key value, and then at each iteration, the key bits that are selected randomly are toggled one by one. The target is to minimize the frequency of differences between the observed and expected responses. Hence, a random key candidate is gradually improved based on the observed test responses. When there is no solution at one iteration, the algorithm resets the key to a new random key value. The main purpose of this attack is to break the very first logic locking technique, i.e., RLL [12]. However, in many cases, it faces a very long execution time with no results. This happens for two main reasons which significantly undermine the success rate of this attack: (1) The key will be initiated randomly and (2) The complexity of the attack will be increased drastically, particularly when the key size is large or the key bits are correlated.

2.4.1.2 OG Combinational Algorithmic (SAT)-Based Attacks

Solving a Boolean *satisfiability* problem is the process of finding the satisfying assignment for a Boolean expression or equation. Many ATPG tools use Boolean *satisfiability* solver, SAT solver, for generating test patterns, and currently, it is one of the fastest ATPG algorithms, particularly for large-size circuits. In 2015, Subramanyan et al. [46] propose a new and powerful attack on logic locking that still gets the benefit of the SAT solver, which is widely used in ATPGs. The engagement of the SAT solver as a means for attacking the locked circuits has got the most attention in recent years for some important reasons, such as the strength, the performance of the attack, and the scalability. As shown in Fig. 2.19, the SAT-based attacks on logic locking, either combinational or sequential, are the largest by count. In this section, we will review the SAT-based attacks that focus on combinational locked netlist.

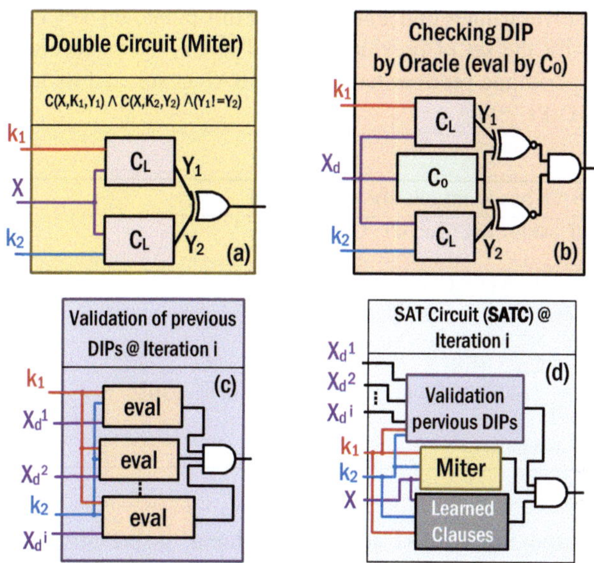

Fig. 2.20 The SAT attack iterative flow [46, 104]: (**a**) building miter (double circuit), (**b**) finding DIP + eval matching with the Oracle, (**c**) Validating DIPs at iteration i for all previously found DIPs, (**d**) SAT circuit (SATC) at iteration i

Conventional SAT Attack

The SAT attack was first introduced by Subramanyan et al. [46]. At the same time, El Massad et al. proposed the same technique, in which a SAT solver is engaged for attacking the combination logic locked netlist [104]. Getting inspired by the miter (distinguisher) circuit that is widely used for formal verification, the SAT attack uses a specific duplication mechanism to break the logic locked netlists. The main steps of the SAT attack have been demonstrated in Fig. 2.20.

As shown in Fig. 2.20a, in the SAT attack, the attacker first duplicates the locked circuit and builds a double circuit (miter). The miter is used for finding an input (x) that for two different key values, i.e., k_1 and k_2, this input generates two different outputs (y_1 and y_2 is XORed). The key values (k_1 and k_2) and the primary input pattern x will be found by a SAT solver query. Such input is referred to as the discriminating/distinguishing input pattern (DIP). Each DIP ($x_d[i]$) will be checked by the oracle (*eval* by applying $x_d[i]$ to c_o), as shown in Fig. 2.20b, assuring that for a previously found DIP, two different keys generate the same output value (part of the correct key pool for current iteration). Each iteration of the SAT attack finds a new DIP, and each adds a new *eval* check to the whole problem. All *eval* checks are then *AND*ed together, as shown in Fig. 2.20c, expanding the constraint to all previously found DIPs. In each iteration, the SAT solver tries to find a new DIP and two key values that satisfy the miter and the all aggregated (*AND*ed) *eval* constraints as shown in Fig. 2.20d. This iterative process continues until the SAT solver cannot find a new DIP. At this point, any key that generates the correct output for the set of previously found x_ds is the correct key. Algorithm 1 provides an algorithmic representation of the SAT attack and its iterative structure for finding all DIPs.

Algorithm 1 SAT-based attack algorithm [46]

1: **function** SAT_ATTACK(Circuit C_L, Circuit C_O)
2:     $i \leftarrow 0$;
3:     $F_0 \leftarrow C_L(x, k_1, y_1) \wedge C_L(x, k_2, y_2)$;
4:     **while** $SAT(F_i \wedge (y_1 \neq y_2))$ **do**
5:         $x_d[i] \leftarrow$ sat_assign$(F_i \wedge (y_1 \neq y_2))$;
6:         $y_d[i] \leftarrow C_O(x_d[i])$;
7:         $F_{i+1} \leftarrow F_i \wedge C_L(x_d[i], k_1, y_d[i]) \wedge C_L(x_d[i], k_2, y_d[i])$;
8:         $i \leftarrow i+1$;
9:     $k_c \leftarrow$ sat_assign(F_i);

The main purpose of the SAT attack was to break the primitive logic obfuscation techniques, including RLL [12], SLL [13], and FLL [45], as described in Sect. 2.3.1. For all these locking techniques, the SAT attack was able to rule out a significant number of key values at each iteration (by finding each DIP), and it was able to break them within a few minutes.

The traditional SAT attack has received significant attention in recent years, and numerous studies demonstrated the limitations/challenges of this powerful attack. Point function techniques (Sect. 2.3.2) show how the strength of DIPs could be reduced to minimize its pruning power. Cyclic locking techniques (Sect. 2.3.3) show the weakness of the SAT attack, while combinational cycles are engaged as a means of logic locking. Routing-based locking techniques (Sect. 2.3.4) and eFPGA IP-level redaction techniques (Sect. 2.3.9) show the effectiveness of complex structures as some fully configurable universal models that could be resilient against the SAT attack and its derivatives. Behavioral timing-based techniques (Sect. 2.3.7), such as delay-based logic locking, show how non-Boolean logic locking techniques cannot be modeled using the SAT attack. Scan-based (Sect. 2.3.5) and FSM-based (Sect. 2.3.6) countermeasures also demonstrate how the limiting access to the DFT structure and sequence-based locking can be engaged to break the SAT attack. Also, other behavioral timing-based techniques, like latch-based and clock-gating locking (Sect. 2.3.7), show how capturing time of scan chain cells can be manipulated that conceal exact read/write of the scan chain for applying the SAT attack. However, since the introduction of the SAT attack, we witness a cat-and-mouse game in this domain, where each defense (attack) is trying to break another attack (defense) by revealing their vulnerabilities.

Approximate-Based SAT Attack

Point function logic locking techniques are the first group of countermeasures that are resilient against the SAT attack. As shown in Fig. 2.8, these techniques will corrupt only a few (e.g., one) PO when the key is incorrect, and the output corruptibility happens for only a limited number of input patterns. Hence, for a limited number of POs (e.g., one PO), the circuit works as the original circuit, and for a rare set of input patterns, the output would be corrupted at rare POs. However, for a wide range of applications, such as image processing engines, this quite low output correction could be ignored at outputs (e.g., missing a few pixels). Hence, unlike the traditional SAT attacks, in which the exactness of the extracted key is guaranteed, in [155, 156], AppSAT and Double-DIP have been proposed, respectively, in which by relaxing this constraint, it is demonstrated that finding approximated key values could minimize the error rate when the circuit is locked using point function techniques.

The overall flow of the AppSAT is demonstrated in Algorithm 2, which is implemented based on the SAT attack and random testing. AppSAT [155] uses the *probably approximate correct* (PAC) model for formulating approximate learning problems. Unlike the traditional SAT attack whose termination condition is when no more DIPs could be found by the SAT solver, the AppSAT termination condition is based on the output error rate of a set of stimuli. AppSAT will be ended in any early step in which the error falls below a certain limit for a randomly selected set of stimuli. If this condition happens, the key value that satisfies the current set of constraints will be recognized as an approximated key with a specified error rate. Some heuristic methods are used in AppSAT for (1) estimating the error of large functions, to avoid any computation complexity, and (2) finding the minimal set of DIPs that lead to the targeted approximated key with the satisfied error. These methods play an important role in terms of the performance of the AppSAT attack.

AppSAT works perfectly fine on the logic locking techniques with low output corruption. Hence, some studies evaluate the robustness of point function techniques when they are combined with the primitive techniques, called compound techniques (Sect. 2.3.2.1). The compound logic locking techniques provide the most benefit of both categories, i.e., the high output corruption achieved by the primitive techniques, as well as the SAT resiliency achieved by the point function techniques. In such cases, AppSAT guarantees that the key of the primitive techniques will be extracted correctly, and the key related to the point function techniques is approximated that meets the error rate requirement.

Double-DIP [156] is an extension of the AppSAT attack in which during each iteration, the discriminating input should eliminate at least two sets of wrong keys. The reason for finding two incorrect sets of keys is for distinguishing between the keys that corresponded to the primitive technique from that of the point function technique. Since point function techniques corrupt the POs for a few input patterns (e.g., *ONE*) when an incorrect key corrupts the POs for more than one set of keys, it corresponds to the primitive logic locking. Otherwise, it corresponds to the point

---

Algorithm 2 AppSAT attack algorithm [155]

1: **function** APPSAT_ATTACK(Circuit C_L, Circuit C_O)
2:   $i \leftarrow 0$;
3:   $F_0 \leftarrow C_L(x, k_1, y_1) \wedge C_L(x, k_2, y_2)$;
4:   **while** $SAT(F_i \wedge (y_1 \neq y_2))$ **do**
5:     $x_d[i] \leftarrow$ sat_assign$(F_i \wedge (y_1 \neq y_2))$;
6:     $y_d[i] \leftarrow C_O(x_d[i])$;
7:     $F_{i+1} \leftarrow F_i \wedge C_L(x_d[i], k_1, y_d[i]) \wedge C_L(x_d[i], k_2, y_d[i])$;
8:     $i \leftarrow i+1$;
9:     every n rounds do
10:      **for each** ($x \in$ Random Patterns) **do**
11:        **if** $C_L(x, k_1, y) \neq C_O(x)$ **then**
12:          $FailedPatterns \leftarrow FailedPatterns + 1$;
13:          $F_{i+1} \leftarrow F_{i+1} \wedge (C_L(x, k_1, y) = C_O(x))$;
14:          $i \leftarrow i+1$;
15:      **if** error $<$ ErrorThreshold **then**
16:        return k_1 as an approximate key
17:   $k_c \leftarrow$ sat_assign(F_i);

---

function technique. The overall flow of the Double-DIP attack has been illustrated in Algorithm 3.

The effectiveness of the Double-DIP attack has been illustrated on the SAR-Lock+SSL, which represents an attack on a compound of point function and primitive techniques. The output result of the Double-DIP attack is an approximated key, in which the key of the primitive technique (SSL) is guaranteed to be correct, and the key of the point function technique (SARLock) is an approximated key that meets the error rate requirement.

In general, considering the exactness of the attack, and assuming that compound logic locking is in place, since approximate-based attacks only guarantee the key corresponding to the primitive logic locking, these attacks could reduce the problem from the compound technique to a single point function technique with an approximated key.

Algorithm 3 Double-DIP attack algorithm [156]

1: **function** DOUBLEDIP_ATTACK(Circuit C_L, Circuit C_O)
2:   $i \leftarrow 0$;
3:   $F_0 \leftarrow C_L(x, k_1, y_1) \wedge C_L(x, k_2, y_2) \wedge C_L(x, k_3, y_1) \wedge C_L(x, k_4, y_2)$;
4:   **while** $SAT(F_i \wedge (y_1 \neq y_2)) \wedge (k_1 \neq k_3)) \wedge (k_2 \neq k_4)$ **do**
5:     $x_d[i] \leftarrow$ sat_assign$(F_i \wedge (y_1 \neq y_2)) \wedge (k_1 \neq k_3)) \wedge (k_2 \neq k_4)$;
6:     $y_d[i] \leftarrow C_O(x_d[i])$;
7:     $F_{i+1} \leftarrow F_i \bigwedge_{j=1}^{4} C_L(x_d[i], k_j, y_d[i])$;
8:     $i \leftarrow i+1$;
9:   $k_c \leftarrow$ sat_assign(F_i);

Bypass Attack

The bypass attack [157] is another attack that gets the benefit of low corruptibility in point function techniques when they are not mixed with the primitive logic locking techniques. The bypass attack instantiates two copies of the obfuscated netlist using two randomly selected keys and builds a miter circuit that evaluates to 1 only when the output of two circuits is different. The miter circuit is then fed to a SAT solver looking for such inputs. The SAT returns with a minimum of two inputs for which the outputs are different. These input patterns are tested using an activated IC (oracle) validating the correct output. Then, a bypass circuit is constructed using a comparator that is stitched to the primary output of the netlist which is unlocked using the selected random key, to retrieve the correct functionality if that input pattern is applied. The bypass attack works well when the SAT-hard solution is not mixed with the traditional logic locking mechanism since its overhead increases very quickly as output corruption of logic locking increases.

Other Attacks on Compound Logic Locking

Due to the robustness of compound logic locking techniques, they received significant attention for a short time after their introduction. Similar to the Double-DIP attack, many studies try to reveal the security vulnerabilities of the compound logic locking techniques. In [158], the bit-flipping attack has been introduced, which relies on the fact that keys corresponding to primitive technique could be separated from the keys of point function technique. This could be achieved based on the hamming distances of outputs of the double circuit. The bit-flipping attack is motivated by the output corruption observation, which shows in primitive techniques, an incorrect key causes substantial corruption at POs. However, the output corruption of point function techniques is very small (e.g., with hamming distance equals with *ONE*). Hence, the calculation of the hamming distance (HD) of the double circuit's output could help to distinguish between two sets of keys. The overall structure of the bit-flipping attack has been described in Algorithm 4. First, for each key bit, some DIPs will be found by the SAT solver, and by checking the HD at the double circuit outputs, the usage (primitive or point function) of the key could be determined. Then, after the separation of the keys into two groups, the bit-flipping attack fixes point function keys, as a random number, and uses a SAT solver to find the correct key values used for the primitive technique. After finding keys of the primitive technique, similar to Double-DIP, the problem from the compound technique is reduced to a single point function technique that could be evaluated using approximate-based attacks for an approximated key, or as described in Sect. 2.4.1.2, bypass attack can recover the correct functionality using a bypass circuitry.

AppSAT-guided removal (AGR) attack is another attack that targets compound logic locking techniques, particularly Anti-SAT as the part of point function [159]. As its name implies, the AGR attack merges the AppSAT attack with a simple

---

Algorithm 4 Bit-flipping attack algorithm [158]

1: **function** BITFLIPPING_ATTACK(Circuit C_L, Circuit C_O)
2:     **for each** $j < $ *Fixed-iteration* **do**
3:         $k_A \leftarrow$ a random key;
4:         **for each** bit b $\in k_A$ **do**
5:             $k_B \leftarrow k_A$ while bit b flipped;
6:             $i \leftarrow 0$; $F_0 \leftarrow C_L(x, k_A, y_A) \wedge C_L(x, k_B, y_B)$;
7:             **while** $SAT(F_i \wedge (y_A \neq y_B))$ **do**
8:                 $x_d[i] \leftarrow$ sat_assign($F_i \wedge (y_A \neq y_B)$);
9:                 $F_{i+1} \leftarrow F_i \wedge (x \neq x_d[i])$;
10:                $i \leftarrow i+1$;
11:                **if** HD $>$ Threshold **then**
12:                    b is in k_1,
13:                    **break**;
            $j \leftarrow j + 1$;
14:     $k_2 \leftarrow$ all key bits / k_1;   ▷ Separation is Done. Fixing k_2.
15:     $k_1 \leftarrow$ SAT_ATTACK (C_L, C_O);   ▷ Find Primitive Keys by SAT.
16:     $C_L^* \leftarrow$ update_netlist($C_L - k_1$)
17:     **return** (BYPASS_ATTACK(C_L^*);

---

structural analysis on the locked netlist (as a postprocessing step). However, unlike AppSAT, the AGR attack recovers the exact correct key. In the AGR attack, the AppSAT is first used to find the approximated key, in which the correctness of the primitive technique keys is guaranteed. Then, AGR targets the remaining key bits that belong to the point function logic locking, such as the Anti-SAT block, through a simple structural analysis.

As demonstrated in Fig. 2.8, one of the main weaknesses of the point function techniques, when they are neither combined with the primitive techniques nor stripped, is that the locking part of the circuit, i.e., flipping and masking circuitry, is fully decoupled from the original parts of the circuit. Hence, a simple structural analysis can recognize the added macros and gates for locking purposes. This is what is called the possibility of removal attack on logic locking discussed in Sect. 2.4.1.3. In AGR, as shown in Algorithm 5, in its postprocessing steps, AGR

Algorithm 5 AGR attack algorithm [159]

1: **function** AGR_ATTACK(Circuit C_L, Circuit C_O)
2:     #*Cand* $\leftarrow$ num_gates(C_L)
3:     **while** (#*Cand* ¿ 1 and !*Timeout*) **do**
4:         AppSAT_Attack();   ▷ 4 times
5:         *Candidates* $\leftarrow \{\}$;
6:         **for each** *gate* $\in C_L$ **do**
7:             **if** *gate$_i$* has the *selected property* **then**
8:                 *Candidates* $\leftarrow$ *Candidates* + 1;
9:     $G \leftarrow$ Find_Max_key_count(*Candidates*);
10:    $C_{Lock} \leftarrow$ remove_**TFI**(C_L, G);   ▷ remove Transitive FanIn of G
11:    **return** C_{Lock};   ▷ C_{Lock}: C_L after removing Anti_SAT block

---

finds the gate (G) at which the keys corresponded to point function are converged (decoupled point function locking part), located at the transitive fan-out of these keys. AGR identifies the candidates for gate G by using structural analysis for all gates in the circuit and then sorts these candidates based on the number of key inputs that converge at a gate and pick the gate G from all candidates, which has the most number of key inputs converge to that gate. Then since the G gate would be the output of the flipping/masking circuitry in the point function techniques, it should have no impact on the functionality when the key is correct. So, the attacker resynthesizes the design with the constant value for the output of G gate (removing all fan-in-cone) and retrieves the correct functionality.

Fault-aided SAT-based attack [160] is another approach working on a specific compound logic locking in which a routing-based locking is combined with a point function technique [58]. In this attack, faults will be inserted at some fault locations in the locked netlist, and then by applying the SAT attack with a tight timeout, the SAT solver helps to find the key, while the fault insertion led to a SAT problem size reduction.

SAT-Based Attacks on Cyclic Locking

In Sect. 2.3.3, we showed the complexity/challenges behind the usage of combinational cycles in the circuit as a means of logic locking [59, 61–66]. SAT attack input must be in directed acyclic graph (DAG) format. Hence, adding cycles will violate this assumption. The incapability of the SAT attack for dealing with combinational cycles is the main motivation behind these techniques, which results in breaking the SAT attack either by (1) trapping it in an infinite loop or (2) forcing it to exit with a wrong key depending on whether the introduced cycles make the circuit stateful or oscillating.

Numerous studies have evaluated the resiliency of these techniques, thereby introducing new attacks on cyclic logic locking [161–163]. In CycSAT [161], the key combinations that result in the formation of cycles are found in a preprocessing step. These conditions are then translated into problem augmenting CNF formulas, denoted as cycle avoidance clauses, the satisfaction of which guarantees no cycle in the netlist. The cycle avoidance clauses are then added to the original SAT circuit CNF and the SAT attack is invoked. CycSAT has been introduced in two variants: (i) CycSAT-I, as shown in Algorithm 6, for the cases whose original circuit is acyclic, and (ii) CycSAT-II on a weaker assumption that the original circuit might be cyclic. However, the complexity that CycSAT-II might deal with is much higher making this variant less effective. Also, since all cyclic-based operations are based on structural analysis on circuits, it still faces exponential complexity when the number of cycles grows exponentially w.r.t. the number of feedbacks.

The lack of scalability to assess all cycles in the preprossessing of the CycSAT attack [161] leads to building incorrect preprocessed circuits, which might keep different types of the cycle even after preprocessing. It results in trapping the SAT attack invoked after the preprocessing step. Hence, the more advanced version of

---

Algorithm 6 CycSAT attack on cyclic locked circuits [59]

1: **function** CYCSAT_ATTACK(Circuit C_L, Circuit C_O)
2:     $W = (w_0, w_1, ...w_m) \leftarrow$ FindFeedback(C_L);
3:     **for each** ($w_i \in W$) **do**
4:         $F(w_i, w_i') \leftarrow$ no_ structural_ path(w_i);
5:     $i \leftarrow 0; NC(\text{k}) = \wedge_{i=0}^{m} F(w_i, w_i')$
6:     $C_L^*(\text{x, k, y}) \leftarrow C_L(\text{x, k, y}) \wedge NC(\text{k})$;
7:     $F_0 \leftarrow C_L^*(\text{x, k}_1, \text{y}_1) \wedge C_L^*(\text{x, k}_2, \text{y}_2)$;
8:     **while** $SAT(F_i \wedge (\text{y}_1 \neq \text{y}_2))$ **do**
9:         $\text{x}_d[\text{i}] \leftarrow$ sat_assign($F_i \wedge (\text{y}_1 \neq \text{y}_2)$);
10:         $\text{y}_d[\text{i}] \leftarrow C_O(\text{x}_d[\text{i}])$;
11:         $F_{i+1} \leftarrow F_i \wedge C_L(\text{x}_d[\text{i}], \text{K}_1, \text{y}_d[\text{i}]) \wedge C_L(\text{x}_d[\text{i}], \text{k}_2, \text{y}_d[\text{i}])$;
12:         $i \leftarrow i+1$;
13:     $k^* \leftarrow$ sat_assign(F_i);

---

cyclic-based SAT attacks was introduced [162, 163] to remedy such shortcoming(s). In BeSAT [162], with a runtime behavioral analysis at each iteration, as demonstrated in Algorithm 7, it detects repeated DIPs when the SAT is trapped in an infinite loop. Also, when SAT solver cannot find any new DIP, a ternary-based SAT is used to verify the returned key as a correct one, preventing the SAT from exiting with an invalid key. icySAT [163] can also produce noncyclic conditions in polynomial time w.r.t. the size of the circuit, avoiding the potentially exponential runtime still witnessed in BeSAT. Also, icySAT improves the attacks on cyclic logic locking techniques for cases whether the original circuit is cyclic if the feedback dependencies are reconvergent, or whenever the types of the cycles are oscillating.

Theory-Based SAT Attack

Since the SAT attack receives the input in conjunctive normal form (CNF), it (or any of its derivatives as described previously) works perfectly fine if the logic locking is of Boolean nature. Hence, a set of recent studies lock the properties of the circuit, which cannot be translated to CNF, such as timing-based or non-CMOS or mixed-signal logic locking as all demonstrated in Sect. 2.3.7.

Azar et al. open a new attack category on behavioral logic locking, which relies on the capability of modeling different theories in satisfiability modulo theory (SMT) solver [106]. The SMT solvers, as the superset of the SAT solvers, can combine the SAT solver with one or more theory solvers, and theory solvers, based on their specification, with richer language sets, can model beyond Boolean nature as well. As a case study, to demonstrate the benefit of the SMT solver as richer means of attack on logic locking, a graph theory solver has been engaged in the SMT attack, showing how delay-based logic locking could be formulated and broken. Based on the structure of SMT solvers, the attack shows how theory solvers could be engaged as a preprocessor (eager approach) or as a coprocessor (lazy approach) to assess and break the problem. One mode known as the *lazy* mode of this attack is illustrated in Algorithm 8. As demonstrated, function *GenTCE* is used and invoked to generate the timing constraints (key-based hold/setup constraints) based on the graph theory solver.

Although the SMT attack demonstrates a general model for attacking behavioral logic locking techniques, Chakraborty et al. also propose TimingSAT that shows a mechanism for modeling delay-based logic locking [164] using the pure SAT solvers. TimingSAT demonstrates the possibility of converting a non-Boolean logic locking problem into its Boolean counterpart. Hence, it consists of two main steps: (1) using a preprocessing mechanism for circuit unrolling approach that helps to capture the timing information in the form of Boolean functions. (2) Then, the locked circuit with captured timing information in Boolean form could be solved using the SAT solver. TimingSAT could be categorized as a simplified formulation of eager approach in SMT, in which with no theory solver, the capture of timing has been accomplished with more challenges using a Boolean-based remapped representation.

The usage of a few theory solvers is investigated in the SMT attack on another side to show the extensibility and richness of the SMT attack on some logic locking techniques, such as compound and cyclic logic locking. Also, further studies on SMT show its capability on other breeds of logic locking as well [165, 166]. In [165], a new SMT attack has been introduced on analog circuits, in which based on SMT formulations, it takes polynomial time (irrespective of the key size) for breaking the analog locking, and it is applicable to the ubiquitous presence in

---

Algorithm 7 BeSAT attack on cyclic locked circuits [162]

1: **function** BESAT_ATTACK(Circuit C_L, Circuit C_O)
2:      $W = (w_0, w_1, \ldots w_m) \leftarrow$ FindFeedback(C_L);
3:      **for each** ($w_i \in W$) **do**
4:          $F(w_i, w_i') \leftarrow$ no_ structural_ path(w_i);
5:      $i \leftarrow 0$;
6:      $NC(k) = \wedge_{i=0}^{m} F(w_i, w_i')$
7:      $C_L^*(x, k, y) \leftarrow C_L(x, k, y) \wedge NC(k)$;
8:      $F_0 \leftarrow C_L^*(x\ k_1, y_1) \wedge C_L^*(x, k_2, y_2)$;
9:      **while** $SAT(F_i \wedge (y_1 \neq y_2))$ **do**
10:          $x_d[i] \leftarrow$ sat_assign($F_i \wedge (y_1 \neq y_2)$);
11:          $y_d[i] \leftarrow C_O(x_d[i])$;
12:          $F_{i+1} \leftarrow F_i \wedge C_L(x_d[i], k_1, y_d[i]) \wedge C_L(x_d[i], k_2, y_d[i])$;
13:          **if** ($x_d[i]$ in DIP) **and** ($C_L(x_d[i], k_1) \neq y_d[i]$)) **then**
14:              $F_{i+1} \leftarrow F_{i+1} \wedge (k_1 \neq \hat{k}_1) \wedge (k_2 \neq \hat{k}_1)$;
15:          **else if** ($x_d[i]$ in DIP) **and** ($C_L(x_d[i], k_2) \neq y_d[i]$) **then**
16:              $F_{i+1} \leftarrow F_{i+1} \wedge (k_1 \neq \hat{k}_2) \wedge (k_2 \neq \hat{k}_2)$;
17:          $i \leftarrow i+1$;
18:      **while** $SAT_{k_1}(F_i)$ **do**   ▷ Correct Key: $\hat{k}_c$
19:          **if** Ternary_SAT(F_i, k_c) **then**
20:              $F_i \leftarrow F_i \wedge (k_1 \neq \hat{k}_c)$
21:          **else**
22:              $k^* \leftarrow \hat{k}_c$;
23:              **break**;

Algorithm 8 SMT attack on DLL (lazy approach) [106]

1: **function** SMTLAZY_ATTACK(Circuit C_L, Circuit C_O)
2:    $C_L^* \leftarrow$ toBOOLEAN(C_L);   ▷ Replace TDK with Buffer
3:    $i \leftarrow 0; F \leftarrow C_L^*(x, k_1, y_1) \wedge C_L^*(x, k_2, y_2)$;
4:    $G_L^* \leftarrow$ toGRAPH(C_L);   ▷ Wires = Edges, Gates = Vertices
5:    $F_T \leftarrow$ GenTCE(G_L^*)   ▷ Theory Learned Clauses
6:    $F_{SMT} \leftarrow F \wedge F_T$;   ▷ SMT Clauses
7:    **while** $SMT(F_{SMT})$ **do**   ▷ $x_d[i], k_1, k_2,$ CC
8:       $y_d[i] \leftarrow C_O(x_d[i])$;
9:       $F \leftarrow F \wedge C_L^*(x_d[i], k_1, y_d[i]) \wedge C_L^*(x_d[i], k_2, y_d[i])$;
10:      $F_{SMT} \leftarrow F \wedge$ CC; $i \leftarrow i+1$;
11:   $k^* \leftarrow$ smt_assign(F_{SMT});

1:    **function** GENTCE(Graph G_L^*)
2:       $Inputs \leftarrow$ find_start_points(G_L^*);
3:       $Outputs \leftarrow$ find_end_points(G_L^*);
4:       $T_{CE}(k) \leftarrow []$;
5:       **for each** $((Sp, Ep) \in (Inputs, outputs)$ **do**
6:          Upper(Sp,Ep)(k) $\leftarrow$!(**distance_leq**(Sp, Ep, t_{cd}));   ▷ Hold
7:          Lower(Sp,Ep)(k) $\leftarrow$ **distance_leq**(Sp, Ep, t_p);   ▷ Setup
8:          $R(Sp,Ep)(k) \leftarrow$ Lower(Sp,Ep)(k) $\wedge$ Upper(Sp,Ep)(k);
9:          $T_{CE}(k) \leftarrow T_{CE}(k) \cup R(Sp,Ep)(k)$;
10:      **return** $T_{CE}(k)$

---

wireless communication networks: Gm-C BPF, LC oscillator, quadrature oscillator, and class-D amplifier, as well as a memristor-based protection technique. Similarly, [166] utilizes SMT-based formulation and algorithm to break RTL logic locking. The algorithm models an RTL design as an RTL finite state machine with datapath (RTL-FSMD), and then it is abstracted out the details of the hardware into a behavioral program on which SMT solving has been invoked.

SAT-Based Attacks on LUT/Routing Locking

LUT-based and specifically routing-based locking, as described in Sect. 2.3.4, significantly increase the complexity of inner calculations of the SAT solver leading to an extremely long runtime per *each iteration* of the SAT attack [79–81]. These techniques rely on building symmetric interconnection into the locked portion of the circuit, extremely increasing the depth of the SAT search tree. The building blocks of the existing solutions in this category are the key-programmable routing blocks, each has its topology, such as crossbar or permutation (logarithmic) network. Having such structures sends the corresponded CNF far away being *under/overconstrained*, and when the SAT problem is a medium-length CNF, it brings difficulties for the SAT solver.

However, some recent studies introduce some SAT-based attacks that narrowly work effectively on this breed of locking [82, 83, 167]. In [83], the authors propose a *canonical prune-and-SAT (CP&SAT)* attack, which exploits a bounded variable addition (BVA) preprocessing step to reduce the size and complexity of the CNF representation of the key-programmable routing blocks used for routing-based

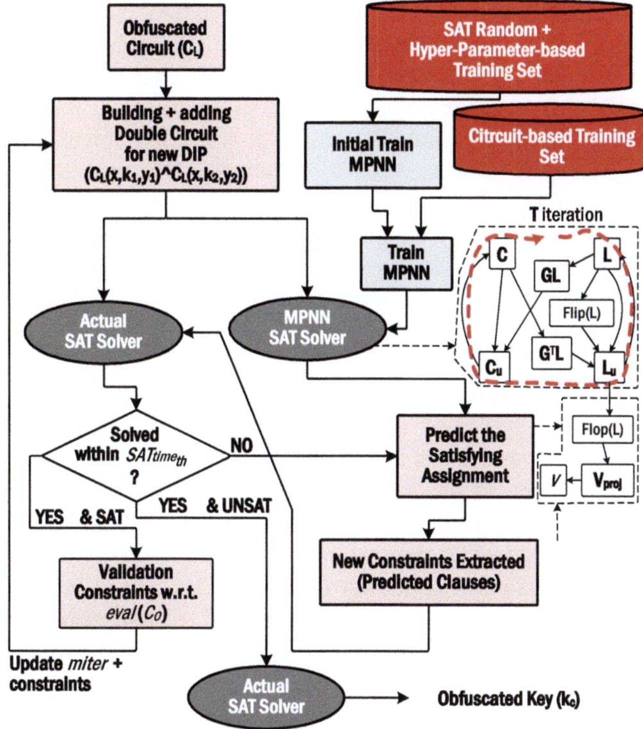

Fig. 2.21 The major steps of NNgSAT attack [167]

obfuscation. After reduction using the BVA, the reduced CNFs (corresponded to numerical bound problems) will be merged again with the circuit's CNF, and the SAT solver could be executed on the reduced CNF version. The main steps of the CP&SAT attack are: (1) It first models routing blocks as the most special case of numerical bound problems;[9] (2) Then by using BVA on the numerical bound problems, helping to reduce the size of the routing blocks as the input to the SAT solver; (3) The reduced problem will be solved by the SAT attack. The approach in [82] proposes the same mechanism with an extension on key-based LUTs. Although these attacks work perfectly fine on routing blocks, their efficiency will be drastically mitigated when routing blocks are twisted with extra logic [83].

In [167], a neural-network-guided SAT attack (*NNgSAT*) has been introduced that examines the capability and effectiveness of a message-passing neural network (MPNN) for solving different types of complex structures, e.g., big multipliers, large routing networks, or big tree structures. In NNgSAT, as demonstrated in Fig. 2.21,

---

[9] Most special case of numerical bounds is when among p variables only *ONE* variable is true, called at-most-1 constraint. It resembles the routing blocks in which for each output, it could be connected to only one of the routing block's input.

MPNN has been engaged and trained on the specific SAT problems to be used as a guide for the SAT solver within the SAT attack. The MPNN-based SAT solver is called in parallel with the actual SAT solver per each SAT iteration. Based on a predefined threshold time, if the actual SAT solver could find the satisfying assignment before the time threshold, the MPNN-based SAT solver will be skipped, and for the next step, both solvers will be called again. However, in those cases that the SAT solver could not find the satisfying assignment before the time threshold, a part of the predicted satisfying assignment by the MPNN-based SAT solver, which has the highest literal votes, will be extracted as a new (guiding) learned constraint, to help the actual SAT solver for finding the precise satisfying assignment.

2.4.1.3   OG Combinational Structural/Functional Attacks

The oracle-guided test-based and SAT-based attacks on combinational logic locking are completely designated for evaluating and revealing the vulnerabilities of the logic locking techniques using the functional attributes of the circuit. However, there exists another set of deobfuscation attacks that try to concentrate on both structural and/or functional properties of the logic locking to break the existing countermeasures. This group of attacks reveals that to have a well-designed and robust logic locking technique, it needs to be evaluated through different forms of analysis, e.g., functional and structural analysis.

Removal Attack

As shown in Fig. 2.8, for implementation of flipping circuit in point function techniques, the locking circuitry is completely decoupled from the original circuit. A removal attack could identify and remove/bypass the locking circuitry to retrieve the original circuit and to remove dependence on key values [159]. The removal attack was first presented to detect and remove SARLock [48]. In the presence of removal attack, different studies investigated the SAT-resistant solutions that are hard to be detected and removed (preventing removal by pure structural analysis), one of the examples of which was Anti-SAT [49].

As we discussed previously, some SAT-based attacks on compound logic locking techniques, such as bit-flipping and approximate-based attacks (Sects. 2.4.1.2, 2.4.1.2), can reduce the logic locking problem to a point function problem. Now with the introduction of the removal attack, such attacks could be integrated with the removal attack to completely break the compound logic locking techniques. For instance, the AGR attack discussed in Sect. 2.4.1.2 integrates AppSAT with removal to break compound logic locking techniques.

Signal Probability Skew (SPS) Attack

The Signal Probability Skew (SPS) attack [168] leverages the structural traces to identify and isolate the Anti-SAT block as a point function technique [49]. *Signal probability skew* (SPS) of a signal x is defined as $s = P_r[x = 1] - 0.5$, where $P_r[x = 1]$ indicates the probability that signal x is 1. The range of s is $[-0.5, 0.5]$. If the SPS of signal x is closer to zero, an attacker has a lower chance of guessing the signal value by random. For a 2-input gate, the signal probability skew is the difference between the signal probability of its input wires. In Anti-SAT, the flipping circuit is constructed using two complementary circuits, g and $\bar{g}$, in which the number of input vectors that make the function g equal to 1 (p) is either close to 1 or $2^n - 1$. These two complementary circuits converge at an AND gate G. Considering this structure, and the definition of SPS, the *absolute difference of the signal probability skew* (ADS) of the inputs of gate G is quite large, noting that the SAT resilience is ensured by this skewed p. After detecting the gate G, removal can be applied. Algorithm 9 shows the SPS attack flow, which identifies the Anti-SAT block's output by computing signal probabilities and searching for the skew(s) of arriving signals to a gate in a given netlist.

Algorithm 9 SPS attack algorithm [168]

1: **function** SPS_ATTACK(Circuit C_L)
2:     $ADS_{arr} \leftarrow \{\}$;
3:     **for each** $gate \in C_L$ **do**
4:         $ADS_{arr}(gate_i) \leftarrow$ Compute_ADS(C_L, $gate_i$);
5:     $G \leftarrow$ Find_Maximum(ADS_{arr});
6:     $Y \leftarrow$ Find_value_from_skew(G);   ▷ Correct value of Anti_SAT output
7:     $C_{Lock} \leftarrow$ remove_**TFI**(C_L, G, Y);   ▷ Transitive FanIn of the gate G
8:     **return** C_{Lock} ▷ C_{Lock}: C_L after removing Anti_SAT block

---

Functional Analysis (FALL) Attack

Since point function techniques are revealed to be vulnerable against structural-based attacks, the stripping functionality (SFLL) is used to invalidate the circuit after removal, as described in Sect. 2.3.2. SFLL is an extended version of point function techniques, in which the original circuit is modified for at least one input pattern (cube) using a *cube stripping unit*, demonstrated in Fig. 2.8b. As shown, Y_{fs} is the output of the stripped circuit, in which the output corresponding to at least one input pattern is corrupted. The restore unit not only generates the flip signal for one input pattern per each wrong key, but it also restores the stripped output to recover the correct functionality on Y. Note that applying removal attack on restore unit recovers Y_{fs}, which is not the correct functionality. In addition, SFLL-HD is able to protect $\binom{k}{h}$ input patterns that are of HD h from the k-bit secret key, and accordingly uses HD checker as a restore unit.

The main aim of the FALL attack is to exploit the resiliency of the SFLL technique [169]. In this attack model, the adversary is assumed to be a malicious foundry that knows the locking algorithm (SFLL) and its parameters (h) in SFLL-HD. The FALL attack is carried out in three main stages and relies on structural and functional analyses to determine the potential key values of a locked circuit:

(i) *Detection of Comparator*: The FALL attack tries to find all nodes which are the results of comparing an input value with key input. It is done by using a comparator identification check. Such nodes ($nodes_{RU}$), which contain these particular comparators, are very likely to be part of the functionality restoration unit. The set of all inputs that appear in these comparators should be in the fan-in-cone of the cube stripping unit. Then, it finds a set of all gates whose fan-in-cone is identical to the members of $nodes_{RU}$. This set of gates must contain the output of the cube stripping unit.

(ii) *Functional Analysis of Candidate Nodes*: The attacker applies functional analysis on the candidate nodes suggested by and collected from step (i) to identify suspected key values. Broadly speaking, the attacker uses functional properties of the cube stripping function used in SFLL, to determine the values of the keys. Based on the author's view, this function has three specific properties. So, per each property, they have proposed a specific attack algorithm. So, three attack algorithms exploit the unateness and hamming distance properties of the cube stripping functions. The input of these algorithms is circuit node c, computed from the first stage, and the algorithm checks if c behaves as a hamming distance calculator in the cube stripping unit of SFLL-HD. If the attack is successful, the return value is the protected cube.

(iii) *SAT-Based Key Confirmation*: They have proposed a SAT-based key confirmation algorithm using a list of suspected key values and I/O oracle access, which verifies whether one of the suspected key values computed from the second stage is correct.

CASUnlock Attack

CASUnlcok [170] is another structural-based attack that is specifically proposed to exploit cascaded logic locking, a.k.a. CASLock [56]. CASLock is a point function technique that proposes a variant cascaded AND-OR tree that builds variant corruptibility for a point function technique, and it is resilient against all previously discussed attacks. This attack works in two ways: (i) By relying on structural traces left after the synthesis process, this attack shows how the flipping circuitry in CASLock can be pinpointed after resynthesis allowing the adversary to recover the original IP. Similar to FALL, it also traces the fan-out of all key inputs and then pinpoints the convergent gate that is the output of the flip signal. (2) It exploits the connectivity of key inputs, thereby enabling the SAT attack to decipher the secret key with only a polynomial number of queries.

EDA-Based (SPI/Valkyrie) Attack

Sparse prime implicant (SPI) attack is another synthesis-based (EDA-based) mechanism that reveals the structural vulnerability in point function techniques [171]. In almost all point function techniques, the approach is relying on the assumption that the underlying EDA synthesis tool used by the semiconductor industry can effectively conceal the structure of flipping/masking/stripping circuitry in the DUT. However, in SPI, a comprehensive exploration has been done through different industrial/academic synthesis tools, including Cadence Genus, Synopsys Design Compiler, Synopsys Synplify, Xilinx Vivado, Mentor Graphics Precision RTL, and ABC, invalidating this assumption. So, similar to what was witnessed in CASUnLock, SPI also confirms that the structural traces left through the EDA tool can be exploited for breaking all point function techniques. SPI is specifically relying on the notion of prime implicants that is the underlying elements in logic optimization and redundancy reduction once the sum of products (SOP) format has been engaged for the circuit representation. In SPI, different specifications and properties have been defined and exploited around this notion to show how protected input patterns in the point function can be revealed. Valkyrie [172] almost follows the same direction, in which a security diagnostic tool has been introduced that checks for structural vulnerabilities. Also, they propose a circuit-recovery attack that shows how an adversary can exploit identified vulnerabilities by the diagnosis tool to recover the original functionality. Similar to the SPI attack, the analysis by Valkyrie invalidates the assumption that the design team can rely on the synthesis tool for absorbing the point function subcircuitry into the original part.

2.4.1.4   Summary of OG Combinational Attacks

As shown in Fig. 2.7, the first *four* logic locking subgroups, i.e., the primitive, the point function, the cyclic-based, and the LUT/routing-based, assume that the scan access for the adversary could be *OPEN*. Hence, the adversary has the capability of targeting each combinational part (CL) directly. This is the main motivation of all functional-oriented oracle-guided attacks on combinational circuits, which rely on I/O query-based techniques. On another side, analysis of the locked designs based on the structural traces reveals that many of the logic locking techniques suffer from this form of vulnerability, as demonstrated by the more recent structural attacks on point function techniques [171, 172]. As reviewed in this section, many of the logic locking techniques in the first *four* logic locking subgroups have been broken using these attacks. Tables 2.12 and 2.13 reflect the current status of oracle-guided combinational attacks. For each attack, these tables try to concisely answer to these *four* questions: (1) How the attack flow works; (2) Which logic locking techniques could be broken using the attack; (3) What is the limitation and challenges of the attack; and (4) What is the existing/potential countermeasure against the attack?

Table 2.12 Overview of oracle-guided attacks on combinational circuits

Attack	Mechanism	Applicable to[a]	Limitation & challenges	Countermeasures
Sensitization [13]	(1) Each key bit is considered as a stuck-at fault, (2) applying fault sensitization, (3) Finding a test pattern that propagates fault to PO/SO, (4) Applying test pattern to the oracle to observe the correct key	some RLL [12]	Only applicable to acyclic combinational circuits, (2) very limited w.r.t. key size and key location	SLL [13], FLL [45]
SAT [46, 104]	(1) Building double (Miter) circuit, (2) Finding DIP using SAT solver for two sets of keys, (3) Comparing with Oracle (ruling out wrong Key(s)), (4) Finding all DIPs, (5) Finding correct key using all DIPs	RLL [12], FLL [45], SLL [13], Lut-based [67]	Circuit must (1) be only acyclic, (2) have open scan access, (3) be only Boolean, (4) have no complex arithmetic units	All except primitive
AppSAT [155], Double-DIP [156]	(1) Running the SAT attack for u iterations, (2) Checking the output error rate for a set of input patterns after u iterations, (3) Returning the key if error rate is below a threshold, (4) Otherwise go back to step 1	Point Function, Point Function + Primitive	(1) Finding the minimal set of input patterns which excite the error, (2) Error rate analysis	G-AntiSAT [54], S-AntiSAT [55], CASLock [56], SFLL-felx/rem [52, 53]
Bit-Flipping [158]	(1) Decoupling keys (point function vs. primitive) based on POs' hamming distance per each DIP, (2) Fixing point-function keys, (3) Finding primitive keys using SAT, (4) Applying Bypass attack for point-function keys	Point Function + Primitive	(1) Not applicable if keys of primitive and point function is twisted, (2) Not applicable if function is stripped	TTLock [51], G-AntiSAT [54], S-AntiSAT [55], CASLock [56], SFLL-felx/rem [52, 53]
CycSAT [161], BeSAT [162], icySAT [163]	(1) Adding cyclic-based constraints to the SAT circuit before invoking the SAT attack (as a pre-processing step), (2) running the SAT attack on a cyclic-defused SAT circuit, (3) Recording/modifying some constraints to avoid infinite loops	cyclic [59, 64]	(1) Adding cycles exponentially w.r.t. the number of feedbacks, (2) Adding stateful/oscillating cycles, (3) asynchronous circuits	SAT-hard cyclic [61–63], Cross-lock [79], Full-lock [80], DF obfuscation [47]

Bypass [157]	(1) Selecting a random key for the obfuscated netlist, (2) Using SAT to find DIPs based on the selected key, (3) Applying DIPs to the Oracle, (4) Adding an extra circuit at the POs to re-flip the incorrect outputs (Bypassing)	Point Function (only low corruptible)	(1) The selected random key might have numerous DIPs, (2) Needs to find all DIPs to consider the attack as an accurate one	TTLock [51], G-AntiSAT [54], S-AntiSAT [55], CASLock [56], SFLL-felx/rem [52, 53]
SMT [106]	(1) Modeling behavioral obfuscation using one theory solver (like graph for timing or RTL (FSMD) or BitVector for hamming distance), (2) Integrating theory solvers with the Boolean SAT solver, (2) Co-solving using theory+SAT solvers	DLL [16]	(1) Theory solvers are limited, (2) theory is hard to be modeled in some behavioral obfuscation techniques	Complex models
NNgSAT [167]	(1) Using message-passing neural network as a classifier to predict the satisfiability, (2) Guiding the actual SAT solver based on the output (prediction) of the message-passing neural network	Cross-lock [79], Full-lock [80], Interconnect [81]	(1) Misguide rate could be high if trained on small set of benchmarks, (2) Accuracy could be low when dynamicity is in place	–
CP&SAT [83]	(1) Simplifying the routing modules using cardinality constraint as a pre-processor, (2) Run the SAT attack on simplified locked netlist	Interconnect [81], Cross-lock [79], Full-lock [80]	Inefficient optimization on strongly twisted logic and routing locking	Interlock [83]

[a] is able to break

Table 2.13 Overview of oracle-guided attacks on combinational circuits (2)

Attack	Mechanism	Applicable to[a]	Limitation & challenges	Countermeasures
Removal [159]	(1) Finding the obfuscation module (wrapper one) using structural analysis, (2) Extracting the original circuit wrapped by isolated obfuscation module	Anti-SAT [49], SARLock [48]	(1) Inefficient if obfuscation module and original design is twisted using fake (irre-movable) logic, (2) When the function is stripped	TTLock [51], G-AntiSAT [54], S-AntiSAT [55], CASLock [56], SFLL-felx/rem [52, 53]
SPS [168]	(1) Computing absolute signal probability skew for each gate, (2) finding the gate which has the maximum skew, (3) determining the correct output based on the location of the selected gate, (4) Revealing the key based on the correct output	Anti-SAT [49]	Inefficient if the sub-circuits of flipping circuitry is (1) not complement of each other, (2) built using LUTs	TTLock [51], G-AntiSAT [54], S-AntiSAT [55], CASLock [56], SFLL-felx/rem [52, 53]
FALL [169]	(1) Finding comparator of flipping circuitry using structural analysis, (2) Finding sub-circuit of restoration unit and cube stripping, (3) finding the suspected key values based on the selected sub-circuit using functional analysis, (4) Verify the key using SAT solver	TTLock [51], SFLL [52]	(1) Hard to find the units through structural analysis when re-synthesis is applied, (2) Hard (almost impossible) if camouflaged gates are used	CASLock [56], SFLL-felx/rem [52, 53]
EDA-based [171, 172]	Analysis of structural traces happening through EDA tools, like synthesis flow	All point functions	–	–

[a] is able to break

2.4.2 Oracle-Guided (OG) on Sequential Circuits

Since the availability of the scan chain undermines the robustness of many logic locking techniques once one of the oracle-guided attacks is in place, as demonstrated in Fig. 2.7, there exist other subgroups, such as scan-based logic locking, scan blockage, and sequential logic locking, in which the availability of the scan chain is targeted to be restricted/blocked. In such scenarios, assuming that the oracle is still available, the adversary access would be limited to the PI/PO of the oracle. Therefore, all of the previously discussed functional-oriented I/O query-based deobfuscation attacks will fail to evaluate and break the locked circuits with limited scan chain access. However, further studies on logic locking show that restricting access still cannot guarantee robustness against state-of-the-art threats. In this section, we will holistically review the attacks that follow such assumptions, and we call them oracle-guided deobfuscation attacks on sequential circuits. This group of attacks follows these assumptions:

It is oracle-guided: The attacker requires to have access to one additional activated/unlocked version of the chip (oracle).

It is on sequential circuits: It implies that having access to the DFT structure, i.e., scan chain pins, is limited/blocked/locked, and the adversary access is limited to PI/PO of the oracle.

It is invasive: The attacker requires to have access to the netlist of the locked circuit (locked GDSII at the foundry or reverse-engineered of chip acquired from the field/market).

2.4.2.1 OG Sequential Unrolling-Based Attacks

In an unrolling-based attack, a derivative of the SAT attack that works on sequential circuits has been engaged to evaluate the robustness of logic locking techniques with restricted scan access. The preliminary version of the SAT-based sequential deobfuscation attack was first introduced in [173]. The sequential SAT attack shows how the SAT solver could still be engaged to break the logic locked circuits with limited scan chain access, while unrolling is used as a preprocessing step. The sequential SAT attack uses an iterative method to prune the search space, similar to the SAT attack. Due to the limited access to internal registers, instead of seeking a DIP in each iteration, it instead finds a sequence of input patterns X denoted as discriminating input *sequence* (X_{DIS}) that can produce two separate outputs for two different keys. To build the sequence using the SAT attack, the sequential SAT gets the benefit of unrolling/unfolding to create a combinational equivalent circuit. Then, the SAT solver will be used for a specific depth to generate the DIS. For instance, Fig. 2.22 shows a sequential circuit unrolled for τ clock cycles. In this case, the SAT could be invoked to return a DIS with a length of τ. In comparison

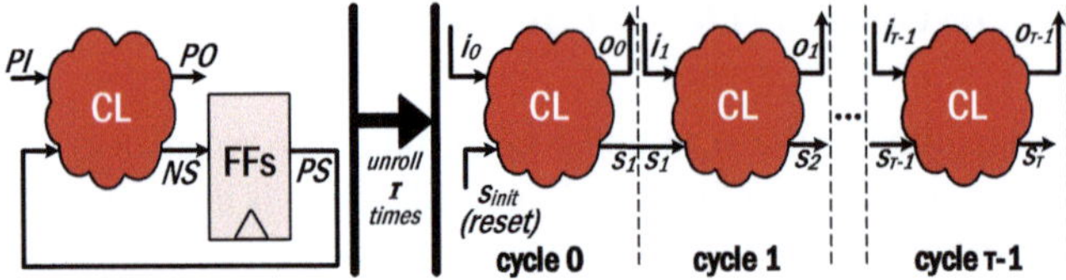

Fig. 2.22 Unrolling the sequential circuit for τ cycles

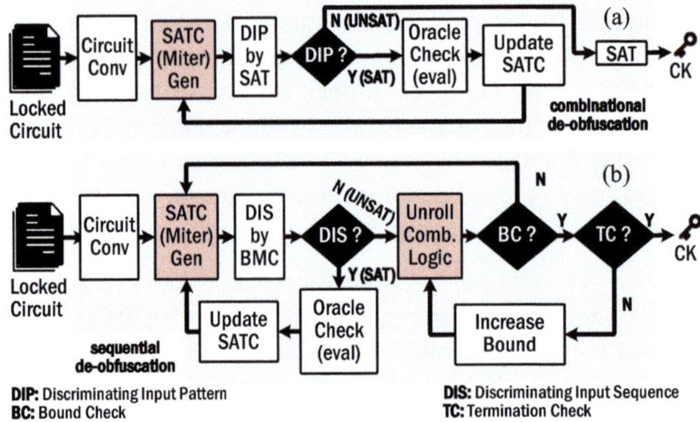

Fig. 2.23 Combinational SAT attack vs. sequential SAT attack

with combinational SAT attack, Fig. 2.23 shows the main steps of the SAT-based sequential deobfuscation, in which unrolling determines the length of DISes.

In an unrolling-based SAT attack, the generation of unrolled instances per clock cycle (as a query) can also be done by a bounded model checker (BMC). Hence, some studies also refer to this group of attacks as BMC attacks on sequential logic locking. So, to accomplish the unrolling step, with determining the boundary, the BMC engine could be invoked to model the locked circuit as an FSM, and the specification could be formalized by temporal logic properties. So, the BMC could be exploited as an alternative approach to do the symbolic model checking before invoking the SAT procedure. Algorithm 10 demonstrates the overall procedure of the primitive sequential SAT attack, once the unrolling is done using the BMC engine. $C(x, k, y)$ indicates the locked circuit generating output sequence y using input sequence x and the key value k, and $C_{BlackBox}(x)$ refers to the output sequence of the oracle for the same input sequence. After building the model from the locked circuit, the attack instantiates a BMC to find the x_{DIS}. Then, the model would be updated with a new constraint to guarantee that the next pair of keys, which will be discovered in the subsequent attack iterations, produce the same output for previously discovered x_{DIS}. The iterations continue until no further x_{DIS} is found within the boundary of b. After reaching the boundary, if the algorithm passes three criteria, the key could be found with one more SAT instantiation. The boundary could be extended if termination conditions are not met. The primitive sequential SAT attack in [173] specified three main termination conditions for this attack:

(1) *Unique Completion (UC):* This condition verifies that the key generated by the algorithm is unique. The attack is successfully ended, and the key is the correct one if there is only one key that meets all previous DISes.

(2) *Combinational Equivalence (CE):* If there is more than one key for all previously found DISes (nonunique key), the attack checks the combinational equivalency of the remaining keys. In this step, the D/Q of FFs are considered as pseudo PO/PI allowing the attacker to treat the circuit as combinational. The resulting circuit is subjected to a SAT attack, and if the SAT solver fails to find a different output or next state for two different keys, it concludes that all remaining keys are correct and the attack terminates successfully.

(3) *Unbounded Model Check (UMC):* If both UC and CE fail, the attack checks the existence of a DIS for the remaining keys using an unbounded model checker. This is an exhaustive search with no limitation on bound (or the number of unrolls). If no DIS is discovered, the existing set of DIS is a complete set, and the attack terminates. Otherwise, the bound is increased and previous steps are repeated.

Algorithm 10 Sequential attack on obfuscated circuits [173]

1: $b = initial\_boundary, Terminated = False$;
2: $Model = C(x, k_1, y_1) \wedge C(x, k_2, y_2) \wedge (y_1 \neq y_2)$;
3: **while** not $Terminated$ **do**
4:       **while** $(x_{DIS}, k_1, k_2) \leftarrow BMC(Model, b) = T$ **do**
5:            $y_f \leftarrow C_{BlackBox}(x_{DIS})$;
6:            $Model = \wedge\, C(x_{DIS}, k_1, y_f) \wedge C(x_{DIS}, k_2, y_f)$;
7:       **if** $\mathbf{UC}(Model, b) \vee \mathbf{CE}(Model, b) \vee \mathbf{UMC}(Model)$ **then**
8:            $Terminated$;
9:       $b = b + boundary\_step$;

As demonstrated in Fig. 2.23, the SAT circuit (SATC) requires an update per each iteration, in which the newly learned clauses will be added to the list of previously found clauses. Hence, with more iteration, the size of the SAT problem will be increased drastically. In sequential SAT, it gets worse because the SAT problem will be expanded in two dimensions, i.e., iterations and unrolling. Hence, sequential SAT attack runs into the scalability issues as it relies on two subroutines which are in **PSPACE** and **NP**, thereby failing to terminate for even moderately small circuits, which contain only a few thousand gates. Hence, some more recent studies investigate different possibilities to mitigate this issue. Shamsi et al. propose a fast sequential deobfuscation attack, called KC2 [174], which implements different dynamic optimization tweaks, such as incremental SAT solving, key-condition sweeping, negative key-condition compression, etc., to improve and accelerate primitive sequential SAT attack. More recently, the open-source toolset RANE [175] has offered a unified framework with unique interfaces and can take advantage of the capabilities (like scalability) of commercial formal verification tools such as Cadence JasperGold and Synopsys Formality for deobfuscation at different stages with fewer difficulties. Although these approaches show significant improvement, more studies still show the lack of scalability induced by the deobfuscation process, especially, for large circuits, e.g., microcontrollers and processors. Another recent study has also explored the effectiveness of different SAT/BMC-related guiding mechanisms, such as restart and initialization, on the scalability of this breed of attacks [176]. In this chapter, warming up the BMC before its invocation has been investigated. The exploration shows that for a significant part of the experience, warm-up as an initialization guide can improve the performance of this type of attack.

2.4.2.2 OG/OL Sequential on FSM Locking

When FSM locking is in place [14, 15, 94, 96–98], unlike all other logic locking techniques, there might exist no wiring, input, or logic, as the explicit declaration of the key (or key gate). Hence, none of the above-mentioned attacks work on FSM locking, where all these attacks assumptions rely on some formulation around key or key gates.

2-Stage (OL/OG) Attack on FSM Locking

To assess the strength of FSM locking techniques with the above-mentioned specification, however, different studies evaluated the possibility of deploying a 2-stage attack on locked FSMs [95, 100]. Algorithm 11 shows the overall flow of 2-stage attacks on FSM locking that is composed of three main steps:

(i) *Topological/Structural Analysis* (described in lines 2–13 of Algorithm 11), which is a detection algorithm to find FFs that are responsible for storing the state values (separating them from datapath FFs). The topological analysis is mostly derived from [177], which identifies FFs whose input contains a combinational feedback path from their output. Then, it reduces the set of possible state FFs by (a) grouping the FFs controlled by the same set of signals and (b) finding strongly connected components (SCCs) using Tarjan's algorithm [95, 100, 178, 179].

(ii) *Functional Analysis* (described in lines 14–21 of Algorithm 11) that finds the state transition graph (STG) based on the list of FFs found in step 1. In the functional analysis stage (stage 2), the attacker attempts to recalculate and extract the STG. This is done by first attempting to find the initial state and then identifying the reachable states by creating a reduced binary decision diagram (BDD) or using a SAT solver.

Algorithm 11 2-stage on FSM locking [95, 100]

1: **function** FSM_EXTRACT(Circuit C_L)
2:     $SFF \leftarrow []$;   ▷ State Flip Flops
3:     $RS \leftarrow$ classify(FFs);   ▷ Classifying FFs into Register Sets
4:     **for each** $set \in RS$ **do**
5:         $set \leftarrow set$ - notSCC(set);   ▷ Keeps Strongly Connected Components
6:         **if** is_splittable(set) **then**
7:             $RS \leftarrow \{RS - set\} \cup$ split(set);
8:     $CLFP \leftarrow$ find_feedback_circuits(C_L, Reg_Sets);
9:     **for each** $set \in RS$ **do**
10:         $set \leftarrow set$ - notInfDep(set);   ▷ Keeps Intersected Influence/Dependence
11:         $set \leftarrow set$ - InputIndependt(set);   ▷ Check Control Metrics
12:         update($CLFP$);
13:         SFF $\leftarrow$ SFF $\cup$ set;
14:     $S_0 \leftarrow$ initial_state(state_regs); $SQ \leftarrow []$;   ▷ State Queue
15:     $SQ \leftarrow SQ \cup S_0$; $STF \leftarrow []$;   ▷ State Transition Table
16:     **while** $SQ \neq []$ **do**
17:         state $\leftarrow SQ$.dequeue();
18:         **for each** DIP **do**   ▷ DIP found by SAT
19:             **if** eval(state_regs, DIP, state) $\notin SQ$ **then**
20:                 SQ.enqueue(nx_state);
21:                 $STF \leftarrow STF \cup \{$state, DIP, nx_state, $PO\}$
         return SQ, S_0, STF;   ▷ States, Initial, Transition Func.

(iii) *Matching/Extracting Original FSM*: In this case, based on the extracted STG, the original part of FSM will be retrieved based on some behavioral specifications. In most FSM locking solutions, the adversary can readily distinguish the original part of the FSM from either extra added states or extra state transitions, leading to extracting the original FSM. In this step, for complex cases that are hard to distinguish between original and extra states, some random-based stimuli will be matched with the oracle.

In this attack, topological analysis is a crucial factor as the success rate of this attack. Because for cases that the topological analysis cannot retrieve a correct set of FFs as the state holders, the returning set might have a subset of actual state FFs with some data FFs, and in this case, behavioral analysis on recalculated STG might lead to an incorrect FSM.

RANE Attack on FSM Locking

API-based invocation of different solvers and formal verification tools in RANE attack allows this framework to formulate multiple threat models and attack flows [175]. Hence, a specific model of FSM locking, i.e., HARPOON-based FSM locking, has been modeled by the RANE attack that shows more scalability versus the 2-stage attacks on FSM. In this model, the initial state has been formulated as the secret (unknown parameter or key variable), and the formal tool has been invoked to find this secret. Then, the formal tool has been called once more to find the unlocking sequence reaching to the initial state. By using these two steps, RANE shows how FSM locking could be still modeled using a similar approach as defined for the primitive sequential SAT attack with a dedicated key variable. Algorithm 12 shows how this attack has been formulated in the RANE framework. Although this new model could provide better scalability compared to 2-stage attacks on FSM, since it is BMC-based (expansion through two dimensions), their experiments still show the scalability issue for larger circuits.

Functional Corruptibility Guided on FSM Locking

In [180], another derivation of SAT-based attack has been studied that is applicable to keyless FSM-based sequential logic locking, called Fun-SAT. In Fun-SAT, the minimum number of unrollings needed to find the correct secret (here the unlocking sequence) will be estimated that will allow the attack to directly jump to the depth close to the final satisfying assignment(s). This method has been realized using bounded-depth function corruptibility (FC) analysis, and the whole attack consists of two major steps, which are FC analysis and SAT solving. Bypassing the gradual unrolling in Fun-SAT can improve the attack performance by up to 90x, showing how the conventional BMC/unrolling suffers from the scalability issue.

ORACALL on Cellular Automata Guided Locking

In [181], a new attack on a specific FSM locking has been introduced, in which the locking of the FSM has been done using cellular automata cells [182], and each FSM transition is protected using a secret key that resembles black holes in traditional

---

Algorithm 12 RANE attack model on FSM locking

―――――――― Finding Secret 1 (init state) ――――――――

1: $Model \leftarrow C_{seq}(x, s_{init1}, y_1) \wedge C_{seq}(x, s_{init2}, y_2)$;
2: **while** $!UC(Model) \wedge !CE(Model) \wedge !UMC(Model)$ **do**
3:     $DIS_i \leftarrow Formal(Model \wedge (y_1 \neq y_2))$;
4:     $y_i \leftarrow C_{BlackBox}(DIS_i)$;
5:     $Model \wedge = C_{seq}(DIS_i, s_{init1}, y_i) \wedge C_{seq}(DIS_i, s_{init2}, y_i)$;

―――――――― Finding Secret 2 (unlocking sequence) ――――――――

6: $Model \wedge = CE_u^0(us_0, s_{rst}, y_{US0}, s_{us1})$;
7: $i \leftarrow 1$;
8: **while** $Formal(Model \wedge (s_{us_i} = \hat{s}_{init})) \rightarrow Fail$ **do**
9:     $Model \wedge = CE_u^i(us_i, s_{usi}, y_i, s_{us_{i+1}})$;
10:     $i \leftarrow i + 1$;
11: **return** $Formal(Model \wedge (s_{us_i} = \hat{s}_{init}))$ ▷ {init state, unlocking sequence}

FSM locking techniques [15]. The overall flow of ORACALL is also similar to other unrolling-based and FSM-based attacks, in which a specific sequence of the pattern will be applied to the circuit as the preprocessing step, and then SAT attack will be called to retrieve the correct key related to each transition.

2.4.2.3 OG Sequential Scan/Leakage-Based Attacks

Similar to attacks described in Sects. 2.4.2.1 and 2.4.2.2, since a subset of defenses target scan locking or blockage (Sect. 2.3.5), a subset of studies eventually evaluated and revealed the vulnerabilities of this breed of locking.

ScanSAT on Scan Locking

ScanSAT aims to break the scan-based logic locking techniques. In scan-based logic locking, since the availability of scan is *locked*, the adversary has only access to the PI/PO. Hence, similar to the primitive sequential SAT attack, KC2, and RANE, ScanSAT [183] is based on the fact that the complex τ-cycle transformation of scan-based locking could be modeled by generating an unfolded/unrolled combinational equivalent counterpart of the scan-locked circuit. In this case, the locking parts added into the scan path become part of the resultant combinational circuit. Hence, the adversary faces a combinational (unrolled) locked circuit with key gates at the pseudo-primary I/Os of the circuit. The combinational equivalent of a locked circuit in Fig. 2.24a is provided in Fig. 2.24b, where the locking on the stimulus and the response are modeled separately as combinational blocks driven by the same scan locking key. In general, ScanSAT models the locked scan chains as a logic locked combinational circuit, paving the way for the application of the combinational SAT

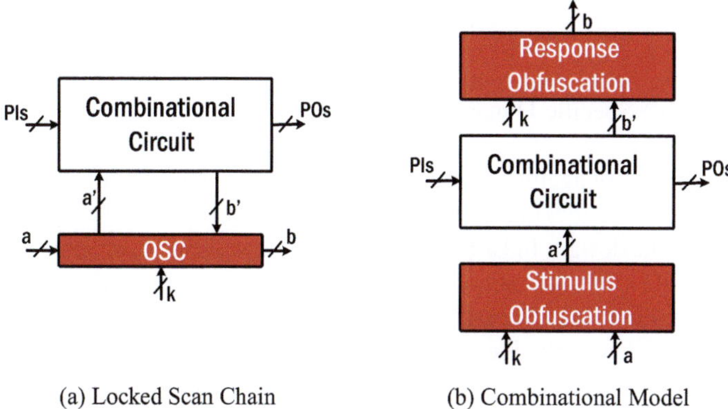

(a) Locked Scan Chain   (b) Combinational Model

Fig. 2.24 Converting a locked scan chain to its combinational counterpart [183]

attack to reveal the key (sequence of unrolled), unlocking the scan chains, and thus, restoring access to the oracle.

In addition to deobfuscating the statically scan-locked circuit, ScanSAT is also able to be applied on the dynamically scan-locked circuit that is locked by DOS architecture [86]. In DOS architecture, an LFSR has been engaged to generate runtime dynamic keys. In oracle-guided attacks, since the output of the netlist evaluation, such as DIPs and DISes, are continuously checked with the oracle, dynamically change of the configuration in the oracle will corrupt this oracle-oriented evaluation and results in failure of such attacks. In ScanSAT, it is assumed that after successfully reverse engineering, the LFSR structure, and consequently its polynomial, is known to the adversary. Hence, finding the seed of LFSR and the update frequency parameter (p as the time interval of updating the key based on the LFSR output), which is the only secret in DOS architecture, would lead to deriving all the keys that are dynamically generated on the chip.

A simple method to identify p is to apply the same stimulus pattern repeatedly from the SI and observe the response through the SO. The point is that after p capture operations, by repeatedly applying the same stimulus, the response would be different because of the updated key; thus, most likely, there will be a noticeable change in the observed response, helping detect the update operation on the key.

After finding p, the same approach that was used for static scan obfuscation would be used in this case. The difference now is that the SAT attack could be executed for at most p iterations (after p iterations, the key is updated). If more than p DIPs are required to identify a dynamic key, the SAT attack needs to be terminated prematurely upon p DIPs. Another SAT attack must be executed subsequently to identify the next dynamic key in the sequence still within p iteration. Since the updated key is generated by the LFSR whose polynomial is known for the adversary, independent SAT attack runs on each dynamic key reveals partial information of the seed; thus, the information from independent SAT attack runs by gradually gathering information about the seed in every run, and finally, by incorporating into the ScanSAT model, the relationship between the seed and the keys would be revealed.

DynUnlock on a Specific Dynamic Scan Locking

As a countermeasure against ScanSAT attack, dynamic encrypt flip-flop (EFF-Dyn) [88] combines scan locking approach from EFF [87] and a PRNG, to introduce dynamicity in the design. In EFF-Dyn, based on the value of scan controlling signal, i.e., scan-enable (SE), the source of the key to the circuit would be changed. In the test mode, the test key must be provided externally, and in case of a mismatch with the locking key embedded in the circuit, there exists a PRNG that updates the key in every clock cycle, thereby controlling the key gates dynamically. However, similar to LFSR, the structure of PRNG and its polynomial would be known for the adversary after successfully reverse engineering. Hence, DynUnlock [184] proposes a similar approach to find the seed of the PRNG in EFF-Dyn.

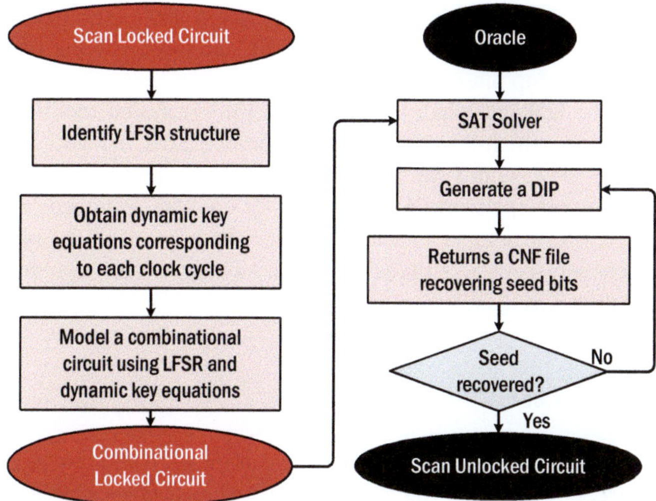

Fig. 2.25 Flowchart for the DynUnlock attack [184]

Assuming that the structure of PRNG is similar to an LFSR, in DynUnlock [184], as demonstrated in Fig. 2.25, it first starts by reverse engineering the LFSR circuit and obtaining the equations corresponding to each clock cycle. Next, it determines the location of key gates inserted between the SFFs. Then it models this sequential logic circuit into a combinational circuit with SFFs replaced with inputs and outputs. Once modeling is complete, the combinational obfuscated counterpart circuit, with seed bits acting as primary key inputs, is fed to a SAT solver, which provides a DIP and its corresponding output pattern. In [184], the authors carry out the attack for just one capture cycle. To recover more bits, they restart the LFSR circuit and obtain a new DIP and its corresponding output pattern from the SAT solver and recover more seed bits. They repeat the restart step until all the seed bits have been recovered, or the remaining seed bits can be brute-forced.

Leakage-Based Attacks on Scan Blockage

Due to the failure of scan obfuscation architectures against sequential SAT attacks, more recent studies evaluate and reveal the effectiveness of the scan chain blockage after activation of the obfuscated circuit. However, the structure of the augmented scan chain and the infrastructure used around them for protection might lead to a different form of security leakage. For instance, the first scan blockage architecture called R-DFS was first introduced in [41], in which the obfuscation key is stored in a custom-designed scan (storage) cell. Based on the value of the SE pin and the new pin called $Test$, the key values could be loaded into SCs either directly from TPM or through SI, and the SO will be blocked after activation to avoid any secret leakage. However, the shift-and-leak attack [38] breaks the R-DFS by exploiting the availability of the shift-in process through the SI, and the capability of reading out the PO through the chip pin-outs in the functional mode.

To remedy the leakage issue, the authors in [38] proposed modification to the R-DFS (mR-DFS), which blocks any shift operation after the obfuscation key is loaded from the TPM, removing the ability of an adversary to apply the shift-and-leak attack. However, the work in [39] illustrates how the architectural drawbacks of the modified infrastructure can lead to another glitch-based shift-and-leak attack, which allows an adversary to still leak the logic locking key through the PO even if the shift operation was disabled.

2.4.2.4   Summary of OG Sequential Attacks

The oracle-guided attacks on sequential circuits mostly rely on three different models to successfully break logic locking techniques: (1) unrolling mechanism that still allows the adversary to get the benefit of satisfiability, and it can be done manually or by using BMC, (2) structural/functional analysis of the locked netlist particularly for FSM-based logic locking, and (3) leakage possibility while the scan chain structure is manipulated. In general, compared to combinational attacks, the biggest shortcoming of sequential attacks is the scalability issue of the attack particularly on large circuits. This is when unrolling or BMC is in place, or some structural/functional analysis has been done, or leakage scenarios are targeted to be modeled. Table 2.14 tries to concisely answer to these *four* questions for attacks on sequential circuits: (1) How the attack flow works; (2) Which logic locking techniques could be broken using the attack; (3) What is the limitation and challenges of the attack; and (4) What is the existing/potential countermeasure against the attack?

2.4.3   Oracle-Less (OL) Attacks

Unlike almost all previous attacks, where access to oracle is one of the basic assumptions in threat modeling, realizing such scenarios might be hard or even impossible in so many cases. For instance, a malicious end-user has stolen one activated chip from the field, and no other copy of the chip is available to the public. Given such scenarios, a significant portion of attacks, which are known as oracle-less attacks, are those assuming there is no/hard possibility for the adversary to get access to an additional activated (unlocked) chip, and the threat model has been defined in a way that the adversary has access to (i) only the reverse-engineered netlist of the chip (locked GDSII at foundry), or (2) only one packaged chip (unlocked/activated) that can go through physical reverse engineering if needed (malicious end-user). Although this group of attacks can be more promising as they do not need the oracle, since there is no reference model, we

Table 2.14 Overview of oracle-guided attacks on sequential circuits

Attack	Mechanism	Applicable to[a]	Limitation & challenges	Countermeasures
Unroll/BMC [173–175], static scanSAT [183]	(1) Unrolling the sequential circuit (FFs are pivot) for u times (u-cycle combinational counterpart), (2) Apply the SAT attack on u-time unrolled circuit, (3) Finding DIPs and unroll more if there is no more DIP in u-time unrolled circuit, (4) Stop based on termination strategies (UC, CE, or UMC)	sequential lock, Encrypt-FF [87], seql [89]	(1) scalability is low (large number of unrolling and/or large circuit), (2) complexity of unrolling is high in multi-clock and clock-gated designs	DOS [86], dyn-EFF [88], latch/clock-based [47, 103, 105]
Dynamic scanSAT [183]	(1) Identify the key update frequency by applying the same stimulus pattern, and observing the response, (2) Extracting the secret seed by applying SAT on unrolled circuit with known update frequency (with updated keys), (3) Extracting original functionality by knowing seed and update frequency	DOS [86]	(1) Hard to apply if different keys generate the partial same response, (2) Impractical for dynamic seed (changing the seed over the time), (3) Impractical if LFSR is replaced with TRNG	Dynamic EFF [88], DisORC [43], DOSC [91]
2-stage [95, 100]	(1) Decoupling state FFs from datapath FFs using structural analysis, (2) Extracting the state transition graph of FSM(s) based on the list of FFs found in step 1, (3) Revealing the original parts of the FSMs	FSM Locking [14, 15, 94]	(1) Hard to apply when datapath and state FFs are strongly connected, (2) Challenging if unreachable states used as trap, (3) Impractical if part of the logic is locked (compound)	DFSSD [96], JANUS/HD [97, 98]
RANE FSM [175], Fun-SAT [180]	(1) setting the initial state of FSM locking as the key, (2) Unrolling the circuit for a specific number of cycles that required to reach initial state, (3) use SAT/BMC to find initial state as well as number of cycles required (unlocking sequence)	FSM Locking [14, 15, 94]	(1) Inapplicable if intermediate states are locked, like adding black holes for intermediate states, (2) Inapplicable if FSM is generated at the run-time	JANUS/HD [97, 98]
DynUnlock [184]	(1) Finding the LFSR/PRNG equations corresponding to each clock cycle, (2) Finding the location of key gates inserted between the SFFs, (3) Replacing SFFs with PIs/POs (de-sequencing), (4) Applying the SAT attack	Dynamic-EFF [88]	(1) Impractical for dynamic seed (changing the seed over the time), (3) Impractical if LFSR/PRNG is replaced with TRNG	–
Shift-and-leak [38]	(1) Finding leaky SFFs which are synthesizable/propagatable, (2) Shifting in state/key into cells, (3) Moving the key into the leaky SFF using shift mode, (4) Observing the key at PO by applying a ATPG-generated pattern	FORTIS [37], R-DFS [41]	(1) Hard to leak the keys if leaky SFFs are very limited (based on the topological sort of the gates), (2) Not applicable if all key storage are placed and mapped ahead of SFFs in the chains	mR-DFS [38], kt-DFS [39], DisORC [43]

[a] is able to break

witness the notion of key *guessing/prediction* because there is no explicit way of confirming the recovered key. Additionally, many of these approaches recover the key partially. So, unlike oracle-guided attacks in which time of the attack is important, here the key coverage percentage is a pivotal metric for the evaluation of the success rate of the attack. Hence, some of these attacks suggest that having an unlocked system is recommended to perform functional tests and to verify the correctness of the extracted key [185]. Based on the structure of oracle-less attacks, they can be categorized into three main subcategories: (i) structural, which could be (a) synthesis-based, (b) ATPG-based, or (c) ML-based, (ii) tampering-based, and (iii) probing-based. In the following, these subgroups will be discovered holistically.

2.4.3.1 OL Structural Synthesis-Based Attacks

Some attack approaches exploit the structural changes induced through the synthesis process. These attacks trace the changes through the synthesis process once the design is fed by different (guessed) key vectors. This breed of attack reveals structure-based shortcomings of logic locking techniques, and the main goal of them is to show that once incorrect (or a specific) key value will be applied to the design, the resynthesis process on the new design is constrained by the key value reveals some information (based on structural analysis) about the correctness of the applied key values.

Desynthesis Attack

As its name implies, the main aim of the desynthesis attack is to eliminate incorrect keys by desynthesizing the locked netlist [186]. In this attack, a hill-climbing approach has been engaged to resynthesize the locked netlist for different key guesses. The key guess that yields the maximum similarity between the locked netlist and its resynthesized versions is considered as the correct key. Figure 2.26 shows a simple example of how the desynthesis attack works based on similarity factor after resynthesis constrained by key value. Ideally, the best scenario that guarantees 100% key recovery is brute force key testing (exhaustively applying all key combinations followed by resynthesis). But, even if the attacker exhaustively tests all key values, for two major reasons, the success rate of such an attack would be very low: (1) The attack uses a heuristic dis-similarity factor and eliminates keys with the smallest dis-similarity (as the incorrect keys). Hence, the efficiency of the algorithm used for checking the dis-similarity crucially affects the success rate. (2) Randomness and imperfection of the commercial synthesis tool per each invocation can be misguiding in different cases, thereby reducing the success rate of this attack.

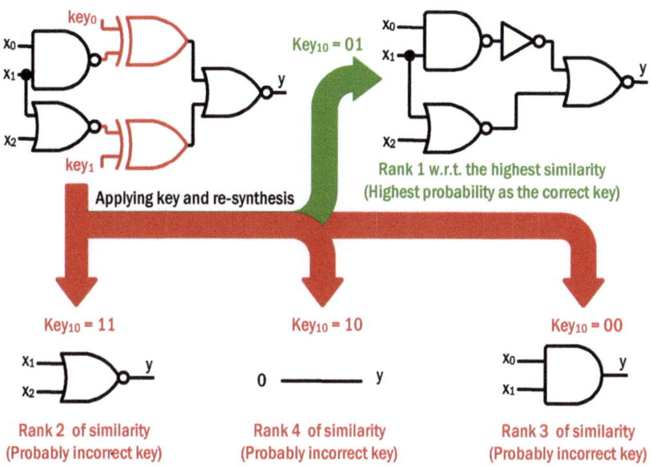

Fig. 2.26 How key-constrained desynthesis works [186]

SWEEP Attack

In SWEEP [185], which is a constant propagation attack, the basic idea of the attack is very similar to the desynthesis attack, which is to assign a constant key value to one key input and synthesize the obfuscated design with that key value. Based on the synthesis report, the attack identifies any structural design features that are correlated to the correct key values. Finally, the correct value of the analyzed key input is identified by comparing them to the synthesis report of the original obfuscated design using a scoring algorithm. SWEEP uses machine learning for the detection and tracing of the features related to the correct key and can be a member of OL structural ML-based attack (Sect. 2.4.3.3) as well. The main steps of the SWEEP attack are (1) training phase and dataset generation, (2) constant propagation, synthesis, and feature extraction from the constrained designs, (3) key correlation analysis that determines feature weighting algorithm, and how the decision will be made, and finally (4) test time which generates the initial predicted key.

Redundancy Attack

The work in [187] proposes a redundancy attack, which is based on the observation that an incorrect key often results in a circuit with significantly more logic redundancy compared to the correct circuit. The redundancy attack determines the likely value of key bits individually by comparing the levels of logic redundancy for each logic value. In the first step, all the key bits are initialized as unspecified until their values are determined. Then the logic value of one key bit is enumerated, and a redundancy identification tool will extract untestable faults for each logic value assignment. The attack assesses the likelihood of a key bit by comparing the changes to the number of untestable faults for a key-bit value of zero and a key-bit value of one. The authors of [187] showed that this attack can recover more than half of the key bits in both RLL and SLL. Similar to previous structural synthesis-based attacks, this attack uses key guessing and reduction after constraining. However, they rely on the fact that an incorrect key can modify the circuit's structure in a way that may reduce the manufacturing test coverage. But this reduction should not happen in the original design, and consequently narrow down the keyspace based on the number of untestable faults per each constraint. Hence, since it is a combination of structural and test-based analysis, it can also be a member of structural ATPG-based attacks (Sect. 2.4.3.2).

Topology-Guided Attack

Topology-Guided Attack (TGA) [188, 189] relies on identifying repeated functions for determining the value of a key bit. The attack is based on the observation that the basic functions in a logic cone are generally repeated multiple times in a circuit, such as basic arithmetic function units, shift registers, counters, etc. These functions are denoted as function units. If one or more key gates are placed in an instance of repeated function unit (FU) during the locking of a circuit, the original netlist can be recovered by searching the equivalent function units (EFUs) with all hypothesis keys. For finding the EFUs in a locked netlist, this attack uses an efficient depth-first search (DFS). If a match is found in the netlist, the hypothesis key becomes the actual key bit. For example, in a four-bit ripple carry adder that consists of eight identical one-bit half adders (HA), HA can be treated as a FU. If one of these HA is locked using an XOR gate, an adversary only needs to find an original HA and then match this with the locked HA to recover the key value.

2.4.3.2   OL Structural ATPG-Based Attacks

ATPG-based attacks are mostly inspired by test concepts to discover the logic locking key values. It is shown that ATPG is a reliable choice for discovering the key values since it is readily available from multiple commercial vendors and it is able to handle large circuits due to its development over the last 50 years. Redundancy attack described previously in Sect. 2.4.3.1 is one of the attacks in this subgroup that tries to guess the incorrect key, while the manufacturing (stuck-at) test coverage for a constrained resynthesized netlist is high.

Differential Fault Analysis Attack

The differential fault analysis (DFA) attack [190] is an ATPG-guided stuck-at fault-based attack, showing how an adversary can determine the logic locking key by injecting faults at the key registers, which hold the key value during normal operation. In this attack model, the first step is to select an input pattern that produces complementary results for both fault-free and faulty circuits. The faulty circuit is the same chip, only with a particular fault inject to keep all the key registers or interconnects to faulty value. Note that the selected input pattern must sensitize only one key bit to the primary output(s). To obtain the specific input test patterns, their method relies on stuck-at faults based on constrained ATPG. Then, the input pattern will be applied to (1) fault-free and (2) faulty circuits, and the responses of them will be gathered. The output responses are XORed to find any mismatch, and based on the output of the XOR gate, the key value can be predicted. The authors of [190] showed that at most $|K|$ test patterns are required to recover the entire secret key of size $|K|$.

CLIC

CLIC-A [191, 192] also uses commercial ATPG to recover key-input values in locked combinational and sequential circuits. There are four methods currently included in CLIC-A. The third and fourth methods are oracle-less, meaning they only require the netlist, which will be discussed in this section. The third method, which is applicable to combinational locking and targeting key-dependent faults [191], targets faults that require multiple key inputs for sensitization. To find these faults, ATPG is first performed on the locked netlist with the unsolved key inputs all constrained to do-not-care values (X). Faults that require the key inputs for detection will be reported as an ATPG failure. Each fault in this set of key-dependent faults is targeted using the second round of ATPG. If a test is found, the generated test is stored, along with the fault to be further analyzed for the key value. The test analysis method used for the generated tests differs depending on the lock type. The fourth method included in CLIC-A solves a key sequence from a sequentially locked circuit [192]. The insight used in the fourth method is that there must exist numerous faults that require the key sequence for detection through sequential ATPG. In this method, the sequential circuit ATPG is typically accomplished by first unrolling the circuit to form a combinational mode. Then CLIC-A targets single stuck-at faults on one combinational frame at a time. This method is specifically effective at solving a key from a circuit locked with an entrance FSM (e.g., [14]).

2.4.3.3 OL Structural ML-Based Attacks

A more recent trend has been opened in the area of logic locking that evaluates the utilization of machine learning (ML) for both defensive and attacking sides. From the attacking point of view, in many of these ML-based attacks, unlike seeking for functional recovery, structural analysis through ML has been done aiming to either find the (correct) key that corresponded to the correct structure or retrieve the original structure by removing the transformations introduced by locking [23, 185, 193–199].

Structural attacks, such as SAIL [193], GNNUnlock [195], and Snapshot [23], aim at discovering the design intent using machine learning. The authors of [193] were motivated based on this observation that obfuscation introduces sparse and local structural changes in a design and the changes are very deterministic. Hence, they proposed the SAIL attack [193], which exposes a vulnerability in logic locking by learning the predictable, localized structural changes that are introduced by the obfuscation process and uses machine learning to learn the deterministic rules applied by commercial CAD tools during synthesis. More precisely, in the SAIL attack, the pre- and postresynthesized locked designs are provided as training data to train the Change Prediction Model (CPM) and the Reconstruction Model (RM). Given a netlist subgraph (considering netlist as a graph) extracted near the selected key input, CPM predicts whether a structural change has occurred. If a change is predicted, RM is utilized to locally revert the structural changes after the resynthesis. SAIL attack is applicable to traditional logic locking techniques (e.g., RLL, FLL, and SLL).

SnapShot [23] is another structural analysis attack on logic locking, which utilizes artificial neural networks to directly predict a key-bit value from a locked synthesized gate-level netlist. The authors in [23] claim that SnapShot can be applied to a wider range of schemes, such as a MUX-based locking scheme. Another example of machine-learning-based attack is GNNUnlock [195], which leverages graph neural networks (GNNs) that learn the common structural features of the protection logic added by point function techniques, such as SFLL-HD [52], TTLock [51], and Anti-SAT [49]. GNNUnlock employs the following techniques: 1) netlist-to-graph transformation to capture each gate's functionality and connectivity in the gate-level netlist, 2) GNN learning on locked circuits, and 3) postprocessing rectification procedure to rectify any potential misclassifications to enhance the accuracy further and remove the identified protection logic effectively.

More recently, some ML-based structural attacks targeted routing-based locking techniques [198, 199]. In [198], the key extraction has been formulated as a link prediction problem, and the prediction has been done using a graph neural network (GNN). Link prediction is a generic problem that can be modeled using GNN, and in this attack, since routing locking will hide the connectivity/wiring, this missing connections are modeled as missing link problems. Then by enclosing subgraph extraction, GNN-based link prediction has been performed. Similar approach has been used in [199], in which another MUX-based routing locking is targeted [200], called D-MUX. Although D-MUX is introduced as a MUX-based routing locking resilient against ML-based attack, the work in [199] still shows that link prediction modeling can break this countermeasure as well. The generality of the link prediction model allows the adversary to formulate different logic locking techniques. For routing-based locking, the target links are connections around MUXes, and for other techniques, it could be either gates or other connections.

2.4.3.4   OL Tampering Attacks

In the IC supply chain, there are potential adversaries who have the capability and access to insert Hardware Trojans into the IC design, which is called tampering attacks. Hardware Trojans are malicious modifications to the original circuitry inserted by adversaries to exploit secret key information of the design, such as logic locking key. The TAAL attack [201] is based on implanting a hardware Trojan in the netlist. In [201], the authors showed that any logic locking techniques that rely on the stored secret key can be broken by inserting a hardware Trojan. The attacking approach of the TAAL attack is to tamper the locked netlist to extract the secret key information and leak the secret key to an adversary once activated. The authors of [201] present three types of TAAL attacks that extract the secret key differently using hardware Trojans placed at different locations in the netlist. In T1 type TAAL attack, an adversary can extract the key from a locked netlist without knowing the details of the logic locking technique used to protect the circuit as it directly leaks the secret key from the tamper-proof memory. In this attack model, the trigger is constructed using a 3-input AND gate along with an inverter placed before one of the AND gate inputs, and the payload is delivered to the primary output of the circuit using a 2-input MUX. Under normal operation, the multiplexer propagates the correct circuit functionality at the output. Once the Trojan gets activated, the output of AND gate becomes 1, which leads to the extraction of the secret key through the multiplexer at the output. T2 type and T3 type TAAL attacks rely on the activation and propagation of the secret key to the primary output and improve the complexity of detecting an attack.

2.4.3.5   OL Probing Attacks

Physical attacks, such as electrical probing attacks and optical probing attacks, can impose the threat of exposing security-sensitive information to an adversary. For example, the electrical probing attack directly accesses the internal wires of a security-critical module and extracts sensitive information in electronic format. Therefore, electrical probing is considered as a contact-based method. Electrical probing attacks can be classified into front-side probing, which is carried out through the upper metal layers, and backside probing, which is mounted through the silicon substrate. Attackers usually deploy focused ion beam (FIB) for probing attacks. FIBs use ions at high beam currents to mill a narrow cavity and get access to the target wire. Active shielding could be considered as a countermeasure against front-side probing attack, in which a shield that carries signals is placed on the top-most metal layer to detect holes milled by FIB; however, it has its limitations [202, 203].

On the other hand, optical probing techniques are often used in backside probing to capture photon emission phenomena during transistor switching. By passively receiving and analyzing the photons emitted from a specific transistor, the signal processed by that transistor can be inferred. In addition to photon emission analysis, laser voltage technique (LVX) and electro-optical frequency modulation (EOFM) are also used during backside attacks. These techniques illuminate the switching transistors and observe the reflected light. The work in [21] shows that optical probing is a threat for logic locking as this method can extract the locking key in a contactless manner; without using invasive methods, like FIBing or circuit edit, and contact-based method, like electrical probing. With that in mind, security against optical analysis mostly concerns protecting the backside of the chip, such as adding a backside polishing detector [202].

2.4.3.6 Summary of Oracle-Less (OL) Attacks

Oracle-less attacks on logic locking mostly focus on the evaluation of design specifications after purposefully manipulating the logic locking part. The manipulating of the logic locking part can be done using different mechanisms: One can apply the constraining of the key value (synthesis-based and ATPG-based), one can insert malicious behavior (tampering-based), one can inject faults, etc. Also, the specification mostly is related to structural specification, testability specification, or ML-based feature extraction. So, in this breed of attacks, the adversary applies the manipulation, gathers the information about the targeted specifications, and decides/calculates/predicts the key value. On the other side, probing attacks, either electrical or optical, can be done on a design, while the key is loaded (or going to be loaded), and no modification or manipulation for observing the logic locking key is required, which makes this model of attack a real threat against all existing logic locking techniques. With the introduction of ML-based and probing-based attacks in recent years, current directions of logic locking faced some changes to be responsive against these tighter and harder threat models. In the following, we will cover some of these directions that require more investigation w.r.t. the existing attack models. Tables 2.15 and 2.16 try to concisely answer to these *four* questions for oracle-less attacks: (1) How the attack flow works; (2) Which logic locking techniques could be broken using the attack; (3) What is the limitation and challenges of the attack; and (4) What is the existing/potential countermeasure against the attack?

2.5 What to Expect from Future Studies

Unlike the very first experiment(s) on logic locking that promise the security of design against reverse engineering, IP piracy, and IC overproduction, we just demonstrate through this chapter that this proactive countermeasure, regardless of its potential, has been challenged continuously for almost two decades; thereby we are witnessing a nonstop cat-and-mouse game between logic locking countermeasures (defenses) as well as deobfuscation approaches (attacks). Considering multiple factors, including but not limited to (i) all threats introduced so far on logic locking, (ii) the shortcoming and architectural drawbacks of existing approaches, (iii) getting

Table 2.15 Overview of oracle-less attacks on logic locking

Attack	Mechanism	Applicable to[a]	Limitation & challenges	Countermeasures
Desynthesis [186]	(1) Applying a random key to the netlist, (2) Re-synthesizing the netlist constrained with the random key, (3) Comparing (structural) the re-synthesized netlist with the raw locked netlist, (4) Guessing the correct key based on the (structural) similarity ratio	RLL [12], FLL [45], SLL [13]	(1) Hard to apply if original gates are combined/twisted with key gates, (2) Depending on the optimizations used in Re-synthesis, it might lead to an incorrect key	Interlock [83], TRLL [43], eFPGA [147, 148]
SWEEP [185], SCOPE [204]	(1) Assigning a key value (0/1) to one key bit, (2) synthesizing w.r.t. the assigned key value, (3) Analyzing the re-synthesized netlist for any design features that are correlated to the correct key value, (4) Indicating the correct key value based on a feature scoring algorithm	Random MUX-based Locking	(1) Low accuracy for large number of key gates (strongly connected MUX-based paths), (3) Complex training phase for larger circuits.	Shielding [205], UNSAIL [206], eFPGA [147, 148]
DFA [190]	(1) Building miter using fault-free circuit and laser-based faulty circuit (fault at key-registers), (2) Finding a DIP using ATPG (sensitize one bit of key), (3) Guessing the key based on the response	Any Locking Technique	(1) Hard to inject fault when backside of die is obstacled using metal coating, (2) More than one laser source is required	–
CLIC-A comb [191]	(1) Finding key-dependent faults via ATPG constrained to don't care values, (2) Targeting each key-dependent fault by ATPG, (3) Analyzing the test generated corresponded to faults for guessing the key value	Primitives (Sect. 2.3.1), point function (Sect. 2.3.2)	(1) Nonscalable for high number of key-dependent faults per cone, (2) Dependent to The availability of don't care	–

(continued)

Table 2.15 (continued)

CLIC-A seq [192]	(1) Unrolling the netlist to build combinational counterpart for u cycles, (2) Finding the faults and guessing the keys as described in CLIC-A comb	FSM locking [14, 15, 94]	(1) Hard to be scalable for large circuit, (2) Not scalable for faults at deep sequence	DFSSD [96]
Redundancy [187]	(1) Assuming all key values as unspecified, (2) Enumerating logic value of one key bit, (3) Extracting untestable faults for each logic value, (4) Indicating the key bit value based on the number of untestable faults	RLL [12], SLL [13], Point Function + Primitive	(1) Hard to apply if original gates are twisted with key gates, (2) Not applicable to MUX-based and LUT-based locking	Interlock [83], TRLL [43], Shielding [205]
TGA [188]	(1) constructing unit functions (repeated) corresponding to the hypothesis key, (2) match unit functions with exisitng unit functions, (3) guessing the key based on the constructed topology	Primitive locking techniques	(1) Low success rate if unique modules are locked, (2) Highly dependent to the topology and structure of the circuit	MUX/LUT locking, eFPGA [147, 148]
SAIL [193]	(1) Train a {change prediction Model} and {reconstruction} model using pre- and post-resynthesized locked designs, (2) Predicting the changes occurred using the trained ML	XOR-based logic locking	Unsuccessful (1) if parts of logic locking is added manually (post-synthesis), (2) on MUX-based and LUT-based locking	eFPGA [147, 148], UNSAIL [206]

[a] is able to break

Table 2.16 Overview of oracle-less attacks on logic locking (2)

Attack	Mechanism	Applicable to[a]	Limitation & challenges	Countermeasures
GNNUnlock [195]	(1) Generating a dataset on targeted locking technique, (2) Translating dataset to graph (netlist to graph), (3) Training the graphNN, (4) Testing on targeted netlist using trained graphNN	Point Function	(1) Training set and test set must be identical, (2) Unsuccessful against dynamic nature (re-configurability)	–
Utangle, MuxLink [198, 199]	(1) Formulating the key-extraction as a link prediction problem, (2) Applying the link prediction using GNN, (3) A post-processing to determine correct vs. false links	InterLock [83], D-MUX [200]	Low success rate if (1) the location of locked paths are distributed, (2) paths are less co-dependent	–
TAAL [201]	(1) Inserting T1/2/3 Trojan into the netlist for extracting/propagating the key, (2) Enabling the Trojan after activation	Any locking technique	(1) Hard to insert Trojan if keys are added with no leakage to outputs, (2) Might be detectable by Trojan detection algorithms	–
Probing [21, 202]	(1) Determining the clock frequency, (2) Localizing key-gate registers/storage, (3) Acquiring optical access for probing, (4) Identifying the key loading period, (5) Extracting key value from EOFM/EOP	Any locking technique	(1) Not applicable if randomness is used during initialization, (2) Localising and simultaneous probing is hard to achieve	Nanopyramid [207], Differential Logic [208]

[a] is able to break

the benefit of cutting-edge science and technologies applicable at this domain, and (iv) further modernization happening in semiconductor industries, it seems that the introduction of a standalone but comprehensive solution for addressing all the vulnerabilities in logic locking is almost beyond the bounds of possibility.

2.5.1 Vulnerabilities/Requirements of Logic Locking

In the following, we first describe state-of-the-art yet most concerning issues around logic locking showing why logic locking has its own certain critical setbacks. Then, we list some of the possible directions that might provide appropriate answers to the existing vulnerabilities. The major questions, yet with no answer, are as follows:

(i) *Lack of Formal Model*: The concept of logic locking with locking/unlocking key resembles that of cryptographic primitives with encryption/decryption key. However, since the introduction of logic locking, there exists no formal definition or formulation as the reference showing what exactly a logic locking countermeasure must approach to be considered (provably) secure. For instance, the point function techniques (Sect. 2.3.2) use the notion of **provably secure**. However, it is only against the SAT attack, and all of them are already broken by exploiting structural specification(s) (Sect. 2.4.1.3). Similarly, configurable logic and routing techniques (Sect. 2.3.4) propose SAT-hard instances that implicitly build provable resiliency with exponential complexity in terms of BDD analysis. However, few recent studies show their vulnerabilities against other structural specification(s) exploited by machine learning (Sect. 2.4.3.3). It is undeniably clear that formalizing the security of logic locking, and defining the logic locking construction that meets the formal model, has been completely elusive, which results in the introduction and evolution of numerous groups but completely independent and in an insular way.

(ii) *Dynamic/Expanding Nature of Threat Modeling*: The applicability of logic locking, as well as the success of the proposed approaches, is heavily dependent on the correct and accurate definition of the threat model. From oracle-guided to oracle-less, from scan available to scan protected, from invasive to noninvasive, and from untrusted foundry to the malicious user at market/field, are some of the notable bullets in the definition of threat model against logic locking. However, over time, the definition[10] might face significant changes that result in invalidating all previous studies. For instance, the notion of **insider threats**, e.g., a rogue employee at a design house, from the design team to verification, integration, etc., or nearly any individual working within any trusted entity but as a malicious actor, can completely invalidate the

---

[10] Here it is all definitions related to IC lifecycle, i.e., from IC specification to disposal.

applicability of existing logic locking techniques, that mostly done at gate level. In addition, unlike almost all attacks that have been done at the gate level, with having this insider threat notion, there exists a possibility of emerging a new thread of attacks on logic locking that have been accomplished at the RTL, whose main aim is to get the benefit of circuit specification(s) at RTL or any high-level presynthesis representation for deobfuscation purposes. This shows the lack of generality and applicability of many existing logic locking techniques, which makes them discredited in the short term with a new backdoor(s) revealed in the new threat model(s).

(iii) *Cutting-Edge Technologies/Devices Against Logic Locking*: As described in Sect. 2.4.3.5, an adversary such as a high-end untrusted foundry with cutting-edge technologies and devices, e.g., microprobing station, scanning electron microscope (SEM), and laser scanning microscope (LSM), should be more than capable of extracting the unlocking key from a chip by contact-based electrical or contactless optical probing. This form of attack is not only limited to logic locking, even cryptographic primitives with proven guarantees are subjected to such attacks. In this case, regardless of the logic locking techniques, with having the information of key management infrastructure (Sect. 2.2.2), the adversary can raid it using either FIB and electrical contact-based probing or contactless optical probing from the backside of the chip, and most modern chips do not have any protection mechanism for the backside of the substrate. Although probing-based attacks on logic locking completely undermine the protection/security promised by logic locking, the security of the key management infrastructure has not been evaluated meticulously, and most of the approaches left this part noninvestigated by just relying on the assumption of having TPM as a reliable solution with no further details.

(iv) *Machine Learning and its Evolution*: Since 2018, the application of machine learning, specifically graph-based neural networks (GNN), has been drastically increased in the domain of logic locking. These approaches rely on some feature extraction that is more related to the structural specification(s) of the logic locked circuit. Then, a training model will help to tune the ML (NN) for the specific logic locking targeted for deobfuscation. Currently, many of the existing logic locking techniques have been broken by a set of ML-based attack models (Sect. 2.4.3.3). These attack models clearly reveal the flaws of existing logic locking techniques, particularly in terms of structural specification(s). Additionally, expanding the threat models and opening new backdoor(s) can also raise the usage of this model of attacks. For instance, assuming the notion of insider threats, and having access to higher abstract(s) of the design, as well as graph-based NN feature extraction can also be done at these higher abstractions making the existing approaches even more prone to another set of ML-based structural analysis.

2.5.2 Possible Research Opportunities in Logic Locking

Following is some of the possible directions that have been already taken with some shallow investigation. However, to get the highest benefit of logic locking, these items must be considered meticulously:

(i) *Security Evaluation Using Formal Models*: Considering that the logic locking research domain is getting bigger, the need for a formal security definition has become inevitable. Some recent studies have evaluated the possibility of defining some security metrics using formalizing the model [52, 209]. However, these models are very limited to the construction of only a specific category (specific threat model), which is combinational construction(s) against the SAT or approximate-based SAT. However, such definitions require more generality and applicability at different threat models and assumptions. Hence, studies that accomplish formal analysis on logic locking that will be followed by the definition of generic and comprehensive security metrics are completely missing, which still needs significant attention by the community. Please note that this process has already been taken in cryptographic primitives, which results in introducing metrics like the indiscernibility of the block ciphers. Although realizing the formalism in logic locking requires a precise definition for what exactly secure logic locking is, which is hard to achieve, it can still help the designers and researchers in the community to pave the road toward what necessarily needs to be focused on [210].

(ii) *Multilayer Logic Locking*: This should not come as a surprise, and there is no single standalone solution for addressing all the above-mentioned vulnerabilities, meeting all requirements, and addressing all challenges in logic locking. Hence, having a methodology that spans over different stages or layers of design and implementation is one valid solution against all threats. One recent study has shown a multilayer of security countermeasures against different threats, known as defense-in-depth [202]. The layers can be a defense against (1) existing attacks on logic locking, (2) attacks around design-for-test infrastructure, and (3) contact-based and contactless probing. The notion of multilayer security can be used for logic locking individually. For instance, the notion of compound logic locking (Sect. 2.3.2.1) or a combination of higher-level logic locking with gate-level logic locking (Sect. 2.3.8) are some of the examples of multilayer logic locking that have been engaged for enhanced security. However, since these compound techniques are introduced to cover and break more attacks (or applicable for a wider range of threat models), they might still suffer from analyses that can lead to breaking them, e.g., compound attacks. So, the composition of multiple logic locking techniques requires algorithmic procedures about the interlocking correlation helping the designers to build a more comprehensive multilayer or compound countermeasures. One intuitive solution for this case is when a primitive logic locking is combined with a scan chain locking/blockage technique, which pushes the complexity toward a two-dimension unrolling-based problem, and the attacker will be limited to only the BMC-based attacks that hugely suffer from scalability issues in large circuits. But note that these techniques might reduce the testability coverage. Additionally, over time, with the advances in formal methods and verification tools, the scalability issue might be mitigated in BMC-based attacks.

(iii) *Logic Locking vs. Zero-Knowledge Computation*: The dynamic nature of threat modeling due to ever-expanding the threats we face through the IC lifecycle weakens the robustness and resiliency promised by the existing logic locking techniques. For instance, the notion of the insider threats and considering even any individual existing within the design house as an untrusted party remind the concept of *zero-knowledge-proof* (ZKP) in cryptographic protocols between two parties, in which a *prover* and a *verifier* work on a statement without leaking any extra information that might expose their belongings related to that statement [211]. In logic locking, this is when the designer wants to send the logic locked design to the integration team, and the designer has no intent to share any information about the design and the logic locking part. In this case, one solution can be enabling logic locking to also serve as a ZKP system. To do this, some requirements must be met: (1) Similar to cryptographic primitives, it requires to support fully indiscernibility, which means that for any individual at any stage of the design, the logic locked design should not provide any additional information compared to the original one. Many of the existing logic locking techniques do not meet these requirements, and they are broken because the locked design, in different ways, can provide hints and additional information to the adversary. One big example is all point function techniques that have already broken via some structural analyses. Or another example is the high success rate of ML-based attacks that rely more on structural-based feature extractions. So, indiscernibility would be a crucial characteristic of a secure logic locking technique, and if the logic locking meets the indiscernibility, none of the structural analyses can be applicable to it. (2) Logic locking must be implemented as early as possible in the design flow, which is at higher levels of abstraction, i.e., RTL or HLS, to cover a wider range of possible threats. Some recent studies propose high-level logic locking (Sect. 2.3.8); however, they just show how logic locking can be migrated from structural gate level to behavioral and semantic level, which can be more helpful for protecting the assets in the design. But, they still leak additional information, which results in being vulnerable to newer attacks, such as SMT [166]. Although the composition of high-level logic locking with some gate-level techniques can also provide robustness against wider threats, there is still the possibility for the adversary to open numerous ways at the higher level representation of the design to break the logic locking. (3) The logic locking part(s) must be **fine-grained**: for reducing overhead, **uniformly distributed**: to eliminate any analysis or reduction, and **with no dependency** with other locked/original part(s) of the design to guarantee that it is not location-dependent.

Some of the existing logic locking techniques meet part(s) of these requirements, but there is still no logic locking that can meet all. For instance, configurable logic and routing techniques add some form of universality that meet these requirements (Sect. 2.3.4), but they still have no full indiscernibility (violating item 1). Also, they are mostly implemented at transistor or layout level (violating item 2). eFPGA-based IP redaction, which can be considered as the superset of LUT-based and routing-based techniques, is another group that meets item 1 (and partially item 2), but they still violate item 3, because they are implemented fully coarse-grained with prohibited overhead ($\sim 1.5 \times$ –$3\times$ for even SoC-size circuits).

References

1. K. Zamiri Azar, H. Mardani Kamali, F. Farahmandi, M. Tehranipoor, *Understanding Logic Locking* (2023)
2. M. Lapedus, *Week In Review: Manufacturing, Test, and Foundry Challenges* (2021). https://semiengineering.com/week-in-review-manufacturing-test-163/
3. M. Tehranipoor, C. Wang, *Introduction to Hardware Security and Trust* (Springer, Berlin, 2011)
4. M. Rostami, F. Koushanfar, R. Karri, A primer on hardware security: models, methods, and metrics. Proc. IEEE **102**(8), 1283–1295 (2014)
5. A.B. Kahng, J. Lach, W. Mangione-Smith, S. Mantik, I.L. Markov, M. Potkonjak, P. Tucker, H. Wang, G. Wolfe, Constraint-based water-marking techniques for design IP protection. IEEE Trans. Comput.-Aided Design Integr. Circuits Syst. **20**(10), 1236–1252 (2001)
6. Y. Alkabani, F. Koushanfar, Active hardware metering for intellectual property protection and security, in *USENIX Security Symposium* (2007), pp. 291–306
7. J. Rajendran, M. Sam, O. Sinanoglu, R. Karri, Security analysis of integrated circuit camouflaging, in *Proceedings of the ACM SIGSAC Conference on Computer & Communications Security* (2013), pp. 709–720
8. J. Lee, M. Tehranipoor, C. Patel, J. Plusquellic, Securing scan design using lock and key technique, in *20th IEEE International Symposium on Defect and Fault Tolerance in VLSI Systems (DFT'05)* (IEEE, 2005), pp. 51–62
9. H. Mardani Kamali, K. Zamiri Azar, F. Farahmandi, M. Tehranipoor, Advances in logic locking: past, present, and prospects. Cryptology ePrint Archive (2022)
10. J. Lee, M. Tebranipoor, J. Plusquellic, A low-cost solution for protecting IPs against scan-based side-channel attacks, in *24th IEEE VLSI Test Symposium* (IEEE, 2006), 6-pp.
11. J. Lee, M. Tehranipoor, C. Patel, J. Plusquellic, Securing designs against scan-based side-channel attacks. IEEE Trans. Depend. Secure Comput. **4**(4), 325–336 (2007)
12. J. Roy, F. Koushanfar, I.L. Markov, EPIC: ending piracy of integrated circuits, in *Design, Automation & Test in Europe Conf. (DATE)* (2008), pp. 1069–1074
13. J. Rajendran, Y. Pino, O. Sinanoglu, R. Karri, Security analysis of logic obfuscation, in *Proceedings of the 49th Annual Design Automation Conference* (2012), pp. 83–89
14. R.S. Chakraborty, S. Bhunia, HARPOON: an obfuscation-based SoC design methodology for hardware protection. IEEE Trans. Comput.-Aided Design Integr. Circuits Syst. **28**(10), 1493–1502 (2009)
15. A.R. Desai, M.S. Hsiao, C. Wang, L. Nazhandali, S. Hall, Interlock- ing obfuscation for anti-tamper hardware, in *Proceedings of the Cyber Security and Information Research Workshop* (2013), pp. 1–8
16. Y. Xie, A. Srivastava, Delay locking: security enhancement of logic locking against IC counterfeiting, in *Design Automation Conference (DAC)* (2017), pp. 1–9

17. J.P. Skudlarek, T. Katsioulas, M. Chen, A platform solution for secure supply-chain and chip life-cycle management. Computer **49**(8), 28–34 (2016)
18. D.P. Affairs, *DARPA Selects Teams to Increase Security of Semiconductor Supply Chain* (2020). https://www.darpa.mil/news-events/2020-05-27
19. C. Helfmeier, D. Nedospasov, C. Tarnovsky, T. Krissler, C. Boit, Christian, J.-P. Seifert, Breaking and entering through the silicon, in *ACM SIGSAC Conference on Computer & Communications Security (CCS)* (2013), pp. 733–744
20. H. Wang, D. Forte, M. Tehranipoor, Q. Shi, Probing attacks on integrated circuits: challenges and research opportunities. IEEE Design & Test **34**(5), 63–71 (2017)
21. M.T. Rahman, S. Tajik, M.S. Rahman, M. Tehranipoor, N. Asadizanjani, The key is left under the mat: on the inappropriate security assumption of logic locking schemes, in *IEEE International Symposium on Hardware Oriented Security and Trust (HOST)* (2020), pp. 262–272
22. A. Chakraborty, N.G. Jayasankaran, Y. Liu, J. Rajendran, O. Sinanoglu, A. Srivastava, Y. Xie, M. Yasin, M. Zuzak, Keynote: a disquisition on logic locking. IEEE Trans. Comput.-Aided Design Integr. Circuits Syst. **39**(10), 1952–1972 (2019)
23. D. Sisejkovic, F. Merchant, L. Reimann, H. Srivastava, A. Hallawa, R. Leupers, Challenging the security of logic locking schemes in the era of deep learning: a neuroevolutionary approach. ACM J. Emer. Technol. Comput. Syst. **17**(3), 1–26 (2021)
24. A. Stern, H. Wang, F. Rahman, F. Farahmandi, M. Tehranipoor, ACED- IT: assuring confidential electronic design against insider threats in a zero trust environment. IEEE Trans. Comput.-Aided Design Integr. Circuits Syst. **41**(10), 3202–3215 (2021)
25. N. Weste, D. Harris, *CMOS VLSI Design: A Circuits and Systems Perspective* (Pearson Education India, 2015)
26. J. Rajendran, O. Sinanoglu, R. Karri, VLSI testing based security metric for IC camouflaging, in *IEEE International Test Conference (ITC)* (2013), pp. 1–4
27. M. Yasin, B. Mazumdar, O. Sinanoglu, J. Rajendran, CamoPerturb: secure IC camouflaging for minterm protection, in *IEEE/ACM International Conference on Computer-Aided Design (ICCAD)* (2016), pp. 1–8
28. M. Li, K. Shamsi, T. Meade, Z. Zhao, B. Yu, Y. Jin, D. Z. Pan, Provably secure camouflaging strategy for IC protection. IEEE Trans. CAD Integr. Circuits Syst. **38**(8), 1399–1412 (2017)
29. B. Shakya, H. Shen, M. Tehranipoor, D. Forte, Covert gates: protecting integrated circuits with undetectable camouflaging, in *IACR Transactions on Cryptographic Hardware and Embedded Systems (CHES)* (2019), pp. 86–118
30. J. Rajendran, O. Sinanoglu, R. Karri, Is split manufacturing secure? in *Design, Automation & Test in Europe Conference & Exhibition (DATE)* (2013), pp. 1259–1264
31. F. Imeson, A. Emtenan, S. Garg, M. Tripunitara, Securing computer hardware using 3D integrated circuit (IC) technology and split manufacturing for obfuscation, in {*USENIX*}*Security Symposium* (2013), pp.495–510
32. K. Vaidyanathan, B. Das, E. Sumbul, R. Liu, L. Pileggi, Building trusted ICs using split fabrication, in *IEEE International Symposium on Hardware-Oriented Security and Trust (HOST)* (2014), pp. 1–6
33. K. Xiao, D. Forte, M. Tehranipoor, Efficient and secure split manufacturing via obfuscated built-in self-authentication, in *IEEE International Symposium on Hardware Oriented Security and Trust (HOST)* (2015), pp. 14–19
34. D. Forte, S. Bhunia, M. Tehranipoor, *Hardware Protection through Obfuscation* (Springer, Berlin, 2017)
35. M. Yasin, J. Rajendran, O. Sinanoglu, R. Karri, On improving the security of logic locking. IEEE Trans. CAD Integr. Circuits Syst. **35**(9), 1411–1424 (2015)
36. Actel Corporation, Design security in nonvolatile flash and antifuse FP-GAs—security backgrounder, in *Technical Report on Quick Logic FPGAs* (2002)
37. U. Guin, Q. Shi, D. Forte, M. Tehranipoor, FORTIS: a comprehensive solution for establishing forward trust for protecting IPs and ICs. ACM Trans. Design Autom. Electron. Syst. **21**(4), 1–20

38. N. Limaye, A. Sengupta, M. Nabeel, O. Sinanoglu, Is robust design-for-security robust enough? Attack on locked circuits with restricted scan chain access, in *IEEE/ACM International Conference on Computer-Aided Design (ICCAD)* (2019), pp. 1–8
39. H.M. Kamali, K.Z. Azar, H. Homayoun, A. Sasan, On designing secure and robust scan chain for protecting obfuscated logic, in *Great Lakes Symposium on VLSI (GLSVLSI)* (Virtual Event, China, 2020), pp. 217–222
40. U. Guin, Z. Zhou, A. Singh, A novel design-for-security (DFS) architecture to prevent unauthorized IC overproduction, in *VLSI Test Symposium (VTS)* (2017), pp. 1–6
41. U. Guin, Z. Zhou, A. Singh, Robust design-for-security architecture for enabling trust in IC manufacturing and test. IEEE Trans. Very Large Scale Integr. Syst. **26**(5), 818–830 (2018)
42. K.Z. Azar, F. Farahmand, H.M. Kamali, S. Roshanisefat, H. Homayoun, W. Diehl, K. Gaj, A. Sasan, COMA: communication and obfuscation management architecture, in *Int'l Symposium on Research in Attacks, Intrusions and Defenses (RAID)* (2019), pp. 181–195
43. N. Limaye, E. Kalligeros, N. Karousos, I.G. Karybali, O. Sinanoglu, Thwarting all logic locking attacks: dishonest oracle with truly random logic locking. IEEE Trans. CAD Integr. Circuits Syst. **40**(9), 1740–1753 (2020)
44. K.Z. Azar, H.M. Kamali, H. Homayoun, A. Sasan, From cryptography to logic locking: a survey on the architecture evolution of secure scan chains. IEEE Access **9**, 73133–73151 (2021)
45. J. Rajendran, H. Zhang, C. Zhang, G.S. Rose, Y. Pino, O. Sinanoglu, R. Karri, Fault analysis-based logic encryption. IEEE Trans. Comput. **64**(2), 410–424 (2015)
46. P. Subramanyan, S. Ray, S. Malik, Evaluating the security of logic encryption algorithms, in *2015 IEEE International Symposium on Hardware Oriented Security and Trust (HOST)* (IEEE, 2015), pp. 137–143
47. M. El Massad, S. Garg, M. Tripunitara, Integrated circuit (IC) decamou-flaging: reverse engineering camouflaged ICs within minutes, in *NDSS* (2015), pp. 1–14
48. M. Yasin, B. Mazumdar, J.J.V. Rajendran, O. Sinanoglu, SARLock: SAT attack resistant logic locking, in *2016 IEEE International Symposium on Hardware Oriented Security and Trust (HOST)* (IEEE, 2016), pp. 236–241
49. Y. Xie, A. Srivastava, Mitigating SAT attack on logic locking, in *International Conference on Cryptographic Hardware and Embedded Systems* (Springer, Berlin, 2016), pp. 127–146
50. K. Shamsi, T. Meade, M. Li, D.Z. Pan, Y. Jin, On the approximation resiliency of logic locking and IC camouflaging schemes. IEEE Trans. Inform. Forensics Secur. **14**(2), 347–359 (2018)
51. M. Yasin, B. Mazumdar, J. Rajendran, O. Sinanoglu, TTLock: tenacious and traceless logic locking, in *IEEE International Symposium on Hardware Oriented Security and Trust (HOST)* (2017), pp. 166–166
52. M. Yasin, A. Sengupta, M.T. Nabeel, M. Ashraf, J. Rajendran, O. Sinanoglu, Provably-secure logic locking: from theory to practice, in *ACM SIGSAC Conference on Computer and Communications Security (CCS)* (2017), pp. 1601–1618
53. A. Sengupta, M. Nabeel, N. Limaye, M. Ashraf, O. Sinanoglu, Truly stripping functionality for logic locking: a fault-based perspective. IEEE Trans. Comput.-Aided Design Integr. Circuits Syst. **39**(12), 4439–4452 (2020)
54. J. Zhou, X. Zhang, Generalized SAT-attack-resistant logic locking. IEEE Trans. Inform. Forensics Secur. **16**, 2581–2592 (2021)
55. Y. Liu, M. Zuzak, Y. Xie, A. Chakraborty, A. Srivastava, Strong anti-SAT: secure and effective logic locking, in *International Symposium on Quality Electronic Design (ISQED)* (2020), pp. 199–205
56. B. Shakya, X. Xu, M. Tehranipoor, D. Forte, CAS-lock: a security-corruptibility trade-off resilient logic locking scheme, in *IACR Transactions on Cryptographic Hardware and Embedded Systems* (2020), pp. 175–202
57. M. Yasin, C. Zhao, J. Rajendran, SFLL-HLS: stripped-functionality logic locking meets high-level synthesis, in *International Conference on Computer-Aided Design (ICCAD)* (2019), pp. 1–4

58. A. Rezaei, Y. Shen, H. Zhou, Rescuing logic encryption in post-SAT era by locking & obfuscation, in *Design, Automation & Test in Europe Conference & Exhibition (DATE)* (2020), pp. 13–18
59. K. Shamsi, M. Li, T. Meade, Z. Zhao, D. Z. Pan, Y. Jin, Cyclic obfuscation for creating SAT-unresolvable circuits, in *Proceedings of the on Great Lakes Symposium on VLSI (GLSVLSI)* (2017), pp. 173–178
60. M.D. Riedel, J. Bruck, The synthesis of cyclic combinational circuits, in *Design Automation Conference (DAC)* (2003), pp. 163–168
61. S. Roshanisefat, H.M. Kamali, A. Sasan, SRCLock: SAT-resistant cyclic logic locking for protecting the hardware, in *Proceedings of the on Great Lakes Symposium on VLSI (GLSVLSI)* (2018), pp. 153–158
62. A. Rezaei, Y. Li, Y. Shen, S. Kong, H. Zhou, CycSAT-unresolvable cyclic logic encryption using unreachable states, in *Asia and South Pacific Design Automation Conference (ASP-DAC)* (2019), pp. 358–363
63. S. Roshanisefat, H.M. Kamali, H. Homayoun, A. Sasan, SAT-hard cyclic logic obfuscation for protecting the IP in the manufacturing sup- ply chain. IEEE Trans. Very Large Scale Integr. Syst. **28**(4), 954–967 (2020)
64. A. Rezaei, Y. Shen, S. Kong, J. Gu, H. Zhou, Cyclic locking and memristor-based obfuscation against CycSAT and inside foundry attacks, in *Proceedings of the Conference on Design, Automation and Test in Europe (DATE)* (2018), pp. 85–90
65. X. Yang, P. Chen, H. Chiang, C. Lin, Y. Chen, C. Wang, LOOPLock 2.0: an enhanced cyclic logic locking approach. IEEE Trans. CAD Integr. Circuits Syst. **41**(1), 29–34 (2021)
66. H. Chiang, Y. Chen, D. Ji, X. Yang, C. Lin, C. Wang, LOOPLock: logic optimization-based cyclic logic locking. IEEE Trans. Comput.-Aided Design Integr. Circuits Syst. **39**(10), 2178–2191 (2019)
67. A. Baumgarten, A. Tyagi, J. Zambreno, Preventing IC piracy using reconfigurable logic barriers. IEEE Design Test Comput. **27**(1), 66–75 (2010)
68. H.M. Kamali, K.Z. Azar, K. Gaj, H. Homayoun, A. Sasan, LUT-lock: a novel LUT-based logic obfuscation for FPGA-bitstream and ASIC-hardware protection, in *IEEE Computer Society Annual Symposium on VLSI (ISVLSI)* (2018), pp. 405–410
69. G. Kolhe, H.M. Kamali, M. Naicker, T. Sheaves, H. Mahmoodi, P.D. Sai Manoj, H. Homayoun, S. Rafatirad, A. Sasan, Security and complexity analysis of LUT-based obfuscation: from blueprint to reality, in *IEEE/ACM International Conference on Computer-Aided Design (ICCAD)* (2019), pp. 1–8
70. S.D. Chowdhury, G. Zhang, Y. Hu, P. Nuzzo, Enhancing SAT-attack resiliency and cost-effectiveness of reconfigurable-logic-based circuit obfuscation, in *IEEE International Symposium on Circuits and Systems (ISCAS)* (2021), pp. 1–5
71. R. Guo, M. Sazadur Rahman, H.M. Kamali, F. Rahman, F. Farahmandi, M. Tehranipoor, EvoLUTe: evaluation of look-up-table-based fine-grained IP redaction, in *2023 Design, Automation & Test in Europe Conference & Exhibition (DATE)* (IEEE. 2023), pp. 1–6
72. S. Rahman, N. Varshney, F. Farahmandi, N.A. Zanjani, M. Tehranipoor, LLE: mitigating IC piracy and reverse engineering by last level edit, in *ISTFA 2023* (ASM International, 2023)
73. V. Betz, J. Rose, FPGA routing architecture: segmentation and buffering to optimize speed and density, in *Proceedings of the ACM/SIGDA Int'l Symposium on Field Programmable Gate Arrays (FPGA)* (1999), pp. 59–68
74. C.J. Alpert, D. Mehta, S.S. Sapatnekar, *Handbook of Algorithms for Physical Design Automation* (CRC Press, Boca Raton, 2008)
75. S. Bose, *Methods and Systems for Placement and Routing*. U.S. Patent 8,332,793, 2012
76. Y. Wang, P. Chen, J. Hu, J. Rajendran, Routing perturbation for enhanced security in split manufacturing. in *Asia and South Pacific Design Automation Conference (ASP-DAC)* (2017), pp. 605–510
77. A. Sengupta, S. Patnaik, J. Knechtel, M. Ashraf, S. Garg, O. Sinanoglu, Ozgur, Rethinking split manufacturing: An information-theoretic approach with secure layout techniques, in *IEEE/ACM International Conference on Computer-Aided Design (ICCAD)* (2017), pp. 329–326

78. S. Patnaik, M. Ashraf, H. Li, J. Knechtel, O. Sinanoglu, Ozgur, Concerted wire lifting: enabling secure and cost-effective split manufacturing. IEEE Trans. CAD Integr. Circuits Syst. **41**(2), 266–280 (2021)

79. K. Shamsi, M. Li, D.Z. Pan, Y. Jin, Cross-lock: dense layout-level inter-connect locking using cross-bar architectures, in *Proceedings of the on Great Lakes Symposium on VLSI (GLSVLSI)* (2018), pp. 147–152

80. H.M. Kamali, K.Z. Azar, H. Homayoun, A. Sasan, Full-lock: hard distributions of SAT instances for obfuscating circuits using fully configurable logic and routing blocks, in *Proceedings of Design Automation Conference (DAC)* (2019), p. 89

81. S. Patnaik, M. Ashraf, O. Sinanoglu, J. Knechtel, Obfuscating the interconnects: low-cost and resilient full-chip layout camouflaging. IEEE Trans. Comput.-Aided Design Integr. Circuits Syst. **39**(12), 4466–4481 (2020)

82. J. Sweeney, M. Heule, L. Pileggi, Modeling techniques for logic locking, in *International Conference on Computer Aided Design (ICCAD)* (2020), pp. 1–9

83. H.M. Kamali, K.Z. Azar, H. Homayoun, A. Sasan, InterLock: an inter-correlated logic and routing locking, in *IEEE/ACM International Con- ference on Computer-Aided Design (ICCAD)* (2020), pp. 1–9

84. A. Saha, S. Saha, S. Chowdhury, D. Mukhopadhyay, B. Bhattacharya, Lopher: SAT-hardened logic embedding on block ciphers, in *Design Automation Conference (DAC)* (2020), pp. 1–6

85. W. Zeng, A. Davoodi, R.O. Topaloglu, ObfusX: routing obfuscation with explanatory analysis of a machine learning attack, in *Asia and South Pacific Design Automation Conference (ASP-DAC)* (2021), pp. 548–554

86. D. Zhang, M. He, X. Wang, M. Tehranipoor, Dynamically obfuscated scan for protecting IPs against scan-based attacks throughout supply chain, in *VLSI Test Symposium (VTS)* (2017), pp. 1–6

87. R. Karmakar, S. Chatopadhyay, R. Kapur, Encrypt Flip-Flop: a novel logic encryption technique for sequential circuits (2018). arXiv preprint arXiv:1801.04961

88. R. Karmakar, H. Kumar, S. Chattopadhyay, Efficient key-gate placement and dynamic scan obfuscation towards robust logic encryption, in *IEEE Transactions on Emerging Topics in Computing* (2019)

89. S. Potluri, A. Aysu, A. Kumar, Seql: secure scan-locking for IP protection, in *International Symposium on Quality Electronic Design (ISQED)* (2020), pp. 7–13

90. H.M. Kamali, K.Z. Azar, H. Homayoun, A. Sasan, SCRAMBLE: the state, connectivity and routing augmentation model for building logic en- cryption, in *IEEE Computer Society Annual Symposium on VLSI (ISVLSI)* (2020), pp. 153–159

91. M. Sazadur Rahman, A. Nahiyan, F. Rahman, S. Fazzari, K. Plaks, F. Farahmandi, D. Forte, M. Tehranipoor, Security assessment of dynamically obfuscated scan chain against oracle-guided attacks. ACM Trans. Design Autom. Electron. Syst. **26**(4), 1–27 (2021)

92. X. Wang, D. Zhang, M. He, D. Su, M. Tehranipoor, Secure scan and test using obfuscation throughout supply chain. IEEE Trans. Comput.-Aided Design Integr. Circuits Syst. **37**(9), 1867–1880 (2017)

93. F. Koushanfar, Active hardware metering by finite state machine obfuscation, in *Hardware Protection through Obfuscation* (2017), pp. 161–187

94. J. Dofe, Q. Yu, Novel dynamic state-deflection method for gate-level design obfuscation. IEEE Trans. Comput.-Aided Design Integr. Circuits Syst. **37**(2), 273–285 (2018)

95. T. Meade, Z. Zhao, S. Zhang, D.Z. Pan, Y. Jin, Revisit sequential logic obfuscation: attacks and defenses, in *IEEE International Symposium on Circuits and Systems (ISCAS)* (2017), pp. 1–4

96. S. Roshanisefat, H.M. Kamali, K.Z. Azar, S.M.P. Dinakarrao, N. Karimi, H. Homayoun, A. Sasan, DFSSD: deep faults and shallow state duality, a provably strong obfuscation solution for circuits with restricted access to scan chain, in *VLSI Test Symposium (VTS)* (2020), pp. 1–6

97. L. Li, A. Orailoglu, JANUS: boosting logic obfuscation scope through reconfigurable FSM synthesis, in *IEEE International Symposium on Hardware Oriented Security and Trust (HOST)* (2021), pp. 1–11

98. L. Li, A. Orailoglu, JANUS-HD: exploiting FSM sequentiality and synthesis flexibility in logic obfuscation to thwart SAT attack while offering strong corruption, in *Design, Automation & Test in Europe Conf. (DATE)* (2022), pp. 1–6

99. M. Sazadur Rahman, R. Guo, H.M. Kamali, F. Rahman, F. Farahmandi, M. Tehranipoor, ReTrustFSM: toward RTL hardware obfuscation-a hybrid FSM approach. IEEE Access **11**, 19741–19761 (2023)

100. M. Fyrbiak, S. Wallat, J. De´chelotte, N. Albartus, S. Bo¨cker, R. Tessier, C. Paar, On the difficulty of FSM-based hardware obfuscation, in *IACR Trans. on Crypto Hardware and Embedded Systems (TCHES)* (2018), pp. 293–330

101. G. Zhang, B. Li, B. Yu, D.Z. Pan, U. Schlichtmann, TimingCamouflage: improving circuit security against counterfeiting by unconventional timing, in *Design, Automation & Test in Europe Conference & Exhibition (DATE)* (2018), pp. 91–96

102. M. Alam, S. Ghosh, S. Hosur, TOIC: timing obfuscated integrated circuits, in *Proceedings of the Great Lakes Symposium on VLSI (GLSVLSI)* (2019), pp. 105–110

103. J. Sweeney, V. Zackriya, V.S. Pagliarini, L. Pileggi, Latch-based logic locking, in *IEEE International Symposium on Hardware Oriented Security and Trust (HOST)* (2020), pp. 132–141

104. K.Z. Azar, H.M. Kamali, S. Roshanisefat, H. Homayoun, C. Sotiriou, A. Sasan, Data flow obfuscation: a new paradigm for obfuscating circuits. IEEE Trans. Very Large Scale Integr. Syst. **29**(4), 643–656 (2021)

105. M.S. Rahman, R. Guo, H.M. Kamali, F. Rahman, F. Farahmandi, M. Abdel-Moneum, M. Tehranipoor, O'Clock: lock the clock via clockgating for SoC IP protection, in *Design Automation Conference (DAC)* (2022), pp. 1–6

106. K.Z. Azar, H.M. Kamali, H. Homayoun, A. Sasan, SMT attack: next generation attack on obfuscated circuits with capabilities and performance beyond the SAT attacks, in *IACR Transactions on Cryptographic Hardware and Embedded Systems (TCHES)* (2019), pp. 97–122

107. J. Knechtel, Hardware security for and beyond CMOS technology: an overview on fundamentals, applications, and challenges, in *International Symposium on Physical Design (ISPD)* (2020), pp. 75–86

108. X. Fong, Y. Kim, K. Yogendra, D. Fan, A. Sengupta, A. Raghunathan, K. Roy, Spin-transfer torque devices for logic and memory: prospects and perspectives. IEEE Trans. Comput.-Aided Design Integr. Circuits Syst. **35**(1), 1–22 (2015)

109. A. Makarov, T. Windbacher, V. Sverdlov, S. Selberherr, CMOS-compatible spintronic devices: a review. Semicond. Sci. Technol. **31**(11), 113006 (2016)

110. S. Baek, K. Park, D. Kil, Y. Jang, J. Park, K. Lee, B. Park, Complementary logic operation based on electric-field controlled spin-orbit torques. Nat. Electron. **1**(7), 398–403 (2018)

111. Q. Alasad, J. Yuan, D. Fan, Leveraging all-spin logic to improve hardware security, in *Great Lakes Symposium on VLSI (GLSVLSI)* (2017), pp. 491–494

112. T. Winograd, H. Salmani, H. Mahmoodi, K. Gaj, H. Homayoun, Hybrid STT-CMOS designs for reverse-engineering prevention, in *Design Automation Conference (DAC)* (2016), pp. 1–6

113. J. Yang, X. Wang, Q. Zhou, Z. Wang, H. Li, Y. Chen, W. Zhao, Exploiting spin-orbit torque devices as reconfigurable logic for circuit obfuscation. IEEE Trans. Comput.-Aided Design Integr. Circuits Syst. **38**(1), 57–69 (2018)

114. G. Kolhe, P.D. Sai Manoj, S. Rafatirad, H. Mahmoodi, A. Sasan, H. Homayoun, On custom LUT-based obfuscation, in *Great Lakes Symposium on VLSI (GLSVLSI)* (2019), pp. 477–482

115. S. Patnaik, N. Rangarajan, J. Knechtel, O. Sinanoglu, S. Rakheja, Advancing hardware security using polymorphic and stochastic spin-Hall effect devices, in *Design, Automation & Test in Europe Conference & Exhibition (DATE)* (2018), pp. 97–102

116. N. Rangarajan, S. Patnaik, J. Knechtel, R. Karri, O. Sinanoglu, S. Rakheja, Opening the doors to dynamic camouflaging: harnessing the power of polymorphic devices, in *IEEE Transactions on Emerging Topics in Computing* (2020)

117. S. Kvatinsky, G. Satat, N. Wald, E. Friedman, A. Kolodny, U. Weiser, Memristor-based material implication (IMPLY) logic: design principles and methodologies. IEEE Trans. Very Large Scale Integr. Syst. **22**(10), 2054–2066 (2013)

118. S. Kvatinsky, D. Belousov, S. Liman, G. Satat, N. Wald, E. Friedman, A. Kolodny, U. Weiser, MAGIC—memristor-aided logic. IEEE Trans. Circuits Syst. II: Express Briefs **61**(11), 895–899 (2014)

119. A. Rezaei, J. Gu, H. Zhou, Hybrid memristor-CMOS obfuscation against untrusted foundries, in *EEE Computer Society Annual Symposium on VLSI (ISVLSI)* (2019), pp. 535–540

120. Z. Chen, D. Farmer, S. Xu, R. Gordon, P. Avouris, J. Appenzeller, Externally assembled gate-all-around carbon nanotube field-effect transistor. IEEE Electron Device Lett. **29**(2), 183–185 (2008)

121. A. Franklin, M. Luisier, S. Han, G. Tulevski, C. Breslin, L. Gignac, M. Lundstrom, W. Haensch, Sub-10 nM carbon nanotube transistor. Nano Lett. **12**(2), 758–762 (2012)

122. A. Todri-Sanial, J. Dijon, A. Maffucci, *Carbon Nanotubes for Interconnects* (Springer, Berlin, 2017)

123. T. Mikolajick, A. Heinzig, J. Trommer, T. Baldauf, W. Weber, The RFET—a reconfigurable nanowire transistor and its application to novel electronic circuits and systems. Semicond. Sci. Technol. **32**(4), 043001 (2017)

124. J. Colinge, A. Kranti, R. Yan, C. Lee, I. Ferain, R. Yu, N. Akhavan, P. Razavi, Junctionless nanowire transistor (JNT): properties and design guidelines. Solid-State Electron. **65**, 33–37 (2011)

125. Y. Bi, K. Shamsi, J. Yuan, P. Gaillardon, G. Micheli, X. Yin, X. Hu, M. Niemier, Y. Jin, Emerging technology-based design of primitives for hardware security. ACM J. Emer. Technol. Comput. Syst. **13**(1), 1–19 (2016)

126. Q. Alasad, J. Yuan, Y. Bi, Logic locking using hybrid CMOS and emerging SiNW FETs. Electronics **6**(3), 69 (2017)

127. Q. Alasad, J. Yuan, Logic obfuscation against IC reverse engineering attacks using PLGs, in *2017 IEEE International Conference on Computer Design (ICCD)* (2017), pp. 341–344

128. F. Rahman, B. Shakya, X. Xu, D. Forte, M. Tehranipoor, Security beyond CMOS: fundamentals, applications, and roadmap. IEEE Trans. Very Large Scale Integr. Syst. **25**(12), 3420–3433 (2017)

129. N. Jayasankaran, A. Borbon, E. Sanchez-Sinencio, J. Hu, J. Rajendran, Towards provably-secure analog and mixed-signal locking against overproduction, in *International Conference on Computer-Aided Design (ICCAD)* (2018), pp. 1–8

130. J. Leonhard, M. Yasin, S. Turk, M. Nabeel, M. Loue¨rat, R. Chotin-Avot, H. Aboushady, O. Sinanoglu, H. Stratigopoulos, MixLock: securing mixed-signal circuits via logic locking, in *2019 Design, Automation & Test in Europe Conference & Exhibition (DATE)* (2019), pp. 84–89

131. K. Juretus, R. Venugopal, I. Savidis, Securing analog mixed-signal integrated circuits through shared dependencies, in *Great Lakes Symposium on VLSI (GLSVLSI)* (2019), pp. 483–488

132. M. Elshamy, A. Sayed, M. Loue¨rat, A. Rhouni, H. Aboushady, H. Stratigopoulos, Securing programmable analog ICs against piracy, in *Design, Automation & Test in Europe Conference & Exhibition (DATE)* (2020), pp. 61–66

133. J. Leonhard, A. Sayed, M. Loue¨rat, MH. Aboushady, H. Stratigopoulos, Analog and mixed-signal IC security via sizing camouflaging. IEEE Trans. Comput.-Aided Design Integr. Circuits Syst. **40**(5), 822–835 (2020)

134. H.M. Kamali, K.Z. Azar, H. Homayoun, A. Sasan, ChaoLock: yet another SAT-hard logic locking using chaos computing, in *International Symposium on Quality Electronic Design (ISQED)* (2021), pp. 387–394

135. J. Leonhard, N. Limaye, S. Turk, A. Sayed, A. Rizo, H. Aboushady, O. Sinanoglu, H. Stratigopoulos, Digitally-assisted mixed-signal circuit security, in *IEEE Transactions on Computer-Aided Design of Integrated Circuits and Systems* (2021)

136. W. Ditto, A. Miliotis, K. Murali, S. Sinha, M. Spano, Chaogates: morphing logic gates that exploit dynamical patterns. Chaos: Interdiscip. J. Nonlinear Sci. **20**(3), 037107 (2010)

137. C. Pilato, F. Regazzoni, R. Karri, S. Garg, TAO: techniques for algorithm-level obfuscation during high-level synthesis, in *Design Automation Conference (DAC)* (2018), pp. 1–6

138. Md. Rafid Muttaki, S. Saha, H. M Kamali, F. Rahman, M. Tehranipoor, F. Farahmandi, RTLock: IP protection using scan-aware logic locking at RTL, in *2023 Design, Automation & Test in Europe Conference & Exhibition (DATE)* (IEEE. 2023), pp. 1–6

139. C. Pilato, A. Chowdhury, D. Sciuto, S. Garg, R. Karri, ASSURE: RTL locking against an untrusted foundry. IEEE Trans. Very Large Scale Integr. Syst. **29**(7), 1306–1318 (2021)

140. M. Zuzak, Y. Liu, A. Srivastava, A resource binding approach to logic obfuscation, in *Design Automation Conference (DAC)* (2021), pp. 235–240

141. R. Muttaki, R. Mohammadivojdan, M. Tehranipoor, F. Farahmandi. HLock: locking IPs at the high-level language, in *Design Automation Conference (DAC)* (2021), pp. 79–84

142. N. Limaye, A. Chowdhury, C. Pilato, M. Nabeel, O. Sinanoglu, S. Garg, R. Karri, Fortifying RTL locking against oracle-less (untrusted foundry) and oracle-guided attacks, in *Design Automation Conference (DAC)* (2021), pp. 91–96

143. C. Karfa, T. Khader, Y. Nigam, R. Chouksey, R. Karri, HOST: HLS obfuscations against SMT attack, in *Design, Automation & Test in Europe Conference & Exhibition (DATE)* (2021), pp. 32–37

144. L. Collini, C. Pilato, A composable design space exploration framework to optimize behavioral locking, in *Design, Automation & Test in Europe Conference & Exhibition (DATE)* (2022), pp. 1–6

145. G. Takhar, R. Karri, C. Pilato, S. Roy, HOLL: program synthesis for higher order logic locking, in *Tools and Algorithms for the Construction and Analysis of Systems (TACAS)* (2022)

146. Md. Rafid Muttaki, R. Mohammadivojdan, H. Mardani Kamali, M. Tehranipoor, F. Farahmandi, Hlock+: a robust and low-overhead logic locking at the high-level language, in *IEEE Transactions on Computer-Aided Design of Integrated Circuits and Systems* (2022)

147. B. Hu, J. Tian, M. Shihab, G. Reddy, W. Swartz, Y. Makris, B.C. Schaefer, C. Sechen, Functional obfuscation of hardware accelerators through selective partial design extraction onto an embedded FPGA, in *Great Lakes Symposium on VLSI (GLSVLSI)* (2019), pp. 171–176

148. P. Mohan, O. Atli, J. Sweeney, O. Kibar, L. Pileggi, K. Mai, Hardware redaction via designer-directed fine-grained eFPGA insertion, in *Design, Automation & Test in Europe Conference & Exhibition (DATE)* (2021), pp. 1186–1191

149. J. Bhandari, A. Moosa, B. Tan, C. Pilato, G. Gore, X. Tang, S. Temple, P. Gaillardon, R. Karri, Exploring eFPGA-based redaction for IP protection, in *International Conference On Computer Aided Design (ICCAD)* (2021), pp. 1–9

150. J. Bhandari, A. Moosa, B. Tan, C. Pilato, G. Gore, X. Tang, S. Temple, P. Gaillardo, R. Karri, Not all fabrics are created equal: exploring eFPGA parameters for IP redaction (2021). arXiv preprint arXiv:2111.04222

151. H.M. Kamali, K.Z. Azar, F. Farahmandi, M. Tehranipoor, SheLL: shrinking eFPGA fabrics for logic locking, in *2023 Design, Automation & Test in Europe Conference & Exhibition (DATE)* (IEEE. 2023), pp. 1–6

152. K.Z. Azar, H. Mardani Kamali, H. Homayoun, A. Sasan, Threats on logic locking: a decade later, in *Proceedings of the 2019 on Great Lakes Symposium on VLSI* (2019), pp. 471–476

153. M. Bushnell, V. Agrawal, *Essentials of Electronic Testing for Digital, Memory and Mixed-Signal VLSI Circuits*, vol. 17 (Springer, Berlin, 2004)

154. S.M. Plaza, I.L. Markov, Solving the third-shift problem in IC piracy with test-aware logic locking. IEEE Trans. Comput.-Aided Design Integr. Circuits Syst. **34**(6), 961–971 (2015)

155. K. Shamsi, M. Li, T. Meade, Z. Zhao, D.Z. Pan, Y. Jin, AppSAT: approximately deobfuscating integrated circuits, in *IEEE International Symposium on Hardware Oriented Security and Trust (HOST)* (2017), pp. 95–100

156. Y. Shen, H. Zhou, Double-dip: re-evaluating security of logic encryption algorithms, in *Proceedings of the on Great Lakes Symposium on VLSI (GLSVLSI)* (2017), pp. 179–184

157. X. Xu, B. Shakya, M. Tehranipoor, D. Forte, Novel bypass attack and BDD-based tradeoff analysis against all known logic locking attacks, in *CHES* (2017), pp. 189–210

158. Y. Shen, A. Rezaei, H. Zhou, SAT-based bit-flipping attack on logic encryptions, in *Design, Automation & Test in Europe Conference & Exhibition (DATE)* (2018), pp. 629–632
159. M. Yasin, B. Mazumdar, O. Sinanoglu, J. Rajendran, Removal attacks on logic locking and camouflaging techniques, in *IEEE Transactions on Emerging Topics in Computing* (2017)
160. N. Limaye, S. Patnaik, O. Sinanoglu, Fa-SAT: fault-aided SAT-based attack on compound logic locking techniques, in *Design, Automation & Test in Europe Conference & Exhibition (DATE)* (2021), pp. 1166–1171
161. H. Zhou, R. Jiang, S. Kong, CycSAT: SAT-based attack on cyclic logic encryptions, in *IEEE/ACM International Conference on Computer-Aided Design (ICCAD)* (2017), pp. 49–56
162. Y. Shen, Y. Li, A. Rezaei, S. Kong, D. Dlott, H. Zhou, BeSAT: behavioral SAT-based attack on cyclic logic encryption, in *Asia and South Pacific Design Automation Conference (ASP-DAC)* (2019), pp. 657–662
163. K. Shamsi, D.Z. Pan, Y. Jin, IcySAT: improved SAT-based attacks on cyclic locked circuits, in *International Conference on Computer-Aided Design (ICCAD)* (2019), pp. 1–7
164. A. Chakraborty, Y. Liu, A. Srivastava, TimingSAT: timing profile embedded SAT attack, in *International Conference on Computer-Aided Design (ICCAD)* (2018), pp. 1–6
165. N.G. Jayasankaran, A.S. Borbon, A. Abuellil, E. Saánchez-Sinencio, J. Hu, J. Rajendran, Breaking analog locking techniques via satisfiability modulo theories, in *IEEE International Test Conference (ITC)* (2019), pp. 1–10
166. C. Karfa, R. Chouksey, C. Pilato, S. Garg, R. Karri, Is register transfer level locking secure? in *Design, Automation & Test in Europe Conference & Exhibition (DATE)* (2020), pp. 550–555
167. K.Z. Azar, H.M. Kamali, H. Homayoun, A. Sasan, NNgSAT: neural network guided SAT attack on logic locked complex structures. in *IEEE/ACM International Conference On Computer Aided Design (ICCAD)* (2020), pp. 1–9
168. M. Yasin, B. Mazumdar, O. Sinanoglu, J. Rajendran, Security analysis of anti-SAT, in *Asia and South Pacific Design Automation Conference (ASP-DAC)* (2017), pp. 342–347
169. D. Sirone, P. Subramanayan, Functional analysis attacks on logic locking. IEEE Trans. Inform. Forensics Secur. **15**, 2514–2527 (2020)
170. A. Sengupta, N. Limaye, O. Sinanoglu, Breaking CAS-lock and its variants by exploiting structural traces (2021). Cryptology ePrint Archive
171. Z. Han, M. Yasin, J. Rajendran, Does logic locking work with {EDA}Tools? in *30th USENIX Security Symposium (USENIX Security 21)* (2021), pp. 1055–1072
172. N. Limaye, S. Patnaik, O. Sinanoglu, Valkyrie: vulnerability assessment tool and attack for provably-secure logic locking techniques, in *IEEE Transactions on Information Forensics and Security* (2022)
173. M. El Massad, S. Garg, M. Tripunitara, Reverse engineering camouflaged sequential circuits without scan access, in *2017 IEEE/ACM International Conference on Computer-Aided Design (ICCAD)* (IEEE. 2017), pp. 33–40
174. K. Shamsi, M. Li, D.Z. Pan, Y. Jinr, KC2: key-condition crunching for sequential circuit deobfuscation, in *Design, Automation & Test in Europe Conference (DATE)* (2019), pp. 534–539
175. S. Roshanisefat, H.M. Kamali, H. Homayoun, A. Sasan, RANE: an open-source formal de-obfuscation attack for reverse engineering of logic encrypted circuits, in *Great Lakes Symposium on VLSI (GLSVLSI)* (2021), pp. 221–228
176. K.Z. Azar, H.M. Kamali, F. Farahmandi, M. Tehranipoor, Warm up before circuit de-obfuscation? An exploration through bounded-model- checkers, in *International Symposium on Hardware Oriented Security and Trust (HOST)* (2022), pp. 1–4
177. Y. Shi, C.W. Ting, B.H. Gwee, Y. Ren, A highly efficient method for extracting FSMs from flattened gate-level netlist, in *IEEE Int'l Symposium on Circuits and Systems (ISCAS)* (2010), pp. 2610–2613
178. R. Tarjan, Depth-first search and linear graph algorithms. SIAM J. Comput. **1**(2), 146–160 (1972)

179. T. Meade, et al. , Netlist reverse engineering for high-level functionality reconstruction, in *Asia and South Pacific Design Automation Conference (ASP-DAC)* (2016), pp. 655–660
180. Y. Hu, Y. Zhang, K. Yang, D. Chen, P.A. Beerel, P. Nuzzo, Fun-SAT: functional corruptibility-guided SAT-based attack on sequential logic encryption, in *IEEE International Symposium on Hardware Oriented Security and Trust (HOST)* (2021), pp. 1–11
181. A. Saha, H. Banerjee, R.S. Chakraborty, D. Mukhopadhyay, ORACALL: an oracle-based attack on cellular automata guided logic locking. IEEE Trans. Comput.-Aided Design Integr. Circuits Syst. **40**(12), 2445–2454 (2021)
182. R. Karmakar, S.S. Jana, S. Chattopadhyay, A cellular automata guided obfuscation strategy for finite-state-machine synthesis, in *Design Automation Conference (DAC)* (2019), pp. 1–6
183. L. Alrahis, M. Yasin, N. Limaye, H. Saleh, B. Mohammad, M. Alqutayri, O. Sinanoglu, Scansat: unlocking static and dynamic scan obfuscation. IEEE Trans. Emer. Topics Comput. **9**(4), 1867–1882 (2019)
184. N. Limaye, O. Sinanoglu, DynUnlock: unlocking scan chains obfuscated using dynamic keys, in *Design, Automation & Test in Europe Conference & Exhibition (DATE)* (2020), pp. 270–273
185. A. Alaql, D. Forte, S. Bhunia, Sweep to the secret: a constant propagation attack on logic locking, in *Asian Hardware Oriented Security and Trust Symposium (AsianHOST)* (2019), pp. 1–6
186. M. El Massad, J. Zhang, S. Garg, M.V. Tripunitara, Logic locking for secure outsourced chip fabrication: a new attack and provably secure defense mechanism (2017). arXiv preprint arXiv:1703.10187
187. L. Li, A. Orailoglu, Piercing logic locking keys through redundancy identification, in *Design, Automation & Test in Europe Conference & Exhibition (DATE)* (2019), pp. 540–545
188. Y. Zhang, P. Cui, Z. Zhou, U. Guin, TGA: an oracle-less and topology-guided attack on logic locking, in *ACM Workshop on Attacks and Solutions in Hardware Security Workshop* (2019), pp. 75–83
189. Y. Zhang, A. Jain, P. Cui, Z. Zhou, U. Guin, A novel topology-guided attack and its countermeasure towards secure logic locking. J. Cryptogr. Eng. **11**(3), 213–226 (2021)
190. A. Jain, T. Rahman, U. Guin, ATPG-guided fault injection attacks on logic locking (2020). arXiv preprint arXiv:2007.10512
191. D. Duvalsaint, X. Jin, B. Niewenhuis, R.D. Blanton, Characterization of locked combinational circuits via ATPG, in *2019 IEEE International Test Conference (ITC)* (2019), pp. 1–10
192. D. Duvalsaint, Z. Liu, A. Ravikumar, R.D. Blanton, Characterization of locked sequential circuits via ATPG, in *2019 IEEE International Test Conference in Asia (ITC-Asia)* (2019), pp. 97–102
193. P. Chakraborty, J. Cruz, S. Bhunia, SAIL: machine learning guided structural analysis attack on hardware obfuscation, in *2018 Asian Hard-ware Oriented Security and Trust Symposium (AsianHOST)* (IEEE. 2018), pp. 56–61
194. P. Chakraborty, J. Cruz, S. Bhunia, SURF: joint structural functional attack on logic locking, in *International Symposium on Hardware Oriented Security and Trust (HOST)* (2019), pp. 181–190
195. L. Alrahis, S. Patnaik, F. Khalid, M. Hanif, H. Saleh, M. Shafique, O. Sinanoglu, Ozgur, GNNUnlock: graph neural networks-based oracle-less unlocking scheme for provably secure logic locking (2020). arXiv preprint arXiv:2012.05948
196. Z. Chen, L. Zhang, G. Kolhe, H.M. Kamali, S. Rafatirad, S.M.P. Dinakarrao, H. Homayoun, C.-T. Lu, L. Zhao, Deep graph learning for circuit deobfuscation. Front. Big Data **4**, 608286 (2021)
197. L. Alrahis, S. Patnaik, M. Shafique, O. Sinanoglu, OMLA: an oracle-less machine learning-based attack on logic locking. IEEE Trans. Circuits Syst. II: Express Briefs **69**(3), 1602–1606 (2021)
198. L. Alrahis, S. Patnaik, M.A. Hanif, M. Shafique, O. Sinanoglu, UNTANGLE: unlocking routing and logic obfuscation using graph neural networks-based link prediction, in *International Conference On Computer Aided Design (ICCAD)* (2021), pp. 1–9

199. L. Alrahis, S. Patnaik, M. Shafique, O. Sinanoglu, MuxLink: circumventing learning-resilient MUX-locking using graph neural network-based link prediction (2021). arXiv preprint arXiv:2112.07178
200. D. Sisejkovic, F. Merchant, L. Reimann, R. Leupers, Deceptive logic locking for hardware integrity protection against machine learning attacks. IEEE Trans. Comput.-Aided Design Integr. Circuits Syst. **41**(6), 1716–1729 (2021)
201. A. Jain, Z. Zhou, U. Guin, TAAL: tampering attack on any key-based logic locked circuits (2019). arXiv preprint arXiv:1909.07426
202. M.T. Rahman, M.S. Rahman, H. Wang, S. Tajik, W. Khalil, F. Farahmandi, D. Forte, N. Asadizanjani, M. Tehranipoor, Defense-in-depth: a recipe for logic locking to prevail. Integration **72**, 39–57 (2020)
203. M. Gao, M. Sazadur Rahman, N. Varshney, M. Tehranipoor, D. Forte, iPROBE: internal shielding approach for protecting against front-side and back-side probing attacks. IEEE Trans. Comput.-Aided Design Integr. Circuits Syst. **42**(12), 4541–4554 (2023). https://doi. org/10.1109/TCAD.2023.3276525
204. A. Alaql, M.M. Rahman, S. Bhunia, SCOPE: synthesis-based constant propagation attack on logic locking. IEEE Trans. Very Large Scale Integr. Syst. **29**(8), 1529–1542 (2021)
205. L. Li, A. Orailoglu, Shielding logic locking from redundancy attacks, in *IEEE VLSI Test Symposium (VTS)* (2019), pp. 1–6
206. L. Alrahis, S. Patnaik, J. Knechtel, H. Saleh, B. Mohammad, M. Al-Qutayri, O. Sinanoglu, UNSAIL: thwarting oracle-less machine learning attacks on logic locking. IEEE Trans. Inform. Forensics Secur. **16**, 2508–2523 (2021)
207. H. Shen, N. Asadizanjani, M. Tehranipoor, D. Forte, Nanopyramid: an optical scrambler against backside probing attacks, in *International Symposium for Testing and Failure Analysis (ISTFA)* (2018), p. 280
208. S. Parvin, T. Krachenfels, S. Tajik, J.-P. Seifert, F.S. Torres, R. Drechsler, Toward optical probing resistant circuits: a comparison of logic styles and circuit design techniques, in *Asia and South Pacific Design Automation Conference (ASP-DAC)* (2022), pp. 1–6
209. K. Shamsi, D.Z. Pan, Y. Jin, On the impossibility of approximation-resilient circuit locking, in *International Symposium on Hardware Oriented Security and Trust (HOST)* (2019), pp. 161–170
210. M. Sazadur Rahman, K. Zamiri Azar, F. Farahmandi, H. Mardani Kamali, Metrics-to-methods: decisive reverse engineering metrics for resilient logic locking, in *Proceedings of the Great Lakes Symposium on VLSI 2023* (2023), pp. 685–690
211. U. Feige, A. Fiat, A. Shamir, Zero-knowledge proofs of identity. J. Cryptol. **1**(2), 77–94 (1988)

Chapter 3
Rethinking Hardware Watermark

3.1 Introduction

The advancement of processing technology has led to rapid growth in integrated circuit (IC) design complexity. There are now more than tens of billions of transistors integrated on a single chip, and the upward trend is expected to increase in the coming years. The continuing device scaling creates a design productivity gap between IC design (which typically increased at about 20% per year) and IC fabrication (which increased at about 40% per year) [1, 2], and this gap is becoming wider. IP reuse emerged as the most important design technology innovation in the past two decades to close this productivity gap and expedite the development of modern SoC products.

The concept of reuse takes advantage of the predesigned and preverified functional blocks, called IP cores. IP reuse and exchange usually take the form of soft, firm, or hard [3–5] IPs. Soft IPs are delivered in the form of synthesizable hardware description language (HDL) codes. Hard IPs are commonly provided in the form of GDSII (Graphical Database System II) files representation of a fully placed and routed design. Firm IPs typically come in the form of entirely placed netlists. These IP cores can be generic, proprietary, or open source. Generally, open-source functions include processor units, ADCs/DACs, RAM, UARTs, ethernet controllers, power modules, media access controllers (MACs), etc. According to a recent report published by the ESD alliance, the semiconductor IP sector has become the largest segment in the electronic design automation (EDA) industry with a total revenue of $1.0529 billion in the 4th quarter of 2020, which is more than 34% of worldwide EDA industry revenue [6]. The flexibility of reusable IPs expedites the creation of SoC products and brings a lot of potential profits to the IP providers. Systems-on-chip (SoCs) combine dozens of intellectual property (IP) cores licensed from different vendors [7, 8]. The reuse of IP cores comes with a risk of IP piracy and overuse. Problems could be in various forms, such as claiming someone else's IP as your own or reselling it, not giving an IP designer credit where

M. Tehranipoor et al., *Hardware Security*,
https://doi.org/10.1007/978-3-031-58687-3_3

it is due and open-source IPs being used for commercial purposes. Therefore, a need exists for provable identification of an IP core within a suspect design. Each IP must have a unique identification that represents the version, ownership rights, design information, and provider. Moreover, the identification can also provide ownership proof, designer information, and IP tracing. The ability to prove the identity of IP is increasing in importance [3, 4, 9–12]. After the IP has been incorporated into an SoC and packaged, designers can still check the identity of the IP.

Researchers have proposed several possible protection methods against illegal IP usage [3, 9, 10, 12–21]. One potential solution for claiming ownership of an IP core is to use watermarks. Watermarking, the process of marking an asset with a known structure, has been proposed to detect IP theft and overuse. Watermarking in hardware IPs is the mechanism of embedding a signature (or a unique code) into an IP core without altering the original functionality of the design. The ownership of the IP can be later verified when the watermark is extracted. The IP watermarking steps are illustrated in Fig. 3.1. Typically, an IP core developer embeds a watermark inside the core and sells the protected IP core in the market. To reduce time-to-market and design complexity, SoC designers use IPs from various IP owners in their designs. The SoCs are fabricated in foundries located around the globe. So, the IP owners have very little control over any illegal usage of their IPs in an SoC [18, 22]. If the IP owner encounters such an issue, they can retain a suspected chip from the market and extract the watermark using their known parameters. If the extracted watermark is matched with the originally embedded watermark into the IP, then the IP owner can easily prove that his/her IP core is illegally used in the SoC. An ideal watermark mechanism is embedded and verified easily and yet does not suffer from high overhead and is resistant to attacks [3].

It is important to note that watermarking is a passive technique that cannot prevent IP infringement, piracy, or overproduction of ICs. It is only suitable for proving intellectual property use. In the past two decades, the semiconductor industry has witnessed legal battles among technological giants in response to intellectual property theft, piracy, and digital rights management. In 2003, Cisco Systems Inc. filed a lawsuit against network equipment manufacturer Huawei Technologies and its subsidiaries, claiming illegitimate copying of its IPs, including source codes, software, documentation, and copyrighted materials [23]. A tough legal battle took place between Intel Corporation and Micron Technology over 3D memory technology development [24]. Semiconductor startup CNEX Labs sued Huawei Technologies for stealing their IP [25]. These copyright infringement activities are not uncommon, and the worst part is that these incidents stay undiscovered in most cases.

Globalization in the semiconductor supply chain granted third parties access to advanced technologies manufactured for critical applications. Outsourcing of semiconductor design tasks, therefore, introduces vulnerabilities into the supply chain that adversaries can exploit. From a global perspective, where IP protection laws vary vastly from one country to another, IP protection and authorship can no longer be limited to passive methods such as patents, copyrights, and watermarks

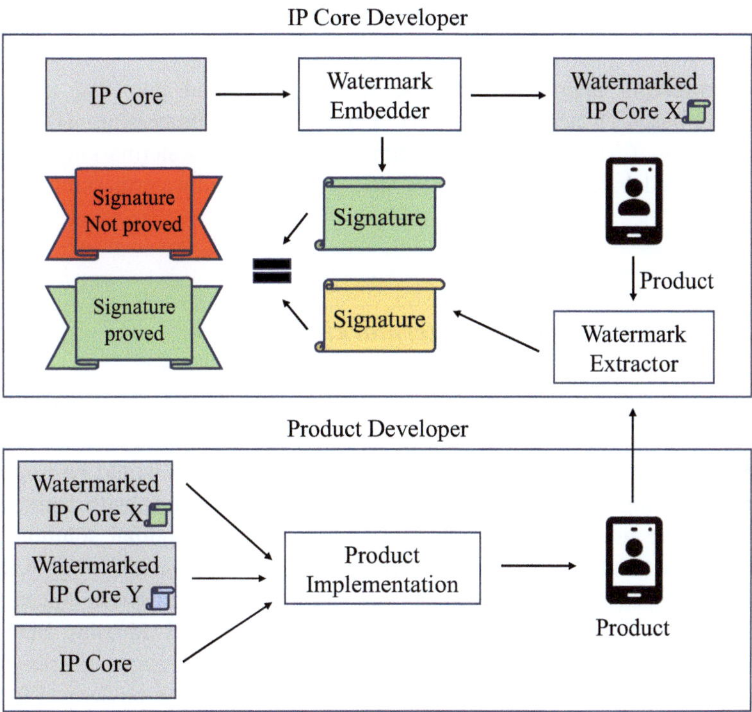

Fig. 3.1 IP watermarking embedding and verification procedures

that merely deter these threats. Furthermore, the existing watermarking techniques [3, 9, 10, 19, 20] are only limited to the specific IP blocks themselves, which are not capable of providing SoC-level detection of adversarial footprints [26]. Therefore, there is a critical need to develop new and innovative watermarking techniques to prevent IP piracy/theft in modern complex SoC environment.

A number of watermarking methods have been reported in the literature for providing proof of IP ownership [27–42]. Researchers have investigated the IP identification/detection methods at various design levels such as system design, behavior design, logic design, and physical design. However, the existing techniques do not explicitly discuss the watermark detection or extraction process in many practical scenarios to prove the ownership of the IP core in complex SoCs. For example, how do the IP designer detect and prove IP overuse in a suspect SoC after the chip has been packaged when the IP designer has access to the chip but not the IP? Therefore, the assumptions to leverage existing approaches are incompatible with the modern threat landscape and must be revisited. In this chapter, we first provide a high-level overview of the state-of-the-art IP watermarking techniques. Then, we provide guidance to shape future research directions to maximize the applicability and impact of watermarking techniques in modern SoCs.

3.2 Existing IP Watermarking Techniques

IP watermarking approaches can be roughly classified into five groups: (i) Constraint-based watermarking, (ii) Digital signal processing (DSP)-based watermarking, (iii) Finite state machine (FSM)-based watermarking, (iv) Test structure-based watermarking, and (v) Side-channel-based watermarking, as shown in Fig. 3.2.

3.2.1 Constraint-Based Watermarking

Several NP-hard optimization problems are used in every phase of the IC design process (i.e., system synthesis, behavioral synthesis, logic synthesis, and physical synthesis). A detailed enumeration of all possible solutions would be impossible due to the complexity of the problems. Heuristic algorithms with some design constraints are used to search for near-optimal or quasi-optimal solutions. This is where constraint-based IP watermarking techniques come into play. Kahng et al. [43] proposed one such approach that is applicable at different stages of the design process. In this approach, NP-hard problems are solved using EDA tools available at that stage. A generic optimizer is used to solve constraint-satisfaction problems (CSP) [43]. As a result, the watermarked design can be derived from the algorithmic constraints added to such a solution.

Figure 3.3 shows the basic conceptual overview of how constraint-based watermarking works. The heuristic algorithm takes the original design specifications. It covers constraints as inputs for design space exploration to select a good solution as the original IP core from an ample solution space. The encrypted authorship message is first converted into a set of embedded constraints. These constraints are then used as additional inputs to the envelope embedding unit. The embedded constraints derived from the author's signature are blended with the cover constraints to generate the stego constraints. At this point, the original problem becomes the stego problem which is fed to the EDA tools to find a near-optimal

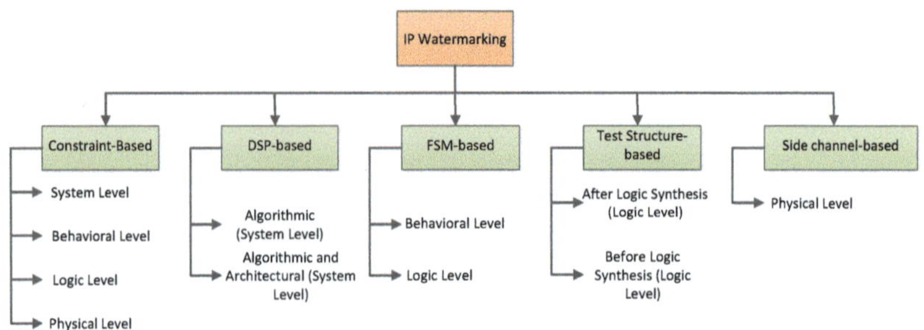

Fig. 3.2 IP watermarking categories and strategies

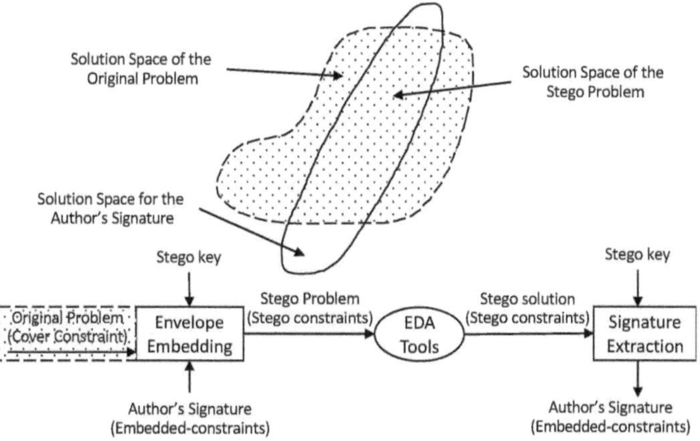

Fig. 3.3 Basic concept of the constraint-based watermarking approach [44]

solution. The final result will be a watermarked IP core that satisfies both the original and embedded constraints. Therefore, this provides a probabilistic proof of authorship, which is expressed in Eq. 3.1 [43].

$$P_c = P(x \le b) = \sum_{i=0}^{b}[(C!/(C-i)!] \cdot i!) \cdot p^{C-i} \cdot (1-p)^i] \tag{3.1}$$

Here, p = probability of satisfying one random constraint by coincidence, P_c = proof of authorship, b = the number of constraints unsatisfied, and C = the number of imposed constraints. P_c should be kept as low as possible when designing constraint-based watermarking strategies so that only the author can satisfy the final solution to the IP design C watermark problem. In addition, the average P_c could be in the range of 10^{-30} [45], making it impossible for anyone (in terms of time and effort) to copy the watermark. Constraint-based watermarking, therefore, allows IP digital rights protection, where it is computationally infeasible to guess the correct solution without a signature, key, etc. Following are the components of a generic constraint-based watermarking procedure as outlined by Kahng et al. [43]:

- Constraint-based watermarking should be a complex optimization problem where achieving an acceptable solution or enumerating enough acceptable solutions grows exponentially with the input size.
- Intellectual property can be defined as the solution to the well-defined optimization problem.
- An optimization problem can be solved by using existing algorithms and/or off-the-shelf software.
- Security requirements are considered as that are similar to those known from currency watermarking.

To solve the stego problem, the heuristic algorithm of the EDA tools uses the stego constraints rather than the design constraints. The constraints can either be incorporated into the optimizer's inputs (i.e., preprocessing) or applied to the optimizer's output (i.e., postprocessing). Preprocessing appears to be the most popular approach. Since its first introduction in [43, 46], many techniques have been proposed for the constraint-based IP watermarking at different abstraction levels, which range from the system synthesis level [27, 46] to the behavioral synthesis level [35, 36], logic synthesis level [37–40], and physical synthesis level [41–43]. In the following, we will review some of the most influential proposals at each level.

3.2.1.1 System Synthesis Level

The nodes of the graph to be partitioned are randomly numbered using integers as illustrated in Fig. 3.4a. The stego constraints associated with watermarks mandate that nodes within a partition remain together. A pair of origin and terminal nodes are selected for each watermark bit. Nodes that have not been selected as the origin node are used to determine the original node. Nodes with the smallest index are selected. The terminal node is determined by the watermark bit. The terminal node is selected when the watermark bit is "1." Alternatively, when the watermark bit is "0," the terminal node is the one with the smallest even index that has not been paired. Consider the letter "A" with the ASCII code "1000001" as a watermark (or a part of the watermark). According to the embedding criteria described above, the pairs of nodes (1, 3), (2, 4), (3, 6), (4, 8), (5, 10), (6, 12), and (7, 13) in Fig. 3.4b should all be in the same partition. According to Fig. 3.4b, one possible watermarked solution is to balance partitioning, so that node differences between two partitions are less than 20%. The graph-partitioning-based watermarking described above can also be extended to a graph coloring problem. In graph coloring, the aim is to find a way to color the vertices of a graph such that no two adjacent vertices are the same color. The graph coloring problem resembles many optimization tasks in the VLSI (Very Large Scale Integration) design flow. As an example, a cache-line code

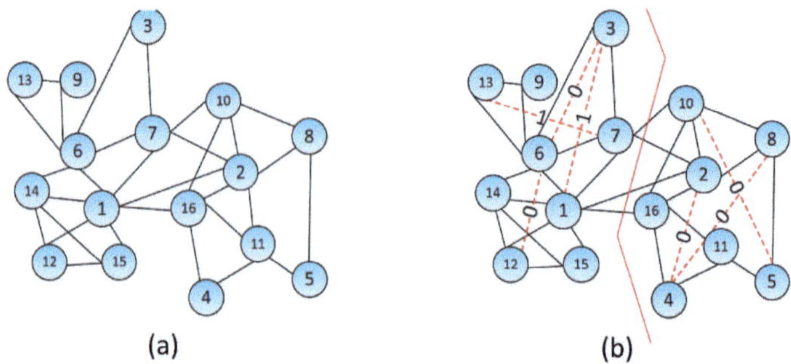

(a)   (b)

Fig. 3.4 (a) Graph to be partitioned with nodes indexed by integers (b) The partitioned graph with embedded watermark [46]

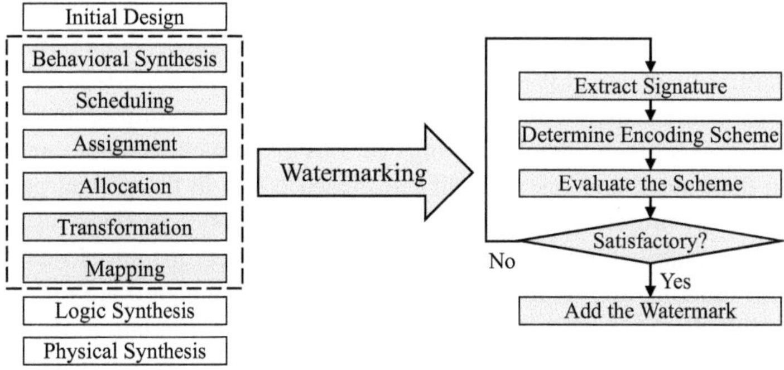

Fig. 3.5 A generic approach for watermark insertion at the behavioral (RTL) synthesis level [36]

optimization problem [27] can be solved by finding a solution with a fixed number of colors, where each color represents one cache line. An additional edge can be added to the graph to represent a watermark.

3.2.1.2 Behavioral Synthesis Level

Task scheduling, resource assignment, transformations, resource allocation, and template mapping are all NP-hard optimization problems that are excellent for embedding the watermark. Koushanfar et al. [36] proposed watermark insertion during register allocation. Figure 3.5 shows a generic approach for watermark insertion at the behavioral synthesis level. After scheduling, the intermediate output generated in one stage needs to be stored in registers and applied to the next stage to maintain the pipelining. The period between the first time a variable is generated and the last time it is used is its lifetime. If two variables exist in two different time frames, then they can share the same register. A graph coloring problem can be solved for an interval graph by allocating registers. The embedded constraints generated by the author's signature can be inserted as an additional edge in the interval graph and solved for the overall solution. Researchers also proposed a do-not-care condition-based solution for constraint-based watermarking insertion [35]. Occasionally, a designer may not care about the output of a truth table for specific inputs, which are termed as "don't care" conditions. IP watermarking can be accomplished with do-not-care conditions that force the output of IPs.

3.2.1.3 Logic Synthesis Level

The logic synthesis process transforms abstract design behavior into a specific implementation comprising logic gates based on the RTL HDL. Multilevel logic minimization and technology mapping are two optimization tasks performed in combinational logic synthesis. According to [38], both tasks are a suitable candidate for constraint-based watermarking. Technology mapping maps the logic network of a design to as few predefined library cells as possible, the complexity of which is NP-hard. Cui et al. [39] proposed an adaptive watermarking technique by modulating some closed cones as part of an original, optimized logic network (master design) for technology mapping. Various disjoint closed cones are analyzed based on their slack and sustainability. If closed cones in the critical path can be more effectively preserved upon remapping using the notion of slack sustainability, then they are qualified to host watermarks. Only qualified disjoint closed cones are selected at random, and templates constrained by the signature are remapped and embedded as the watermark. By using this parametric formulation, designers can take advantage of a design's headroom to increase the signature length or strengthen the watermarks. A similar approach has been proposed where watermarking constraints have been imposed only to a careful selection of noncritical paths based on the design specification [40]. A watermarking technique based on the Schmitt Trigger insertion at logic synthesis level has been proposed [37]. The signature is presented as a BASE64 hash code, an MD5 hash code, and an ASCII string to create a unique sequence of watermarked bits.

3.2.1.4 Physical Synthesis Level

The logic synthesis process produces a gate-level netlist and circuit layout derived from the mapped gate-level netlist. Based on the mapped gate-level netlist, physical synthesis produces an optimized netlist and circuit layout. Floorplanning, clock tree synthesis, placement, scan chain reordering, routing, physical sign-off are common steps of physical synthesis. In [43], several watermarking techniques based on these tasks are presented. Authorship signatures are used to determine path timing constraints. These timing constraints are replaced with "sub-path" constraints to insert watermarks. The synthesis solution may not satisfy both subpath constraints under the original timing constraint [47]. As indicated by Kahng et al. [43], there is a slight chance of satisfying both subpath constraints when the original constraint is satisfied. A strong proof of authorship can be achieved by constraining hundreds of timing paths. Similar to routing-based IP protection techniques [48, 49], standard-cell routing-based watermarking approach at physical synthesis level has been proposed [43]. According to this approach, the watermarking constraints are based on the (per-net) costs associated with the routing resources. In other words, if the watermark bit is "1," the network will incur unusual costs if traffic is redirected in the wrong direction and/or vice versa. Thus, a watermark can be verified by examining the distribution of resources in specific nets. By constraining the placement of selected logic cells in rows with the specified row parity, the watermark can also be inserted [43]. In other words, some cells are constrained to be placed in even-index rows, while others are constrained to be placed in odd-index rows. The typical placement instance has tens of thousands of standard cells, and this method provides an authorship proof with a long watermark. Still, some cells may incur a higher routing cost due to watermarking constraints. Therefore, watermarked logic cells should be carefully chosen. Routing grid-based watermark insertion has been proposed [41] where each grid cell in the layout picture is assigned a judging parameter value, and the picture is partitioned with small grids of appropriate size. A watermarking signature is then drawn on the grided layout. Text messages or meaningful symbols may be used as signatures. The cells are scattered using some algorithm, such as the pseudo-random coordinate transformation (pReT) algorithm, to generate a new distribution of the original watermarking cells. As a result, the watermarking can be spread out over the entire layout, making it difficult to tamper with the watermark by changing just a tiny part of the layout. By combining directed NBTI (negative bias temperature instability) aging [50] and delay logic, Zheng et al. [42] presented a new approach to IC digital watermarking. Their work demonstrates that delay logic measures relative speed differences between competing logic paths due to variation in the process.

3.2.2 Digital Signal Processing Watermarking

Watermarking using digital signal processing (DSP) is introduced at the algorithmic level of the design flow in [28] and [29]. The primary goal of both techniques is to enable designers to make slight modifications to the decibel (dB) requirements of filters without compromising their operation. According to [28], the designer of a high-level digital filter needs to first encode a single character (7 bits) as a watermark. The high-level filter's design is then broken into seven segments, each of which is employed as a modulation signal for one of the bits. It entails breaking the filter into seven segments and using each portion as a carrier signal with little dB change if the bit is "1" or none if the bit is "0."

The authors of [29] divided the problem into two parts. Watermarking has been implemented at both the algorithmic and architectural levels to improve robustness. They have used a similar approach to [28] on the algorithmic level, where seven bits are added. However, they employed a static technique at the architectural level to watermark the transpose of the finite impulse response (FIR) filter [29]. Their solution is based on employing different filter building block architectures according to the bits that need to be embedded. A pipelined structure is used to create the transposed form FIR filter, with each pipeline stage consisting of basic multiplier–adder–delay block functions. Watermarking at the architectural level is based on making circuit modifications utilizing the fundamental blocks of each pipeline step without modifying the filter's transfer function. Two of these fundamental block changes are shown in Fig. 3.6. Both structures are capable of encoding a two-bit code: 00, 01, and 10.

Both methods have a low embedding overhead and a low design overhead. However, as long as the DSP filter is not completely hidden in the design, they are immediately recognizable. Both techniques can be applied at the DSP

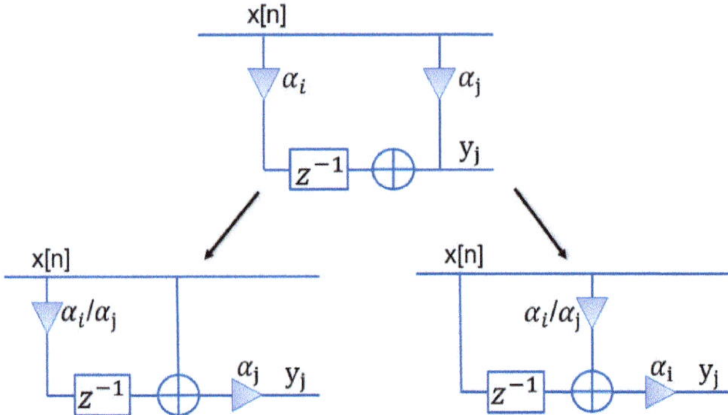

Fig. 3.6 Architecture-level circuit transformations [29]

algorithmic level, which is a relatively high level in the design flow. The authors did not take into account the robustness of their technique or any other criteria. Furthermore, the methods rely on a relatively low data rate, just one character (7 bits), making them unfeasible for application in an industrial setting. The watermark is especially susceptible to design variations at lower levels because of the low bit rate. Furthermore, such a watermark is quite vulnerable to masking attacks since even minor modifications in the filter function may readily hide or even delete the watermark.

3.2.3   FSM-Based Watermarking

Similar to FSM-based protection techniques [51, 52], the FSM-based watermark is embedded at the behavioral level by introducing additional FSM states or transitions which does not interfere with the normal chip functionality. FSM-based watermarking is a fairly researched area of hardware security: One reason behind the popularity of this technique is that FSMs in a design provide an easy way to extend the state space to hide watermark into it. In general, FSM-based watermarking techniques should meet the following requirements:

- Watermarking states or transitions should be well hidden into the existing state space so that the newly introduced states/transitions do not become easy target to the removal attack.
- Watermarking states or transitions must not change the original functionality of the IP.
- Newly introduced states/transitions should have reasonable impact on the original design in terms of overhead.
- It should not be affected by design optimization techniques during synthesis or physical design.

FSM watermarking techniques embed watermarking signature into the original design by modifying the state transition graph (STG). State transition graphs consist of states and transitions, and thus, FSM watermarking techniques can be classified into two types: transition-based watermarking [53–55] and state-based watermarking [30, 31]. State-based watermarking schemes involve adding new states or changing state encoding. On the other hand, transition-based watermarking schemes add new transitions to the FSM or utilize unused transitions. The following sections review some of the related solutions of each type.

3.2.3.1 State-Based FSM Watermarking

The first state-based watermarking strategy was proposed by Oliveira [30], where the watermark is inserted as a new property. The strategy involves manipulation of the STG by introducing redundancy so that it exhibits the chosen watermarking property. The user needs to choose a copyright text of arbitrary length of string. The string is then encrypted and hashed to generate the signature. The technique proposed in [30] breaks this signature into a combination of input sequence. Extra states and transitions are added into the STG in such a way that the input sequence, when applied, satisfies the property. The property itself is purely topological and does not depend on the state encoding. In this approach, modification of the STG is performed by duplicating existing STG. Next, one or more states and transitions are copied from the original STG and used to connect the previously duplicated STG to the original STG. This approach of STG modification adds large overhead to the design. Moreover, this approach is weak against the state minimization attack as it will remove all redundant states. Also, it uses a counter to detect expected input sequence. This is a weak point as identifying and removing this counter will render the watermarking scheme useless.

Lewandowski et al. [31] proposed another state-based watermarking scheme by embedding a watermark signature via state encoding. In this approach, a watermark graph is first constructed from the binary representation of the signature. The binary string is divided into blocks of a chosen size. Each node of the watermark graph will represent a block of the binary string. Then the algorithm inserts directed edge from one node to the next. These steps are repeated until entire binary string is encoded into the nodes of the watermark graph. Once the watermark graph is built, it is then embedded into the design STG. The proposed algorithm looks for isomorphism between the two STGs using a greedy heuristic. If the algorithm fails to find a subgraph isomorphic to the watermark STG, a subgraph that is most similar to the watermark graph is chosen and extra transitions are added to it to make it isomorphic to the watermark graph. Then, the nodes of watermark graph are mapped into the nodes of the subgraph, and the state encoding of the subgraph is set by the state encoding of watermark graph. This approach of IP watermarking is susceptible to state recoding attack. State recoding attack can change the state encoding of the watermarked FSM and corrupt/change the watermark signature.

3.2.3.2   Transition-Based FSM Watermarking

One of the earliest papers on FSM watermarking is [53], where the authors proposed an algorithm to extract the unused transitions in an FSM. The algorithm visits each state of the FSM to look for unused transitions and uses them for watermarking. If the algorithm fails to find any unused transition in the design, new input–output pins are inserted to expand the STG.

The state transition diagram shown in Fig. 3.7a is the original FSM. The input is 2-bit and output is 1-bit. In this example, the watermark length is 2-bit and it is represented as ((00/1), (11/0)). Figure 3.7b shows the watermarked FSM without augmenting the input. If the current FSM is completely specified, then the input is extended by 1-bit and then watermarking transitions are added as shown in Fig. 3.7c.

Utilizing existing transitions as well as unused transition for embedding watermark in an FSM was proposed in [54]. The authors break the watermarking process into three parts: generating the signature, embedding it into the FSM, and detecting the watermark. Figure 3.8 shows the watermarking framework with

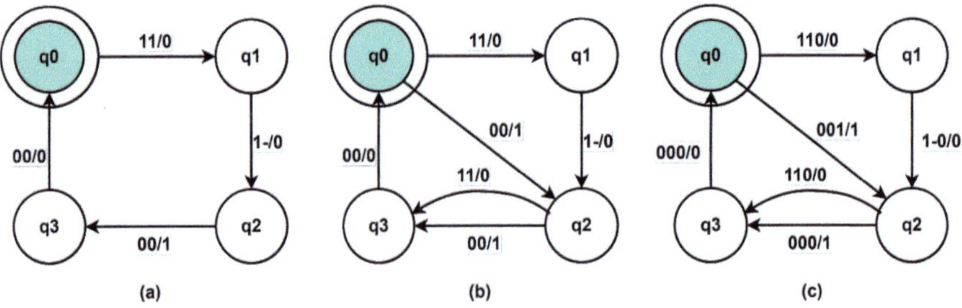

Fig. 3.7 FSM watermarking example from [53]: (**a**) Original FSM, (**b**) Adding new transitions, and (**c**) Extending input and adding new transitions

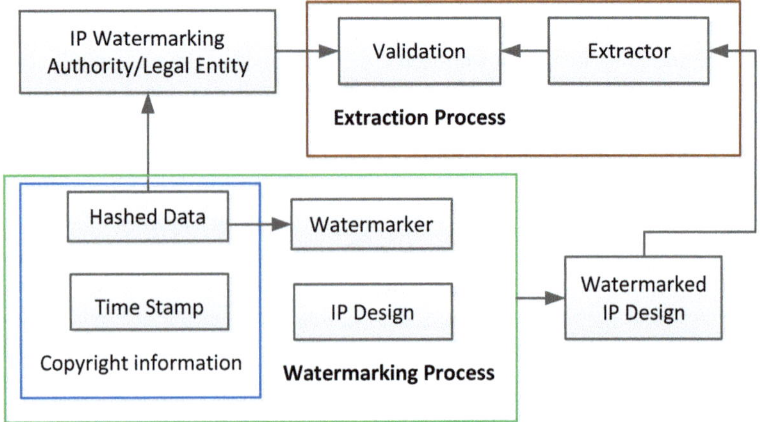

Fig. 3.8 Watermarking with third party keeping a time-stamped signature [55]

third party keeping a time-stamped signature. The technique utilizes a third party, namely watermarking authority, to generate the time-stamped watermark signature. This signature is stored in the database and reused during extraction process to provide authorship proof. For the watermark insertion, the paper [54] proposes two algorithms. The first algorithm works based on the input bits comparison. It first generates random inputs and associates watermark signature with these random inputs. Then it looks for a match between this randomly generated input and existing input of a randomly chosen state of the STG. If there is a match between the inputs, the output of the state is compared with the signature. If the output matches, the state will be used as watermarking state. If there is no match between the outputs, a new transition is added to the state. And, if there is no match found with the inputs of the STG, the inputs are extended using an extra bit and the newly generated states are used for watermarking. This algorithm adds new input bits even if the STG is not completely specified which adds unnecessary overhead. Another algorithm works based on the mapping of signature bits into existing FSM outputs. The algorithm visits different states and tries to match signature bits with each state's output. If there is a match, that state is used as watermarking state. But if there is no match, it tries to add new transition using free state space. If there is no free state space, the algorithm extends the input similar to the other approach. Both of these algorithms perform poorly when the existing FSM has outputs of large size. In this case, probability of matching/mapping with the signature bits becomes far less.

In [56], the authors have proposed a watermarking scheme inspired by the stealthy nature of hardware Trojans (HTs). This technique is interesting since it provides very low overhead in the target IP and very good resiliency against the known attacks since Trojans are very difficult to detect by existing testing methods [57, 58]. In this approach, the designer of the IP can verify the watermark by loading the test vector twice through the scan chain (which triggers the watermark) and by observing the payload output. The payload response is normal when the trigger test vector is loaded for the first time. However, when the trigger test vector is loaded for the second time, the payload bit is flipped in the response. These normal and HT-affected payload responses form the proof of ownership. However, in [56], hardware Trojans have been inserted in gate-level abstraction only. It can be enhanced to design and embed watermarking in other abstraction levels like RTL and physical design level [4, 58]. This proposition requires further investigation and research efforts to materialize properly.

The problems with the watermark signature mapping into the existing FSM output are addressed in [55]. The proposed FSM watermarking technique also uses existing transitions for watermarking. But the signature is not directly mapped into the outputs of states rather the watermark appears at certain points in the output sequence under the specific input. These specific positions are generated using a pseudo-random number generator [59, 60]. Similar to the previous approaches, if existing transitions cannot be used for watermarking, new transitions are introduced. If the FSM is completely specified, the FSM input is extended.

None of the FSM watermarking techniques discussed above discusses about the watermark extraction process when the IP is integrated deep into an SoC. In this case, the output of the watermarked FSM may not be directly observable using the primary outputs of the SoC. This problem is addressed in [61], where the authors proposed a hybrid scheme combining FSM watermarking with test-based watermarking. Additionally, the synthesis process can introduce security risks in the implemented circuit by inserting extra do-not-care states and transitions [62]. The IP is watermarked in two phases: adding watermark into existing FSM and reordering the scan cells when the IP is integrated in an SoC. The scan chain provides an easy watermark extraction process. This approach also does not consider a scenario where the IP is illegally used in a different SoC and that the test access may not be available. The attacker, in this scenario, may not provide open access to the IP or the access may be restricted which blocks the designer from proving their authorship.

3.2.4 Test Structure-Based Watermarking

The core concept for testing-based watermarking techniques is embedding a water-mark into a test sequence at the behavior design level. After integrating the IPs into the full SoCs, test signals have to be traceable. Using this fact, the authors combined this test sequence with the watermark generating circuit, so that any IP in the chip may be observed and tested even after the chip has been packaged [18, 63]. In the test mode, the selected IP sends output test patterns and watermark sequences. We can determine the identity of the IP provider according to the watermark sequence. Several postfabrication watermark verification techniques based on test structures have been proposed in literature. Despite utilizing the sophisticated sequential automated test pattern generation (ATPG), scan design accomplishes the testability of sequential design by achieving the controllability and observability of the flip-flops present in the design [64]. For decades, scan chain-based testing methodology has been the backbone of design-for-test (DFT) in the EDA industry, and it remains as the most prevalent technique. D-type flip-flops (DFFs) are substituted by scan flip-flops to enable a design having scan chain capabilities (SFFs). SFFs are made up of a DFF, a multiplexer, and two new signals: scan-data SD and test control TC. As demonstrated in Fig. 3.9, SFFs can be chained by connecting the Q output of one SFF to the SD input of the next SFF. The Q-SD connection style [20] is used for

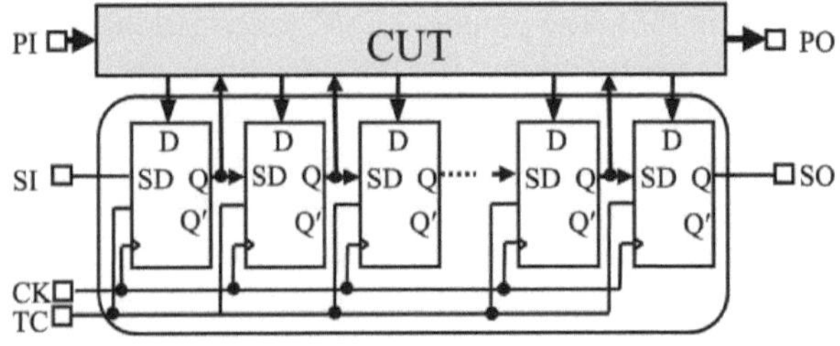

Fig. 3.9 Scan chain architecture used in DFT synthesis flow [20]

this. The core under test (CUT) is switched between normal and testing modes via TC. Test data from the new primary input SI will be applied to the scan chain in the testing mode, and response data will be gathered from the new primary output SO [20].

In [20], a scheme was proposed to insert watermark by controlling the connection style between two connected scan cells, to minimize various overhead due to watermarking. As a result, under the validation vector, the scan chain will output an output response that contains watermark information at certain positions. This watermarking can be implemented by alternating the local routing, and only negligible overhead on the test power is incurred, as shown in the middle part of Fig. 3.10. It can be applied to protect hard IP cores. If some hard IP cores provide some space for one to change the local routing, the watermark can be easily removed as the watermarked connection style may conflict with that determined by the test power optimization. An alternative scheme was proposed that the dummy scan cells will be inserted in scan chain when the optimized connection style conflicts with the watermarked style. Such dummy scan cells are merged in the common scan cells but play no role in testing, and they also enable all connection patterns to satisfy the optimization criteria, as shown in the lower part of Fig. 3.10. Although some extra scan cells are incurred, it is robust against the possible removal attack. Designers can determine the amount of resilience they want from the flexible framework, reducing design overhead.

An approach for watermarking at the gate level while selecting the chain of scan registers for sequential logic test generation was suggested in [32]. The IP's watermark is transformed into user-specific constraints for selecting the scan register chain. The watermark is inserted using a set of standards for uniform circuit ordering. For this methodology, the watermark verification method is relatively straightforward and can be readily conducted by introducing a particular set of test vectors and returning a set of scan chain outputs uniquely tied to the watermark signature. However, ordering the scan cells is an NP-hard problem. In order to minimize the test power, a similar method was proposed in [33], which introduces the addition of the constraints originated by the owner's digital signature. Fault coverage and test application time remain unchanged since only the order of the scan cell gets changed. In all of these proposed schemes, the watermark pattern can be observed in test mode. As a result, field verification of the ownership of the IP is possible. However, all of these schemes are susceptible to removal attacks since the watermarked test core structure is independent of the functional logic. The attacker can easily redesign the test structure for completely removing or corrupting the watermark partially.

A persuasive technique called synthesis-for-testability (SFT) was proposed in [34] for IP watermarking. The watermark is embedded as hidden constraints to the scan chain ordering mechanism very similar to [33]. However, the SFT watermarking mechanism inserts the watermark into the chain before synthesis, but the DFT-based watermarking methods proposed in [32, 33] are performed after logic synthesis. The most promising fact of this technique is that test functions can be

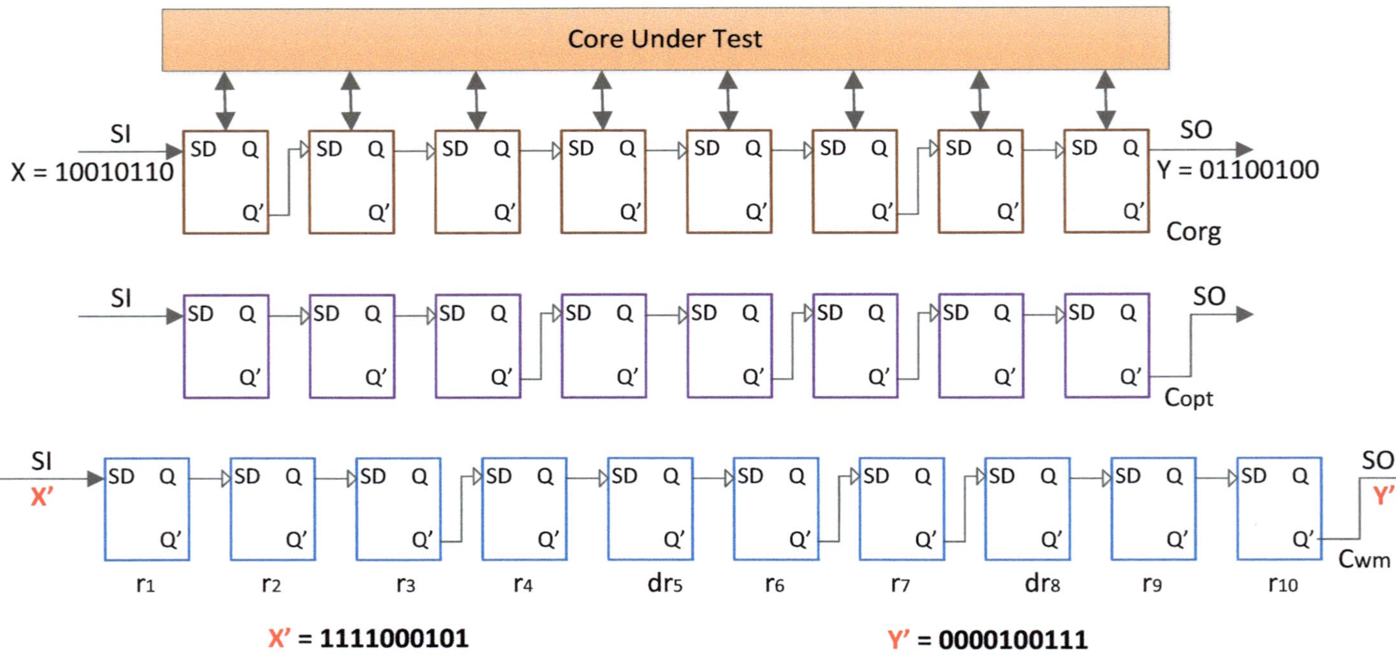

Fig. 3.10 DFT-based watermarking scheme [20]

mingled with core functions, making the attempts to remove or alter the watermark a thousand times harder. If the attacker attempts to modify or remove the watermark now, there is a high possibility that design specification and optimal characteristics will get impacted. This method is quite promising, but in terms of wide acceptability, it has a long way to go since the SFT technique has not been widely adopted as a testing method in the current industry.

3.2.5 Side-Channel-Based Watermarking

Side-channel analysis can utilize the physical information leaked from a cryptographic device and is frequently used to retrieve the secret keys and their security issues [65–68]. The side-channel-based watermark is that instead of leaking out secret information, the side channel is engineered to contain a watermarking signal. The main idea is to use a side channel, e.g., power consumption, to embed a watermark into an IP core. Then the verifier uses that side-channel information to extract the watermark and prove ownership. In [69], a power signature-based watermarking technique was introduced for protecting IP cores, which uses the power supply pins to extract the watermark in FPGAs. It utilized the fact that supply voltage to the IP core depends on the switching of shift registers located in the IP core. The main idea is to convert the watermark into a specific voltage signal using a pattern generator and then decode the extracted signature to prove ownership. Figure 3.11 shows a watermark verification via power analysis.

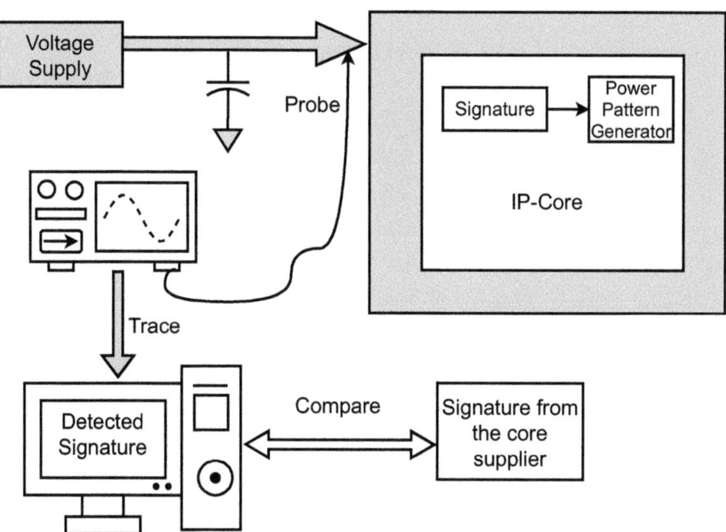

Fig. 3.11 Watermark verification via power analysis [69]

As the signature is hidden below the noise floor, it is difficult to perform removal attack, but this method has reliability issues while decoding the watermark signature. Also, modern SoCs contain tens of IP cores; hence, there is a high probability that in a functional chip the power consumption of the other IP cores can alter the watermark signature of the target IP. To solve this issue, [70] proposes a technique that involves embedding ring oscillators into IP core to make it more distinguishable during parallel testing and operation. A group of ring oscillators (ROs) can cause a huge amount of switching activity which results in a significant amplification of the power consumption. Thus, an additional RO block controlled by an input vector can lead to a successful core detection [71], but it increases the area overhead of circuit, and could be susceptible to removal attacks.

In [72], authors propose a watermark verification scheme using a correlation analysis based on the measurement of the power consumption of an IC. In addition, the authors of [73] discussed two different power-based watermark approaches such as spread spectrum-based watermark and input-modulated watermark. In the spread spectrum-based watermark, a PRNG (pseudo-random number generator) is used to generate a bitstream and then a leakage circuit maps the bitstream to a physical power consumption. The leakage circuit produces additional leakage or no additional leakage depending on the bitstream. In the input-modulated watermark, a leakage generating circuit is implemented in such a way that it results in power consumption when a known input pattern is applied. To implement this watermark, the verifier should have knowledge about some bits of the IP core that can vary for different measurements. Both of these techniques have low overhead, but it is possible to reverse engineer the chips and remove the watermark.

3.3 Assessment of Existing Watermarks

3.3.1 General Requirements for Hardware IP Watermarking

The general requirements for IP watermarking technology are almost similar to the requirements of multimedia watermarking technology. However, multimedia watermarking technology has more freedom to alter the cover media and embed the watermark [3]. Such alteration is restricted in hardware IP watermarking technology because the watermarked IP core must remain functionally correct. Based on the requirements for a hardware IP watermarking technique presented in [3], the following design criteria are briefly outlined:

- *Credibility*: The watermark should be promptly detectable for the proof of authorship. No third party (i.e., other than IP owner) should be able to claim the watermark by chance, i.e., the probability of occurrence of a particular watermark on a nonwatermarked IP core must be very low. The watermark insertion is half the process only, and tracking and detection is the remaining important process in any watermarking technique. It is advantageous to ease the detection of watermark and enable the origin of fraudulence to be traced after possible attacks.

- *Low Overhead*: The performance of the IP core in terms of area, speed, and power should not be degraded/affected after embedding watermark. Both the computational time and cost needed for the watermark embedding should be kept low.
- *Resiliency (Robustness)*: The robustness measures based on the strength of the hidden signature against different types of attacks. The watermark of the IP core should be impossible or difficult to remove or tamper without the complete knowledge of the software or design.
- *Invisibility*: The embedded watermark should not affect the functionality of the IP core. The watermark of the IP core should not be readily detectable by third parties. It should be well-concealed such that only the IP owner should be able to reveal it.
- *Granularity*: The watermark implementation should be distributed in a hardware design in order to protect the watermark from removal or masking.

In Table 3.1, performance comparison of existing watermarking techniques in terms of credibility, overhead, resiliency, invisibility, granularity, and resiliency to attacks (i.e., masking, forging, tampering, removal) has been illustrated.

3.3.2 Attack Analysis for IP Watermarking

In general, there exist four main types of attacks: removal, tampering, forging, and reverse attacks. The shared prerequisite of these attacks is that they should not degrade the design performance. That is, an evidently deteriorated design is not what an attacker wants to steal.

3.3.2.1 Removal Attacks

For removal attacks, the main goal of the adversary is to remove the watermark entirely. This task is normally very difficult to succeed with the shared prerequisite [74]. The complexity and feasibility of this attack depend on the type of watermarking approach as well as on how well the watermarking scheme is hidden inside the original design. This task is usually very difficult to succeed with the shared prerequisite. There is currently no known metrics which can be used to determine how deeply a watermark is hidden into a design. In general, the watermark should be tangled with the existing features or functionalities of the design in such a manner that confuses the attacker to perform removal attack.

Table 3.1 Performance evaluation and comparison of the existing watermarking techniques

Watermarking	Abstraction level	Existing techniques	Performance evaluation[a]	Resistance to attacks[b]
	System synthesis	DSP code partitioning [46]	Low/Low/High/Low/Low	Low/Low/High/Low
		Cache-line coloring [27]	Low/Low/High/Low/Low	Low/Low/High/Low
	Behavioral synthesis	Do-not-care-based [35]	Low/Low/Medium/Low/High	Low/Low/Low/Low
		Register allocation [36]	Low/Low/Low/Low/Low	Low/Low/High/Low
	Logic synthesis	Schmitt trigger [37]	Medium/Medium/Low/Low/Low	Low/High/High/Low
		Combinational logic [38]	Low/High/Low/Low/Low	High/Low/Low/Low
Constraint-based		Tech. mapping [39]	Low/High/Low/Low/Low	High/Low/Low/Medium
		Critical path [40]	Medium/Medium/Low/Low/Low	High/Low/High/Medium
	Physical synthesis	Path-time constraint [43]	Medium/High/Low/Low/Low	Low/Low/Low/Medium
		Standard-cell routing [43]	Low/High/Low/Low/Low	High/Low/Low/High
		Row-based placement [43]	Low/High/Low/Low/Low	High/Low/Low/High
		Routing grid [41]	Medium/High/Low/Low/Low	High/Low/High/High
		FPGA [42]	Medium/High/Low/Low/Low	High/Low/High/Medium
DSP-based	Algorithmic	Character encoding[28]	Medium/Low/Low/Low/Low	Low/Low/Medium/Medium
	Architectural	Character encoding and static[29]	Medium/Low/Low/Low/Low	Low/Low/Medium/Medium
	State-based	Watermark as a property [30]	Low/High/Low/Low/Low	Low/Low/Medium/Low

Category	Subcategory	Method	a	b
FSM-based		State encoding-based [31]	Medium/High/Medium/Low/Low	Low/Low/Low/Low
	Transition-based	Unused transition-based [53]	Medium/Medium/Medium/Low/Medium	Medium/High/Medium/Low
		Existing transition-based [54]	Medium/Low/High/Medium/Medium	High/Low/Medium/Medium
		Robust watermarking [55]	High/Low/High/Medium/High	High/High/Medium/High
	Hybrid	FSM- and test-based [61]	High/Low/High/Medium/High	High/High/High/Medium
Test-based	After logic synthesis	Scan chain reordering [32]	Medium/Low/Low/Low/Low	Low/Low/Low/Low
		Scan reordering [33]	Medium/Low/Low/Low/Low	Low/Low/Low/Low
	Before logic synthesis	Synthesis-for-testability [34]	Medium/Low/High/High/Low	High/High/High/Medium
Power-based	Physical Synthesis	Power trace analysis [69]	Medium/Low/Low/High/Low	High/Low/Low/Medium
		Viability-based power [70]	Medium/High/High/Low/Low	Low/Low/Low/Low
		Input modulated [73]	Medium/Low/High/High/High	High/Low/High/High
		Spread spectrum-based [73]	Low/Low/High/High/High	High/Low/High/High

[a] Credibility/overhead/resiliency/invisibility/granularity
[b] Removal/forging/tampering/RE

3.3.2.2   Tampering Attacks

Performing a successful removal attack is typically very difficult. As a result, the attacker can tamper with the watermarked design to hide the watermark's presence, known as masking attack. The minimal number of watermark bits that must be changed to result in a successful masking attack varies depending on the detection system. A probability of masking is defined as the chance that an attack would modify or erase enough information to make the watermark invisible without degrading the design's performance excessively [74].

3.3.2.3   Forging Attacks

In forging attacks, the adversary implants his own watermark in the watermarked IP to claim his ownership to the design. The attacker may redo the watermark insertion process using his/her own signature or simply perform a ghost search for inserting the watermark. A ghost search is a way of creating an ostensibly genuine but distinct watermark depending on the detection mechanism of the targeted watermarked design and using it as the adversary's signature. The likelihood of a successful ghost search is the same as the probability of coincidence, as defined in [74].

3.3.2.4   Reverse Engineering (RE) Attacks

Another type of attack on embedded watermark of an IP is reverse engineering attack. This attack is not so common since it depends on a large number of factors. However, reverse engineering is a strong attack on embedded watermark. Firstly, it needs to be ensured that the attacker has sufficient knowledge, facilities, time, and money to make it possible to perform reverse engineering an IP. Secondly, given the attacker has sufficient knowledge, capabilities, time, and money, the attacker needs to assess whether more profit will be earned or not by selling the illegal copies of the IP than the cost of performing such reverse engineering attacks. Finally, removal of the watermark is as difficult as completely designing a specific functionality. Reverse engineering attacks are not impossible; nevertheless, it is implausible to occur in normal designs. It targets complex expensive designs and security-critical designs dedicated to cryptographic applications for instance. Moreover, all available IP watermarking solutions do not work everywhere, and more robust security solutions are required such as ciphering the FPGA bitstream [75]. It needs to be kept in mind that watermarking is a solution against illegal copying of IPs specifically, but not against IP reverse engineering which is considered to be a serious threat.

3.3.2.5 Countermeasures

Security analysis and countermeasures of an IP watermarking technique against masking and removal attacks are dependent on the watermark insertion and detection mechanism used, and they differ from case to case. In [3], authors reported a countermeasure to defend against the forging attack. If the watermarked design is forged by simply the addition of watermark, the IP owner is able to provide an IP core with only his watermark, while the attacker has only the IP core with both watermarks. The IP core clearly belongs to the IP owner. The FSM-based watermarking scheme proposed in [55] provides good resiliency against the removal and masking attacks since the watermark bits are dispersed arbitrarily and hidden in the existing transitions of the FSM. This method to the state assignments of pseudo-input variables makes it almost infeasible to attack the watermarked FSM. The length of the watermark verification pattern can be altered without reducing the watermark strength. Very similar to scan protection techniques [76, 77], the scan chain-based watermarking technique proposed in [20], based on ordering of the connections between some pairs of scan flip-flops in the scan chain, provides high watermark strength. However, this method is highly vulnerable to removal attacks.

Resiliency against the removal attack increases if the watermarking scheme is inseparable from the design and the attacker cannot distinguish it from the original design. One way to achieve this is by watermarking different parts of the design separately. Kirovski et al. [78] proposed that multiple watermarks can be inserted into different locations of the IP. Rather than inserting a large watermark, the authorship signature is broken down into a series of small watermarks, each of which is randomly supplemented into a different region of the design and can be verified separately of the rest of the design. According to the authorship information, the small watermarks are translated into sets of extra constraints and allocated to pseudo-randomly selected locations. Since just a particular location of the design is required to decode the stego constraints owing to the localized watermark in that location, the approach allows parts of the watermarked IP to be secured individually. Furthermore, because the local watermarks are independent of one another, an attacker would have to change a significant portion of the IP to remove the copyright information. Charbon [45] proposed that multiple watermarks can be inserted at multiple levels of the design, establishing the concept of hierarchical watermarking. Multiple watermarks implanted in different abstraction levels, each independent of the other, give more powerful protection for the IP. The authorship information can only be removed if the attacker is able to remove all of the watermarks in each design level. One issue with the hierarchical watermarking is that the design overhead becomes large with an increased number of watermarked levels. A similar concept to localized watermarking can be used to solve this problem. In [79], the authors have proposed another interesting approach for implementing secure DSP circuits for consumer electronics applications, considering a double line of defense: (1) key-based structural obfuscation for preventing attacks and (2) physical-level watermarking for detecting attacks. This methodology seems interesting since it provides almost zero overhead with very high tamper resistance. The key-based structural obfuscation (i.e., act as the first line of defense) ensures security against malicious intents for counterfeiting and Trojan insertion. Embedded watermark in the physical level (i.e., acting as the second line of defense) helps detect fake ICs or IPs and removes them from the supply chain.

3.4   Comprehensive Threat Models

In order to develop appropriate watermarking approach to provide proof of IP ownership, understanding an attacker's objective, the assets she has access to, the capabilities she can use to exploit the vulnerability, and the different attack methods she may employ are important. To exploit vulnerabilities, the adversary must identify the objectives, assets, and capabilities available. The IP owner, design house, SoC integrator, untrusted foundry, third-party design service provider, outsourced semiconductor assembly and tests (OSATs), and end-users can be potential antagonists against watermarking [13, 19, 80]. Based on their capabilities, we have developed a threat model consisting of the objective, asset holding, and capability for each of the entities mentioned above in Table 3.2. The following discussion defines these entities and briefly discusses their objective, capabilities, and asset holding in the threat model against watermarking.

In the supply chain of the SoC design, the "IP owner" is the person who owns the legitimate rights to the intellectual property. Various abstraction levels can be used to describe the design, from layout to HDL code. Owners of the IP provide them to design houses and permit them to use the predesigned IP cores under certain conditions. The "design house" is the entity that owns the whole system-on-chip. It purchases licenses of different third-party IPs (3PIP) from IP providers and integrates them with in-house IPs to design the complete SoC [19, 80]. In some cases, the design house outsources the SoC integration part to other companies, termed "SoC integrators," which are responsible for designing the final product for the design house. They are responsible for connecting the design house, the manufacturer, and the market. An SoC integrator can tamper, remove, or reverse engineer the watermark from the 3PIP cores if they have access to both the soft and hard IP cores as well as knowledge of each IP's functionality. Additionally, the design is subjected to a thorough functional analysis to identify bugs. Rogue designers may be able to tamper with DFT structures such as scan chains. In addition to these tools, the SoC integrator can use state-of-the-art reverse engineering software to extract netlists. These tools enable the integrator to analyze the implementation and functionality of each gate within the core. Malicious SoC designers attack 3PIP with the primary purpose of stealing IP. A rogue design house may remove the original watermark of the 3PIP owner and insert its own watermark to claim ownership proof. As a result, 3PIP vendors have always had trust issues with SoC integrators [18, 19, 63, 80, 81].

Table 3.2 Threat model for watermarking at different phases of design

Attributes	Design house	SoC integrator	Design service Provider (e.g., DFT)	Foundry	OSATs	End-user
Asset	RTL to layout	RTL to layout	Netlist/Layout	Layout	IC	IC
Objective	Piracy reuse counterfeit overproduction	Piracy reuse counterfeit overproduction	Piracy reuse	Piracy reuse overproduction	Piracy reuse overproduction	Piracy
Capability	Tampering removal reverse eng.	Tampering removal reverse eng.	Tampering removal reverse eng.	Removal reverse eng.	Reverse eng.	Reverse eng.

The SoC integrator hires third-party design service providers to perform specific design, implementation, and verification tasks [82–85]. Their access to gate-level netlists and the device's scan chain makes them capable of launching several attacks [18, 63, 86–88]. Their capability may also include netlist reverse engineering and access to failure analysis (FA) labs. The goal for attacking the hardware for a third-party service provider is IP piracy and reuse.

Due to the increased demand in consumer electronics, fabless semiconductor companies and integrated device manufacturers (IDMs) outsource their fabrication and manufacturing test services to offshore foundries and OSATs (outsourced semiconductor assembly and test) [22, 80]. The foundry requires the physical layout (GDSII) to manufacture the device, which makes them a significant suspect for IP infringement in the supply chain. They may have access to the manufacturing test patterns as well. In addition to the latest FA tools, each foundry is also capable of reverse engineering. Another asset available to the foundry is access to the DFT structures that will detect and analyze any failure in the die [13, 63]. A foundry equipped with the capabilities mentioned above can reverse engineer the chip and localize the key-storage element, key-delivery unit, key gates, interconnect, and DFT distribution, bypassing a design's security. Researchers are being urged to reexamine the threat of IP piracy by end-users due to advancements in reverse engineering over the past few years. End-users have access to the unlocked chip and any related documentation. She can access any FA tools in any industrial or academic FA lab and learn about the chip's functionality, algorithm, and implementation [13, 63].

IP authors use watermarking schemes to embed authorship information into the IP. When developing different IP watermarking schemes, designers often restrict themselves within the boundary of the IP. The IP could be used legally or illegally within an SoC with some other IPs. Verifying the watermark signature embedded into the IP will require access to the IP within the SoC [70]. Watermark designers often consider that they have access to the IP. This assumption is valid but only within a specific context. If the IP owner contracts with an SoC integrator, the IP owner can assume that some form of IP access will be given for authorship verification. IP owners can also force the SoC integrator for this access to authorship verification when forming the contract. Providing the access will benefit the SoC design house as proof of ownership of the IPs used in the SoC will eventually provide proof of ownership of the SoC (i.e., prove ownership within the contract). But the assumption may not be valid in the context of IP theft. The rogue SoC design house, which stole the IP, is not within any legal contract with the IP author and is not bound to provide easy access to the IP within the SoC to prove ownership (i.e., prove ownership outside contract). Hence, we should expect the rogue SoC integrator to actively block any path or access to the authorship verification scheme. The majority of the existing watermarking techniques rely on a challenge or specific input pattern to generate the watermark signature containing proof of ownership. When input access to the watermarked IP for the specific pattern remains ambiguous or blocked, the watermarked scheme will be rendered useless.

Table 3.3 Assumptions for different types of threat models

Watermarking methods	Access to IP	Access to IC	Prove ownership	Abstraction level	Threat model
Digital watermark	Yes	Yes/No	Within contract	RTL	Removal RE
	Yes/No	Yes	Outside contract	RTL/Gate	Removal RE
Universal watermark	No	Yes	Outside contract	RTL/Gate	Removal RE
Active watermark	Yes/No	Yes	Outside contract	RTL/Gate	Removal RE
WM for Trojan detection	Yes/No	Yes	Outside contract	RTL/Gate	Removal RE

Based on the threat models discussed above for different entities involved in the supply chain and to meet the requirements for next-generation watermarking in resisting IP piracy, we envision four emerging watermarking methods. A summary of the threat model and assumptions of these emerging watermarking techniques are outlined in Table 3.3. Further details on these envisioned watermarking techniques are discussed in the following section.

3.5   Futuristic Solutions in Watermarking

In this section, we provide a roadmap for further research on IP watermarking. Designing a robust and efficient IP watermarking method is necessary to make reusable IP watermarking secure against piracy/theft, tampering, and reverse engineering in practical scenarios. In the past two decades, the literature introduced several watermarking solutions for providing proof of ownership for mostly single-module IPs as described in Sect. 3.2. However, the assumptions made to leverage these approaches are incompatible with the modern SoC threat landscape (see Table 3.3) and must be addressed. The following aspects should be taken into consideration as high-level guidance for potential future research directions to maximize the applicability and impact of watermarking techniques in modern SoC.

3.5.1   Digital Watermark

For digital watermarking approach, the ability of an IP owner to detect and prove IP overuse in a suspect SoC can be coarsely defined by three scenarios: i) the IP owner has physical access to the SoC and access to the IP, ii) the IP owner has physical access to SoC but lacks access to the IP, and iii) IP owner does not have physical access to the SoC but has access to the IP through cloud environment. Here, each watermark evaluation scenario is further defined and explored.

3.5.1.1 Access to the IP and IC

In this scenario, the IP owner has access to the IP and the IC. So, the IP owner can easily verify the watermark. This is an optimistic scenario for the IP owner because if the IP is stolen, the attacker may not allow the IP owner an easy access to the IP. Nonetheless, this scenario is prevalent in cases where the IP owner and the SoC developer are within a contract, and the SoC developer has agreed to provide access to the IP owner to be able to extract the watermark.

In this scenario, the IP owner has physical access to the SoC and control of the IP. A software code, IP primary inputs and outputs (I/Os) and/or the scan chains, may be available for use in watermark detection and verification. Additionally, time-dependent information with physical access only (e.g., power or electromagnetic side-channel information) is also available for use. Most existing watermarking techniques (see Sect. 3.2) assume full access to the SoC and the target IP, resulting in the highest opportunity of watermark generation and extraction from a silicon design. However, it is improbable that a skilled attacker misusing a pirated IP would allow easy access to the IP through standard I/O and/or scan chain operation. Watermarks that aim to address this unlikely scenario should consider the following design requirements:

- The watermark can be implemented in register-transfer level (RTL), gate level, or layout level. If integrated into scan chain, the watermark can be embedded at the gate level during DFT insertion.
- The watermark must be detectable with primary I/O, scan chain access, or especially developed software code that runs on the processing unit.
- It should not be affected by circuit modification when optimized for power, area, and/or performance.
- Although embedded in the IP, the watermark should not be affected by logic synthesis, place and routing process, timing closure, and power closure.
- The watermark logic must be disjoint from the IP functionality. Therefore, during the watermark insertion, IP functionality must not be altered.
- The performance of the IP, where the watermark is inserted, should not be impacted. Therefore, performance overhead after watermark insertion should be negligible.
- The watermark insertion, generation, and extraction process need to be stealthy. The watermark should be extremely hard to detect, change, or remove.

3.5.1.2 Access to the IC But Not the IP

This is a more likely scenario where the IP owner has access to the IC but lacks access to the IP. We believe this is a more practical case because the attacker, who steals the IP or overuses it, should not be expected to allow the IP owner easy access to the IP. In general, this is a scenario where the IP is pirated by an attacker or a rogue SoC integrator without any contract with the IP owner. In this case, the IP owner has access to the suspect IC only and not the input–output of the embedded IP. As a result, only the inputs and outputs of the IC, incorporating the test structure, are available for proving ownership of the IP. Existing watermarking techniques (as described in Sect. 3.2) do not explicitly discuss the watermark extraction process in the scenario where the IP owner has access to the IC but lacks access to the IP. In this scenario, access to the IP may not be available when the IP is illegally used in a different IC due to the fact that attacker knowingly blocks access to the stolen IP. It may be possible to extract the watermark signature from the IP if the watermarking signature is side-channel-based [69]. However getting the signature could be more difficult for some techniques such as FSM-based and test-based [89] because the attacker may actively block the observability. Hence, attention and innovative research is required to design watermarks that prove ownership outside of the contract. In this case, following design requirements should be considered while designing and embedding watermark signatures:

- As access to the IP is not granted in this scenario, the watermark signature should be detectable without direct access to the IP.
- The watermark signature extraction could be side-channel-based [69] when access to the IP is no longer viable. Therefore, watermark signature generation circuitry must be resilient against removal and reverse engineering attacks.
- The watermark can be implemented in RTL, gate-level, or layout-level abstraction. However, the change in abstraction level must not create any additional vulnerability to the underlying watermarking technique.
- Integrated circuits go through several design, implementation, integration, and optimization steps during the IC design flow. The security of the watermarking technique should not be impacted by the SoC design flow.

3.5.1.3   Access to the IP But Not the IC

In this scenario, we assume that the verifying party (e.g., IP owner) has control over the IP but does not have physical access to the SoC. In other words, IP owner is not able to use I/Os or scan chains to apply inputs to the SoC to extract the embedded watermark. Therefore, one possible option could be accessing to the IP through cloud environment. Another possible option could be to activate the watermark circuit externally using electromagnetic (EM) pulse, laser, or alike. Moreover, IP owner can use the side-channel concept [69, 73] to detect the watermark sequencing when externally activated during verification. There has been a significant amount of research regarding remotely activated hardware Trojans reported in the literature such as Power Distribution Network (PDN) Trojan based on side channel [90]. This technique can be used to activate the watermark generation circuit remotely. Furthermore, other techniques such as EM fault injection (EMFI) and laser fault injection (LFI) can be used to activate the watermark generation circuit externally [91]. Another possible option is to obtain access to the IP watermark through embedded HOST IP in the SoC. Such access requires co-operation between IP owner and the SoC designer. This case is similar to the previous one mentioned in Sect. 3.5.1.1. In case of utilizing EM, laser, etc., watermark designer should consider the following requirements:

- The watermark can be implemented in RTL, gate-level, or layout-level abstraction.
- A controller engine is needed inside SoC to extract watermarks from IPs and transfer them to the cloud environment through HOST IP, Ethernet, or WiFi.
- In other scenarios, the watermark must be detectable only when it is externally activated (i.e., using EM, laser, or optical probe, etc.)
- Similar to the previous scenarios, the watermark should not be affected by circuit synthesis, placement, routing, and optimization process.
- Additionally, the watermark circuitry must not alter the functionality of the IP core.
- The watermark signature should be easy to detect but difficult to remove.

3.5.2 Universal Watermark

In designing a universal watermark, the IP owner should be able to detect and prove IP overuse in a suspect SoC when the IP owner has physical access only to global I/O signals (i.e., clock, Vdd, or reset) of the IC but lacks control of the IP. In this scenario, we assume that the IP owner does not have any legal contract with the SoC integrator, i.e., the SoC integrator has pirated or reverse-engineered the IP. In general, access through scan is not readily available because of multiple scan chains connected to decompressor and compactor [76, 92], making it difficult to locate and access an IP deeply embedded into the SoC. For example, a rogue SoC integrator/an attacker may change scan cell order in the IP or deactivate the JTAG interface from the board. Then, the IP owner cannot access the IP or cannot get correct stimuli into the scan chain if cells are reordered. In that case, IP watermark detection and verification is not possible through standard I/O or scan chain operation. Most of the existing watermarking techniques (see Sect. 3.2) require access to the IP to prove ownership. Therefore, we need an alternative solution for watermark design, detection, and verification without any IP access. In this scenario, watermark designers should consider the following design requirements:

- Considering the conditions for universal watermark, the signature must be detectable without primary I/O or scan chain access.
- The watermark should be detectable only when activated using global I/O signals (i.e., clock, Vdd, or reset) of the IC, possibly with side-channel information.
- The watermark should be intertwined with the existing design in such a way that the watermark circuitry is resistant against removal and tampering attacks.
- The watermark can be implemented in RTL, gate level, or layout level based on the abstraction level of integration (soft, firm, or hard IP).

- The watermark should not be affected by synthesis, integration, design, and optimization process.
- The watermark signature should be easy to detect by the IP owner but hard to identify or remove by the SoC integrator who is out of contract.

We provide an example solution for IP owners (e.g., verifying party) to detect and prove IP overuse in a suspect SoC when the IP owner has physical access to SoC but not access to her IP. In this context, an optical analysis such as photon emission or electro-optical methods can be utilized during watermark verification. First, watermarked signature can be embedded to the IP core at the netlist level. A suspect SoC can be evaluated by activating the watermark generation circuit using global signals (i.e., clock, Vdd, or reset) for the watermark signature generation. Then, we can use the electro-optical frequency map (EOFM) probing system [93] to extract the watermark from retrieving a reference map of the IC. In this approach, the optical contactless probing is used to measure magnitude and phase of EOFM markers for identifying the IP core layout localization and bit sequences that represent watermark signature. Based on this information, the watermark signature can be reconstructed, and ownership can be verified by comparing reconstructed watermark signature with the inserted watermark signature. Although the proposed approach may sound innovative and promising, it has a number of challenges before it can be used in practice. Some of those challenges include (1) reliable measurement of photon emissions, (2) addressing errors in the collected bit sequences that represent watermark signature, (3) ensuring attacker will not be able to manipulate the IP such that the signature is impacted.

3.5.3 Active Watermark

An IP watermark has always been considered "passive" as it does not stop the attacker from stealing the IP. The watermarked IP also does not change its functionality if it has been stolen and used in another SoC. The IP owners can only prove their authorship if they can have access to the IP in the SoC. If the embedded watermark is of active nature, i.e., it prevents IP piracy or changes IP functionality, it would deter IP theft in a very effective way. Hence, we envision an active watermark, when inserted into an IP, should be aware of its surrounding in an SoC. When the IP owner is making a contract with a design house, they can plan for an IP ecosystem encircling the watermarking strategy. The watermarked IP and the neighborhood IPs form an ecosystem, which should function correctly only when they maintain a certain functionality within the same SoC. For example, the SoC integrator can make the communication between the watermarked IP and the neighborhood IPs dependent on the correct signature generated by the watermarked IP. The watermarked IP will also get a response/feedback from its neighborhood IPs so that it is aware of its surrounding. So, the SoC would function correctly only when the watermarked IP generates the correct signature. Therefore, this kind of watermarking strategy can make IP theft much more difficult. Moreover, if the watermarked IP is stolen and used in a different SoC, the IP would not function correctly as it cannot verify its neighborhood IPs. This forces the attacker to steal not only the watermarked IP but also the neighborhood IPs. Furthermore, the attacker also needs to place these IPs in an SoC properly and maintaining their interdependent relationship within the same SoC. All these should make stealing the watermarked IP near impossible [94]. When developing active watermarking techniques, the designers should also consider the availability of access to the IP. Assuming easy access to the IP even when the IP is stolen may not be correct in reality as discussed in earlier sections.

3.5.4   Watermark for Trojan Detection

Watermarking techniques add some degree of overhead to the design, and IP designers may need to relax the design constraints to accommodate this overhead. These added features remain hidden, unused, and nonfunctional until they are activated or decoded to provide proof of authorship [58]. It will be an outstanding accomplishment if the features added for watermarking schemes can also thwart other security vulnerabilities. One possible approach could be to use the watermarking features to detect Trojans in the design. Although this idea mostly remains unexplored in the literature, Shukry et al. [95] proposed a watermarking strategy that can also detect Trojans in the design. The proposed watermarking scheme is particularly sensitive to manipulative design modifications, to the point where any change in the original design will impact the watermark output, hence called fragile watermarking (see Fig. 3.12). Thus, the watermarking strategy can effectively detect purposeful tampering attacks and Trojans insertion. The authors selected system FSMs to insert the fragile watermark because they are usually in charge of system control. Hardware Trojans usually modify them to change how the system works. Although the fragile watermarking technique is an excellent way to detect design alterations, it cannot detect a skillfully-inserted Trojan to manipulate the function output. In [96], a challenge–response-based watermarking technique has been proposed. This method is very sensitive to small modifications in the design, thus providing great resiliency against tampering. It also exhibits a high watermark verification accuracy with reliable detection of malicious alterations or hardware Trojans. In both of these cases, it is assumed that both the IP and IC are accessible for watermark verification. Although these research efforts perform an excellent job showing that detecting Trojans with watermarking techniques is feasible, this problem domain requires more research and development to make it more practical.

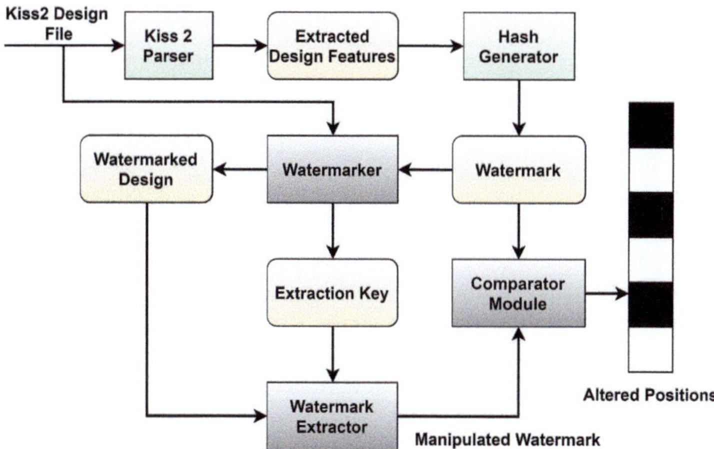

Fig. 3.12 Fragile FSM watermarking for Trojan detection [95]

3.5.5 *Side-Channel-Based Watermark*

In the side-channel-based watermark, a unique signal is embedded into the side channel of the device that serves as the watermark [97]. Common examples of side-channel parameters that can be exploited to insert or verify the watermark include power consumption, execution time, and electromagnetic emission. So far, most of the IP watermarking techniques use the power consumption signature of the IP core to watermark the IP (see Sect. 3.2.5). Other side-channel measures such as execution time, electromagnetic radiation, or contactless optical techniques can also be utilized to insert or verify watermarks. Since the embedded watermark system can cause power or area overhead, the watermark should not always be ON. Instead, it should be activated nondeterministically and for a brief amount of time. Furthermore, in the power signature-based watermark methods, the IP developers utilized the testing mode of the SoC to detect unauthorized use of their cores because they assumed that the inputs and the outputs of each IP core are not available at the finished SoC. Therefore, the IP developers by putting SoC into test mode and observing its power signature, if the core under question is contained in the SoC, will observe a very high correlation between the power signature of the core and that of the SoC [66, 67, 70]. However, using the power signature of a designer core as a watermark in the modern SoC cannot provide definite results on core detection because the power consumed by the rest cores in the SoC alters the designer core's signature. Therefore, the watermark verification in modern SoC using power signature analysis needs a further study.

3.5.6 Emerging Topics in AMS Watermarking

Analog and mixed-signal (AMS) chips are the most counterfeited parts in the semi-conductor industry. Different types of attacks such as removal, RE, counterfeiting, etc., can be performed at the system level, circuit level, or layout level assuming that the attacker has access to both IP and SoC. SoCs may contain different IPs including memory, analog, and digital IPs [59, 98]. However, there are not many techniques that have been proposed to protect IPs in this domain compared to the digital domain. Existing analog and mixed-signal watermarking techniques [99–101] require modifying the original layout. These watermarking approaches are somewhat difficult to verify and also prone to removal attacks. In the watermarking scheme proposed in [99], the watermark is embedded into the number of fingers of the transistors at the layout level. It can cause performance degradation due to a lack of proper implementation. The watermark extraction requires reverse engineering of a suspected chip which can be time-consuming. The analog watermarking technique requires addressing the unique characteristics of analog and mixed-signal designs while considering the functional differences between various classes of designs. It should be universal and applicable to AMS chips to prove the ownership, distinguish the IPs in the SoC, and detect IP piracy and overuse. For example, an encrypted secret bitstream can be converted into an analog waveform using quantization and error correction. Then, this watermarked output can be generated from an analog chip by applying a secret input dependent on that chip behavior and transfer function. In this way, the designer does not need to modify the original design for the watermarked waveform generation. They need to develop an algorithm based on the chip behavior, which can perform an exhaustive search for input signals to generate the watermarked output from that chip.

3.5.7 Automatic Cost-Effective IP Watermarking

Watermarking of embedded IPs should be capable of preserving the original functionality of the IP core with minimal area overhead. Therefore, IP watermarking approaches should imply a low overhead on the design process and the final watermarked product. Moreover, the watermark extraction time and cost should be kept low. Thus, the watermark embedding and reconstruction should be as fast, cost-effective, easy to evaluate, and ubiquitous as possible to be of any practical use. Furthermore, the watermark generation, embedding, and extraction should be easily integrated into any IP cores in modern SoCs, such as soft, firm, or hard. Therefore, it requires sufficient generic IP watermarking techniques with the very low area and timing overheads. Such techniques must be implemented in automatic design flows using CAD tools for rapid component tagging in a designer-friendly process.

3.5.8 Watermark Security Assessment

Several attacks against IP watermarking have been reported in the literature, such as removal attack, tampering attack, forging attack, and reverse engineering attack (see Sect. 3.3.2). However, few researchers have presented defense mechanisms to these attacks in the literature (see Sect. 3.3.2.5). Most of these defense mechanisms do not have high practical significance because they cannot provide comprehensive protection for IP watermarking against piracy. In this context, watermark methods should be integrated with advanced security techniques for secure IP protection. For example, in [102], authors have proposed the traceable IP protection algorithm in the Blockchain environment to improve the robustness of copyright information, where an IP core can be protected by inserting a watermark. As an attacker intends to destroy the content of IP core, first she should attack the watermark in the IP core, given that copyright requires verification. Moreover, if IP watermark techniques are ever to be deployed in critical security systems, they must resist removal, forging, and reverse engineering attacks. So far, very little has been done in this area. This is an area that requires significant research and development. Furthermore, there is a need to develop computer-aided design (CAD) frameworks for the robustness assessment of IP watermarking against all types of attacks, such as removal, forging, tampering, reverse engineering, etc.

References

1. T. Nie, Performance evaluation for IP protection watermarking techniques, in *Watermarking: Volume 2* (2012), p. 119
2. P.E. Calzada, Md. Sami U.I. Sami, K.Z. Azar, F. Rahman, F. Farahmandi, M. Tehranipoor, Heterogeneous integration supply chain integrity through blockchain and CHSM. ACM Trans. Design Autom. Electron. Syst. **29**(1), 1–25 (2023)
3. C.-H. Chang, M. Potkonjak, L. Zhang, Hardware IP watermarking and fingerprinting, in *Secure System Design and Trustable Computing* (Springer, Cham, 2016), pp. 329–368
4. M. Tehranipoor, H. Salmani, X. Zhang, *Integrated Circuit Authentication* (Springer, Cham, 2014), pp. 978–3
5. K.Z. Azar, H.M. Kamali, F. Farahmandi, M. Tehranipoor. *Understanding Logic Locking* (Springer, Berlin, 2023)
6. ESD Alliance, *ESD Alliance Electronic Design Market Data* (2021). https://www.semi.org/en/news-media-press/semi-press-releases/esda-q4-reports
7. Md. Sami U.I. Sami, F. Rahman, F. Farahmandi, A. Cron, M. Borza, M. Tehranipoor, Invited: end-to-end secure SoC life-cycle management, in *2021 58th ACM/IEEE Design Automation Conference (DAC)* (2021), pp. 1295–1298
8. J. Villasenor, M. Tehranipoor, Chop shop electronics. IEEE Spectr. **50**(10), 41–45 (2013)
9. M. Potkonjak, G. Qu, F. Koushanfar, C.-H. Chang, 20 Years of research on intellectual property protection, in *2017 IEEE International Symposium on Circuits and Systems (ISCAS)* (2017), pp. 1–4. https://doi.org/10.1109/ISCAS.2017.8050602
10. W. Hu, C.-H. Chang, A. Sengupta, S. Bhunia, R. Kastner, H. Li, An overview of hardware security and trust: threats, countermeasures, and design tools. IEEE Trans. Comput.-Aided Design Integr. Circuits Syst. **40**(6), 1010–1038 (2021). https://doi.org/10.1109/TCAD.2020.3047976

11. J. Wurm, Y. Jin, Y. Liu, S. Hu, K. Heffner, F. Rahman, M. Tehranipoor, Introduction to cyber-physical system security: a cross-layer perspective. IEEE Trans. Multi-Scale Comput. Syst. **3**(3), 215–227 (2017)

12. D. Forte, S. Bhunia, M.M. Tehranipoor, *Hardware Protection Through Obfuscation*, 1st edn. (Springer, Berlin, 2017). ISBN: 3319490184

13. M. Tanjidur Rahman, M. Sazadur Rahman, H. Wang, S. Tajik, W. Khalil, F. Farahmandi, D. Forte, N. Asadizanjani, M.M. Tehranipoor, Defense-in-depth: a recipe for logic locking to prevail. Integration **72**, 39–57 (2020)

14. D. Mehta, N. Mondol, F. Farahmandi, M. Tehranipoor, AIME: watermarking AL models by leveraging errors. in *2022 Design, Automation Test in Europe Conference Exhibition (DATE)* (2022). Preprint

15. H.M. Kamali, K.Z. Azar, F. Farahmandi, M. Tehranipoor, Advances in logic locking: past, present, and prospects (2022). Cryptology ePrint Archive

16. P. Mishra, S. Bhunia, M. Tehranipoor, *Hardware IP Security and Trust* (Springer, Berlin, 2017)

17. N. Nalla Anandakumar, M.S. Hashmi, M. Tehranipoor, FPGA-based physical unclonable functions: a comprehensive overview of theory and architectures. Integration **81**, 175–194 (2021). ISSN: 0167–9260.

18. X. Wang, D. Zhang, M. He, D. Su, M. Tehranipoor, Secure scan and test using obfuscation throughout supply chain. IEEE Trans. Comput.-Aided Design Integr. Circuits Syst. **37**(9), 1867–1880 (2017)

19. B. Colombier, L. Bossuet, Survey of hardware protection of design data for integrated circuits and intellectual properties. IET Comput. Digit. Tech. **8**(6), 274–287 (2014)

20. A. Cui, G. Qu, Y. Zhang, Dynamic watermarking on scan design for hard IP protection. IEEE Trans. Inform. Forensics Secur. **10**(11), 2298–2313 (2015)

21. S. Bhunia, M. Tehranipoor, *Hardware Security: A Hands-on Learning Approach* (Morgan Kaufmann, Los Altos, 2018)

22. Md. Tauhidur Rahman, D. Forte, Q. Shi, G.K. Contreras, M. Tehranipoor, CSST: preventing distribution of unlicensed and rejected ICs by untrusted foundry and assembly, in *2014 IEEE International Symposium on Defect and Fault Tolerance in VLSI and Nanotechnology Systems (DFT)* (IEEE. 2014), pp. 46–51

23. https://www.computerworld.com/article/2578617/cisco-sues-huawei-over-intellectual-property.html. *Cisco Sues Huawei Over Intellectual Property* (Computer World)

24. https://www.mazzarellalaw.com/blog/2018/12/intel-files-trade-secret-theft-lawsuit/. *Intel Files Trade Secret Theft Lawsuit* (Mazzarella & Mazzarella LLP)

25. https://techcrunch.com/2019/05/22/semiconductor-startup-cnex-labs-alleged-huaweis-deputy-chairman-conspired-to-steal-its-intellectual-property/. *Semi-Conductor Startup CNEX Labs Alleged Huawei's Deputy Chairman Conspired to Steal Its Intellectual Property* (TechCrunch)

26. J. Robertson, M. Riley, The big hack: How China used a tiny chip to infiltrate us companies. Bloomberg Businessweek **4**(2018) (2018)

27. G. Qu, M. Potkonjak, Analysis of watermarking techniques for graph coloring problem, in *Proceedings of the 1998 IEEE/ACM International Conference on Computer-Aided Design* (1998), pp. 190–193

28. R. Chapman, T.S. Durrani, IP protection of DSP algorithms for system on chip implementation. IEEE Trans. Signal Process. **48**(3), 854–861 (2000). https://doi.org/10.1109/78.824679

29. A. Rashid, J. Asher, W.H. Mangione-Smith, M. Potkonjak, Hierarchical watermarking for protection of DSP filter cores, in *Proceedings of the IEEE 1999 Custom Integrated Circuits Conference (Cat. No.99CH36327)* (1999), pp. 39–42. https://doi.org/10.1109/CICC.1999.777240

30. A.L. Oliveira, Techniques for the creation of digital watermarks in sequential circuit designs. IEEE Trans. Comput.-Aided Design Integr. Circuits Syst. **20**(9), 1101–1117 (2001). https://doi.org/10.1109/43.945306

31. M. Lewandowski, R. Meana, M. Morrison, S. Katkoori, A novel method for watermarking sequential circuits, in *2012 IEEE International Symposium on Hardware-Oriented Security and Trust* (2012), pp. 21–24. https://doi.org/10.1109/HST.2012.6224313

32. D. Kirovski, M. Potkonjak, Intellectual property protection using watermarking partial scan chains for sequential logic test generation, in *High Level Design, Test Verification* (1998)

33. A. Cui, C.-H. Chang, Intellectual property authentication by watermarking scan chain in design-for-testability flow (2008), pp. 2645–2648. https://doi.org/10.1109/ISCAS.2008.4542000

34. C.-H. Chang, A. Cui, Synthesis-for-testability watermarking for field authentication of VLSI intellectual property. IEEE Trans. Circuits Syst. I: Regul. Papers **57**(7), 1618–1630 (2010). https://doi.org/10.1109/TCSI.2009.2035415

35. G. Qu, L. Yuan, Secure hardware IPs by digital watermark, in *Introduction to Hardware Security and Trust* (Springer, Berlin, 2012), pp. 123–141

36. F. Koushanfar, I. Hong, M. Potkonjak, Behavioral synthe-sis techniques for intellectual property protection. ACM Trans. Design Autom. Electron. Syst. **10**(3), 523–545 (2005)

37. T.-B. Huynh, T.-T. Hoang, T.-T. Bui, A constraint-based watermarking technique using Schmitt Trigger insertion at logic synthesis level, in *2013 International Conference on Advanced Technologies for Communications (ATC 2013)* (IEEE. 2013), pp. 115–120

38. D. Kirovski, Y.-Y. Hwang, M. Potkonjak, J. Cong, Protecting combinational logic synthesis solutions. IEEE Trans. Comput.-Aided Design Integr. Circuits Syst. **25**(12), 2687–2696 (2006)

39. A. Cui, C.H. Chang, S. Tahar, IP watermarking using incremental technology mapping at logic synthesis level. IEEE Trans. Comput.-Aided Design Integr. Circuits Syst. **27**(9), 1565–1570 (2008)

40. S. Meguerdichian, M. Potkonjak, Watermarking while preserving the critical path, in *Proceedings of the 37th Annual Design Automation Conference* (2000), pp. 108–111

41. M. Ni, Z. Gao, Constraint-based watermarking technique for hard IP core protection in physical layout design level, in *Proceedings. 7th International Conference on Solid-State and Integrated Circuits Technology, 2004*, vol. 2 (IEEE, Piscataway, 2004), pp. 1360–1363

42. J.X. Zheng, M. Potkonjak, Securing netlist-level FPGA design through exploiting process variation and degradation, in *Proceedings of the ACM/SIGDA international symposium on Field Programmable Gate Arrays* (2012), pp. 129–138

43. A.B. Kahng, J. Lach, W. Mangione-Smith, S. Mantik, I.L. Markov, M. Potkonjak, P. Tucker, H. Wang, G. Wolfe, Constraint-based water-marking techniques for design IP protection. IEEE Trans. Comput.-Aided Design Integr. Circuits Syst. **20**(10), 1236–1252 (2001)

44. M. Tehranipoor, C. Wang, *Introduction to Hardware Security and Trust* (Springer, Berlin, 2011)

45. E. Charbon, Hierarchical watermarking in IC design, in *Proceedings of the IEEE 1998 Custom Integrated Circuits Conference (Cat. No. 98CH36143)* (IEEE, Piscataway, 1998), pp. 295–298

46. I. Hong, M. Potkonjak, Techniques for intellectual property protection of DSP designs, in *Proceedings of the 1998 IEEE International Conference on Acoustics, Speech and Signal Processing, ICASSP'98 (Cat. No. 98CH36181)*, vol. 5 (IEEE, Piscataway, 1998), pp. 3133–3136

47. K.Z. Azar, H.M. Kamali, H. Homayoun, A. Sasan, SMT attack: next generation attack on obfuscated circuits with capabilities and performance beyond the SAT attacks, in *IACR Transactions on Cryptographic Hardware and Embedded Systems (TCHES)* (2019), pp. 97–122

48. H.M. Kamali, K.Z. Azar, H. Homayoun, A. Sasan, Full-lock: hard distributions of SAT instances for obfuscating circuits using fully configurable logic and routing blocks, in *Proceedings of Design Automation Conference (DAC)* (2019), p. 89

49. H.M. Kamali, K.Z. Azar, H. Homayoun, A. Sasan, InterLock: an intercorrelated logic and routing locking, in *IEEE/ACM International Conference on Computer-Aided Design (ICCAD)* (2020), pp. 1–9

50. H. Dogan, D. Forte, M.M. Tehranipoor, Aging analysis for recycled FPGA detection, in *2014 IEEE International Symposium on Defect and Fault Tolerance in VLSI and Nanotechnology Systems (DFT)* (2014), pp. 171–176

51. M.S. Rahman, R. Guo, H.M. Kamali, F. Rahman, F. Farahmandi, M. Tehranipoor, ReTrustFSM: toward RTL hard-ware obfuscation-a hybrid FSM approach. IEEE Access **11**, 19741–19761 (2023)

52. H.M. Kamali, K.Z. Azar, H. Homayoun, A. Sasan, SCRAMBLE: the state, connectivity and routing augmentation model for building logic encryption, in *IEEE Computer Society Annual Symposium on VLSI (ISVLSI)* (2020), pp. 153–159

53. I. Torunoglu, E. Charbon, Watermarking-based copyright protection of sequential functions. IEEE J. Solid-State Circuits **35**(3), 434–440 (2000). https://doi.org/10.1109/4.826826

54. A.T. Abdel-Hamid, S. Tahar, E.M. Aboulhamid, Finite state machine IP watermarking: a tutorial, in *First NASA/ESA Conference on Adaptive Hardware and Systems (AHS'06)* (2006), pp. 457–464. https://doi.org/10.1109/AHS.2006.40

55. A. Cui, C.-H. Chang, S. Tahar, A.T. Abdel-Hamid, A robust FSM watermarking scheme for IP protection of sequential circuit design. IEEE Trans. Comput.-Aided Design Integr. Circuits Syst. **30**(5), 678–690 (2011). https://doi.org/10.1109/TCAD.2010.2098131

56. M. Shayan, K. Basu, R. Karri, Hardware trojans inspired IP watermarks. IEEE Design Test **36**(6), 72–79 (2019). https://doi.org/10.1109/MDAT.2019.2929116

57. H. Salmani, M. Tehranipoor, J. Plusquellic, A layout-aware approach for improving localized switching to detect hardware Trojans in integrated circuits, in *2010 IEEE International Workshop on Information Forensics and Security* (IEEE, Piscataway, 2010), pp. 1–6

58. C. Lamech, R.M. Rad, M. Tehranipoor, J. Plusquellic, An experimental analysis of power and delay signal-to-noise requirements for detecting Trojans and methods for achieving the required detection sensitivities. IEEE Trans. Inform. Forensics Secur. **6**(3), 1170–1179 (2011)

59. Md. Tauhidur Rahman, K. Xiao, D. Forte, X. Zhang, J. Shi, M. Tehranipoor, TI-TRNG: technology independent true random number generator, in *2014 51st ACM/EDAC/IEEE Design Automation Conference (DAC)* (IEEE, Piscataway, 2014), pp. 1–6

60. N. Nalla Anandakumar, S.K. Sanadhya, M.S. Hashmi, FPGA-based true random number generation using programmable delays in oscillator-rings. IEEE Trans. Circuits Syst. II: Express Briefs **67**(3), 570–574 (2020)

61. A. Cui, C.-H. Chang, L. Zhang, A hybrid watermarking scheme for sequential functions, in *2011 IEEE International Symposium of Circuits and Systems (ISCAS)* (2011), pp. 2333–2336. https://doi.org/10.1109/ISCAS.2011.5938070

62. A. Nahiyan, K. Xiao, K. Yang, Y. Jin, D. Forte, M. Tehranipoor, AVFSM: a framework for identifying and mitigating vulnerabilities in FSMs, in *2016 53rd ACM/EDAC/IEEE Design Automation Conference (DAC)* (IEEE, Piscataway, 2016), pp. 1–6

63. Y.-C. Fan, Testing-based watermarking techniques for intellectual-property identification in SOC design. IEEE Trans. Instrum. Measur. **57**(3), 467–479 (2008)

64. M.L. Bushnell, V.D. Agrawal, *Essentials of Electronic Testing for Digital, Memory and Mixed-Signal VLSI Circuits* (Springer, Berlin, 2002). https://doi.org/978-0-306-47040-0

65. P.C. Kocher, J. Jaffe, B. Jun, Differential power analysis, in *Advances in Cryptology - CRYPTO '99, 19th Annual International Cryptology Conference, Santa Barbara, California, USA, August 15–19, 1999, Proceedings*, vol. 1666. Lecture Notes in Computer Science (Springer, Berlin, 1999), pp. 388–397

66. N. Ahmed, M. Tehranipoor, V. Jayaram, Supply voltage noise aware ATPG for transition delay faults, in *25th IEEE VLSI Test Symposium (VTS'07)* (2007), pp. 179–186

67. J. Ma, J. Lee, M. Tehranipoor, Layout-aware pattern generation for maximizing supply noise effects on critical paths, in *2009 27th IEEE VLSI Test Symposium* (2009), pp. 221–226

68. N. Nalla Anandakumar, SCA resistance analysis on FPGA implementations of sponge based MAC-PHOTON, in *8th International Conference, SECITC 2015, Bucharest, Romania, June 11–12* (Springer, Cham, 2015), pp. 69–86

69. D. Ziener, J. Teich, FPGA core watermarking based on power signature analysis, in *2006 IEEE International Conference on Field Programmable Technology* (2006), pp. 205–212. https://doi.org/10.1109/FPT.2006.270313

70. G. Blanas, H.T. Vergos, Extending the viability of power signature-based IP watermarking in the SoC era, in *2016 IEEE International Conference on Electronics, Circuits and Systems (ICECS)* (2016), pp. 281–284
71. X. Zhang, M. Tehranipoor, Design of on-chip lightweight sensors for effective detection of recycled ICs. IEEE Trans. Very Large Scale Integr. Syst. **22**(5), 1016–1029 (2013)
72. C. Marchand, L. Bossuet, E. Jung, IP watermark verification based on power consumption analysis, in *2014 27th IEEE International System-on-Chip Conference (SOCC)* (2014), pp. 330–335. https://doi.org/10.1109/SOCC.2014.6948949
73. G.T. Becker, M. Kasper, A. Moradi, C. Paar, Side-channel based watermarks for integrated circuits, in *2010 IEEE International Symposium on Hardware-Oriented Security and Trust (HOST)* (IEEE, Piscataway, 2010), pp. 30–35
74. A.T. Abdel-Hamid, S. Tahar, E.M. Aboulhamid, A survey on IP watermarking techniques. Des. Autom. Embedd. Syst. **9**(3), 211–227 (2004)
75. L. Bossuet, G. Gogniat, W. Burleson, Dynamically configurable security for SRAM FPGA bitstreams, in *18th International Parallel and Distributed Processing Symposium, 2004. Proceedings* (2004), p. 146. https://doi.org/10.1109/IPDPS.2004.1303128
76. K.Z. Azar, H.M. Kamali, H. Homayoun, A. Sasan, From cryptography to logic locking: a survey on the architecture evolution of secure scan chains.IEEE Access **9**, 73133–73151 (2021)
77. S. Roshanisefat, H.M. Kamali, K.Z. Azar, S.M.P. Dinakarrao, N. Karimi, H. Homayoun, A. Sasan, DFSSD: deep faults and shallow state duality, a provably strong obfuscation solution for circuits with restricted access to scan chain, in *VLSI Test Symposium (VTS)* (2020), pp. 1–6
78. D. Kirovski, M. Potkonjak, Local watermarks: methodology and application to behavioral synthesis. IEEE Trans. Comput.-Aided Design Integr. Circuits Syst. **22**(9), 1277–1283 (2003). https://doi.org/10.1109/TCAD.2003.816208
79. A. Sengupta, M. Rathor, Enhanced security of DSP circuits using multi-key based structural obfuscation and physical-level watermarking for consumer electronics systems. IEEE Trans. Consum. Electron. **66**(2), 163–172 (2020). https://doi.org/10.1109/TCE.2020.2972808
80. U. Guin, D. DiMase, M. Tehranipoor, A comprehensive framework for counterfeit defect coverage analysis and detection assessment. J. Electron. Testing **30**(1), 25–40 (2014)
81. K.Z. Azar, F. Farahmand, H.M. Kamali, S. Roshanisefat, H. Homayoun, W. Diehl, K. Gaj, A. Sasan, COMA: communication and obfuscation management architecture, in *International Symposium on Research in Attacks, Intrusions and Defenses (RAID)* (2019), pp. 181–195
82. K.Z. Azar, M.M. Hossain, A. Vafaei, H. Al Shaikh, N.N. Mondol, F. Rahman, M. Tehranipoor, F. Farahmandi, Fuzz, penetration, and AI testing for SoC security verification: challenges and solutions (2022). Cryptology ePrint Archive
83. M.M. Hossain, A. Vafaei, K.Z. Azar, F. Rahman, F. Farahmandi, M. Tehranipoor, SoCFuzzer: SoC vulnerability detection using cost function enabled fuzz testing, in *2023 Design, Automation & Test in Europe Conference & Exhibition (DATE)* (IEEE, Piscataway, 2023), pp. 1–6
84. H. Al-Shaikh, A. Vafaei, M. Md Mashahedur Rahman, K.Z. Azar, F. Rahman, F. Farahmandi, M. Tehranipoor, Sharpen: SOC security verification by hardware penetration test, in *Proceedings of the 28th Asia and South Pacific Design Automation Conference* (2023), pp. 579–584
85. N.N. Mondol, A. Vafaei, K.Z. Azar, F. Farahmandi, M. Tehranipoor, RL-TPG: automated pre-silicon security verification through reinforcement learning-based test pattern generation, in *Design, Automation and Test in Europe (DATE)* (IEEE, Piscataway, 2024), pp. 1–6
86. S. Roshanisefat, H.M. Kamali, H. Homayoun, A. Sasan, RANE: an open-source formal de-obfuscation attack for reverse engineering of logic encrypted circuits, in *Great Lakes Symposium on VLSI (GLSVLSI)* (2021), pp. 221–228
87. K.Z. Azar, H.M. Kamali, H. Homayoun, A. Sasan, NNgSAT: neural network guided SAT attack on logic locked complex structures, in *IEEE/ACM International Conference On Computer Aided Design (ICCAD)* (2020), pp. 1–9

88. K.Z. Azar, H.M. Kamali, F. Farahmandi, M. Tehranipoor, Warm up before circuit de-obfuscation? An exploration through bounded-model- checkers, in *International Symposium on Hardware Oriented Security and Trust (HOST)* (2022), pp. 1–4

89. P.P. Sarker, U. Das, M.B. Monjil, H.M. Kamali, F. Farahmandi, M. Tehranipoor, GEM-water: generation of em-based watermark for SoC IP validation with hidden FSMs, in *ISTFA 2023* (ASM International, 2023), pp. 271–278

90. F. Schellenberg, D.R.E. Gnad, A. Moradi, M.B. Tahoori, An inside job: remote power analysis attacks on FPGAs, in: *2018 Design, Automation Test in Europe Conference Exhibition (DATE)* (2018), pp. 1111–1116. https://doi.org/10.23919/DATE.2018.8342177

91. J. Lee, S. Narayan, M. Kapralos, M. Tehranipoor, Layout-aware, IR-drop tolerant transition fault pattern generation, in *2008 Design, Automation and Test in Europe* (2008), pp. 1172–1177

92. P. Srinivasan, R. Farrell, Hierarchical DFT with combinational scan compression, partition chain and RPCT, in *2010 IEEE Computer Society Annual Symposium on VLSI* (2010), pp. 52–57

93. A. Stern, D. Mehta, S. Tajik, F. Farahmandi, M. Tehranipoor, SPARTA: a laser probing approach for trojan detection, in *2020 IEEE International Test Conference (ITC)* (2020), pp. 1–10. https://doi.org/10.1109/ITC44778.2020.9325222

94. Z. Ibnat, M. Sazadur Rahman, M.M. Rahman, H.M. Kamali, M. Tehranipoor, F. Farahmandi, ActiWate: adaptive and design-agnostic active watermarking for IP ownership in modern SoCs, in *2023 60th ACM/IEEE Design Automation Conference (DAC)* (IEEE, Piscataway, 2023), pp. 1–6

95. S.M. Hussein Shukry, A.T. Abdel-Hamid, M. Dessouky, Affirming hardware design authenticity using fragile IP watermarking, in *2018 International Conference on Computer and Applications (ICCA)* (2018), pp. 1–347. https://doi.org/10.1109/COMAPP.2018.8460369

96. A. Nair, Patanjali SLPSK, C. Rebeiro, S. Bhunia, SIGNED: a challenge-response based interrogation scheme for simultaneous watermarking and trojan detection (2020)

97. U. Das, M Sazadur Rahman, N. Nalla Anandakumar, K.Z. Azar, F. Rahman, M. Tehranipoor, F. Farahmandi, PSC-watermark: power side channel based IP watermarking using clock gates, in *2023 IEEE European Test Symposium (ETS)* (IEEE, Piscataway, 2023), pp. 1–6

98. U. Guin, X. Zhang, D. Forte, M. Tehranipoor, Low-cost on-chip structures for combating die and IC recycling, in *2014 51st ACM/EDAC/IEEE Design Automation Conference (DAC)* (IEEE, Piscataway, 2014), pp. 1–6

99. D.L. Irby, R.D. Newbould, J.D. Carothers, J.J. Rodriguez, W.T. Holman, Low level watermarking of VLSI designs for intellectual property protection, in *Proceedings of 13th Annual IEEE International ASIC/SOC Conference (Cat. No. 00TH8541)* (IEEE, Piscataway, 2000), pp. 136–140

100. R.D. Newbould, D.L. Irby, J.D. Carothers, J.J. Rodriguez, W.T. Holman, Mixed signal design watermarking for IP protection. Integr. Comput.-Aided Eng. **10**(3), 249–265 (2003)

101. U. Das, Md Rafid Muttaki, M.M. Tehranipoor, F. Farahmandi, ADWIL: a zero-overhead analog device watermarking using inherent IP features, in *2022 IEEE International Test Conference (ITC)* (IEEE, Piscataway, 2022), pp. 155–164

102. L. Xiao, W. Huang, Y. Xie, W. Xiao, K.-C. Li, A blockchain-based traceable IP copyright protection algorithm. IEEE Access **8**, 49532–49542 (2020). https://doi.org/10.1109/ACCESS.2020.2969990

Chapter 4
SoC Security Verification Using Fuzz, Penetration, and AI Testing

4.1 Introduction to SoC Security Verification

With the ever-increasing adoption and utilization of the integrated circuit (IC) supply chain horizontal model through the last decade, a wide range of parties and entities have to be involved to contribute and accomplish part(s) of the various stages of design, fabrication, testing, packaging, and integration, which results in forming expansive globalization and distribution of the semiconductor supply chain. Globalization by outsourcing significantly reduces the logistic/manufacturing cost and time-to-market and allows companies access to advanced nodes, specialized resources, and cutting-edge technologies [1]. However, the main impact of globalization is the loss of control/trust that raises a wide range of threats in the supply chain, including IP piracy, IC overproduction, malicious functionality insertion, etc., that can induce catastrophic consequences with huge financial/reputation loss [2].

Over the past decade, as demonstrated in Fig. 4.1, the same trend has been witnessed for (i) the involvement of numerous entities, (ii) complexity and heterogeneity of modern SoCs, (iii) re-usage of existing third-party IPs, (iv) the security threats variety, and (v) spending growth against the threats. As demonstrated, this trend faces a significantly higher rate in recent years showing why urgent action is required for automation of the SoC security verification in the current time frame. While the SoC designs and architectures are getting larger, more complex, and more heterogeneous, with a variety of IPs that have several highly sensitive assets such as encryption keys, device configurations, application firmware (FW), e-fuses, one-time programmable memories, and on-device protected data, guaranteeing the security of these assets against any unauthorized access or any attack is paramount. As demonstrated in Fig. 4.2, a typical SoC architecture comes with numerous security IPs, such as cryptographic cores (encryption/decryption), True Random Number Generator (TRNG) modules, Physical Unclonable Function (PUF) units, one-time memory blocks, etc. With either generation, propagation, or usage of the assets by these IPs, the distribution of these security assets may happen across the

M. Tehranipoor et al., *Hardware Security*,
https://doi.org/10.1007/978-3-031-58687-3_4

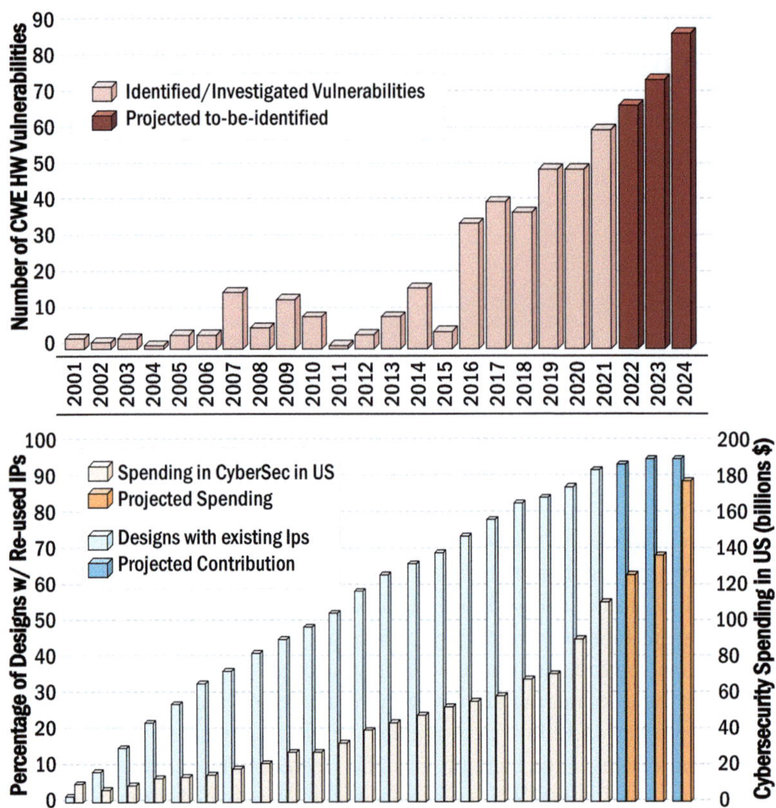

Fig. 4.1 IP reuse vs. threats vs. security-related spending

SoC architecture, and since some of these IPs are from third-party vendors, any unauthorized or malicious access to these assets can result in company trade secrets for device manufacturers or content providers being leaked [3].

The involvement of third-party vendors, which compromises the trustworthiness of the entire SoC, is not the only source of vulnerability in the SoC architecture. Security vulnerabilities can also emerge as the consequence of design/implementation/integration drawbacks through different stages of the SoC design flow, which makes the designing of a secure SoC more challenging. These vulnerabilities can also emerge in different forms, such as information leakage, access control violation, side-channel or covert channel leakage, presence or insertion of malicious functions, exploiting test and debug structure, and fault-injection-based attacks [4–8]. Some of these security vulnerabilities may be introduced unintentionally by a designer during the transformation of a design from specification to implementation. This is because the designers at the design house spend most of their design, implementation, and integration efforts to meet area, power, and performance criteria. For verification, they mostly focus on functional correctness [9]. Additionally, the

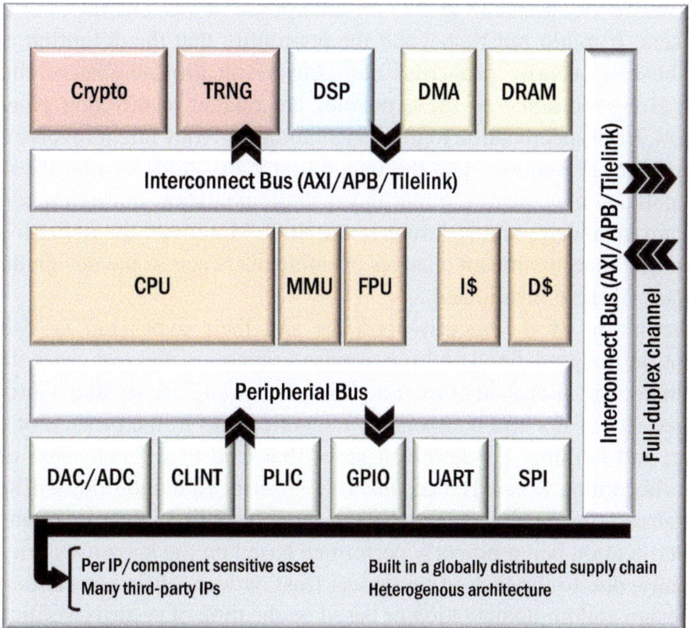

Fig. 4.2 An SoC design with the integration of a wide variety of IPs

flaws in the computer-aided design (CAD) tools can unintentionally result in the emergence of additional vulnerabilities in SoCs. Moreover, rogue insiders in the design houses can intentionally perform malicious design modifications that create a backdoor to the design. An attacker can utilize the backdoor to leak critical information, alter functionalities, and control the system.

It should be noted that many of the existing security countermeasures introduced in the literature or widely used in the industry, such as hardware obfuscation, watermarking, metering, camouflaging, etc. [10–16], have nothing to do with such SoC-level vulnerabilities, as many of these security vulnerabilities originate precisely from unexpected interactions between layers and components, and traditional techniques fail at catching these cross-layer problems or do not scale to real-world designs. Therefore, apart from the existing hardware security countermeasures that might be applied to the design, the security verification is required to be evaluated meticulously, particularly for the SoC-level vulnerabilities [17, 18].

Considering that the different IP components of an SoC have their own highly sensitive security assets that should be accessed/exploited by some other components, most system design specifications include a (limited) number of security policies that define access constraints and permissions to these assets at different phases during the system execution. As SoC complexity continues to grow and time-to-market shrinks, verification of these security policies has become even more difficult to ensure that the design under investigation does not violate these policies, whose violation can lead to the emergence of vulnerabilities or backdoors for attackers. It might not be beyond the possibility that the definition of a set of well-established security properties can help verification engineers alleviate the problem. However, some of these policies are related to different phases of the design flow, from specification to implementation, and they might involve the design house, different IP vendors, and the integration team. Additionally, the realization of these policies may require a multilayer implementation through a combination of hardware, firmware, and software in the SoC. Moreover, the definition of these policies might face significant changes or refinements across the design flow, which makes many of them invalidated.

The definition of the security policies and their expansion is also entirely dependent on the predefined and pre-assessed scenarios and test cases that consist of prohibited and illegal actions. For SoCs becoming larger and more complex with larger sets of IPs and components integrated into it, the process of detection, gathering, and building all these test cases that lead to the definition of security policies is becoming worse. Hence, *unknown* security vulnerabilities will still appear in SoC transactions, leading to breaches of confidentiality, integrity, or authenticity, despite verification being properly performed based on the known security policies. Additionally, due to the lack of reciprocal trust between different entities involved in SoC design and implementation or based on the time of security verification (i.e., pre- or post-verification), the access of the security verification engine/tool will vary to the system, from full access with knowledge about all internal operations, wires, registers, etc. to *NO* access to the internal knowledge of the system. The access differs case by case; however, in all cases, it will affect the outcome of security verification, in terms of security policies conformance, performance, the complexity of security verification flow, etc.

A literature review on the software testing domain reveals that the procedure of software testing has been suffering for almost two decades from the same challenges and difficulties. Software testing is the main way of verifying software, from specification to release, against the defined requirements and accounts for about half of the cost and time of development [19, 20]. For instance, considering the main objective of software testing and verification, which is systematically evaluating the software, in carefully controlled circumstances, the scope of software testing is heavily dependent on the knowledge and internal access, which categorizes the software testing into three main breeds, namely (i) white box with full knowledge of code, (ii) black box with no knowledge relevant to the internal structure of the code, and (iii) gray box with limited knowledge of the internal structure. In this case, the automation of verification depends on the breadth of testing. Recently, in software testing, with the emergence of automated and semi-automated techniques, like the usage of self-guidance and self-refinement concepts (e.g., artificial intelligence, machine learning, the mutation-based approaches like fuzzing [21]), the testing procedure has been evolved tremendously in this domain. These techniques are highly successful in detecting software vulnerabilities since they are automated and scalable to large codebases, do not require the knowledge of the underlying system, and are highly efficient in detecting many security vulnerabilities.

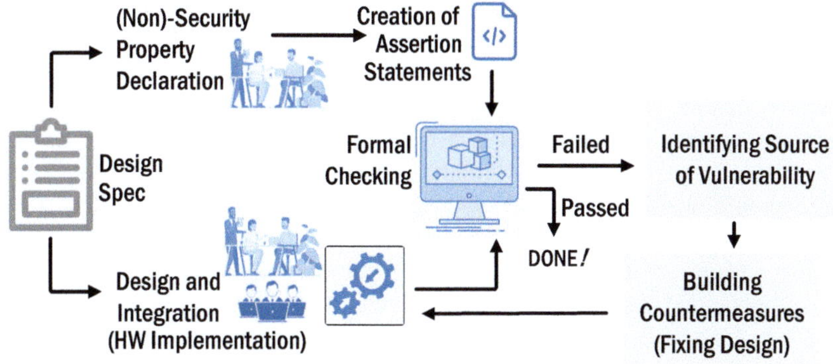

Fig. 4.3 Property-driven functional verification

The utilization of existing techniques for security verification of modern SoCs is mainly limited to the expert review, and they do not provide acceptable scalability [22]. The top view of such techniques has been demonstrated in Fig. 4.3. In such solutions, the conventional formal verification techniques, in spite of significant recent advances in automated formal technologies (functional verification) such as satisfiability (SAT) checking and satisfiability modulo theories (SMT), that are also widely used for the evaluation of IP protection techniques (hardware obfuscation) [8, 23, 24] cannot promise the desired scalability for the security verification of modern SoCs, and the gap between the scale and complexity of modern SoC designs and those which can be handled by formal verification techniques has continued to grow. Similarly, symbolic execution and model checking [25] are suffering from scalability for verification at the SoC level. As an instance, the work in [4, 26] proposed a technique to analyze the vulnerabilities of FSM, called AVFSM. However, apart from the FSM, an SoC contains other modules (exception handlers, test and debug infrastructures, random number generators, etc.) which must be inspected during security verification. The authors in [27] provided a method to write the security-critical properties of the processor. They found that the quality of security properties is as good as the developer's knowledge and experience. Moreover, there is a lack of comprehensive threat model definition, which must be considered while developing security properties. There are some other approaches that have developed security properties and metrics by considering only a small subset of vulnerabilities (e.g., the vulnerability in hardware crypto-systems [28], side-channel vulnerabilities [29, 30], and abstraction level limitations like the behavioral model [31]).

Also, since many of the security vulnerabilities in the SoC originate precisely from unexpected interaction between layers and components, identifying novel methodologies is hard because researchers often do not have enough access to all parts of the system, which is particularly true for proprietary hardware microarchitectures. The authors in [32] demonstrate how commercially available tools can fail to detect security-relevant RTL bugs including some that originate from cross-modular and cross-layer interactions. The work in [33] presented a methodology to infer security-critical properties from known processor errata. However, manufacturers only release errata documents enumerating known bugs. Unknown security vulnerabilities can exist in an SoC design that are not listed in available errata documents. Another approach for finding security bugs is information flow tracking (IFT) techniques. The authors in [34–38] utilize IFT and statistical models to map hardware security properties. However, this technique requires design instrumentation and tainting all the input variables, which require more computational time and memory resources. Hence, IFT and statistical modeling, which requires expert knowledge of the design, become more complex with increasing design complexity. There is an increasing need for methodologies and frameworks for security verification of modern SoCs that are scalable to large and complex designs, highly automatic, effective, and efficient in detecting security-critical vulnerabilities.

Based on the trend of software verification testing techniques and their efficacy for the evaluation of software specifications and requirements, it is evident that the same but futuristic trend will potentially happen in the area of SoC security verification. However, there definitely exist numerous limitations and challenges, in terms of the concept migration, implementation, assumptions, metrics, and the outcome. Hence, with such a gap, and due to notable lack of detailed and comprehensive evaluation on SoC security verification, in this chapter, we will examine and reevaluate the principles and fundamentals of SoC security verification through automated and semi-automated architectures. Moving forward, with the ever-increasing complexity and size of SoCs and with the contribution of more and more IPs and less and less trustworthiness between components, SoC security verification through a more closed environment, like the gray- and black-box models, will get more attention. Hence, in this chapter, with more focus on such models, and by trying to get the benefit of semi-automated or automated approaches, like AI or ML, fuzz testing, and penetration testing, we provide a comprehensive overview of SoC security verification as follows [39–43]:

1. We first identify the source of vulnerabilities indicating the necessity of an automated verification framework for the SoC security verification.
2. We then define the assumptions for SoC security verification based on different factors like the designer's desire, followed by the requirements of the framework, such as scalability, high coverage, etc.
3. We examine the possibility of engaging software approaches, with more specific concentration on self-guided or self-refinement approaches, such as fuzz, penetration, and AI testing for SoC security verification.
4. We discuss about the future research directions and challenges for implementing an automated verification framework to identify security vulnerabilities based on the self-refinement approaches.

Figure 4.4 first shows the main steps of a modern IC supply chain which is plunged in globalization with the involvement of multiple IPs. As also demonstrated

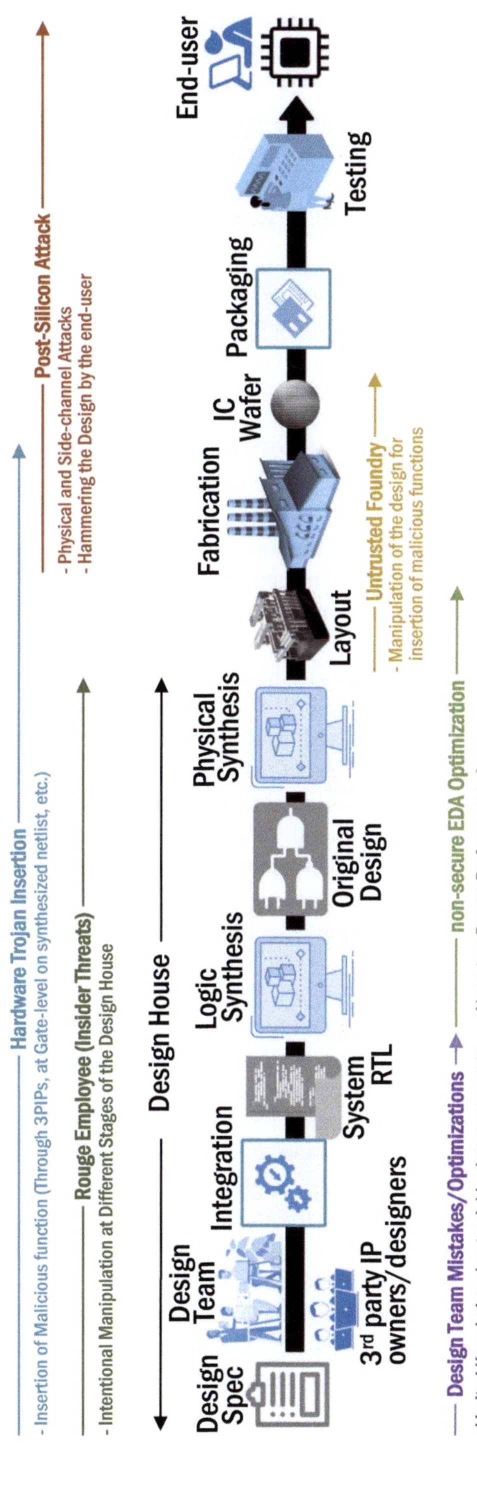

Fig. 4.4 Potential threats within SoC design flow

in Fig. 4.4, an SoC design can encounter security vulnerabilities during different stages of its design and lifecycle; each is excited from a unique source. Beginning from the very early stages of the IC design to its fabrication, the following are the main sources of such security vulnerabilities in the SoC design and implementation:

(**SV1**) *Inadvertent designer mistakes*: The development of multiple IPs/components may be distributed between different design teams and third-party IPs. This can result in a non-clear definition over the interaction between these IPs, non-sophisticated exception handling for inter-component communications, limited knowledge about the behavior of neighboring components, (communication) protocol malfunctioning, and lack of understanding of security problems due to the high complexity of the designs and a variety of assets. Hence, it may result in different forms of vulnerabilities, such as secret information leakage or losing the reliability of the SoC.

(**SV2**) *Rogue employee (insider threat)*: Unlike (**SV1**), deliberate malfunction can be invoked by the rogue employee(s) that pose significant security threats to the security of the whole SoC.

(**SV3**) *Untrusted third-party IP vendors*: Similar to (**SV2**), with violation of rules/protocols of communication between their IPs and other components, or unauthorized interactions with no monitoring, third-party IPs can pose the same security issues.

(**SV4**) *EDA optimizations*: Almost all efforts in the development and improvement of CAD tools have been directed to optimize and increase the efficiency of synthesis, floor planning, placement, and routing in terms of area, power, throughput, delay, and performance. These CAD tools are not well equipped with the understanding of the security vulnerabilities [26] of integrated circuits (ICs) and can, therefore, introduce additional vulnerabilities in the design. As an instance, the CAD tools can potentially but unintentionally merge trusted blocks with the untrusted ones or move a security asset to an untrusted observable point, which opens the possibility of different attacks like eavesdropping on the side-channel analysis.

(**SV5**) *Security compromising stages of IC supply chain*: For modern SoCs, design-for-test (DFT) and design-for-debug (DFD) are designed to increase the observability and controllability for post-silicon verification and debug efforts. However, increasing the observability and preserving security are two conflicting factors. Test and debug infrastructures may create confidentiality and integrity violations. For example, in [31], scan chains have been exploited to leak security-critical information, and numerous studies through the last decade have evaluated the utilization of DFT/DFD infrastructure for attacking the design [13]. Similar to (**SV4**), many of these vulnerabilities are emerged unintentionally due to the accomplishment of different stages of the IC design flow.

(**SV6**) *Impact of hardware Trojan insertion*: This case is a derivation of (**SV2**) and (**SV3**), in which SoC designs are also prone to many maliciously introduced threats, such as hardware Trojans. These hardware Trojans can be inserted by untrusted entities involved in a number of design and supply chain stages including third-party IP (3PIP) vendors and rogue in-house employees causing sensitive information leakage, denial of service, reduction in reliability, etc. in the SoC. Also, insider threats are particularly dangerous since they have full observability and access to the whole design and source files. When a chip is deployed into the final design and Trojan was inserted during stages, an attacker can monitor physical characteristics of a design (such as delays, power consumption, transient current leakage) to recover secret information.

(**SV7**) *Lack of trustworthiness in EDA Tools*: While it's generally assumed that CAD tools are trusted nodes within the IC design flow, updating the SoC design infrastructure to support cloud-based development and employing fully distributed processes with remote EDA tools challenges this assumption. In such cases, software tools cannot be considered trusted anymore (the current assumption is that even for cloud-based systems, it is just assumed that the cloud infrastructure is secure, which is not always correct), and the SoC designs are accordingly prone again to numerous malicious threats.

(**SV8**) *Untrusted fabrication site*: The consequence of this case is similar to (**SV7**). In this case, the fab site as an untrusted entity, with full knowledge about the layout of the design, can apply manipulation before the fabrication for further usage after the fabrication.

(**SV9**) *Efforts on SoC design optimization*: Most of the designers are always concentrating on minimizing the overhead in terms of area, power, and performance. However, in many of these optimization cases done by the designer, the outcome can possibly incur forms of security issues, such as sharing memory banks between non-trusting processes in multi-tenanted systems that were the main consequence of Spectre and Meltdown attacks [44, 45].

(**SV10**) *Hammering by the end-user*: This group of threats can be related to either logical or physical attributes and specification of the components integrated into the SoC, such as memory. Similar to rowhammer attack on DRAMs, logic-level hammering and repeating of different sequences of action(s) around the targeted component might lead to the reveal of some information, and then by repeating the scenario (sequence of actions), some form(s) of information breach can happen in the SoC.

Table 4.1 provides a top view of these sources of vulnerabilities and their characteristics. It clearly shows that there is a critical need to verify the security of the SoC at each of the stages and verify the trustworthiness of an SoC. However, if the vulnerabilities reach the post-silicon stage, there would be limited flexibility (almost none) in changing or fixing them. Moreover, the cost of fixing the design is significantly higher as we advance through the later stages of the design. Furthermore, vulnerabilities that reach the manufacturing stage will cause revenue loss. Therefore, it is essential to develop efficient security verification approaches to ensure the security and trustworthiness of SoC designs with more concentration at the pre-silicon stage.

Table 4.1 Overview of source of SoC security vulnerabilities

Cat.	Source of vulnerability	Design flow stage	Examples of vulnerabilities	Threat model
(SV1)	Inadvertent designer mistakes	Software, firmware, boot loader, register file, cache, register-transfer (RT) level, high-level language (HLL)	(1) Insecure implementation of controller circuit (FSM) or boot loader, (2) incorrect synchronization or protocol handshaking between IPs (master/slave), (3) incorrect mutual exclusion of write/execute operation leading to illegal access.	Insecure boot mode, inter-IP illegal accesses, information breach, malfunctioning of security operations
(SV2)	Rogue employee at design house (insider threat)	Software, firmware, boot loader, register file, cache, RTL, HLL	(1) Manipulating hardware, firmware, software, which facilitates obtaining the security assets after fabrication, including logic analyzer module insertion, disabling security permissions/policies, etc.	Insecure boot mode, inter-IP illegal accesses, information breach, malfunctioning of security operations
(SV3)	Untrusted third-party IP vendors	HLL, RTL, Gate level	(1) Continuous watching/monitoring the bus for obtaining information by the IP, (2) incorrect protocol handshaking that leading to illegal actions (illegal memory write/read).	Inter-IP Illegal actions, bypassing IP-based checks (security checks)
(SV4)	EDA opti-mizations	Logical or physical synthesis flow	(1) Insecure optimization of the design, such as sharing (merging between trusted and untrusted region), insecure control flow, insecure data flow	Information breach, inter-IP illegal accesses, flaws in security policies implementation
(SV5)	Security compromis-ing stages of IC supply chain	Gate level, design service provider (DFT/DFD)	(1) Opening backdoor for attacks through test/debug infrastructures, (2) reading internal values of the design	Information breach
(SV6)	Impact of hardware Trojan insertion	Gate level	(1) Manipulating hardware that facilitates obtaining the security assets after fabrication, (2) insertion of Trojan for malfunctioning	Information breach, inter-IP privacy violation, malfunctioning
(SV7)	Lack of trust-worthiness in EDA Tools	Logical and physical synthesis flow	**(SV2)** + **(SV6)**	**(SV2)** + **(SV6)**
(SV8)	Untrusted fabrication site	GDSII at fabrication site	**(SV6)**	**(SV6)**

(continued)

Table 4.1 (continued)

Cat.	Source of vulnerability	Design flow stage	Examples of vulnerabilities	Threat model
(SV9)	Efforts on SoC design optimization	RTL, HLL	Incorrect optimization with open corner cases that leads to security vulnerabilities	Information breach, inter-IP illegal accesses, insecure protocol implementation
(SV10)	Hammering by the end-user	Post-silicon over the fabricated chip	Applying continuous and repeating tests on specific target based on physical or logical reflects	Information breach

4.1.1   Security Assets in SoC

For any IP component, firmware, software, etc., involved and integrated into the SoC, there exists a set of information or data, whose leakage can lead to catastrophic consequences with huge financial/reputation loss. This sensitive information or data are known as security-critical values, a.k.a. *security assets*, that must be protected against any potential form of threat. Any successful retrieval of such information or data in an illegal way might result in trade secret(s) loss for device manufacturers or content providers, identity theft, and even destruction of human life. These assets are usually known to the designers, and they are defined based on the specifications of the design. As an instance, encryption/decryption or private key in cryptographic primitives is assets, and the location and usage of them are known for the designers through the SoC design and implementation [46]. The following gives us some insight about the main primitives in an SoC that must be considered as the security assets:

(SA1) On-device key: (*Secret/private key(s) of an encryption algorithm*) These assets are stored on a chip mostly in some form of nonvolatile memory. If these are breached, then the confidentiality requirement of the device will be compromised.

(SA2) Manufacture firmware: (*Low-level program instructions, proprietary firmware, and protected states of the controller(s)*) These assets may have intellectual property and system-level configuration values, and compromising these assets would allow an attacker to counterfeit the device.

(SA3) On-device protected data: (*Sensitive user data + meter reading*) Leakage of these assets is more related to identity theft, and an attacker can invade someone's privacy by stealing these assets or can benefit himself/herself by tampering these assets.

(SA4) Device configuration: (*Service/resource access configuration*) These assets determine which particular services or resources are available to a particular user, and an attacker may want to tamper these assets to gain illegal access to the resources.

(**SA5**) **Entropy**: (*Random numbers generated for cryptographic primitives*) These
assets are directly related to the strength of cryptographic algorithms embedded
into the SoC, e.g., initializing vector or cryptographic key generation. Successful
attacks on these assets would weaken the cryptographic strength of a device.

The choice of security assets varies design by design and abstraction layer by
an abstraction layer. Mainly, the declaration of security assets is heavily dependent
on the *security policies* that is defined by the designers of a different component
integrated into the SoC. Hence, apart from these general security assets listed here,
which are known to the hardware designers, the security assets begin to expand
within the SoC due to the interaction of different IPs. Consequently, with the
increase of the list of security assets, SoC security verification through traditional
methodologies, e.g., formal-based and satisfiability-based approaches, becomes
almost impractical.

4.1.2 Security Policies/Properties in SoC

Based on the source of vulnerabilities, the threat model per each source, and
security assets defined for the design under investigation, a set of requirements
will be defined, whose realization will assist the design team to guarantee the
protection of the security assets against the given threat models. Different threat
model categorization can be evaluated for SoC verification, such as (i) overwriting
or manipulating confidential data by the unauthorized entity (integrity violation),
(ii) unauthorized disclosure of confidential information/data (confidentiality violation), and (iii) malfunctioning or disruption of function/connectivity of a module
(availability violation) [47].

For SoC security verification, (*security policies/properties*) define a mapping
between the requirements and some design constraints. Then, to fulfill the desired
protection level, these constraints must be met by building some infrastructures, and
these infrastructures are built by the IP design team(s) or SoC integration team. The
definition of security policies/properties is dependent on different actions/behaviors
located at multiple stages or abstraction layers, and they might also be updated or
refined through various stages [48, 49]. As an instance, the requirement defined as
*switching the chip from functional to test mode should not leak any information
related to the security assets* in a typical SoC can be mapped to constraint
(policy/property) defined as *An asynchronous reset signal assertion for scan chain
is required for secret/private key registers while the chip's mode is switching from
the functional to the test mode.*

Per each requirement, for mapping to a security policy/property, details and
underlying conditions must be considered meticulously, and then these conditions/-
constraints must be met to guarantee the protection of the asset(s). It is evident that
the definition of security policies/properties may vary depending on multiple factors,
like architecture and components of the SoC, interconnections and bus interface,

state of the execution (e.g., boot time, normal execution, test time), the state in the development life cycle (e.g., manufacturing, production, test/debug), etc. Below we provide a categorization for general policies that are required to be considered for security purposes in SoCs. This categorization covers both system-level and lower-level security policies/properties.

(**SP1**) *Definition of access restriction policies*: This set of policies/properties is one of the main requirements that define the constraints regarding how different components, either hardware, framework, or software, can access security assets. The definition of access restricting policies/properties directly/indirectly affects other policies/properties as well. For instance, given specific restricted access will change the data flow or control flow in the SoC. So, other policies/properties can be met/violated based on a newly defined access restriction policy/property.

(**SP2**) *Definition of data/control flow restricting policies*: In many cases, particularly security assets related to cryptographic primitives, by observing the responses of the components under investigation to a sequence of actions, the security assets can be retrieved with no direct access. In such cases, even though (**SP1**) has been defined and established properly, since there exists an insecure form of data/control flow, the confidentiality would be violated. Unlike (**SP1**), the realization of this group of policies/properties requires highly sophisticated protection techniques with advanced formulation/model. So, to keep the complexity of security policies at a reasonable degree, this policy/property is more efficient to be employed for high-critical assets with high confidentiality requirements.

(**SP3**) *Definition of HALT/OTS/DOS restricting policies*: This policy/property indicates the liveness of components throughout the execution of different operations. This mostly can be done by checking the status signals per each request for the component, to make sure that there is no halt, out of service (OTS), or denial of service (DOS) that violates the system availability requirements. Policies/properties related to the malfunctioning can also be part of availability violation (as the component does not provide the correct functionality or correct timing behavior).

(**SP4**) *Definition of insecure sequence execution restricting policies*: An authorization is always required once a component needs to get access to a security asset in the SoC. However, the flow between the authorization and getting access must be flawless with no possibility of changes that invalidates the access control. One of the examples for this group of policies is time-of-check to time-of-use (TOCTOU), which shows in the middle of authorization and usage the changes of the state of the security asset can lead to some invalid/illegal actions which are happening only when the resource is in an unexpected state.

(**SP5**) *Definition of insecure boot restricting policies*: The process of the boot may involve multiple critical security assets, including the definition of access restriction policies, cryptographic and on-device keys, firmware, etc., and any security leakage can lead to multiple vulnerabilities at different layers. Policies for protecting the boot can be defined individually related to (**SP1-4**) or unified on a set of actions/requirements.

(**SP6**) *Definition of inter-component integrity policies*: This policy is more likely low-level, i.e., IP-level inter-communication, showing that any (secure) communication between two components must be kept intact and with no changes done by a third component.

(**SP7**) *Definition of inter-component confidentiality policies*: Similar to (**SP6**), at low-level, any inter-communication between two components must be kept fully secure and confidential between these two components, and there should be no possibility for any third component to receive any part of inter-component communicated data.

(**SP8**) *Definition of inter-component authenticity policies*: It is a low-level policy verifies the authenticity of the component requesting a security asset.

The definition of security assets, security policies, and building a clear relationship between them are indispensable preliminary steps of SoC security verification. In the context of security verification, a security policy/property must be a complete statement that can check assumptions, conditions, and expected behaviors of a design [50], which is more likely a formal description of the design behavior/specification. The coverage of security policies/properties can be considered as a metric for the security assessment of an SoC, and the violation of the policies/properties implies that the design should be fixed.

4.1.3 Security Policy/Property Languages in SoC

To build the automated SoC security verification, a straightforward constraint definition is required to verify that based on the specification of the security policies/properties, the design adheres to those properties. To meet such requirements, there should be a unified language for the declaration of the security properties, and the security verification framework must be able to convert this language to hardware implementation and testing. Due to the dynamic nature of the threat model, the language must be rich and amenable to be expanded as needed and to specify characterizations like sensitivity levels, affectability, and observability. As an instance, security policies/properties determine the secure vs. insecure region(s), and accordingly, the security language must be able to check policies like confidentiality, as the sensitive data from the secure region should not leak to any insecure region.

These days, designers mostly use one of the powerful assertion languages such as *property specification language* (PSL) [51] and *SystemVerilog Assertions* (SVA) [52] to describe interesting behavioral events of a design. These languages use temporal logic representations such as Linear Temporal Logic (LTL) [53] and Computational Tree Logic (CTL) [54]. Languages based on LTL and CTL usually describe design behaviors and properties in four layers: Boolean expression, sequence, property specification, and assertion directive layers. These layers can be used on top of different HDL languages including Verilog and VHDL.

4.1.4   *Pre-silicon vs. Post-silicon Verification in SoC*

SoC security verification can be done either before or after the fabrication stage, called pre-silicon and post-silicon verification, respectively. Generally, in the pre-silicon verification, the verification target is typically a model of the design (a representation of the design at a specific design stage, like postsynthesis netlist, or the generated layout) than an actual silicon artifact. The pre-silicon verification activities consist of code and design reviews, simulation and testing, as well as formal analysis. These tests are running at different corner cases with constrained inputs. One of the biggest advantages of pre-silicon verification is its high observability as the design representation is available and any internal signal/wire/register can be observed and verified. However, since it is mostly simulation basis at MHz speed limited by the simulator performance, it takes a lot of time for verification of all policies/properties.

On the other side, post-silicon verification can be considered as one of the most crucial yet expensive and complex forms of the SoC verification, in which a fabricated but pre-production silicon of the targeted SoC (initial prototypes of the chips that are fabricated and are used as test objects) will be picked as the verification vehicle, and then a comprehensive set of tests will be executed on it. The goal of post-silicon verification is to ensure that the silicon design works properly under actual operating conditions while executing real software and identify (and fix) errors that may have been missed during pre-silicon verification. In post-silicon verification, since the silicon is used as the verification vehicle, tests can run at target clock speed enabling the execution of long use cases at a much smaller time slot. However, it is considerably more complex to control or observe the execution of silicon than that of a pre-silicon verification, as the access and observability of many internal nodes will be lost.

4.1.5   *Adversarial Model in SoC*

To ensure that an asset is protected, the design team needs, in addition to the security policies/properties, to define and determine comprehension of the power of the adversary. The notion of the adversary can vary depending on the asset being considered. For instance, regarding the cryptographic and on-device keys as the security asset, the end-user would be an adversary, and the keys must be protected against the end-user. As another example, the content provider (and even the system manufacturer) may be included among adversaries in the context of protecting the private information of the end-user. So, based on the definition of security assets, source of vulnerabilities, and security policies/properties, the adversary model also needs to be clearly defined helping to have a stronger SoC security verification mechanism. Hence, rather than focusing on a specific class of users as adversaries, it is more convenient to model adversaries corresponding to each policy and define protection and mitigation strategies with respect to that model.

4.1.6 Verification Model in SoC

The SoC security verification can be done at different stages each focusing on different sets of security policies/properties. Targeting the design at different stages will affect the model defined for the verification. In addition, the model can also be related to the adversarial model as the source of threats. For instance, for the untrusted foundry with access to all masking layer information, to malicious end-user that has access to the fabricated chip mostly as a black box, the security verification can be modeled differently. Hence, based on the access provided for the SoC verification framework, it can be generally divided into three main categories: white, gray, and black. The definition of these verification models in SoC is very close to their definition at the software level. In the white verification model, all internal wires, signals, nodes, registers, etc. are fully observable allowing the security properties/policies to be implemented in detail and very specifically based on the requirement. The black verification model treats the SoC as almost a black box, and only the general ports and potentially scan pins are available for testing. The gray model is any verification model that stands between white and black models, and the level of access to internal parts may be different per each case.

4.1.7 Scope of Security Verification in SoC

With the involvement of multiple abstraction layers in a complex and heterogeneous SoC, based on the security policies/properties associated with each abstraction layer, the SoC security verification can be categorized into three main domains:

(**SC1**) *Low-Level (Hardware-Level)*: Once the security vulnerabilities arise from the underlying hardware, such as RTL and gate-level netlist, the scope of verification needs to be concentrated at logic-level (low-level), such as hardware Trojan insertion, counterfeiting, fault injection, etc.

(**SC2**) *Platform-Level (System-Level)*: In this category, the vulnerabilities associated with the inter-IP communication and system-level bugs can be exploited mostly by untrusted third-party components during runtime. As an example, any confidentiality and integrity violation for any inter-component communication that can lead to information leakage or unexpected actions can be considered as system-level vulnerabilities that require system-level security policies/properties definition.

(**SC3**) *Software-Level (Framework-Level)*: This group refers to vulnerabilities arising from the intercommunication between hardware parts and the software or the framework. Additionally, network-based vulnerabilities, like communication of an embedded computing unit with off-chip modules or cloud, can be categorized as a member of this group. In this group, the definition of security policies/properties would be more at the higher level of abstraction combined with checking flags/status at hardware levels.

4.2 SoC Security Verification: Challenges

Based on what we learned so far, for the automation of SoC security verification, we need to accomplish some steps: (i) identification of the source of vulnerabilities (**SV**s), which helps to define the threat models (adversarial model), (ii) indicating the security assets per component (**SA**s), (iii) definition of security policies/properties (**SP**s) based on the threat model and the chosen security assets, (iv) formalizing the security policies/properties using the unified language with consideration of the security domain (**SC**s), and (v) implementation and testing. To accomplish these steps, there exist challenges showing why new approaches like the self-refinement technique, i.e., fuzz, penetration, and AI testing, are needed for the SoC security verification. The following section covers some of the biggest challenges that cannot be solved using the existing approaches, like formal satisfiability-based techniques, model checking, information flow tracking, etc.:

(i) *Preciseness*: For almost all aforementioned steps, the course of action(s) decided and accomplished by the designer(s) requires the highest precision to make the whole SoC security verification procedure a successful practice. Precisely evaluation of **SV**s, understanding the threat models, precisely choosing **SA**s, precisely defining **SP**s, and formalizing using the selected language directly and significantly affect the outcome of the framework. For SoCs getting more complex and more heterogeneous, precisely indicating exacts **SV**s, **SA**s, and **SP**s becomes more and more challenging. The designer(s) needs to know all underlying information about all components, modules, frameworks, their intercorrelation, handshaking, their corresponding security levels, etc., which is almost impractical in modern SoCs.

(ii) *Multi-stage/layer verification*: The definition of **SV**s, **SA**s, and **SP**s is fully dependent on the stage of the design flow and the abstraction layer of the design. There is no guarantee that when the security verification is passed in one stage of the design, then the same **SP**s will be passed in another abstraction layer. Changes and refinements per each stage, and moving from one abstraction layer to another one, might arise new vulnerabilities. A simple example for this case is the transitions done by synthesis tools, like high-level synthesis (HLS) and other computer-aided design (CAD) tools, which may add new data/control flow that can be exploited leading to new vulnerabilities [26, 55]. This is why the invocation of SoC security verification is required at different stages and different layers of abstraction.

(iii) *Verification vs. Dynamicity*: Based on how the **SV**s, **SA**s, and **SP**s are defined, dynamicity might happen during runtime, meaning that there are potentially new **SA**s introduced when a specific set of operations are executed on the original **SA**s. This will propagate the original ones, and based on the dependency/relation to other variables, some other variables may preserve critical information that must be considered as new **SA**s. Additionally, the threat models will be updated over time, resulting in new **SV**s, which lead to the introduction of new **SP**s. Also, protecting the design against one **SV** may make it vulnerable to the other one. For example, protecting a design against information leakage may create side-channel leakage that can be exploited by an attacker to retrieve sensitive information.

(iv) *Unknownness*: As we mentioned previously, the utilization of conventional techniques and tools, i.e., formal satisfiability-based techniques, model checking, information flow tracking, etc., for SoC security verification is mainly limited to the expert review, and they do not provide acceptable scalability. This is getting worse when the side effect of dynamicity comes to the action, which is the introduction of unknown vulnerabilities. With unknown vulnerabilities, there is no precise definition for **SV**, **SA**, and **SP**, and they might inadvertently be caught either during testing by the designer(s) or through attacks, e.g., hammering, by adversaries. This is when the self-refinement techniques' contribution plays an important role, by using smart hammering and testing, to detect such vulnerabilities before releasing the SoC into the field/market.

4.3 SoC Security Verification: Assumptions

To have a successful SoC security verification solution, there exist some fundamental assumptions that must be considered meticulously as the most basic requirements of the proposed solution:

(i) *Time of Verification*: Since moving toward the final stages of SoC design, implementation, and testing makes the evaluation and investigation much harder, it is paramount to accomplish the verification, particularly for primary **SA**s and **SP**s at the earliest possible stage. Assuming that the vulnerabilities reach the post-silicon stage, the verification model will change from a low-level white model to a chip-level black model, which makes the verification much harder. Moreover, the cost of fixing the design is significantly higher as we advance through the later stages of the design. Furthermore, vulnerabilities that reach the manufacturing stage will cause revenue loss. According to the rule-of-ten for product design [56], modifying a design at the later stages of the SoC design flow is ten times costlier than doing in the previous steps.

(ii) *Evolutionary Mechanism*: Almost all existing security verification frameworks that are relying on conventional formal-based, satisfiability-based, or model-checking-based tools tend to provide a binary response to the **SP**(s) associated with a set of **SA**s. So, the analysis behind the verification is very limited to a specific set of events, and the verification solutions just indicate that no flow occurs or that there is at least one that violates the defined **SP**(s). However, they do not provide any indication in between, e.g., predicting that we are approaching a vulnerability or not (the majority part of **SP**(s) will be satisfied/passed). Providing evolutionary response (like providing feedback) can help the framework by itself to mutate the test cases based on the collected data and in a smart way and narrow the search space for reaching to potential vulnerabilities. For building such structure, a definition of some security metrics or coverage metrics is needed as well to guide the test cases to approach the vulnerability and eventually get the global minimal.

(iii) *Hammering-Based Verification*: As mentioned in Sect. 4.2, due to the dynamic nature of threat model as well as the propagation of **SA**s that results in the introduction of newer **SA**s, unknown security vulnerabilities will appear in SoC transactions, leading to breaches of confidentiality, integrity, or authenticity, even while the verification has been done appropriately based on the known security policies. Hence, it is crucial for a security verification framework to be capable of finding both known and unknown vulnerabilities in the SoC, even though there exists no clear/precise definition for the actual vulnerabilities in the targeted SoC. This is when evolutionary-based verification comes into action and may help to detect such unknown vulnerabilities by a smart hammering and helping to avoid such scenarios before moving to the next design stage.

(iv) *Hardware–Software Co-verification*: Some hardware vulnerabilities in SoC designs are not explicitly vulnerable unless triggered by the software [32]. In some cases, it might be possible to formalize such vulnerabilities via a complex sequence checking at the hardware level, but if the verification mechanism allows doing either hardware–software co-analysis or software-level verification, the definition of **SP**(s) can become more straightforward. So, the SoC security verification framework should handle such interactions as well to ensure the security of the SoC.

(v) *Verification at a Different Level of Access*: As the contribution of proprietary third-party IPs in modern SoCs is getting more and more, it decreases the visibility of the verification engineer from the internal specifications of the SoC. Therefore, the security verification framework should be able to verify the security of the SoC considering a gray-box or black-box model.

4.4 SoC Security Verification: Flow

Considering the aforementioned challenges and assumptions, to get the benefit of self-refinement techniques for SoC security verification, the followings are the major steps that must be considered as fundamentals of the SoC security verification for both known and unknown vulnerabilities:

(**Step 1**) *Risk Assessment*: The main purpose of this step is to identify the **SA**(s). The **SA**(s) will be determined based on the ownership, domain, usage, propagation, static or dynamic nature, etc., and the outcome of this step would be a semantic definition as the requirements of verification activities.

(**Step 2**) *Definition of Adversarial Entry Region*: Per each defined **SA**, this step defines the most intuitive adversarial actions (**SV**s) around the **SA** that might lead to governing the assets, such as different entry point candidates, data/control relevancy between assets, untrusted region(s), etc.

(**Step 3**) *Definition of Security Policies/Properties (Known)*: Based on **SA**(s) and **SV**(s), each vulnerability is converted to a set of rules, and then those rules will be converted to a set of properties (**SP**s). For widely arisen vulnerabilities that can be categorized as known vulnerabilities, the formalism can be done by definition of assertions that monitor illegal events/sequences. The reachability of the unified language plays an important role in this step (e.g., LTL [53] and CTL [54]) that paves the way for converting the **SP**s to hardware implementation.

(**Step 4**) *Definition of Security Policies/Properties (Unknown)*: Unlike known vulnerabilities, in case of unknown vulnerabilities (e.g., any possible scenario that leads to security asset leakage), the policy/property can be defined in the form of a cost function, which represents the evolutionary behavior of the vulnerability and tries to approach a certain state/location of the design in a different way to trigger unknown vulnerabilities. Cost functions can be considered as a relaxed or higher-level version of assertion-based **SP**s, and by using self-refinement tools, as discussed hereinafter, self-mutation can help to move toward testing data that excite the conditions leading to the specific class of vulnerability under investigation.

(**Step 5**) *Hardware Implementation of Security Verification Model*: For both known and unknown cases, by using the unified languages, all **SP**s must be converted to hardware implementation. For known cases, it is more likely assertion-based scenarios that check the sequence of events for a specific incident. For unknown cases, it is based on instrumentation and mutation, which helps to build the evolutionary mechanism.

(**Step 6**) *Security Verification and Testing*: This step involves running verification, testing, and refinement-based tools based on the definition of **SP**s and their hardware implementation that lead to verifying the security of the SoC. This step can be done at different stages of the IC design, from high level to GDSII as pre-silicon, and after fabrication on initial prototypes as post-silicon.

In the following sections, we will discuss how self-refinement techniques and tools, like fuzz, penetration, and AI testing, can be engaged along with the consideration of cost function definition for finding both known and unknown vulnerabilities of the SoC.

4.5 SoC Security Verification: Fuzzing

The words *fuzz testing* and *fuzzing* represent randomized testing of programs to find out the anomalies and vulnerabilities. Fuzzing is usually an automatic or semi-automatic approach that is intended to cover numerous predefined (instrumented) corner cases of invalid inputs to trigger the existing vulnerabilities in a program. Fuzzing was first applied by Miller et al. [57] to generate random inputs, which were provided to Unix to find the specific inputs that cause crashes.

Generating random inputs without any feedback from the design under investigation is referred to as blind fuzz (close to random) testing which suffers from low coverage, especially when more complex input models are introduced. This has led to additional stages being introduced to the fuzzing platforms. Generally, three main steps are involved in a fuzzing that will be invoked iteratively on the program:

(i) *Test scheduling*, which relates to the problem of ordering the initial seeds of each fuzz iteration in a way that leads to full coverage as fast as possible.

(ii) *Mutation steps*, which incorporate a variety of methods ranging from a simple crossing of seeds to optimized genetic algorithms in order to produce new seeds for further exploit generation.

(iii) *Selection*, where the useful seeds are pruned out of all generated seeds. This process requires metric evaluation as the prominent way of deciding which seeds result in better coverage of the whole design under test [58].

Among the existing and widely used fuzzer tools, American fuzzy lop (AFL) [21] is one of the most popular software fuzzers that uses an approximation of branch coverage as its metric, while another famous fuzzer named Hongfuzz [59] accounts for the unique basic blocks of code visited. Based on the benchmarks used by different fuzzers, the baseline of fuzzing, crash type, coverage mode, and seed formulation, past fuzzing mechanisms divided into numerous groups, whose details and comparison can be found in [60]. The following first shows how the verification model can change the way fuzzer acts on the targeted program, and then we will investigate how the fuzzer can be engaged for SoC security verification. As discussed previously, the purpose of using such self-refinement approaches is to overcome the scalability issues of formal verification methods [61].

4.5.1 Formal Definition of Fuzz Testing

Based on the verification model, fuzzing-based techniques can be classified into three categories: white box, gray box, and black box, which are defined based on the availability of information during verification or runtime phases. This information may include the source code of the hardware or software designs, detail information about the security specification and functionalities, code coverage, control and data flow graphs, and execution (simulation or emulation)-related information such as

CPU utilization, memory usage, etc. The following describes these three categories of fuzzing.

4.5.1.1 Black-Box Fuzzing

Black-box testing does not take any information from the program under test. This approach also does not obtain the input format of the program, rather it generates random inputs that are mutated from a given seed data provided as arguments or a file. This approach can use some predefined rules, such as bit flipping, varying input length (bits), sign reverse, etc., to generate mutated inputs. Some recent black-box fuzzing-based approaches use grammar or input-specific information to obtain partially valid inputs [62]. Black-box fuzzing is very convenient to use when the design specification is unknown and little information is available about the internal parts of the design under test. On the contrary, it is very challenging to generate test cases for a program with a very large number of execution paths due to the lack of diversity of mutated inputs, which may usher the failure of reaching the corner cases. Again, due to the inherent blindness, black-box fuzzing-based approaches struggle with the code coverage in practical use cases of finding vulnerabilities.

4.5.1.2 White-Box Fuzzing

Unlike black-box fuzzing, in white-box fuzzing, all of the information of the target design is transparent and available for the verification engineers to use. This approach was first introduced by Godefroid et al. [63]. White-box fuzzing utilizes as much information as needed to guide the seed generation effectively. This approach analyzes symbolic constraints for all conditional statements in the program in order to develop the path constraints for all possible executions. Integrating a coverage-maximizing metric to white-box fuzzing can accelerate the approach to find the inputs triggering the vulnerabilities in the program [63]. Although theoretically it seems that white-box fuzzing can generate inputs to cover all of the possible execution paths, practically it is very challenging to achieve due to too many execution paths and time restrictions of running a program.

4.5.1.3 Gray-Box Fuzzing

Gray-box fuzzing is a hybrid approach that mixes black-box and white-box fuzzing. Gray-box fuzzing obtains partial information of the design under verification. For example, while in white-box fuzzing we could instrument the whole design for attaining code coverage, here we might be only able to instrument the final binary and not the code itself, or in case of an SoC design, we might have access to the bus interface without being able to probe anything else. A directed information feedback can largely assist in guiding the mutation engine so that it can cover more control paths and hence find the vulnerabilities in a short time. The common method of mutation strategies for this type of fuzzing is genetic algorithm [64], taint analysis [65], etc. Taint analysis assists to focus on mutating the inputs which have a larger impact on the targeted vulnerabilities. As this fuzzing approach possesses some information about the design, it can implement a targeted feedback system to trigger the vulnerabilities and explore new paths in the program, which significantly improves the detection capability and coverage of the fuzzing approach.

4.5.2 Fuzzing Hardware Like Software

The complexity of hardware/software security co-verification is caused by complications that arise from interfacing numerous sub-modules from both sides of the table. The approaches presented to this day have failed to propose a scalable method with sufficient coverage that captures the hardware and software vulnerabilities in a unified platform simultaneously. Fuzzing has proved to be a powerful tool for detecting vulnerabilities in real-world software programs that encompass huge code bases. This has opened the door for many researchers to investigate the possibility of applying the same mechanisms to a full system that includes hardware, firmware, and software components.

4.5.2.1 Hardware to Software Abstraction

In order to have a coherent model of the system so the previous fuzzing efforts can be reused, one method is to translate the hardware to the software world. Utilizing state-of-the-art fuzzing tools for hardware verification is a promising concept which does not enforce the development of a new platform for RTL verification. In order to do that, RTL designs are translated to equivalent software models by a hardware translator, such as Verilator [66], which generates C++ programs of the given RTL designs. The generated C++ classes are instantiated by a wrapper which acts as the main function and stimulates the program under test for simulation. The wrapper, which is a test-bench in hardware terms, is designed in a way that holds functionalities and security properties that act as a cost function which provides feedback information for future iterations. The generated C++ programs can also be instrumented to increase the visibility of the internal simulations. Instrumentation can also help to achieve code coverage, line coverage, branch coverage, etc. to evaluate the overall coverage during simulation.

A metric that we call the cost function is introduced to measure the extent to which the security of an SoC design has been compromised and how close are the test scenarios to activating a targeted vulnerability. In that sense, the cost function is helping to build the evolutionary mechanism for a vulnerability that guides fuzzing toward detecting the vulnerability in a significantly shorter time period compared to blind fuzzing. The cost function may also include general metrics such as code

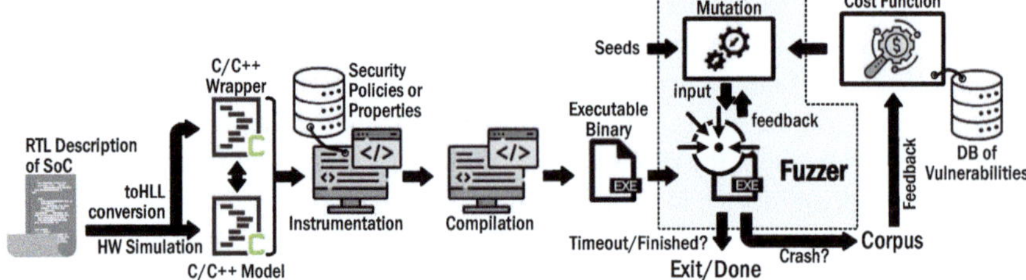

Fig. 4.5 A fuzzing framework on software level of RTL

coverage, line coverage, branch coverage, etc., which can be obtained through instrumentation of RTL designs or translated software models. The cost function works in tandem with the mutation engine of the fuzzer to assist in generating better mutated test cases that can trigger various corner cases.

Figure 4.5 shows how the transformation of RTL code to an executable binary that encompasses instrumentation takes place. Also this figure represents how the cost function interacts with the fuzzer outputs as well as the mutation engine. The special characteristic that is worth mentioning here is the fact that the cost function is selected based on the vulnerability instance that is currently being targeted. This means that instead of a general feedback from the fuzzer outputs, one can make more suitable decisions for further analysis because now the feedback is tailored for the targeted bug in the running verification session.

4.5.2.2 Prior Art HW-to-SW Fuzzing

This section reviews the studies that incorporate hardware-to-software abstraction conversion for realizing software fuzzing for hardware verification and the prominent deficiencies of each method are mentioned. The authors in [67] proposed a mutational coverage-guided fuzzing-based framework in order to resolve modern SoC verification challenges, such as co-verification of hardware and firmware, as well as scalability. The framework includes adversarial behavior and coverage metrics to evaluate the security properties written in their proposed logic HyperPLTL (Hyper Past-time Linear Temporal Logic). However, the primary disadvantage of this technique is that it requires a huge amount of technical expertise, particularly for building **SP**s, and hence incurs higher chances of erroneous and low coverage results which could eventually limit the scope of bugs detection. Tripple et al. [61] developed software models of hardware and then applied fuzzing on the software leveraging Google's OSS-fuzz. The main problem with this approach is due to its dependence on general metrics such as code coverage rather than a dedicated cost function targeted at hardware model properties.

Moghimi et al. [68] utilized fuzzing by mutating the seeds of existing Meltdown variants in order to discover various Meltdown-type attack variants. The authors provided randomly mutated inputs to the associated faulty loads and used the cache as proof of covert channel for side-channel leakage. Oleksenko et al. in [69] developed a dynamic testing methodology, SpecFuzz, for identifying speculation execution vulnerabilities (e.g., Spectre). SpecFuzz instruments the program and runs the simulation for speculative execution in software-level traversing all possible reachable code paths that may be triggered due to branch mispredictions. During simulated execution of the program, speculative memory accesses are visible to integrity checkers, which is combined with traditional fuzzing techniques. SpecFuzz can detect potential Spectre like vulnerabilities, but it is only useful for this certain type of attacks that exploit speculation in CPU pipeline.

DifFuzz proposed in [70] is a fuzzing-based approach in order to detect side-channel vulnerabilities related to time and space. This technique analyzes two copies of the same program which have different secret data while providing the same inputs to them both. The authors developed a cost metric which estimates the resource consumption, such as the number of executed instructions and memory footprint, between secret-dependent paths for both programs. However, this technique is primarily dependent on the details of microarchitectural implementation information. The more visibility into the microarchitectural state, the more coverage is achieved. Unfortunately, obtaining good visibility is not always available for many hardware designs which binds these techniques to low accuracy. Yuan Xiao et al. in [71] developed a software framework leveraging fuzzing concepts to identify Meltdown-type vulnerabilities in the existing processors. The authors build up the code using some templates, which are executed to find out the vulnerabilities. The authors leveraged cache-based covert channel and differential tests to gain visibility into the microarchitectural state changes, which eventually helps to analyze the attack scenarios.

Ghaniyoun et al. [72] proposed a pre-silicon framework for detecting transient execution vulnerabilities called IntroSpectre. IntroSpectre is developed on top of Verilator. The authors resolved the challenges of lacking visibility into the microarchitectural state by integrating it into the RTL design flow which makes it identify unreachable potential side-channel leakages. The authors utilize the fuzzing approach to generate different attack scenarios consisting of code gadgets and analyze the logs obtained from simulation to identify the potential transient execution vulnerabilities.

4.5.2.3 Limitations and Challenges

There are many rudimentary differences between fuzzing a software program and an RTL hardware model, the first one of which is due to the difference in input arguments. A digital circuit has the notion of input ports that take different values in each cycle, unlike software that reads its inputs from a variety of sources including arguments, files, or through OS system-calls. Fuzzing requires a solid definition of the format of the input, so it can do meaningful mutations and generate new passable tests. So the actual input to be fuzzed should be thoroughly explained to the fuzzer when working with a translated hardware design. The format of the input has a great impact on how well the mutation engine performs.

The second issue to address is due to the difference in software versus hardware coverage metrics. Many fuzzers depend on instrumentation to obtain coverage metrics and direct the seed generation toward uncovered sections of the design. While some metrics such as branch and line coverage can approximately be mapped to each other in both hardware and software models, other metrics such as FSM coverage or toggle rate are not translatable [61]. The traditional hardware verification uses these metrics to target specific classes of vulnerabilities, and any effort in software domain should comply with these previous platforms as well.

The third dominant issue with fuzzing on the translated hardware is with regard to the cost function. Software fuzzers look for crashes, exceptions, and memory checks as the vulnerabilities that could exist in a software model. These scenarios are not convertible to hardware designs, particularly low-level or platform-level vulnerabilities, because these forms of errors like exceptions do not exist in the hardware realm. Hardware is inherently different when it comes to targeting vulnerabilities because software crashes can still happen even when the hardware is fully functional and secure. Fuzzing the translated model without introducing our own notion of hardware vulnerabilities through a cost function will result in the fuzzer expending its resources on detecting bugs induced by the translator rather than the hardware model itself [73]. Another problem related to this issue is due to the changes in variables and functions when performing the translation. This makes the translation of properties from hardware RTL model to software a crucial task that has not been investigated.

4.5.3 *Direct Fuzzing on Hardware RTL*

The verification faces additional challenges with enlarging and growth of the design's size and scope, e.g., the verification for a full-scale complex and hetero-geneous SoC. The primary challenge relevant to the SoC security verification is the lack of end-to-end verification methods which can resemble the behavior of every hardware component and the runtime implications of software components or framework(s) executing on it, i.e., hardware/software co-verification with maximum coverage. Again, scalability is the biggest challenge in modern SoC verification due to the complexity and massive size of the designs. In order to tackle scalability, an automated and systematic verification platform is required. However, automation in SoC verification is very difficult for several reasons. Firstly, the verification engineer faces extreme challenges to precisely assess the security policies/properties of an SoC design due to the enormous number of components that interact with each other. Identifying the security policies/properties in a comprehensive fashion largely influences the quality of SoC verification. Secondly, the verification engineer faces challenges in modeling the attack scenarios that may happen in the SoC. These

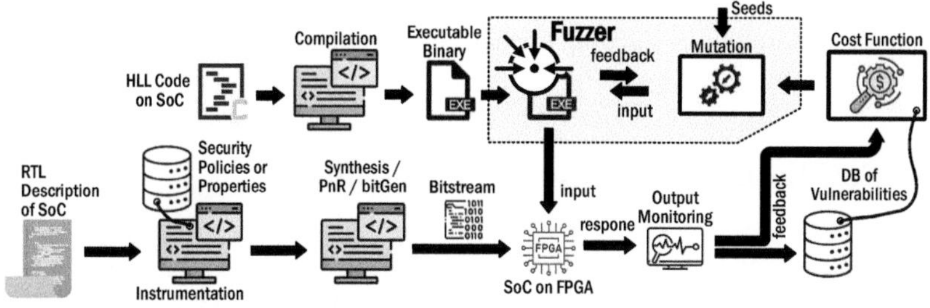

Fig. 4.6 A framework for direct fuzzing of RTL

attacks may include side-channel based, direct hardware or software exploitable attacks that could run on the SoC. It is very challenging to prepare different attack models from the specifications for different untrusted third-party IPs [74], untrusted OEM firmware [75], and untrusted software [25]. Again, the verification engineer may not get a complete manual for third-party IPs. Hence, SoC verification is still a burdensome task that demands significant research to develop a robust and scalable solution.

In such an environment the emulation-based fuzzing methods, with more concentration on gray-box fuzzing mode, can greatly improve the performance of the simulation-based approaches by running the design on an FPGA and interfacing the fuzzer to the prototype design under test in a more realistic manner. Figure 4.6 represents a general platform for interfacing FPGA with a test generator fuzzer. The difference of this method with what was discussed in Fig. 4.5 is mostly in the bitstream generation as opposed to a binary under test and the fact that the test cases are introduced to the actual design through a direct memory access channel. The instrumentation in this approach is outputted through the probe analyzers available at FPGA monitoring implementation, and it helps the cost function to better guide the mutation engine.

Direct fuzzing on FPGA-accelerated simulation incurs some major challenges such as resetting the FPGA, which is required to take the program to a particular known state to decrease the verification time or defining the branch coverage, which is needed to track the verification coverage. In order to solve the first challenge, memory snap-shooting techniques can be used to reset the program and set the desired state as a preparation for each test intended to be performed with new fuzzing inputs. In order to estimate the branch coverage, branches are mapped to multiplexers which output one of the two input values in each cycle.

Kevin Laeufer et al. in [76] proposed a coverage-directed fuzzing approach for RTL testing on FPGAs leveraging ideas from the software testing community (RFuzz). The authors proposed a new approach for coverage metrics, which uses multiplexer as branch coverage. RFuzz employs FPGA. In general, fuzzing can be a promising solution for developing an end-to-end mechanism for full system verification, and it requires a minimum amount of adjustment since it is already well-developed in other areas of research, standalone simulation-based fuzzing, or FPGA-based interfacing still cannot cover a wide variety of threat models and source of vulnerabilities showing further investigation is inevitable in this domain.

4.6 SoC Security Verification: Penetration

Penetration testing (PT) is a methodology to assess the vulnerabilities in an application or a network and exploit those weaknesses to gain access to the resources. Since accessing security-critical resources of a computing system necessitates the exploration of exploitable vulnerabilities, vulnerability assessment (VA) is an indispensable precursor to PT. In the literature, therefore, VA and PT are often considered a single framework to assess the security of a network or application [77].

4.6.1 Higher Abstraction Penetration Testing

Motivated by the ever-growing attack surface of software, the secure software development lifecycle has been adopted by leading software companies over the last two or three decades. Security issues are considered, analyzed, and attempted to be discovered at even early stages of development [78]. Consequently, in the software development lifecycle, VAPT has become a well-defined methodology to weed out bugs and vulnerabilities. In fact, there are governmental and private organizations that accredit individuals and groups based on their ability [77, 79]. Open Web Application Security Project (OWASP) is a non-profit organization, for example, that has published and sanctioned a detailed step-by-step process of performing penetration testing of web applications and firmware [80]. Open Source Security Testing Methodology Manual (OSSTMM) also offers detailed guidelines on how to conduct VAPT on different systems [81]. CREST, Tigerscheme, and the Cyberscheme are some other professional bodies that provide industrial certifications and qualifications. CREST in particular provides the intelligence-driven red team penetration testing framework to assess the robustness of an operational team against cyberthreats [82]. Mitre sponsors the maintenance of two detailed databases of known software vulnerabilities—the CVE (Common Vulnerabilities and Exposure List) and the CWE (Common Weakness Enumeration). These two databases contain extensive information on most common software vulnerabilities, example implementations, and even discuss potential mitigation solutions. Therefore, VAPT is a well-delineated methodology in software and network security that has been implemented in practice successfully over the course of the last two or three decades.

The main objective of a penetration test pertaining to an application or network is to find out exploitable vulnerabilities. Colloquially, VAPT has also been described as ethical hacking [83]. This terminology is inspired by the fact that often, the best approach to finding unanticipated vulnerabilities in a computer system is to *hack* into the system from the outside looking in. This is essentially a *simulated attack*, in which the penetration test engineer tries to anticipate the course of action that malicious actors might adopt while trying to compromise the security of the system. An analogy is often used in penetration testing colloquialism, where it is compared to hiring a thief to break into one's own house to find the loopholes in the implemented security system. The term penetration in this context, therefore, stands for acquiring legitimate access to resources illegitimately. We present the typical steps in a typical VAPT approach in the following:

(i) *Reconnaissance*: In this step, the tester gathers extensive knowledge on the application, network, or computing system to be pen tested. The principal goal of this step is to get familiarized with the particular technology, protocols, software versions, IP addresses, and the configuration being used by the system. Websites, job postings, and even social engineering can be used to gather information [84]. *Httrack*, *Harvester*, and *Whois* are some tools that can help in this step.

(ii) *Scanning*: In the scanning step, the system or network is at first probed to find points of entry. An example is *Nmap* program to find open ports in a network. Subsequently, a commercially available tool like *Nessus* is used to find known vulnerabilities in the system.

(iii) *Exploitation*: From the list of known vulnerabilities, the tester comes up with an *attack* or exploitation plan. The goal of this step is to identify the sequence of actions that can be taken to gain access to unprivileged resources in the system. *Metasploit* is an open-source tool that is frequently used in academic and professional settings to realize this step. Effective exploit management (search, upgrade, and documentation) or a large number of payloads (tasks that are done after successful exploitation of the target system) are available in Metasploit. In general, payloads can be either simple and focused on a single activity (e.g., user creation) or complex and comprehensive and provide more advanced functionality.

(iv) *Post Exploitation*: The post exploitation step documents the steps taken to gain access to non-permitted resources (if successful). It might also involve the tester attempting to escalate privileges already gained in the system. The documentation of the steps taken gives valuable insight into the weakness of the system.

4.6.2   Formal Definition of Penetration Testing

Similar to other testing approaches, based on the depth and level of access as well as verification model, there are three types of penetration testing:

4.6.2.1 Black-Box Pen Testing

The testers do not have any prior access to any resources on the test target when performing black-box penetration testing. They are expected to figure out all of the minutiae of the system, as well as any flaws, depending on their previous experience and individual expertise. The tester's primary goal is to audit the external security boundary of the test target; as a result, the tester replicates the activities and procedures of an actual attacker who may be located at a location other than the test target's boundary and who does not know anything about the target. OSSTMM makes a distinction within what would be typically referred to as black-box PT between blind and double blind testing. In double blind testing, the target is not notified ahead of time of the audit, whereas in blind testing it is informed ahead of time.

4.6.2.2 White-Box Pen Testing

Contrary to black-box PT, the testers are provided with all of the internal information about the system. This is meant to simulate an attack from an internal threat like a malicious employee. White-box PT offers higher granularity of testing while at the same time offering the benefit of not relying heavily on trial and error as is common in black-box PT.

4.6.2.3 Gray-Box Pen Testing

Gray-box PT is somewhere in between black-box and white-box methodologies in terms of the information available to the testers.

4.6.3 Penetration Testing on Hardware: Definition

Compared to the software domain, hardware penetration testing is in its infancy. The term has been used as a stand-in for merely vulnerability assessment or for post-silicon testing and debugging in hardware security literature. The authors in [85], for example, equate penetration testing of hardware to post-silicon debugging and testing. The examples they provide amount to testing the software layer being run on the hardware and not the hardware itself.

In keeping with the stated goals of software PT, we believe that the principal objective of hardware PT should be to discover vulnerabilities in the hardware that can be exploited. However, certain differences must be noted. Firstly, vulnerabilities in hardware can be purely hardware-oriented such as malicious modification of the hardware description to compromise the integrity and confidentiality of the device

[5], side-channel leakage [86], and fault injection [87, 88]. Alternatively, they can be the source of cross-layer vulnerabilities which can be exploited through the software layer of the computing stack. Spectre and Meltdown vulnerabilities, which leverage the weakness in hardware implementation of speculative execution, are examples of these types of vulnerabilities [44, 45]. Secondly, the after-deployment *simulated attack* scenario to probe for vulnerabilities provides limited benefits when translated to hardware. In contrast to software that can be updated with a patch once a vulnerability has been discovered, hardware cannot be easily patched (especially ASICs). A pre-silicon testing methodology would be much more beneficial to the designers and verification engineers.

In light of the challenges and foregoing differences with the software domain, we define pre-silicon hardware penetration testing as a testing methodology that propagates the effects of vulnerability to an observable point in the design in spite of cross-modular and cross-layer effects present in the design. In contrast to randomized testing which develops test patterns without the knowledge of the vulnerability it is seeking to detect, hardware penetration testing assumes a gray- or black-box knowledge of the specification of the design and a gray-box knowledge of the bug or vulnerability it is targeting. The gray-box knowledge of the bug implies that the tester has knowledge of the type of bug or vulnerability it is, and how it might impact the system but not the precise location of its origin or the precise point in the design where it might manifest in a complex SoC. Penetration in this context, therefore, refers to the propagation of a vulnerability from an unobservable point in the design to an observable point.

4.6.4 Penetration Testing on Hardware: Framework

In this section, we demonstrate how a binary particle swarm optimization (BPSO)-based penetration testing framework can be used as a promising solution for the SoC security verification domain.

4.6.4.1 Binary Particle Swarm Optimization (BPSO)

The particle swarm optimization (PSO) is an evolutionary computation technique motivated by the behavior of organisms. PSO has been widely employed in a range of optimization situations due to its simplicity and ease of implementation. The PSO method is initialized by randomly placing a population of individuals, called particles, in the search space and then searching for the optimal solution by updating individual generations. At each iteration, for the jth index in the ith particle of the swarm, the position and velocity of the particle are updated through the following equations:

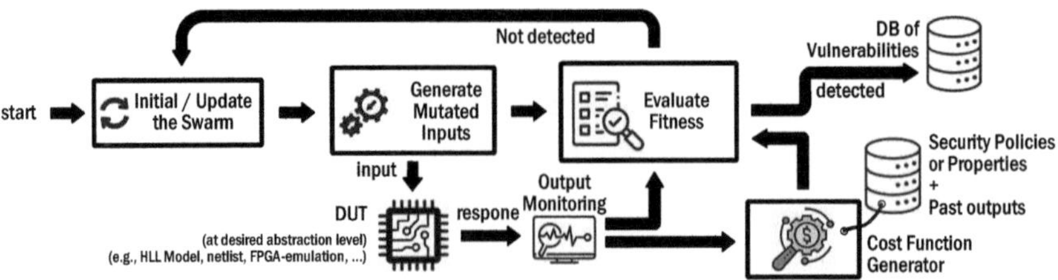

Fig. 4.7 A BPSO-based hardware penetration testing framework for detecting RTL vulnerabilities

$$v_{i,j}(t+1) = w v_{i,j}(t) + c_1 R_1 (p_{besti,j} - x_{i,j}(t))$$
$$+ c_2 R_2 (g_{besti,j} - x_{i,j}(t)) \tag{4.1}$$

$$x_{i,j}(t+1) = x_{i,j}(t) + v_{i,j}(t+1) \tag{4.2}$$

R_1 and R_2 are uniformly distributed random numbers between 0 and 1, c_1 and c_2 are acceleration coefficients, and w is called positive inertia constant.

Kennedy and Eberhart modified the continuous variable PSO algorithm for binary spaces in [89], which came to be known as the Binary PSO (BPSO) algorithm. The adaptation for binary spaces is done through a set of constraints. The constraints are imposed on equation 4.2 in the following form:

$$x_{i,j}(t+1) = \begin{cases} 0, & \text{if } rand() \geq S(v_{i,j}(t+1)) \\ 1, & \text{otherwise} \end{cases} \tag{4.3}$$

where $S(.)$ is the sigmoid function used for transforming the velocity.

4.6.4.2　BPSO-Based Hardware Penetration Testing

Based on our earlier discussions on hardware penetration testing, the framework shown in Fig. 4.7 can be a BPSO-based pen testing approach on hardware that can be applicable for SoC security verification. At the core of the vulnerability detection process is the cost function generator. This generator (ideally automatically) generates a mathematical function that describes the vulnerability that the tester is attempting to detect. The input to the design is described in terms of a binary vector which is mutated upon based on the evaluation of cost function for a generation of input test vectors. For each generation of the swarm, the algorithm tries to minimize (or maximize) the cost function (which helps to build the evolutionary mechanism). The observed output can be any observable point in the SoC including hardware signals during functional simulation, memory address contents available from simulation or emulation, observable points or signals created by RTL instrumentation or scan chain insertion, and even the output of a user space program. The cost function generator generates the mathematical description of the vulnerability based on the observable output point, a description of security policy, and an append-only database of observed past outputs for a particular sequence of inputs applied or actions performed. The cost function generator must rely on keeping a record of outputs attained for *a sequence* of inputs since only a sequence of inputs can trigger certain hard-to-detect vulnerabilities. For example, the AES-T1100 Trojan described in the Trust-Hub database [90] gets activated upon the application of a predefined sequence of plaintext.

In order to apply such BPSO-based pen testing framework, there are three prerequisites that must be met:

(i) The tester should possess a preliminary knowledge of vulnerability in addition to the impact it might have on observable output and how it can lead to the violation of predefined security policies of the device. We argue that this is a reasonable supposition since a significant portion of RTL hardware vulnerabilities have well-studied effects. For example, hardware Trojans can cause integrity, confidentiality, and availability violations in a circuit [9, 91]. Security unaware design practices can lead to a design having unanticipated leakage of information and assets to an observable point or to an unauthorized 3PIP in the design [92–95]. Furthermore, there are open-source databases (e.g., Common Weakness Enumeration) that catalog commonly found vulnerabilities in hardware along with the impact they might have. The framework's primary application would be to test vulnerabilities for which the tester has a high-level working knowledge of how they can result in a security policy violation.

(ii) The tester should have access or visibility to certain points in the design anticipated to be affected by the triggering of the vulnerability. The encryption key used in the crypto core of an SoC is an asset. To test whether a vulnerability exists in the design which can lead implicit or explicit flow of this asset to a PO, the observable point in the design would be the PO. On the other hand, if we are testing to check if the vulnerability enables flow of the asset to an unauthorized 3PIP, the observable point should be the SoC bus through which this type of transaction might take place.

(iii) The tester should have reasonable (but not necessarily exact) knowledge of how to trigger the vulnerability. This in turn would dictate the input to mutate on. For example, let us consider the debug unit vulnerability described in [96] where the password check of the JTAG interface could potentially be bypassed by resetting the unit. In this case, the hardware debug signals exposed to the outside world would be the relevant inputs. For the key asset scenario described earlier, the input to mutate would be physical signals of the crypto core exposed to the outside world or a user space program that can access AES resources. Similarly, for triggering software-exploitable vulnerabilities, the input to mutate would be the data and control flow of a user space program.

This stands in contrast to random and blind fuzz testing, which find vulnerabilities by applying random inputs to the design which in turn can unexpectedly lead the design to a nonfunctional or vulnerable state. The BPSO algorithm is suited for vulnerability detection at the pre-silicon level since any input to a digital design can be considered in terms of binary vectors. Discernibly hardware signals are binary quantities. The user space programs that run on modern SoCs can also be described in terms of binary vectors by translating the associated program into corresponding assembly instructions. Additionally, to incorporate sequential inputs, each particle in the swarm can be considered as input vectors applied at different clock cycles.

4.6.4.3   Validity of Gray-Box Assumptions

Since the BPSO-based penetration test framework assumes knowledge on the part of the tester and the availability of RTL code, the readers might assume that this violates the gray-box testing goals of the framework. We now discuss why the prerequisite knowledge assumed earlier does not necessarily violate gray-box assumptions especially in the context of SoC verification. Modern SoCs can contain tens of different third-party IPs, many of which are too complex with their implementation details abstracted away by integrating CAD tools. Even during functional verification, the verification engineer would only have a high-level knowledge of the vulnerability and functionality of the integrated 3PIP and not the minutiae of RTL implementation. In such cases, it can become extremely challenging for the verification engineer to trigger the vulnerability with existing verification tools due to complex transactions occurring inside the SoC, implicit timing flows, or unanticipated security unaware design flaws. We also note that formal verification tools such as Cadence JasperGold®, even with definitive knowledge of the impact a vulnerability might have, can throw false positives or suffer from state explosion problems [97].

4.7   SoC Security Verification: AI Testing

Pre-silicon verification is an essential but time-consuming and tedious part that consumes about 70% of the total time allocated for hardware design flow [98]. Similar to fuzz and pen testing, machine learning (ML) can be used for the SoC security verification to make the process automatic and evolutionary-based. ML can be used in different aspects of the verification process, such as generating new test cases that cover more functional states, producing new stimuli and input patterns for increasing coverage, analyzing the test results, etc. The biggest challenge of using ML will be selecting appropriate data, models, and ML techniques through trial and error to get the desired automation and coverage that the state-of-the-art techniques are unable to provide.

4.7.1 Higher Level Machine Learning

In general, three commonly used words—machine learning (ML), deep learning (DL), and neural network (NN)—are sub-fields of AI. While intelligent devices use AI to replicate human thought processes, ML is the mechanism to develop its intelligence. ML uses statistical models and extracts patterns from data to train a system without direct instructions. ML facilitates an intelligent approach to continue learning and improving through feedback. Even though it is common to use machine learning and deep learning interchangeably, they are not the same. DL is a sub-field of ML, whereas NN is a sub-field of DL. Figure 4.8 shows the simplified version of the machine learning workflow. While applying machine learning, the first step is to formulate the problem statement. After that, collecting appropriate data is an essential stage to train the model to solve the problem like a human. The overall performance of the ML model depends on how representative the training data is. ML model learns from some extracted features of the relevant data. The higher the amount of data, the better the training would be. Hence, ML data must be the adequate amount (the more the data, the better) and must have appropriate depth and feature (if inherently 2D data is represented by 3D data, the training method will not be effective). Data also have to be representative unbiased and should include every corner case possible. Once the developers have enough data, they can move forward to train different ML models to see which model works best for their problem statement and collected data.

Figure 4.9 shows different tasks a machine learning model can perform (classification, anomaly detection, etc.). Depending on the task, developers have to choose a method of training, for example, supervised or unsupervised learning. Once the developers have selected the task, training method, and the model itself, they can then fine-tune the best model for their application and deploy it for usage.

Using ML to alleviate some manual interventions by verification engineers has become a recent research trend. In these research works, the authors have used ML to automate different aspects of hardware verification. For example, Hughes et al. [99] used a combination of supervised and reinforcement learning to generate constrain-random stimulus, which will ensure to hit the hard-to-hit combination in highly complex functional design space. Hutter et al. [100] used AI to automatically tune the decision-making procedure of bounded model checking SAT solvers, which in result would boost the verification procedure. Sometimes verification engineers need to recreate a failure to trace back the input that causes system failure. However, stimulating the origin of a system failure is time-consuming and computationally

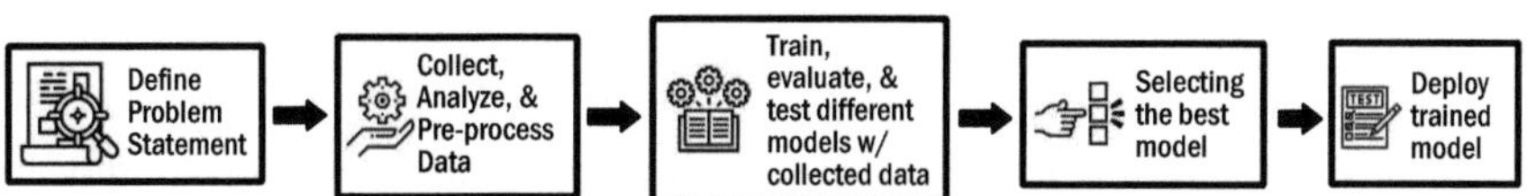

Fig. 4.8 Applied machine learning workflow (general application)

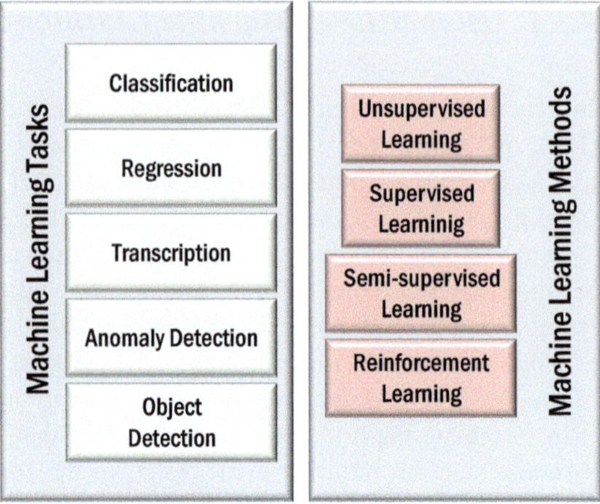

Fig. 4.9 Applied machine learning tasks and methods

draining. Gaur et al. [101] proposed an ML model for this debugging purpose. The ML model will be trained to calculate the switching probability of the design output, which can be used to simulate system failures with negligible overhead.

4.7.2 AI for Hardware Verification

When trying to use ML for hardware verification, the ML workflow shown in Fig. 4.8 has to be followed. Figure 4.10 illustrates the workflow on how ML can be used to enhance hardware verification process. In the following subsections, different requirements and possible challenges of using this workflow for hardware verification are discussed in detail.

4.7.2.1 Requirements/Workflow for Using ML in Verification

Hardware verification is done in every abstraction level of the hardware design flow. For example, after materializing a design's concept and architectural specification, behavioral verification is done on the RTL. Next, functional verification is done at a gate level, transistor level, or during DFT insertion. Finally, the synthesized layout goes through a physical verification process. Therefore, while implementing ML for automating verification, the first requirement is to select the level of abstraction, meaning which method (behavior, function, or physical verification) to use and which level (RTL, gate, transistor, or layout) to use. This is the first step of using ML in pre-silicon verification as shown in Fig. 4.10.

Next, verification engineers must choose what part of the verification to automate through ML. For example, ML can generate stimuli, generate new test cases to increase code or branch coverage, and produce new guided, constrained-random,

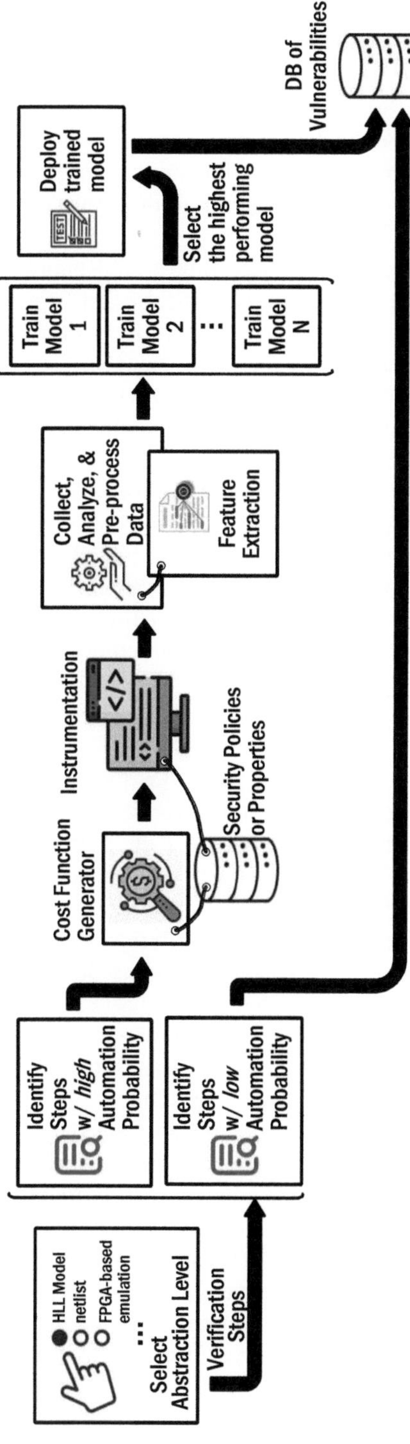

Fig. 4.10 Workflow of hardware verification using applied ML

or entirely random inputs to hit more functional or behavior nodes. Therefore, it is required to select the abstraction level and the role of ML in automation as a problem statement as the first part of ML workflow. As shown in Fig. 4.10, verification engineers have to divide the verification steps into two parts. The first part can be automated through ML, and the second part is the state-of-the-art verification steps that do not need further optimization. Verification engineers have to build problem statements and cost functions corresponding to the first part and leave the second part.

One factor that goes into consideration while establishing a problem statement is if the verification engineer has full access to the design (white-box setting) or has only access to primary input and primary output of the design (black-box setting). Traditionally design engineer and verification engineer are two different entities, and verification engineer does not require any knowledge of the design itself. Also, if the IP came from a 3PIP vendor, white-box knowledge of the IP is inaccessible, as often 3PIPs are encrypted. However, the level of access to the design information (white-box/gray-box/black-box setting) is crucial for collecting training data for the machine learning model. As shown in Fig. 4.10, to get the appropriate training data, the verification engineer may need to instrument the IP. These training data have to fulfill the requirement of being comprehensively representative of the complete problem statement, unbiased, and should cover all corner cases. After collecting required training data and analyzing the data structure, it is mandatory to extract features that represent the problem statement the best. These data features will be used to statistically model the problem statement and give human-like predictions using ML models.

Depending on the data available on the IP (as the model can have one of the black-, gray-, or white-box approaches), the verification engineer must select the best ML method (shown in Fig. 4.9) suited for the predefined problem statement and collect data features. Selecting an appropriate model will be a trial and error process because the model will work differently for different problem statements and data features. For example, in Fig. 4.10, different ML models (model 1, model 2,..., model N) are trained using the same training data, the performance of all these different ML models is compared on the basis of the same cost function, and the best performing model will be selected for automation. While selecting the appropriate method, the verification engineer must also consider the computational power, platform, and resource available for training.

After analyzing, optimizing, and evaluating the trained model, the automated part of the verification process will be integrated with the verification steps that do not require further optimization. This integrated part will produce an accelerated ML-based hardware verification method.

One of the critical requirements of using ML in verification is forming an evaluation matrix and objective function to measure the performance of the ML model. This objective function and evaluation matrix must be dynamic, comprehensive, and analog so that reinforcement feedback can be provided for increasing model accuracy.

4.7.2.2   Challenges of ML-Based Verification

The accuracy and performance of the ML model depend hugely on the quality and quantity of collected data, and collecting an adequate number of relevant, unbiased, comprehensive data is always a challenge. It takes numerous man-hours, computational resources, and manual interventions to collect these data. However, most of the collected data are noisy, biased, and not comprehensive in real-life cases. For this reason, preprocessing and data sorting imposes a challenge for hardware verification. Extracting appropriate features from collected data is another demanding task. While building a model, verification engineers need to spend most of their time analyzing the data to get the appropriate specification and feature depth of their model that represents the problem statement the best. Selecting the wrong data feature will result in poor model performance. This is why careful selection of data features is an essential task.

Efficient model selection has one of the highest impacts. The model structure and depth must be coherent and complementary with the data structure and must train itself from the given data features. As there are many ML methods and models to choose from, training different models and comparing different methods can give the best-suited model for a particular hardware verification problem. How and where to instrument the code, what platform to use, and what properties to check are also very challenging questions from the verification perspective. Testing, developing, and implementing appropriate objective functions are challenging in real life. For example, one objective function may be appropriate for a certain scenario and irrelevant for another. Developing a generalized, scalable, and appropriate objecting function poses a challenge from this perspective.

4.8   Future of SoC Security Verification

In this section, we provide possible directions for further research on SoC security verification using fuzz, penetration, and AI testing. Considering the scalability issue that is ever-increasing by facing with larger and more complex SoCs, introduction of reliable and efficient verification techniques using these self-refinement mechanisms is inevitable. In spite of extensive research efforts in developing scalable and automated security verification techniques over the years, there are still many challenges remain to design secure and trustworthy SoCs. While formal verification tools like JasperGold can be used for known vulnerabilities, here we demonstrated how, by appropriately definition of security policies/properties and evolutionary-based cost-functions, smarter approaches such as fuzz, penetration, and AI testing that are capable of generating smart test cases can extend the scope for both known and unknown SoC security vulnerabilities and could be a promising direction in SoC security verification. Although recent years show some preliminary usage of these techniques, as discussed previously, many of these techniques still need significant improvement, in terms of performance as well as coverage. In addition, depending on the security vulnerability, source of vulnerability, and security policy/property, in many cases there will be a need for developing hybrid approaches combining the inherent advantages of different security verification methods to detect a wide variety of security vulnerabilities in emerging SoCs. In addition, while design-time security verification techniques can detect certain types of vulnerabilities, it is infeasible to remove all possible vulnerabilities during pre-silicon security verification. Due to observability constraints in fabricated SoCs, post-silicon security verification (on initial prototypes or FPGA-based emulation) approaches should be considered as well. So, the future research needs to employ both post-silicon and pre-silicon security verification. Finally, security verification tools need to check for various security vulnerabilities across different phases in the design cycle. Specifically, per each technique, the following draws some of the possible future research directions:

4.8.1 Fuzz Testing

While utilizing fuzz testing for SoC verification is a new concept, still the majority of studies explore the best-fitted conventional methods adopted from the software realm. The threat model of software designs is truly different from the hardware. Future work in this field should address this issue by introducing equivalent cost functions and threat surfaces that can be measured and evaluated. Information flow analysis, static code analysis, and dynamic heuristics developed based on simulation flows have proved to be good candidates for the development of metrics, especially in a white-box scenario.

The second problem yet not addressed properly is due to the fact that most recent studies focus on constructing a general approach that works for all possible vulnerabilities in the design. The metrics mentioned above can be more efficient in detecting hardware vulnerabilities if they consider the scope of the vulnerability targeted at each session of testing. For example, a full exploration of the system while generating many test cases for full system security assurance is not necessary when the memory interface is the only untrusted entity in the system. This can help with the scalability of the fuzzing approach, while a general method might be more suitable for detecting new vulnerabilities.

The automation required for developing an end-to-end fuzzing approach is another issue that needs further investigation. There still exists a gap for easy-to-use, plug-and-play software based on fuzz testing that can be incorporated as is into other verification tools.

4.8.2 Pen Testing

The efficacy of the pre-silicon hardware Pen Testing framework, such as the previously mentioned BPSO-based architecture, is dictated by how effectively the associated cost functions can encapsulate the vulnerability being targeted. The effectiveness of the cost function, in turn, is contingent upon the tester's ability to identify corresponding inputs and effects of the vulnerability. As we mentioned previously, there is an ever-growing database of vulnerabilities that the community understands how to trigger (at a high level) and what impacts they might have. However, due to the modular design practices prevalent in the industry today, the visibility and accessibility within the design is getting reduced. This means gaining access to the signals or points of interest may be a challenge, especially taking time constraints into consideration. For example, the designer may anticipate that the impact of a vulnerability may be visible through the common bus used in the SoC. However, in a pre-silicon setting, the time taken to simulate the design to an appropriate number of clock cycles such that the vulnerability is triggered may become unacceptably high. In such cases, FPGA emulation of the design can be considered as a promising approach to speed up the process.

We note that the previously mentioned BPSO-based framework assumes no particulars on how the observable point is observed. It can be through simulation, emulation, or any other approach preferred by the tester based on time and cost considerations. So long as the tester can provide the algorithm with observed outputs, the algorithm would be able to mutate the input based upon the feedback provided by evaluation of the cost function. Manual formulation of cost functions can become nonscalable if the designers want to test a large variety of vulnerabilities across different platforms and architectures. The best approach to tackle this challenge is to devise an automatic cost function generation methodology based on a general description of the type and scope of the vulnerability as well as microarchitectural implementation details.

4.8.3   AI Testing

State-of-the-art verification processes require hours and hours of manual intervention but fail to achieve desired coverage goals. Moreover, so many different IPs (hard, soft, and firm IPs) are integrated from so many different vendors in a practical system that developing scalable and reusable test cases becomes a daunting task. ML has the potential to overcome both of these issues faced by the traditional verification process. However, ML has its own challenges because the usage of ML for hardware verification is still in its early stages. Once verification engineers overcome the initial challenges of trial and error and begin to explore machine learning for each verification stage, they will establish the specifications for the collected data, identify the best features, determine the appropriate model structure, and assemble everything needed to construct an automated framework. From these established structures and using transfer learning of ML, different IPs with different specifications can be automated to increase verification coverage. Of course, there will always be unique cases where manual interventions will be needed. However, with the help of reinforcement learning, the rate of human intervention requirement will reduce exponentially. One of the possible future direction would be building of this kind of unbiased, scalable, reusable ML model which will increase verification coverage and drastically decrease manual efforts required by the existing process. The best way to implement all of these is to tackle one abstraction level at a time, starting from the RTL. Verification engineers have to generate an automated method to check the scope of using ML in each step of the verification process and start utilizing ML for the promising steps. Also, generating an evolving cost function with dynamic behaviors is another task. After collecting data and training the model, verification engineers have to check the performance of the whole verification step, with and without ML. If the ML model outperforms the traditional approach with lower overhead, then the ML approach should be established as the standard procedure.

References

1. A. Yeh, Trends in the global IC design service market, in *DIGITIMES* (2012)
2. M. Rostami, F. Koushanfar, R. Karri, A primer on hardware security: models, methods, and metrics. Proc. IEEE **102**(8), 1283–1295 (2014)
3. K.Z. Azar, M.M. Hossain, A. Vafaei, H. Al Shaikh, N.N. Mondol, F. Rahman, M. Tehranipoor, F. Farahmandi, Fuzz, penetration, and AI testing for SoC security verification: challenges and solutions, in *Cryptology ePrint Archive* (2022)
4. A. Nahiyan, K. Xiao, K. Yang, Y. Jin, D. Forte, M. Tehranipoor, AVFSM: a framework for identifying and mitigating vulnerabilities in FSMs, in *2016 53nd ACM/EDAC/IEEE Design Automation Conference (DAC)* (IEEE, 2016), pp. 1–6
5. M. Tehranipoor, F. Koushanfar, A survey of hardware Trojan taxonomy and detection. IEEE Des. Test Comput. **27**(1), 10–25 (2010)
6. G.K. Contreras, A. Nahiyan, S. Bhunia, D. Forte, M. Tehranipoor, Security vulnerability analysis of design-for-test exploits for asset protection in SoCs, in *2017 22nd Asia and South Pacific Design Automation Conference (ASP-DAC)* (IEEE, 2017), pp. 617–622
7. H. Al-Shaikh, M.B. Monjil, K.Z. Azar, F. Farahmandi, M. Tehranipoor, F. Rahman, QuardTropy: detecting and quantifying unauthorized information leakage in hardware designs using g-entropy, in *2023 IEEE International Symposium on Defect and Fault Tolerance in VLSI and Nanotechnology Systems (DFT)* (IEEE, 2023), pp. 1–6
8. K.Z. Azar, H.M. Kamali, H. Homayoun, A. Sasan, SMT attack: next generation attack on obfuscated circuits with capabilities and performance beyond the SAT attacks, in *IACR Transactions on Cryptographic Hardware and Embedded Systems (TCHES)* (2019), pp. 97–122
9. K. Xiao, D. Forte, Y. Jin, R. Karri, S. Bhunia, M. Tehranipoor, Hardware Trojans: lessons learned after one decade of research. ACM Trans. Des. Autom. Electron. Syst. (TODAES) **22**(1), 1–23 (2016)
10. Y. Alkabani, F. Koushanfar, Active hardware metering for intellectual property protection and security, in *USENIX Security Symposium* (2007), pp. 291–306
11. J. Rajendran, M. Sam, O. Sinanoglu, R. Karri, Security analysis of integrated circuit camouflaging, in *Proceedings of the ACM SIGSAC Conference on Computer & Communications Security* (2013), pp. 709–720
12. D. Forte, S. Bhunia, M. Tehranipoor, *Hardware Protection Through Obfuscation* (Springer, 2017)
13. K.Z. Azar, H.M. Kamali, H. Homayoun, A. Sasan, From cryptography to logic locking: a survey on the architecture evolution of secure scan chains. IEEE Access **9**, 73133–73151 (2021)

14. H.M. Kamali, K.Z. Azar, F. Farahmandi, M. Tehranipoor, Advances in logic locking: past, present, and prospects, in *Cryptology ePrint Archive* (2022)

15. U. Das, M.S. Rahman, N.N. Anandakumar, K.Z. Azar, F. Rahman, M. Tehranipoor, F. Farahmandi, PSC-watermark: power side channel based IP watermarking using clock gates, in *2023 IEEE European Test Symposium (ETS)* (IEEE, 2023), pp. 1–6

16. P.P. Sarker, U. Das, M.B. Monjil, H.M. Kamali, F. Farahmandi, M. Tehranipoor, GEM-water: generation of EM-based watermark for SoC IP validation with hidden FSMs, in *ISTFA 2023* (ASM International, 2023), pp. 271–278

17. M.M.M. Rahman, S. Tarek, K.Z. Azar, M. Tehranipoor, F. Farahmandi, Efficient SoC security monitoring: quality attributes and potential solutions, in *IEEE Design & Test* (2023)

18. M.M.M. Rahman, S. Tarek, K.Z. Azar, F. Farahmandi, EnSAFe: enabling sustainable SoC security auditing using eFPGA-based accelerators, in *2023 IEEE International Symposium on Defect and Fault Tolerance in VLSI and Nanotechnology Systems (DFT)* (IEEE, 2023), pp. 1–6

19. D. Beyer, T. Lemberger, Software verification: testing vs. model checking, in *Haifa Verification Conference* (Springer, 2017), pp. 99–114

20. C. Calcagno, D. Distefano, J. Dubreil, D. Gabi, P. Hooimeijer, M. Luca, P. O'Hearn, I. Papakonstantinou, J. Purbrick, D. Rodriguez, Moving fast with software verification, in *NASA Formal Methods Symposium* (Springer, 2015), pp. 3–11

21. *American Fuzzy Lop (AFL) Fuzzer* (2018). http://lcamtuf.coredump.cx/afl/ (visited on April 2018)

22. W. Chen, S. Ray, J. Bhadra, M. Abadir, L.-C. Wang, Challenges and trends in modern SoC design verification. IEEE Des. Test **34**(5), 7–22 (2017)

23. K.Z. Azar, H.M. Kamali, H. Homayoun, A. Sasan, NNgSAT: neural network guided SAT attack on logic locked complex structures, in *IEEE/ACM International Conference On Computer Aided Design (ICCAD)* (2020), pp. 1–9

24. K.Z. Azar, H.M. Kamali, F. Farahmandi, M. Tehranipoor, Warm up before circuit de-obfuscation? An exploration through bounded-model-checkers, in *International Symposium on Hardware Oriented Security and Trust (HOST)* (2022), pp. 1–4

25. P. Subramanyan, S. Malik, H. Khattri, A. Maiti, J. Fung, Verifying information flow properties of firmware using symbolic execution, in *2016 Design, Automation & Test in Europe Conference & Exhibition (DATE)* (IEEE, 2016), pp. 337–342

26. A. Nahiyan, F. Farahmandi, P. Mishra, D. Forte, M. Tehranipoor, Security-aware FSM design flow for identifying and mitigating vulnerabilities to fault attacks. IEEE Trans. Comput.-Aided Des. Integr. Circuits Syst. **38**(6), 1003–1016 (2018)

27. B. Kumar, A.K. Jaiswal, V.S. Vineesh, R. Shinde, Analyzing hardware security properties of processors through model checking, in *2020 33rd International Conference on VLSI Design and 2020 19th International Conference on Embedded Systems (VLSID)* (IEEE, 2020), pp. 107–112

28. B. Yuce, N.F. Ghalaty, P. Schaumont, TVVF: estimating the vulnerability of hardware cryptosystems against timing violation attacks, in *2015 IEEE International Symposium on Hardware Oriented Security and Trust (HOST)* (IEEE, 2015), pp. 72–77

29. J. Demme, R. Martin, A. Waksman, S. Sethumadhavan, Side-channel vulnerability factor: a metric for measuring information leakage, in *2012 39th Annual International Symposium on Computer Architecture (ISCA)* (IEEE, 2012), pp. 106–117

30. A. Nahiyan, J. Park, M. He, Y. Iskander, F. Farahmandi, D. Forte, M. Tehranipoor, Script: a cad framework for power side-channel vulnerability assessment using information flow tracking and pattern generation. ACM Trans. Des. Autom. Electron. Syst. (TODAES) **25**(3), 1–27 (2020)

31. H. Salmani, M. Tehranipoor, Analyzing circuit vulnerability to hardware Trojan insertion at the behavioral level, in *2013 IEEE International Symposium on Defect and Fault Tolerance in VLSI and Nanotechnology Systems (DFTS)* (IEEE, 2013), pp. 190–195

32. G. Dessouky, D. Gens, P. Haney, G. Persyn, A. Kanuparthi, H. Khattri, J.M. Fung, A.-R. Sadeghi, J. Rajendran, {HardFails}: insights into {software-exploitable} hardware bugs, in *28th USENIX Security Symposium (USENIX Security 19)* (2019), pp. 213–230

33. R. Zhang, N. Stanley, C. Griggs, A. Chi, C. Sturton, Identifying security critical properties for the dynamic verification of a processor. ACM SIGARCH Comput. Archit. News **45**(1), 541–554 (2017)

34. W. Hu, A. Althoff, A. Ardeshiricham, R. Kastner, Towards property driven hardware security, in *2016 17th International Workshop on Microprocessor and SOC Test and Verification (MTV)* (IEEE, 2016), pp. 51–56

35. M. Tiwari, J.K. Oberg, X. Li, J. Valamehr, T. Levin, B. Hardekopf, R. Kastner, F.T. Chong, T. Sherwood, Crafting a usable microkernel, processor, and I/O system with strict and provable information flow security, in *2011 38th Annual International Symposium on Computer Architecture (ISCA)* (IEEE, 2011), pp. 189–199

36. R. Kastner, J. Oberg, W. Huy, A. Irturk, Enforcing information flow guarantees in reconfigurable systems with mix-trusted IP, in *Proceedings of the International Conference on Engineering of Reconfigurable Systems and Algorithms (ERSA)* (The Steering Committee of The World Congress in Computer Science, Computer… 2011), p. 1

37. W. Hu, J. Oberg, A. Irturk, M. Tiwari, T. Sherwood, D. Mu, R. Kastner, On the complexity of generating gate level information flow tracking logic. IEEE Trans. Inf. Forensics Secur. **7**(3), 1067–1080 (2012)

38. J. Oberg, W. Hu, A. Irturk, M. Tiwari, T. Sherwood, R. Kastner, Information flow isolation in I2C and USB, in *2011 48th ACM/EDAC/IEEE Design Automation Conference (DAC)* (IEEE, 2011), pp. 254–259

39. M.M. Hossain, A. Vafaei, K.Z. Azar, F. Rahman, F. Farahmandi, M. Tehranipoor, SoCFuzzer: SoC vulnerability detection using cost function enabled fuzz testing, in *2023 Design, Automation & Test in Europe Conference & Exhibition (DATE)* (IEEE, 2023), pp. 1–6

40. H. Al-Shaikh, A. Vafaei, M.M.M. Rahman, K.Z. Azar, F. Rahman, F. Farahmandi, M. Tehranipoor, Sharpen: Soc security verification by hardware penetration test, in *Proceedings of the 28th Asia and South Pacific Design Automation Conference* (2023), pp. 579–584

41. M.M. Hossain, K.Z. Azar, F. Farahmandi, M. Tehranipoor, TaintFuzzer: SoC security verification using taint inference-enabled fuzzing, in *International Conference On Computer Aided Design (ICCAD)* (IEEE, 2023), pp. 1–9

42. N.N. Mondol, A. Vafaei, K.Z. Azar, F. Farahmandi, M. Tehranipoor, RL-TPG: automated pre-silicon security verification through reinforcement learning-based test pattern generation, in *Design, Automation and Test in Europe (DATE)* (IEEE, 2024), pp. 1–6

43. N. Farzana, M.M. Hossain, K.Z. Azar, F. Farahmandi, M. Tehranipoor, FormalFuzzer: formal verification assisted fuzz testing for SoC vulnerability detection, in *Asia and South Pacific Design Automation Conference (ASP-DAC)* (IEEE, 2024), pp. 1–6

44. P. Kocher, J. Horn, A. Fogh, D. Genkin, D. Gruss, W. Haas, M. Hamburg, M. Lipp, S. Mangard, T. Prescher, et al., Spectre attacks: exploiting speculative execution, in *2019 IEEE Symposium on Security and Privacy (SP)* (IEEE, 2019), pp. 1–19

45. M. Lipp, M. Schwarz, D. Gruss, T. Prescher, W. Haas, A. Fogh, J. Horn, S. Mangard, P. Kocher, D. Genkin, et al., Meltdown: reading kernel memory from user space, in *27th USENIX Security Symposium (USENIX Security 18)* (2018), pp. 973–990

46. E. Peeters, Soc security architecture: current practices and emerging needs, in *2015 52nd ACM/EDAC/IEEE Design Automation Conference (DAC)* (IEEE, 2015), pp. 1–6

47. S.J. Greenwald, Discussion topic: what is the old security paradigm? in *Proceedings of the 1998 Workshop on New Security Paradigms* (1998), pp. 107–118

48. N. Farzana, F. Rahman, M. Tehranipoor, F. Farahmandi, SoC security verification using property checking, in *2019 IEEE International Test Conference (ITC)* (IEEE, 2019), pp. 1–10

49. N. Farzana, F. Farahmandi, M. Tehranipoor, SoC security properties and rules, in *Cryptology ePrint Archive* (2021)

50. P. Mishra, M. Tehranipoor, S. Bhunia, Security and trust vulnerabilities in third-party IPs, in *Hardware IP Security and Trust* (Springer, 2017), pp. 3–14

51. M. Gruninger, C. Menzel, The process specification language (PSL) theory and applications. AI Magazine **24**(3), 63 (2003)

52. S. Vijayaraghavan, M. Ramanathan, *A Practical Guide for SystemVerilog Assertions* (Springer Science & Business Media, New York, 2005)
53. A. Pnueli, The temporal logic of programs, in *18th Annual Symposium on Foundations of Computer Science (SFCS 1977)* (IEEE, 1977), pp. 46–57
54. A. Cimatti, E. Clarke, F. Giunchiglia, M. Roveri, NuSMV: a new symbolic model checker. Int. J. Softw. Tools Technol. Transf. **2**(4), 410–425 (2000)
55. C. Dunbar, G. Qu, Designing trusted embedded systems from finite state machines. ACM Trans. Embedded Comput. Syst. (TECS) **13**(5s), 1–20 (2014)
56. D.M. Anderson, *Design for Manufacturability: How to Use Concurrent Engineering to Rapidly Develop Low-Cost, High-Quality Products for Lean Production* (CRC Press, New York, USA, 2020)
57. B.P. Miller, L. Fredriksen, B. So, An empirical study of the reliability of UNIX utilities. Commun. ACM **33**(12), 32–44 (1990)
58. J. Wang, Y. Duan, W. Song, H. Yin, C. Song, Be sensitive and collaborative: analyzing impact of coverage metrics in greybox fuzzing, in *22nd International Symposium on Research in Attacks, Intrusions and Defenses (RAID 2019)*. Chaoyang District, Beijing: USENIX Association, Sep 2019, pp 1–15. ISBN: 978-1-939133-07-6. https://www.usenix.org/conference/raid2019/presentation/wang
59. *honggfuzz* (2018). http://honggfuzz.com/ (visited on April 2018)
60. G. Klees, A. Ruef, B. Cooper, S. Wei, M. Hicks, Evaluating fuzz testing, in *Proceedings of the 2018 ACM SIGSAC Conference on Computer and Communications Security* (2018), pp. 2123–2138
61. T. Trippel, K.G. Shin, A. Chernyakhovsky, G. Kelly, D. Rizzo, M. Hicks, Fuzzing hardware like software, in *arXiv preprint arXiv:2102.02308* (2021)
62. J. De Ruiter, E. Poll, Protocol state fuzzing of {TLS} implementations, in *24th {USENIX} Security Symposium ({USENIX} Security 15)* (2015), pp. 193–206
63. P. Godefroid, M.Y. Levin, D.A. Molnar, et al., Automated whitebox fuzz testing, in *NDSS*, vol. 8 (2008), pp. 151–166
64. R.L. Seagle Jr., A framework for file format fuzzing with genetic algorithms (2012)
65. M.M. Hossain, F. Farahmandi, M. Tehranipoor, F. Rahman, BOFT: exploitable buffer overflow detection by information flow tracking, in *2021 Design, Automation & Test in Europe Conference & Exhibition (DATE)* (IEEE, 2021), pp. 1126–1129
66. https://www.veripool.org/verilator/
67. S.K. Muduli, G. Takhar, P. Subramanyan, Hyperfuzzing for SoC security validation, in *Proceedings of the 39th International Conference on Computer-Aided Design* (2020), pp. 1–9
68. D. Moghimi, M. Lipp, B. Sunar, M. Schwarz, Medusa: microarchitectural data leakage via automated attack synthesis, in *29th {USENIX} Security Symposium ({USENIX} Security 20)* (2020), pp. 1427–1444
69. O. Oleksenko, B. Trach, M. Silberstein, C. Fetzer, SpecFuzz: bringing Spectre-type vulnerabilities to the surface, in *29th {USENIX} Security Symposium ({USENIX} Security 20)* (2020), pp. 1481–1498
70. S. Nilizadeh, Y. Noller, C.S. Pasareanu, DifFuzz: differential fuzzing for side-channel analysis, in *2019 IEEE/ACM 41st International Conference on Software Engineering (ICSE)* (IEEE, 2019), pp. 176–187
71. Y. Xiao, Y. Zhang, R. Teodorescu, SPEECHMINER: a framework for investigating and measuring speculative execution vulnerabilities, in *arXiv preprint arXiv:1912.00329* (2019)
72. M. Ghaniyoun, K. Barber, Y. Zhang, R. Teodorescu, IntroSpectre: a pre-silicon framework for discovery and analysis of transient execution vulnerabilities, in *2021 ACM/IEEE 48th Annual International Symposium on Computer Architecture (ISCA)* (IEEE, 2021), pp. 874–887
73. A. Tyagi, A. Crump, A.-R. Sadeghi, G. Persyn, J. Rajendran, P. Jauernig, R. Kande, TheHuzz: instruction fuzzing of processors using golden-reference models for finding software-exploitable vulnerabilities (2022)
74. A. Basak, S. Bhunia, T. Tkacik, S. Ray, Security assurance for system-on-chip designs with untrusted IPs. IEEE Trans. Inf. Forensics Secur. **12**(7), 1515–1528 (2017)

75. P. Subramanyan, D. Arora, Formal verification of taint-propagation security properties in a commercial SoC design, in *2014 Design, Automation & Test in Europe Conference & Exhibition (DATE)* (IEEE, 2014), pp. 1–2

76. K. Laeufer, J. Koenig, D. Kim, J. Bachrach, K. Sen, RFUZZ: coverage-directed fuzz testing of RTL on FPGAs, in *2018 IEEE/ACM International Conference on Computer-Aided Design (ICCAD)* (IEEE, 2018), pp. 1–8

77. S. Shah, B.M. Mehtre, An overview of vulnerability assessment and penetration testing techniques. J. Comput. Virol. Hacking Tech. **11**(1), 27–49 (2015)

78. H.H. Thompson, Application penetration testing. IEEE Secur. Privacy **3**(1), 66–69 (2005)

79. W. Knowles, A. Baron, T. McGarr, The simulated security assessment ecosystem: does penetration testing need standardisation? Comput. Secur. **62**, 296–316 (2016)

80. https://owasp.org/www-project-web-security-testing-guide/latest/3-The%5C_OWASP%5C_Testing%5C_Framework/1-Penetration%5C_Testing%5C_Methodologies (visited on 02 Oct 2022)

81. https://www.isecom.org/OSSTMM.3.pdf (visited on 02 Oct 2022)

82. https://www.crest-approved.org/what-is-star-fs/index.html (visited on 02 Oct 2022)

83. R. Baloch, *Ethical Hacking and Penetration Testing Guide* (Auerbach Publications, Boston, 2017)

84. T. Dimkov, A. Van Cleeff, W. Pieters, P. Hartel, Two methodologies for physical penetration testing using social engineering, in *Proceedings of the 26th Annual Computer Security Applications Conference* (2010), pp. 399–408

85. H. Khattri, N.K.V. Mangipudi, S. Mandujano, HSDL: a security development lifecycle for hardware technologies, in *2012 IEEE International Symposium on Hardware-Oriented Security and Trust* (IEEE, 2012), pp. 116–121

86. M. He, J. Park, A. Nahiyan, A. Vassilev, Y. Jin, M. Tehranipoor, RTL-PSC: automated power side-channel leakage assessment at register-transfer level, in *2019 IEEE 37th VLSI Test Symposium (VTS)* (IEEE, 2019), pp. 1–6

87. H. Wang, H. Li, F. Rahman, M.M. Tehranipoor, F. Farahmandi, Sofi: security property-driven vulnerability assessments of ICs against fault-injection attacks. IEEE Trans. Comput.-Aided Des. Integr. Circuits Syst. **41**(3), 452–465 (2021)

88. M.R. Muttaki, M. Tehranipoor, F. Farahmandi, FTC: a universal fault injection attack detection sensor, in *IEEE International Symposium on Hardware-Oriented Security and Trust (HOST)* (IEEE, 2022)

89. J. Kennedy, R.C. Eberhart, A discrete binary version of the particle swarm algorithm, in *1997 IEEE International Conference on Systems, Man, and Cybernetics. Computational Cybernetics and Simulation*, vol. 5 (IEEE, 1997), pp. 4104–4108

90. H. Salmani, M. Tehranipoor, R. Karri, On design vulnerability analysis and trust benchmarks development, in *2013 IEEE 31st International Conference on Computer Design (ICCD)* (IEEE, 2013), pp. 471–474

91. B. Shakya, T. He, H. Salmani, D. Forte, S. Bhunia, M. Tehranipoor, Benchmarking of hardware Trojans and maliciously affected circuits. J. Hardw. Syst. Secur. **1**(1), 85–102 (2017)

92. X. Zhang, M. Tehranipoor, Case study: detecting hardware Trojans in third-party digital IP cores, in *2011 IEEE International Symposium on Hardware-Oriented Security and Trust* (IEEE, 2011), pp. 67–70

93. X. Zhang, H. Salmani, Integrated circuit authentication: hardware Trojans and counterfeit detection (2014)

94. A. Nahiyan, M. Sadi, R. Vittal, G. Contreras, D. Forte, M. Tehranipoor, Hardware Trojan detection through information flow security verification, in *2017 IEEE International Test Conference (ITC)* (IEEE, 2017), pp. 1–10

95. S. Bhunia, M. Tehranipoor, *Hardware Security: A Hands-on Learning Approach* (Morgan Kaufmann, San Francisco, 2018)

96. S. Gogri, P. Joshi, P. Vurikiti, N. Fern, M. Quinn, J. Valamehr, Texas A&M Hackin' Aggies' security verification strategies for the 2019 Hack@ DAC competition. IEEE Des. Test **38**(1), 30–38 (2020)

97. F. Farahmandi, Y. Huang, P. Mishra, *System-on-Chip Security* (Springer, Cham, 2020)
98. H. Kaeslin, *Digital Integrated Circuit Design: From VLSI Architectures to CMOS Fabrication* (Cambridge University Press, New York, 2008)
99. W. Hughes, S. Srinivasan, R. Suvarna, M. Kulkarni, Optimizing design verification using machine learning: doing better than random, in *arXiv preprint arXiv:1909.13168* (2019)
100. F. Hutter, D. Babic, H.H. Hoos, A.J. Hu, Boosting verification by automatic tuning of decision procedures, in *Formal Methods in Computer Aided Design (FMCAD'07)* (IEEE, 2007), pp. 27–34
101. P. Gaur, S.S. Rout, S. Deb, Efficient hardware verification using machine learning approach, in *2019 IEEE International Symposium on Smart Electronic Systems (iSES) (Formerly iNiS)* (IEEE, 2019), pp. 168–171

Chapter 5
Runtime SoC Security Validation

5.1 Introduction

As the demand for more advanced technologies grows, systems-on-chip (SoCs) are becoming more complex and capable of supporting more features. The ever-increasing complexity and strict time-to-market make it challenging to perform a thorough security verification of the SoC. At the same time, IP reuse has become a common trend among SoC designers to meet strict time-to-market requirements. This introduces potentially untrusted third-party IPs (3PIPs) into the supply chain. Moreover, in the horizontal model of the IC supply chain, there is an increasing risk of introducing security vulnerabilities during the design process or at the foundry. Furthermore, the EDA tools used for logic/physical synthesis are highly tuned for performance/power/area efficiency, allowing for unintended security vulnerabilities. Hence, a security monitoring mechanism is a must for today's SoCs [1, 2].

Over time, we witnessed numerous research into hardware-oriented security and trust solutions. However, most of this research focuses on IP-level security solutions, e.g., hardware obfuscation, watermarking, metering, camouflaging, side-channel analysis, etc. [3–6]. These solutions are not applicable to SoC security vulnerabilities because vulnerabilities in the SoC exploit inter-IP (cross-layer) connections/transactions [1, 7, 8]. Few recent studies, however, focus on SoC security monitoring and vulnerability detection. A centralized monitoring engine that implements security rules (policies) and acts as the policy server was studied in [9–11]. In such solutions, a security wrapper customized for security policies is added to each security-critical IP for monitoring events, configurations, and data, as well as coordinating with the centralized controller to ensure policy compliance. However, these wrappers follow a fixed/static architecture and thus cannot be stay tuned for newer possible backdoors after the fabrication, so they cannot address dynamic vulnerabilities.

M. Tehranipoor et al., *Hardware Security*,
https://doi.org/10.1007/978-3-031-58687-3_5

Using design-for-debug (DfD) infrastructures to monitor IP-level (intra-IP) and SoC-level (inter-IP) signals to detect SoC vulnerabilities was studied in [12, 13]. This approach also uses a centralized policy engine that monitors signals/events captured by the DfD infrastructures. However, certain challenges must be tackled when focusing on security and observing internal signals through the debug infrastructure. For example, repurposing the DfD (for vulnerability detection) should not interfere with its use for debugging. Furthermore, IP providers must map security-critical signals/events to the DfD instrumentation for the IP to adapt to this DfD-based security solution. Assuming this in modern SoC designs with multiple third-party IPs is very difficult.

Additionally, as an instance of a commercial solution, an industry standard solution for debugging and monitoring is developed by UltraSoC [14]. This solution utilizes distributed and hierarchical monitoring units across the design with optionally inserted logic filters/analyzers. However, the solution is more suited for debugging than security monitoring as it mostly checks for static access control and data flow.

Considering the limitations of existing security monitoring approaches, in this chapter, we define a set of solid quality attributes (*"ilities"*) needed for reliable and efficient security monitoring. These *ilities* are (ILITY-1) *scalability* (overhead and implementation effort), (ILITY-2) *sustainability*, (ILITY-3) *adaptability*, (ILITY-4) *observability*, and (ILITY-5) *distributibiity*. We provide a meticulous definition for these *ilities* needed for security monitoring. Then, we open up four different research avenues as potential solutions to fulfill the requirement for security monitoring. With a holistic evaluation, we demonstrate how each method may bring advantages for the designer or raise challenges.

5.2 Background of Security Vulnerabilities and Monitoring

The design objectives of an SoC are traditionally driven by (design/implementation/fabrication) cost, performance, and time-to-market constraints while leaving security largely unaddressed. Furthermore, security-aware design practices do not yet exist. Thus, there can be different sets of vulnerabilities in an SoC that may come from various sources. Security vulnerabilities may come from unintentional design mistakes or intentional (malicious) modifications (by rogue insiders or third parties). There can also be security and trust issues in the design that make the SoC vulnerable to different attacks, e.g., side-channel attacks, fault injection attacks, information leakage, and Trojans. In addition, many products have DfT and DfD that improve controllability/observability [15, 16]. However, if there is no proper access control to these structures, attackers may misuse them for controlling and observing the secrets in the SoC. Moreover, the CAD tools used in the design process are unaware of potential security vulnerabilities. Hence, protecting assets in an SoC becomes more challenging. We can define an asset as any information that should be protected from an adversary. In general, these security vulnerabilities enable adversaries to gain unprivileged access to these assets or disrupt the correct/secure operation of the SoC.

A security policy is a rule that should be maintained to protect the confidentiality, integrity, and availability (CIA) of security assets against a security vulnerability. Hence, security policies are at the core of any security monitoring engine. Depending on where the vulnerabilities reside and the logic required to detect them, various SoC transactions/events must be tracked to build a reliable security monitoring technique. From a security monitoring point of view, security policies can be classified into two categories, i.e., static and dynamic. Static policies are rules that are specific to a design strategy and are typically fixed for an SoC. On the other hand, dynamic policies are rules that are defined to prevent attacks or unwanted behaviors that may or may not occur at runtime.

High-level security policies need to be mapped/built into security monitors. If security monitors are static and are not upgradeable, security policies can be implemented as synthesizable RTL codes and integrated into IP or SoC designs. However, a security policy that maps to RTL implementation cannot be changed after chip fabrication. On the other side, recent studies [9–11, 13] implement security monitors as configurable wrappers with a central monitor or policy server. The security policies are mapped into the central monitor with firmware upgrades [9] or as configuration bits [10, 11, 13]. The way security policies are mapped into the security monitoring engine significantly affects the coverage (number of detected vulnerabilities) as well as the lifespan (over time with consideration of the newer attacks) of the design equipped with security monitors.

5.3 "ilities" Needed for Security Monitoring

To detect and address security vulnerabilities both in the design phase (static) and in-field (dynamic), we need a security monitoring approach that meets a set of quality attributes at the same time, which are as follows:

(ILITY-1) Scalability: Scalability of an SoC security monitoring means how suitable it is for a fixed set of policies and can support more types of security policies defined for different vulnerabilities in an SoC. As the area overhead of security monitoring increases with more security policy implementations, the more common logic (reuse) for different security policies (larger the room for adding security policies), the more scalable the security monitoring becomes.

(ILITY-2) Sustainability: SoC security monitoring sustainability means that the monitoring technique is reconfigurable to address dynamic and runtime vulnerabilities (e.g., zero-day attacks). Security policy upgradability is a desired feature of security monitoring, which is not comprehensively supported by the existing practices. If the policies cannot be upgraded in-field, the monitoring technique becomes static because it can only check for the vulnerabilities considered during the design process (not for unforeseen newer threats).

(ILITY-3) Adaptability: Adaptability of an SoC security monitoring indicates how it can be used in different SoCs with minimum adjustments or modifications. Having a security monitoring solution that is optimized and efficient for a specific SoC but is not suitable for other SoCs is not desirable. It implies that the signals/events of a security monitor should be designed in a way that migrating from one design to another can be done in a plug-and-play manner.

(ILITY-4) Observability: It refers to how security-critical interconnects/signals are selected so that more policies can be tracked/implemented with minimal observation. Limited observability means that security monitoring requires a large number of signals for a few policies, thereby limiting the performance and the number of policies that can be implemented (lower scalability). Thus, the security monitor must have efficient observability in different parts of the SoC, especially those responsible for security-critical functions.

(ILITY-5) Distributability: Distributability means that the monitoring approach has distributed monitoring units across the design (hierarchical monitoring). Some of the existing monitoring solutions [9–11] use centralized monitoring where the monitor connects to the system bus and sits at a fixed location in the SoC. Unfortunately, vulnerabilities can occur anywhere in the SoC, and the effect of these vulnerabilities is not always visible on the system bus. Thus, monitoring only the system bus will severely limit the effectiveness of the security monitor. Moreover, connecting the central monitor with signals and interconnects across the SoC is not practical due to the resulting complexity of IO placement and routing (PnR). Thus, the monitoring technique should be divided and distributed across the SoC to increase the observability of transactions and events and to eliminate PnR complexity.

5.4 Possible Avenues of SoC Security Monitoring

In summary, none of the existing solutions meet these *ilities* simultaneously. In this section, we explore four avenues of security monitoring solutions, each of which fulfills different requirements and is suitable for different scenarios. We will also try to find a solution that closely meets the general requirements outlined above.

5.4.1 Solution 1: Synthesizable Assertions

Assertions are commonly used in (functional) RTL/netlist verification [17]. Considering that they can also be implemented as synthesizable codes, known as synthesizable assertion (SA), they can be distributively integrated into the SoC to check for security policies. SA-based checkers can be inserted either at the IP level or at the SoC level, as shown in Fig. 5.1. For IP-level implementation, the assertions

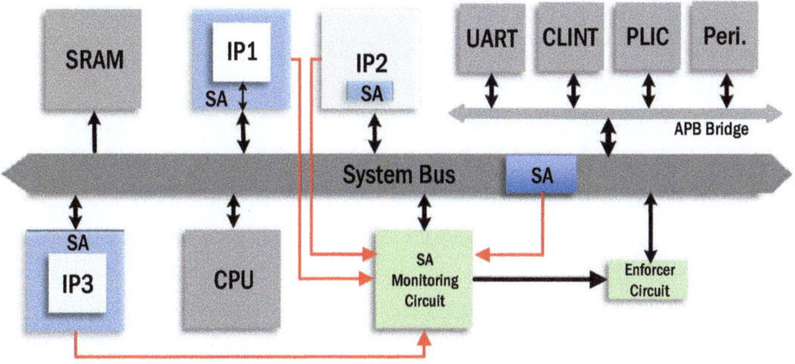

Fig. 5.1 Security monitoring with synthesizable assertion-based checkers

will be a part of the IP itself. Thus, assertions will have access to the signals inside the IP. So, these assertions can be used to check for different properties of the IP for different vulnerabilities. An example of such an SA-based property checker would be to check for illegal state transitions in an IP controller FSM. For SoC-level implementation, the assertions will be outside of the IPs but in the system interconnect or in the glue logic for IP integration. The SAs will have access to system bus transactions and/or IP I/Os. So, these assertions can check for system-level security policies. In this case, the system integrator identifies the security policies based on the assets. The SA-based monitors will help to identify improper handling of these assets. An assertion-based monitor for access control policy is an example of SoC-level implementation.

As the SA-based checkers will be distributed across the SoC, there should be an assertion monitoring circuit that will check the responses from the SA-based checkers (Fig. 5.1). The SA monitoring circuit consists of a simple logic that collects responses from the distributed checkers and generates interrupts based on a predefined priority. Whenever an assertion is violated, enforcement activities such as stopping the current process, blocking the transaction, and disabling the IP address should be carried out. Thus, the interrupt signals should be sent to the processor or a separate enforcer circuit, if available.

Scalability of Solution 1: An SA-based monitoring system is scalable in that any static security property can be assessed with minimal overhead. A new SA-based checker can be introduced into the SoC regardless of other SAs. As the SA-based checkers work independently, a larger number of SAs will not degrade the performance.

Sustainability of Solution 1: As its name implies, SAs are policies or rules implemented statically. Also, SAs cannot be modified/changed after chip fabrication. Hence, SA-based monitors cannot check for dynamic vulnerabilities. Similarly, synthesizable assertions cannot be used to address zero-day attacks as they are not upgradeable in-field.

Adaptability of Solution 1: The SA-based monitors check static security policies and properties that are mostly specific to an SoC architecture or IP functionality. As these monitors are not upgradable, there is limited scope to use these monitors in another SoC without changing the RTL code. This creates portability issues and requires designer intervention.

Observability of Solution 1: The distributed SA-based checkers provide high observability both at the IP level and at the SoC level. Therefore, synthesizable assertions can be a suitable choice for any static property checking.

Distributability of Solution 1: SA-based checkers are independent and can be distributed across the SoC. Static security policies are implemented using these monitors, and the vulnerability detection status is sent to the SA monitoring unit. Thus, this monitoring technique has high distributability.

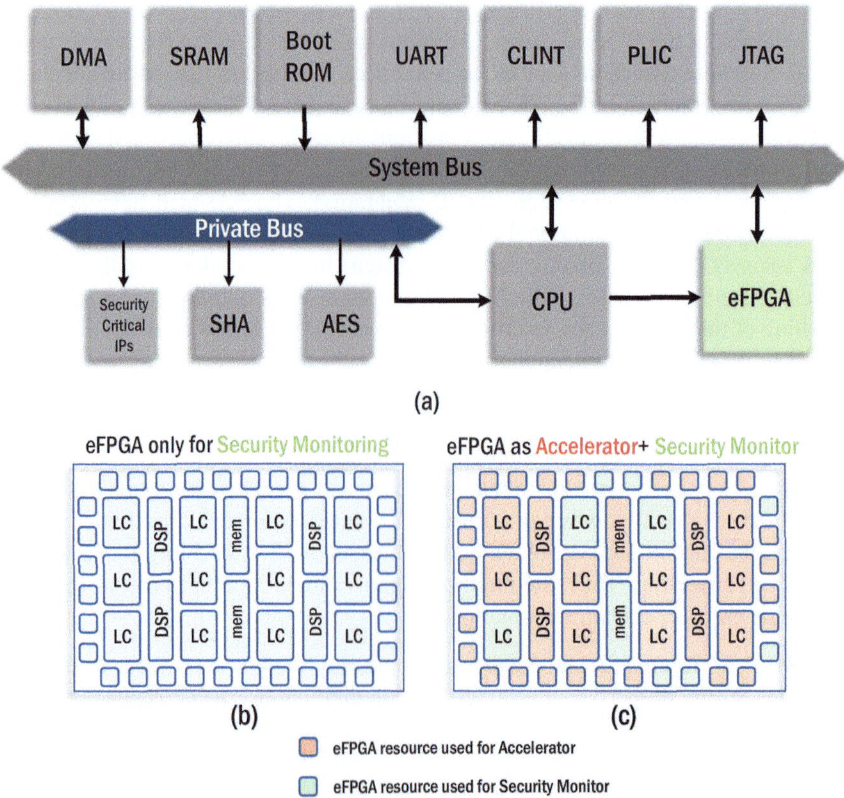

Fig. 5.2 Security monitoring with reconfigurable eFPGA fabric: (**a**) connecting eFPGA to the SoC, (**b**) coarse-grained eFPGA used only for security monitoring, and (**c**) security monitoring using unutilized eFPGA resources

5.4.2 Solution 2: Coarse-Grained eFPGA-Based Monitoring

Solution 2 focuses on mitigating shortcomings of Solution 1 (Sect. 5.4.1), i.e., lack of sustainability/adaptability. To enable these *ilities*, reconfigurability is the key, and for this purpose, eFPGA would be a solution to be integrated into the SoC for implementing policies (Fig. 5.2a), which provides runtime reconfigurability for updating the security policies over time. In such a scenario, with the coarse granularity of eFPGA architecture, security policies are more related to scenarios that monitor bust transactions and status registers. The implementation of this approach starts with identifying security policies based on the threat model, complexity of implementation, and resource utilization. The security policies are then translated into RTL codes and eventually into a bitstream to program the eFPGA. As the eFPGA will be used only for security monitoring (Fig. 5.2b), in Solution 2, a large number of security policies must fit into the eFPGA. However, the actual number of security policies depends on the size of the eFPGA fabric, the types of policies, and their implementation strategies.

The eFPGA must be connected to the private bus or the system bus within the SoC depending on the security policies being implemented (Fig. 5.2a). If security policies require monitoring many signals from different parts of the SoC, then the I/O pads required for the eFPGA fabric become much larger. Hence, based on the type of policies (and signal needed for monitoring), a subset of them may be possible to be implemented via Solution 2 (due to limited I/O pads). With reconfigurability, eFPGA-based security monitoring can be used to address dynamic vulnerabilities like Rowhammer, ransomware, and DMA attacks. Dynamic vulnerabilities cause/exploit unspecified behavior of the SoC, and detecting them typically requires more complex logic. Fortunately, an eFPGA fabric can provide enough resources to implement detection logic for dynamic vulnerabilities.

Scalability Solution 2: Using eFPGA only for security would allow for a larger number of policies to be implemented and monitored, thus making this solution scalable. However, adding eFPGA fabrics into the SoC is not an easy option, mostly due to the size of the eFPGA and eventually PnR challenges. Although the size of the eFPGA fabric depends on the number of security policies and their implementations, the inherent architecture of the eFPGA makes it bulkier than most ASIC modules. Hence, the overhead is very high which often limits the usage of eFPGA only for security.

Sustainability Solution 2: With reconfigurability provided by the eFPGA fabric, new security policies can be added in-field, which makes this solution sustainable for addressing dynamic vulnerabilities and zero-day attacks.

Adaptability Solution 2: Since monitoring relies on bus transactions and shared registers, this solution can easily be integrated and then reprogrammed for a different SoC, which makes it a highly adaptable solution for monitoring.

Observability Solution 2: eFPGA-based monitoring has limited observability as the security policies have access only to the interconnect(s) that are connected to the eFPGA. It is challenging to connect all security-critical signals from different parts of the design to the eFPGA due to PnR challenges and lack of enough IO pads.

Distributability Solution 2: With coarse granularity, where the eFPGA is placed at a specific location in the SoC (Fig. 5.2a), it does not have any distributed monitoring capability. This approach is similar to a centralized monitoring solution. Thus, the distributability of this solution is low.

5.4.3 Solution 3: Multi-application eFGPA Used for Monitoring

Although Solution 2 (Sect. 5.4.2) is sustainable/adaptable, its overhead is a significant shortcoming that must be addressed. Over time, we witnessed that reconfigurable logic, particularly eFPGA is a promising solution for numerous applications on modern SoCs, e.g., for acceleration, power efficiency, configurable computation systems, etc. [18–21]. So, it is highly possible that for performance/power efficiency, eFPGA could be used in SoCs. In such cases, while these workloads take a fair share of the eFPGA resources, there is still some unused portion left. In Solution 3, we assume eFPGA is used for 1+ applications mentioned above, and then security policies will be implemented in the unused portion of the eFPGA, as shown in Fig. 5.2c. In this case, the eFPGA fabric size and the number of IOs will remain the same, and we have a strict resource constraint to implement security monitoring. Hence, unlike the previous solutions, more effort is needed for identifying suitable security policies, ranking these policies, and optimizing their implementations.

In this case, since eFPGA is primarily engaged for other non-security applications, we just consider unutilized resources for implementing security policies. Hence, the overhead is almost gone as we target to increase the utilization ratio of eFPGA by having one more security-oriented application. Similar to Solution 2 (Sect. 5.4.2), the eFPGA here will be connected to the system bus as it will contribute to the SoC functionality (Fig. 5.2a). So, the security policies are related to the system bus. Sorting out the security policies and ranking them (based on the type of policies, overhead, and signals needed for monitoring) is a crucial step in Solution 3.

Scalability Solution 3: Since the eFGPA is for other applications, there exists limited room in the eFPGA to add security policies. However, it is possible to replace the already implemented policy to fit new policies, which makes a trade-off between upgradability and area overhead in this solution. Thus, the scalability of this solution depends on several parameters, e.g., the size of the eFPGA fabric, accelerator implementation, the utilization ratio of the eFPGA for other applications, targeted security policies, and their implementations.

Sustainability Solution 3: With eFPGA in place, monitoring can support reconfigurability. However, it is limited by the number of resources available for security monitoring. Thus, the sustainability of this solution is less than Solution 2.

Adaptability Solution 3: Since eFPGA is used for multiple applications, from one SoC to another, rearranging the eFGPA with the same policies might be much more challenging (as the resource utilization will be changed based on the application mapped on eFPGA). Hence, for the security application, the adaptability of this solution is also lower than Solution 2.

Observability Solution 3: Depending on the application implemented in the eFPGA, it will be connected to an interconnect. Thus, the security monitor will have access to this interconnect. However, routing complexity and a limited number of eFPGA IO restrict the observability of this solution to other parts of the design (similar to Solution 2).

Distributability Solution 3: The security monitoring approach is completely based on a shared eFPGA that sits at a specific location in the SoC. Hence, the distributability of this monitoring mechanism is low.

5.4.4 Solution 4: Distributed Monitoring Units + eFPGA

Although Solutions 2/3 may provide some of the quality attributes like sustainability/adaptability, they are virtually poor solutions in terms of observability/distributability. Additionally, the scalability (particularly overhead) needs meticulous consideration that which security policies are the most suited. Solution 4 opens a new potential solution, in which we still get the benefit of eFPGAs for reconfigurability, while with a hierarchical implementation (a combination of static/dynamic implementation), scalability/observability/distributability issues will be significantly mitigated. Solution 4 introduces additional distributed monitors, which we term Security Status Monitors (SSMs) (Fig. 5.3). The SSMs are identical to the performance monitors in modern SoCs. The eFPGA connects to the SSMs and acts as a control hub, as shown in Fig. 5.3. These SSMs mainly focus on IP-internal and system bus transactions [22].

SSMs act as a bridge for encapsulating data needed to be transferred to eFPGA for checking the policies. In fact, the security policies implemented in the eFPGA are based on the information provided by the SSM. So, some information filtering/profiling logic can be offloaded to the SSMs. With the inclusion of SSMs, the logic needed for the dynamic part (inside eFPGA) becomes minimal. SSMs gather security-critical per IP (region), do some pre-computation (pre-check), and then forward them to the eFPGA for final checks. In this case, since the logic to be mapped on eFPGA is minimal, the eFPGA could be used either only for security or multiple applications (unutilized resources for security monitoring). In either case, since the logic becomes minimal, scalability/observability/distributability issues will be relaxed.

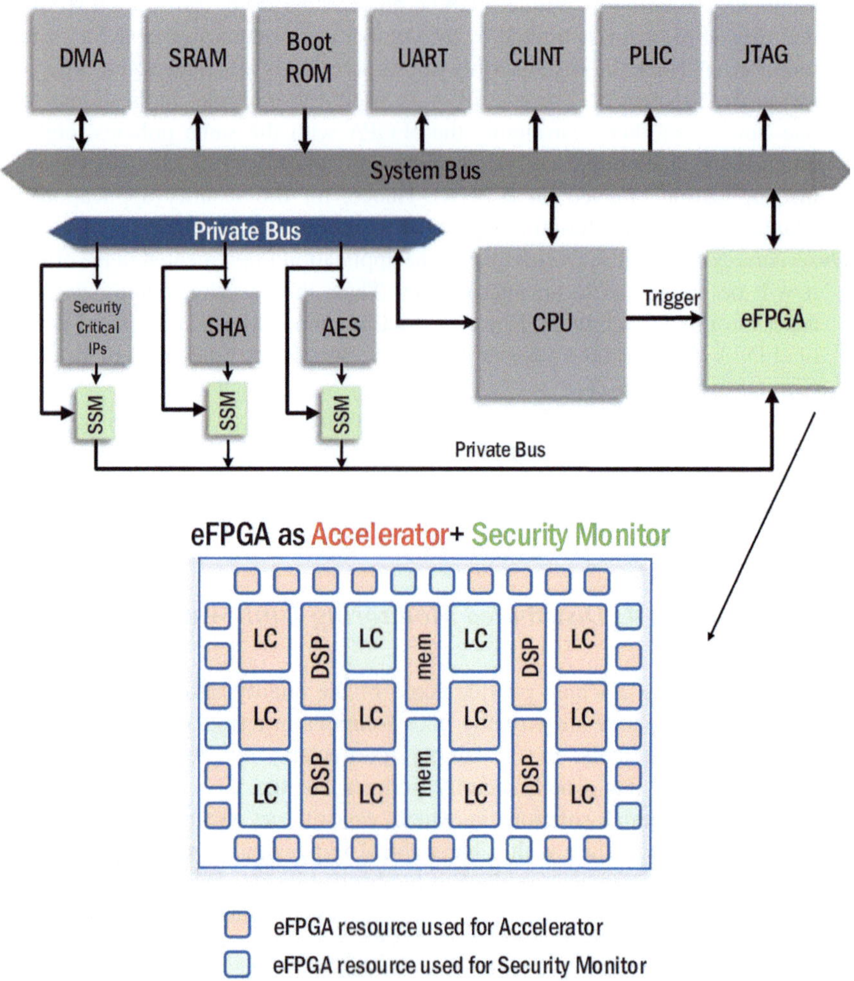

Fig. 5.3 Distributed SSMs with eFPGA-based security monitoring

Scalability Solution 4: Having hierarchical implementation improves the scalability of this solution. More security policies can be checked in this solution with the help of distributed status monitors. Although the addition of SSM may increase the area overhead compared to Solution 3, the SSMs should incur low area overhead since they implement simple static policies and event detection logic.

Sustainability Solution 4: In this solution, security policies implemented in the eFPGA can still be configured at runtime to address unforeseen vulnerabilities or zero-day attacks. Also, distributing the monitoring mechanism creates more space and opportunities for addressing runtime threats.

Adaptability Solution 4: SSMs are static, which make it less adaptable than Solution 3. This is because the distributed SSMs will connect to SoC-level and IP-internal logic, mostly specific to an SoC. However, simple modifications to the SSM IO logic would allow the SSMs to be reused in different SoCs.

Observability Solution 4: SSMs as the bridge between IPs and eFPGA increase the observability of this solution as they monitor security-critical signals both at the IP level and at the SoC level. Hence, with the help of SSMs, it is possible to perform IP-internal monitoring and policy checking.

Distributability Solution 4: With the hierarchical structure, this solution has better distributability than Solutions 2/3. The SSMs not only perform event detection and data filtering but can also implement static policies similar to Solution 1.

5.5   Distributed Monitoring with Centralized Checker: A Case Study

When the circuit size targeted to be mapped into an eFPGA fabric is getting larger, the fabric size will explode, and the utilization ratio decreases significantly. Furthermore, as the utilization of larger eFPGA fabrics decreases, the number of resources available (unutilized) increases. For instance, in a 28×28 eFPGA fabric, there are more than 2K LUTs unutilized, while only 14 LUTs are unutilized in a 4×4 eFPGA fabric. Since eFPGAs will be mostly used for acceleration purposes (e.g., to model a complex compute-intensive kernel migrated from the CPU to the eFPGA), which requires a large fabric, it opens the possibility of reutilizing unused resources for another (non-computation-intensive) application. So, *EnSAFe* enables the security monitoring of the entire system in a dynamic manner on these unutilized resources, which is typically not a resource-intensive application and can be fit on the unutilized part. Figure 5.4 demonstrates the outcome of *EnSAFe* in a RISC-V-based SoC architecture. As shown, it consists of two main components: (1) *eFPGA for acceleration/security*, which realizes both acceleration and dynamic monitoring together, and (2) *security status monitor (SSM)*, which is responsible for collecting required security-critical implications in a compressed manner. Enabling the security monitoring on eFPGA fabric allows us to upgrade the security policies in-field at runtime by reprogramming the eFPGA, which helps address new vulnerabilities or to prevent zero-day attacks.

In this model, the eFPGA is assumed to be connected to the system bus, as it will perform hardware acceleration. So, the eFPGA can directly monitor the system bus transactions. It also enables the CPU to configure and communicate with the eFPGA when necessary (triggering). In this case, based on the policy, the eFPGA will be triggered to monitor the system bus for a specific time of operation. The CPU sends configuration bits to the eFPGA, which will be used as the trigger to start/stop

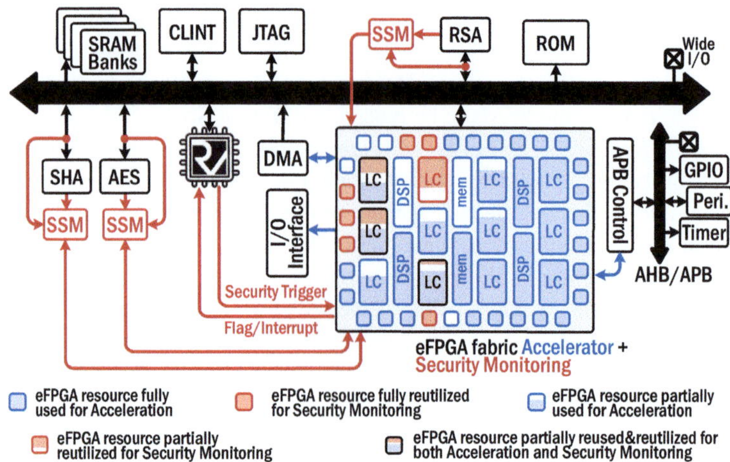

Fig. 5.4 Security monitoring via the eFPGA and security status monitor (SSM)

monitoring. In *EnSAFe*, we use an enforced memory mapping in eFPGA so that no unauthorized IP can access the eFPGA at any time during the SoC lifecycle.[1]

To complete the monitoring procedure, we propose a new monitoring module, called SSM. The SSMs, as shown in Fig. 5.4, are added per IP (security-critical ones), and they follow a very generic definition to minimize the customization required per IP.[2] The SSM concentrates mostly on inter-IP transactions. The SSMs will also take care of a limited amount of intra-IP signaling, which is difficult to achieve by only observing the system bus transactions. This, in turn, simplifies the policy implementation and reduces resource utilization of the eFPGA. Based on the generic specification of the IP (main behavior, timing diagram, handshaking protocol, etc.), they filter out and store critical data of interest. Then, this data will be sent in a compressed format to the eFPGA (if triggered for monitoring). The compression will be accomplished using a predefined protocol per IP (e.g., compressed AXI (AXI4-Lite) signaling, function, and timing description). SSMs are a static part of the SoC, and all the security policies in EnSAFe are written based on the information available through the SSMs (and main bus). Although it seems that having SSMs static may limit the reconfigurability, our case studies show these signals can cover a wide range of vulnerabilities.

Figure 5.5 shows an overall view of the SSM architecture. SSMs have an identical connection to IPs (as IPs use the same handshaking, e.g., AXI). Each SSM is connected to the eFPGA fabric directly[3] through dedicated (bidirectional) wiring (to unutilized I/O pins). SSMs consist of a limited number of buffers to temporarily

---

[1] Since multiple components may be authorized to engage eFPGA for applications, *EnSAFe* regulates them by defining specific security policies.

[2] SSMs resemble performance monitors used in today's modern SoCs [23].

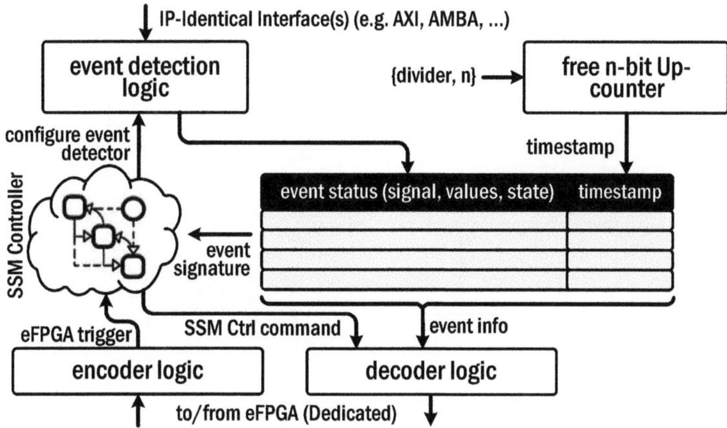

Fig. 5.5 Architecture of a security status monitor (SSM)

store IP-related information/data based on the event detection logic output. The events are defined based on the I/O values of IPs and their generic specifications. The event detection logic looks for these events or event sequences in the IO signal values (address, data, valid, ready, etc.). The event detection logic consists of a set of configurable registers, comparator(s), and tiny FSM(s). If the event to be detected occurs only in one clock cycle and does not depend on any sub-events, we do not need an FSM in the event detection logic. In this case, a simple comparator will suffice. On the other hand, if the event to be detected has precedent sub-events that occur over multiple clock cycles, we use a tiny FSM to detect such events. The tiny FSM(s) uses configurable register contents to determine which event/value to look for in the IO signals. Thus, the event detection logic is implemented in a commandable manner so that the eFPGA can send various commands for detecting multiple events. A free up-counter is used in SSMs for building the timestamp for recording the events. The SSMs will monitor and collect data with timestamps so that the eFPGA will be able to tell at which time the vulnerability occurred. The security policies implemented within the eFPGA use the event information from the SSMs to detect security vulnerabilities.

Figure 5.6 shows the overall flow of the *EnSAFe* framework. As shown, after generating the eFPGA fabric for the accelerator (based on the targeted eFPGA architecture), resource evaluation has been done to check the availability of unutilized resources. Afterward, based on the availability of unutilized resources, security-critical IP specifications, and potential threat models, we create security

---

[3] As there will be multiple SSMs in the SoC, each for a security-critical IP, the communication between the eFPGA and the SSMs through a shared private bus may create a bottleneck. Since eFPGA for acceleration occupies a significant portion of die size, our experiments show slight delay for routing dedicated wires between IPs' SSMs and eFPGA.

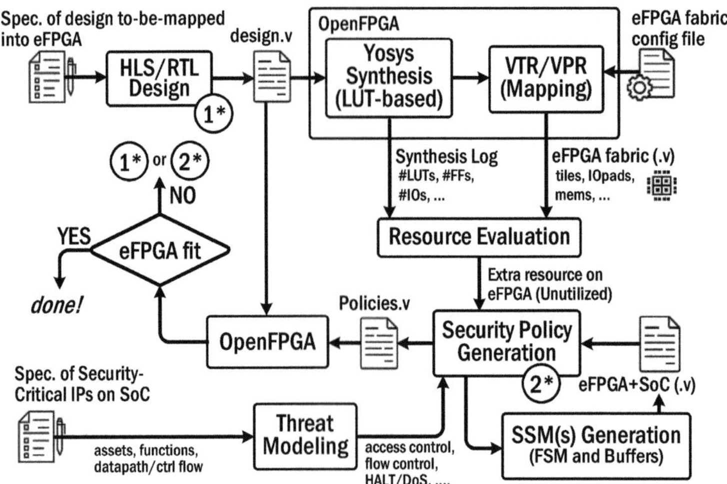

Fig. 5.6 Overall flow of the EnSAFe framework

policies and SSM subcircuits. The SSMs' subcircuits will be integrated back into the SoC, and the policies' subcircuit will be integrated with the accelerator for one more time mapping on eFPGA fabric.

If the routing and placement algorithms used in the eFPGA automation tool (VTR/VPR) fail to remap the accelerator and security policies together, the *EnSAFe* framework should reduce the logic size. There exist two valid options for the designer in this case: (1) back to redesigning the accelerator part (if the security is the priority) and (2) back to a relaxed selection of policies (if the performance/throughput is the priority). Although possible, our experiments show that for the case studies investigated so far via the *EnSAFe* framework, no failure happened for remapping.

Furthermore, as shown in Fig. 5.6, the security policies (to monitor security-critical transactions of various IPs inside the SoC) are based on the threat models and assets associated with security-critical IPs. Threat models are carefully selected based on the nature of vulnerabilities (per IP) and their impact on the SoC. Security policies use the data collected from the SSMs and system bus monitoring to detect security vulnerabilities. We implemented the security policies in synthesizable HDL, primarily as tiny FSMs, which can be mapped into the eFPGA fabric and updated as needed (in the field).

Once the security policies are mapped into the eFPGA and the SSMs are added and connected for collecting data, in case of any security policy violation, either the eFPGA or the CPU can take care of the policy enforcement. However, to keep resource utilization for the security portion at a minimum, we have considered the CPU as the enforcement entity. So, as shown in Fig. 5.4, depending on the nature of the policy, the eFPGA raises a flag or sends interrupts to the CPU. The CPU can take any necessary actions from there, depending on the severity of the violation.

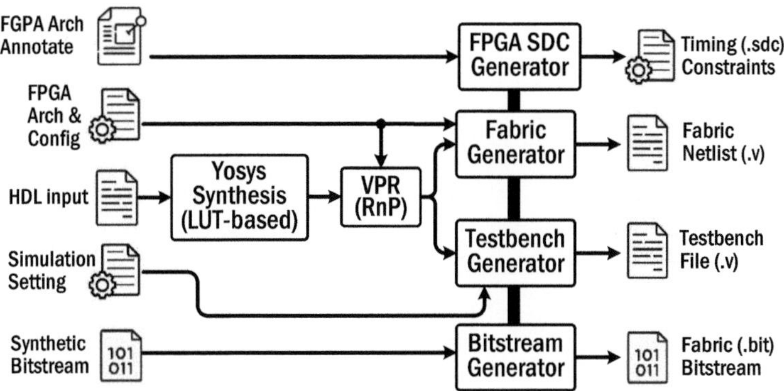

Fig. 5.7 Custom eFPGA design flow based on OpenFPGA [25]

These enforcement actions include but are not limited to stopping execution of the current process, blocking transactions from/to suspect IP, and sending "disable" configuration bits to an IP to isolate it from the SoC.

To assess the performance and efficiency of the proposed *EnSAFe* framework, we targeted Ariane SoC [24] to be enabled with both acceleration and security monitoring. Ariane SoC consists of a 64-bit, 6-stage, in-order RISC-V processor, and several (peripheral) IPs like UART, SPI, Ethernet, GPIO, etc., all connected using AXI crossbar. We augmented the SoC by adding a 128-bit Advanced Encryption Standard (AES) crypto module and a 64-bit Direct Memory Access (DMA) module. Additionally, to have a fair comparison, for the acceleration, we inserted an eFPGA with approximately equal size to the eFPGA used in Arnold SoC [18]. For building the eFPGA for the integration, we engaged OpenFPGA framework [25], which provides a complete Verilog to bitstream generation flow with customizable eFPGA architectures (suited for the design to be mapped). Using the framework summarily shown in Fig. 5.7, the OpenFPGA framework creates a tightly organized set of programmable tiles, block RAMs, DSP blocks (if needed), and I/O banks. We integrated the generated eFPGA fabric into the Ariane SoC through the system AXI4 crossbar. Additionally, for SSMs, direct connections to (unutilized) I/O banks are added. For overhead comparison, all experiments went through Synopsys Design Compiler using open SkyWater 130 nm process [26]. Also, the test of the security monitoring scenarios has been done using Vivado (for synthesis and integration) and OpenOCD (JTAG-enabled debugging). Furthermore, the test is accomplished (with functional verification) using an emulation model on Genesys 2 Kintex-7 FPGA Development Board [27].

We have considered three use case scenarios for eFPGA-based security monitoring proof of concept. They involve three different vulnerabilities and their detection approach using the proposed eFPGA-based monitoring architecture. Table 5.1 lists and describes these targeted vulnerabilities.

Table 5.1 Vulnerabilities targeted for monitoring in EnSAFe

Vulnerability	Description	Category	Security policy monitor	SSM event	FPGA-based resource needed for the policy
Rowhammer Attack	Repeated reading/writing and recharging of a DRAM row for illegal access to security asset(s) stored in RAM	CWE-1260	Repeated rd/wr req should not exceed a threshold within a limited time	DDR3 controller interface watch	61 LUTs + 68 FFs
AES Information Leakage	A Trojan leaks the AES secret key through the common bus in the SoC (triggering with a specific plaintext)	AES-T1300, CVE-2018-8933, CVE-2014-0881	No cipher key (partial or full) should be visible through the primary output	Cipher key and Ciphertext observation	46 LUTs + 2 FFs
AES Denial of Service	A Trojan tracks incoming patterns (plaintext) and HALT will be occurred once a specific pattern is observed	AES-T500, BASICRSA-T200, BASICRSA-T400	The valid_out signal must become high after n number of clock-cycle (once plaintext is refreshing)	AES FSM watch (counter-based)	9 LUTs + 10 FFs
DRAM Access Control Violation	A malicious IP tries to illegally access the secure DRAM region	CWE-1257, CWE-1190, CVE-2022-37302	Identify illegal access to the secure memory region by monitoring read/write requests	DDR3 controller interface watch	N/A

(Vulnerability 1) Rowhammer Attack: Rowhammer attack [28] is performed by repeated reading/writing and recharging of a DRAM row. This attack targets the security assets that have been stored in RAM banks during the execution of different applications on SoC. The repeat of read/write will redirect access to different rows of the RAM on the same bank, allowing the adversary to catch values not requested for (illegal access). To model this attack, we defined the security policy as follows: **repeated read/write accesses should not exceed a certain threshold within a limited time**. In this case, SSM dedicated to RAM (DDR3) will monitor DRAM bus continuously to find high-frequency read/write (count of read/write + timestamp). The event detection logic is parameterized for the number of memory banks, threshold value for counting, and time window (clock cycles) and can look for the rowhammer attack simultaneously in multiple banks (up to eight banks). The threshold value corresponds to a physical attribute of the DRAM which defines the minimum frequency of row buffer recharge that may result in rowhammer effect. We experimentally chose this value for the DRAM during policy implementation. However, the threshold value does not have any cost impacts on the SSM. In the following snippet, the security policy for rowhammer detection is formulated at a high level.

```
1 if time_window_start
2     count = 0;
3 else if read/write_addr_within_bank
4     count = count + 1;
5 if count > threshold
6     interrupt = 1;
```

(Vulnerability 2) AES Information Leakage: This vulnerability is Trojan-enabled, and when triggered, the cipher key or parts of the cipher key will be leaked through the primary outputs (within the ciphertext). In this specific Trojan, the trigger for this Trojan is the specific pattern in the plain text, and the payload is the cipher key leakage. The SSM dedicated to the AES module will check the AES data, including plaintext, cipher key, and the ciphertext for catching this vulnerability.[4] By sending the data collected and encoded in the SSM, the policy added into the eFPGA will detect whether the Trojan is triggered or not. In this case, the policy states that **no cipher key (or its parts) should be visible through the primary output** (shown in the following snippet).

```
1 for all ciphertext_bytes do
2     if ciphertext_byte = cipherkey_byte
3         interrupt = 1;
```

(Vulnerability 3) AES Denial of Service: Similar to vulnerability 2, this vulnerability is also Trojan-enabled, in which the Trojan is triggered by a sequence of specific patterns in the plaintext. If the trigger matches, the whole process goes into a denial of service (DoS). AES module cannot generate a *valid_out* signal while in a DoS situation (HALT in operation). In this case, the policy is that **the**

---

[4] As the cipher key is a security-critical entity, comparing the output with each byte of the key is performed inside the SSM.

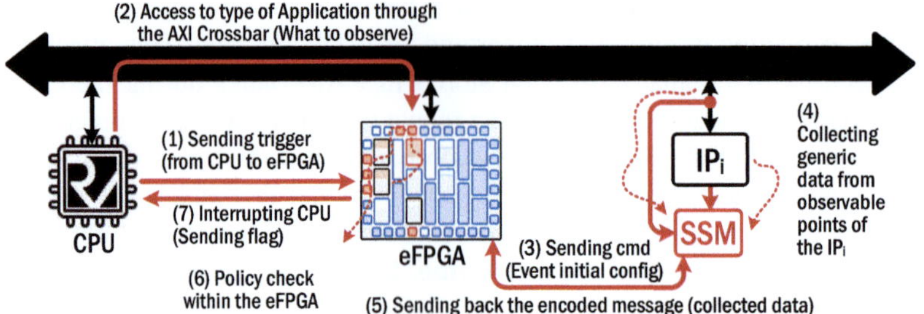

Fig. 5.8 Steps to Detect Vulnerability in the SoC via *EnSAFe*

valid_out signal must become high after n **number of clock cycle**, where n is
the number of clock cycle needed to complete the encryption (high level shown
in the following snippet).

```
1  for all clock_cycles do
2      if valid_input
3          count = 0;
4      else if !valid_out
5          count = count+1;
6      if count > n
7          interrupt = 1;
```

In all use case scenarios, the *EnSAFe* framework follows an identical procedure.
Figure 5.8 shows the overall steps of detecting the vulnerabilities/threats. Per each
use case scenario, the CPU has the right to send a trigger signal to the eFPGA to
start monitoring. A set of commands will be sent to the SSM from eFPGA that
determines which event to be monitored. The SSM gathers all the signals into the
event tracker buffers and sends the encoded data back to the eFPGA. Then, the
eFPGA, with minimal logic resources, will complete the monitoring for detecting
potential vulnerabilities.

5.5.1   Experimental Evaluation

As discussed previously, for larger eFPGAs, more unutilized resources are available.
Table 5.1 shows what resources are required per each vulnerability to implement
the policy into the eFPGA. As shown (in the last column), only tens (10–70)
of LUTs/FFs are required per policy. Our security policies (which behave like
synthesizable assertions) can easily fit into the unutilized resources. With this
low overhead, we can develop a more comprehensive security policy engine
that monitors a wide range of policies within the eFPGA. Table 5.2 provides a
comparative illustration between the resources required for modules mapped to
the eFPGA in Arnold SoC vs. that of security policies. The resources needed for
acceleration are more than tenfold. This is why the eFPGA fabric for accelerators is
huge with a low utilization rate, and it allows us to reuse them with no overhead for

Table 5.2 Resource utilization (postsynthesis) for implementing different use cases (acceleration vs. security) on the eFPGA fabric

Resources	Acceleration use cases					Security policies		
	Custom I/O	Wide I/O	BNN	CRC	Elliptic	V1	V2	V3
GPIO	72	192	384	64	192	———— 18[*1] ————		
LUT	355	565	1465	122	4064	61	46	9
FF	240	550	935	41	2122	68	2	10
eFPGA size	———— 32×32[*2] ————				28×28	———— 2×2[*3] ————		

[*1]: 18 is the total IO needed for all vulnerabilities (together). V_i: vulnerability i
[*2]: 32×32 is the eFPGA fabric for all I/Os, BNN, and CRC
[*3]: 2×2 is the eFPGA fabric for all security policies together

Table 5.3 Area distribution of some main components of the SoC

Module	Area [μm^2]	Cell count	Normalized ratio (w.r.t CPU)
CVA6 (CPU)	1,734,757	216,922	1
eFPGA for sec. policy	380,903	42,424	0.219
eFPGA for acceleration	79,715,267	11,085,611	46.004
AES core	902,325	110,375	0.520
SSM (3)	656,750	85,154	0.37
Direct memory access	167,780	13,572	0.096
SPI peripheral	12,827	1,269	0.007

security purposes. Additionally, as listed in Table 5.2, the size of fabric needed for each application (separately) reveals how security portion could be small compared to the acceleration part(s). For instance, for Arnold (with more than 6K LUTs, 4K FFs, and DSPs), the fabric size becomes 32×32, while for the security policies of Table 5.1, the eFPGA fabric size is only 2×2, which shows a ratio of 256:1. Future work will expand the list of use cases to demonstrate that even the inclusion of more security policies does not impact the ratio significantly.

Table 5.3 shows a simple breakdown of some RISC-V Ariane SoC modules' area overhead. In this experiment, a dedicated eFPGA fabric per application (one for security and one for accelerator) is implemented, integrated, and tested. All the ratios are normalized w.r.t. the size of the CPU to provide a big picture of eFPGAs' sizes. As shown, the eFPGA for acceleration is almost 46x bigger than the CPU alone. This is consistent with the output of Arnold SoC [18], where eFPGA occupies ~80% of the die size. However, the eFPGA used for the security policies of targeted vulnerabilities is almost 5x smaller than the CPU. This, again, confirms that once the eFPGA fabric is in place, in which the utilization is also low, the unutilized part can be reused for other non-compute-intensive applications, like security monitoring, which has almost no impact on the overhead. Additionally, the area overhead of the SSM reported in Table 5.3 shows that adding these monitoring modules per each IP (based on the inter-IP transactions and important intra-signaling) does not incur considerable overhead. The main part of the SSMs is the event buffer, and tuning this part (compressing and decompressing) affects the overhead significantly. However, for the targeted vulnerabilities, based on the size and the depth of buffer used for detecting the vulnerabilities, the overhead is kept reasonable. In this experiment, we implemented security monitoring for the three vulnerabilities (rows 1–3) in Table 5.1. When the chip is fabricated and in-field, there could be unforeseen vulnerabilities. To address these unforeseen vulnerabilities, we have to reprogram the eFPGA with a revised security policy. Row 4 in the Table 5.1 denotes such an unforeseen vulnerability where a malicious IP tries to access the secure memory region in the DRAM. The SSM used for rowhammer detection monitors the DRAM interface and the eFPGA is connected to the system bus (Fig. 5.8). Therefore, we can reprogram the eFPGA with a new security policy that uses the same SSM to address this access control violation.

5.6 Takeaways and Future Possibilities

As the threats around modern SoCs are evolving faster than we imagined, we need an approach that can be dynamically effective against unforeseen attacks. While reconfigurability is necessary for every security monitoring solution, the solutions presented here also consider other requirements. Ideally, we want a solution that is scalable, reconfigurable, and adaptable under duress and can also cover a significant portion of the security-critical entities inside an SoC. Table 5.4 includes the quality attributes of an ideal security monitoring solution and summarizes how the four solutions discussed in Sect. 5.4 contribute to the requirements. As discussed, Solution 1 is very effective for general-purpose SoCs containing limited security-critical assets and less susceptible to dynamic threats. On the other side, Solutions 2/3/4 are more likely to be used in higher security-critical SoCs depending on the severity of security requirements. Applications like autonomous vehicles, space operations, and safety-critical IoTs require the highest level of security, where area overhead can be overlooked for security. In this case, as reflected in Table 5.4, Solution 2 is more suitable as it provides more reconfigurability in the policies. In cases where security is as vital as area constraints, Solutions 3/4 can be implemented. For SoCs having several crypto IPs (i.e., Security Engines and Security Enclaves) that require IP-internal security monitoring, Solution 4 may be the best option.

Although the security monitoring solutions discussed in this chapter can provide security for a wide range of SoC vulnerabilities, there is still room for improvement. The complex functionality of SoCs often creates additional opportunities for attackers to perform zero-day attacks. Future directions for these monitoring solutions can include extensive research into detecting these dynamic threats in a more comprehensive way and how efficient solutions may be customized for these dynamic threats. For example, in Solution 4, how SSMs can be customized for improving the utilization of eFPGA to have the support of more security policies (improving scalability).

Table 5.4 Analysis of the four security monitoring solutions to meet the quality attributes

Solutions	Quality attributes					
	Scalability	Sustainability	Adaptability	Observability	Distributability	Area overhead
Synth. assertions (Sect. 5.4.1)	HIGH	N/A	LOW	HIGH	HIGH	LOW
Coarse-grained eFPGA-based monitoring (Sect. 5.4.2)	HIGH	HIGH	HIGH	MEDIUM	LOW	VERY HIGH
Multi-application eFGPA also used for monitoring (Sect. 5.4.3)	LOW to MEDIUM	MEDIUM	MEDIUM	MEDIUM	LOW	VERY LOW
Distributed monitoring units + eFPGA (Sect. 5.4.4)	MEDIUM to HIGH	MEDIUM to HIGH	LOW to MEDUIM	HIGH	MEDIUM to HIGH	LOW

References

1. K.Z. Azar, M.M. Hossain, A. Vafaei, H. Al Shaikh, N.N. Mondol, F. Rahman, M. Tehranipoor, F. Farahmandi, Fuzz, penetration, and AI testing for soc security verification: challenges and solutions, in *Cryptology ePrint Archive* (2022)
2. M.M.M. Rahman, S. Tarek, K.Z. Azar, M. Tehranipoor, F. Farahmandi, Efficient SoC security monitoring: quality attributes and potential solutions, in *IEEE Design & Test* (2023)
3. Y. Alkabani, F. Koushanfar, Active hardware metering for intellectual property protection and security, in *USENIX Security Symposium* (2007), pp. 291–306
4. J. Rajendran, M. Sam, O. Sinanoglu, R. Karri, Security analysis of integrated circuit camouflaging, in *Proceedings of the ACM SIGSAC Conference on Computer & Communications Security* (2013), pp. 709–720
5. K.Z. Azar, H.M. Kamali, H. Homayoun, A. Sasan, From cryptography to logic locking: a survey on the architecture evolution of secure scan chains. IEEE Access **9**, 73133–73151 (2021)
6. H.M. Kamali, K.Z. Azar, F. Farahmandi, M. Tehranipoor, Advances in logic locking: past, present, and prospects, in *Cryptology ePrint Archive* (2022)
7. M.M. Hossain, A. Vafaei, K.Z. Azar, F. Rahman, F. Farahmandi, M. Tehranipoor, SoCFuzzer: SoC vulnerability detection using cost function enabled fuzz testing, in *2023 Design, Automation & Test in Europe Conference & Exhibition (DATE)* (IEEE 2023), pp. 1–6
8. H. Al-Shaikh, A. Vafaei, M.M.M. Rahman, K.Z. Azar, F. Rahman, F. Farahmandi, M. Tehranipoor, Sharpen: Soc security verification by hardware penetration test, in *Proceedings of the 28th Asia and South Pacific Design Automation Conference* (2023), pp. 579–584
9. A. Basak, et al., A flexible architecture for systematic implementation of SoC security policies, in *IEEE/ACM International Conference on Computer-Aided Design (ICCAD)* (2015), pp. 536–543
10. A.P.D. Nath, S. Ray, A. Basak, S. Bhunia, System-on-Chip security architecture and CAD framework for hardware patch, in *2018 23rd Asia and South Pacific Design Automation Conference (ASP-DAC)* (IEEE, 2018), pp. 733–738
11. S.K. Saha, C. Bobda, FPGA accelerated embedded system security through hardware isolation, in *2020 Asian Hardware Oriented Security and Trust Symposium (AsianHOST)* (2020), pp. 1–6. https://doi.org/10.1109/AsianHOST51057.2020.9358258
12. X. Wang, et al., IIPS: infrastructure IP for secure SoC design. IEEE Trans. Comput. **64**(8), 2226–2238 (2014)
13. A. Basak, et al., Exploiting design-for-debug for flexible SoC security architecture, in *ACM/EDAC/IEEE Design Automation Conference (DAC)* (2016), pp. 1–6
14. *UltraSoC Embedded Analytics for SoCs*. https://www.linkedin.com/company/ultrasoc-technologies/?originalSubdomain=uk
15. K.Z. Azar, H.M. Kamali, H. Homayoun, A. Sasan, Threats on logic locking: a decade later, in *Proceedings of the 2019 on Great Lakes Symposium on VLSI* (2019), pp. 471–476
16. K.Z. Azar, H.M. Kamali, H. Homayoun, A. Sasan, SMT attack: next generation attack on obfuscated circuits with capabilities and performance beyond the SAT attacks, in *IACR Transactions on Cryptographic Hardware and Embedded Systems (TCHES)* (2019), pp. 97–122
17. N. Farzana, M.M. Hossain, K.Z. Azar, F. Farahmandi, M. Tehranipoor, FormalFuzzer: formal verification assisted fuzz testing for SoC vulnerability detection, in *Asia and South Pacific Design Automation Conference (ASP-DAC)* (IEEE, 2024), pp. 1–6
18. P.D. Schiavone, D. Rossi, A. Di Mauro, F.K. Gürkaynak, T. Saxe, M. Wang, K.C. Yap, L. Benini, Arnold: an eFPGA-augmented RISC-V SoC for flexible and low-power IoT end nodes. IEEE Trans. Very Large Scale Integr. (VLSI) Syst. **29**(4), 677–690 (2021)
19. H.M. Kamali, K.Z. Azar, K. Gaj, H. Homayoun, A. Sasan, LUT-lock: a novel LUT-based logic obfuscation for FPGA-bitstream and ASIC-hardware protection, in *IEEE Computer Society Annual Symposium on VLSI (ISVLSI)* (2018), pp. 405–410

20. H.M. Kamali, K.Z. Azar, H. Homayoun, A. Sasan, Full-lock: hard distributions of SAT instances for obfuscating circuits using fully configurable logic and routing blocks, in *Proceedings of Design Automation Conference (DAC)* (2019), p. 89
21. H.M. Kamali, K.Z. Azar, F. Farahmandi, M. Tehranipoor, SheLL: shrinking eFPGA fabrics for logic locking, in *2023 Design, Automation & Test in Europe Conference & Exhibition (DATE)* (IEEE, 2023), pp. 1–6
22. M.M.M. Rahman, S. Tarek, K.Z. Azar, F. Farahmandi, EnSAFe: enabling sustainable SoC security auditing using eFPGA-based accelerators, in *2023 IEEE International Symposium on Defect and Fault Tolerance in VLSI and Nanotechnology Systems (DFT)* (IEEE, 2023), pp. 1–6
23. H. Kyung, et al., Performance monitor unit design for an AXI-based multi-core SoC platform, in *ACM Symposium on Applied Computing* (2007), pp. 1565–1572
24. F. Zaruba, et al., The cost of application-class processing: energy and performance analysis of a Linux-ready 1.7-GHz 64-bit RISC-V core in 22-nm FDSOI technology. IEEE Trans. Very Large Scale Integr. (VLSI) Syst. **27**(11), 2629–2640 (2019)
25. X. Tang, et al., OpenFPGA: an open-source framework for agile prototyping customizable FPGAs. IEEE Micro **40**(4), 41–48 (2020)
26. SkyWater Technology, *SkyWater Open Source PDK*. https://github.com/google/skywater-pdk (2020)
27. Digilent Genesys 2, *Genesys 2 FGPGA Development Board*. https://digilent.com/reference/programmable-logic/genesys-2/start (2016)
28. O. Mutlu, J.S. Kim, RowHammer: a retrospective. IEEE Trans. Comput.-Aided Des. Integr. Circuits Syst. **39**(8), 1555–1571 (2020). https://doi.org/10.1109/TCAD.2019.2915318

Chapter 6
Large Language Models for SoC Security

6.1 Introduction

The omnipresence of systems-on-chip (SoCs) in contemporary digital systems emphasizes its significant role in the modern era. As SoCs permeate a wide range of devices, from mobile phones and wearable technology to the core systems of autonomous machinery, their role in today's technological infrastructure is undeniable. The broad adoption of SoCs not only reflects their versatility but also brings to the fore critical security considerations, especially as they handle and process sensitive data across various platforms. SoCs, by their design, consist of multiple intellectual property (IP) cores, each with distinct functionalities and security requirements. This integration of functionalities and the complex interplay between IP cores render SoCs vulnerable to a range of security threats. Such threats include but are not limited to information leakage [1, 2], side-channel attacks [3–5], access control violations [6], and hardware Trojans [7], especially within third-party IPs, a testament to the necessity for exhaustive security verification within the design lifecycle of the system.

The rigorous nature of this verification process clashes with the rapidly accelerating demand for the mass production of electronic devices, creating a tension that complicates the maintenance of robust security measures. This pressure is compounded by the market's urgency to minimize the time-to-market for new products, pitting thoroughness against expedience [8]. This dynamic often results in a precarious balance, where insufficient verification can lead to expensive post-production corrections. Current security solutions [6, 9–19], while extensive in their approach, struggle with scalability in light of the growing complexity of SoC architectures. They also lack the flexibility to adapt to new designs and the ever-changing landscape of hardware security threats and do not comprehensively cover all aspects of hardware vulnerabilities.

M. Tehranipoor et al., *Hardware Security*,
https://doi.org/10.1007/978-3-031-58687-3_6

On the other hand, the advent of LLMs [20–25] has precipitated a remarkable phase in the field of natural language processing (NLP), significantly broadening the horizons of what artificial intelligence can achieve with human language. These advanced models have not only excelled in traditional linguistic tasks but also transcended expectations, demonstrating unprecedented success in complex reasoning and problem-solving and unlocking enhanced capabilities, leading to performance breakthroughs in tasks like text generation, summarization, and machine translation. The proficiency of LLM extends beyond mere linguistic manipulation; now they are instrumental in deciphering and generating nuanced language through advanced forms of understanding [26] and reasoning [27]. This ability is particularly evident in their ability to engage in sophisticated tasks such as logical reasoning [27], analogical thinking [28], and even engaging in discourse that requires a deep understanding of context and subtleties of language. These models, especially prominent iterations such as GPT-4 [23], have demonstrated a formidable capacity to adapt to a variety of tasks through zero-shot and few-shot learning [22] methods, allowing them to perform tasks without extensive task-specific data training.

Recognizing the intricacies and myriad challenges inherent in SoC security, the integration of LLMs into this field is not just opportune, but potentially revolutionary. The qualities of pattern recognition, natural language understanding, advanced reasoning, knowledge transfer, and automation capabilities make LLMs a promising candidate for enhancing the security and verification processes of SoC designs. The versatility and sophistication demonstrated by LLMs in linguistic processing, coding, and complex reasoning tasks suggest a rich vein of capabilities that could be harnessed to enhance SoC security. The prospect of applying LLM-driven tools in the domain of SoC promises a substantial leap forward, moving from conventional methods to more advanced and intelligent solutions. Figure 6.1 presents a comprehensive illustration of the potential applications of LLM in SoC security.

Initial research endeavors [29–32] have begun to scratch the surface, addressing individual security challenges within hardware systems. These exploratory steps suggest the ability of LLMs to contribute meaningfully to this field. Despite these promising developments, the current volume of research on the application of LLMs in SoC security remains limited, suggesting that the full scope of their potential remains largely untapped. This chapter seeks to bridge this research gap by providing a comprehensive exploration of the possibilities that LLMs hold for SoC security verification. It explores how these advanced models can be adapted and applied to the unique context of SoC, proposing a paradigm where security verification is not only about identifying and rectifying vulnerabilities but also about preemptively crafting more secure systems through intelligent design and analysis. This approach could redefine the landscape of SoC security, offering new directions and methodologies informed by the emergent intelligence of LLMs.

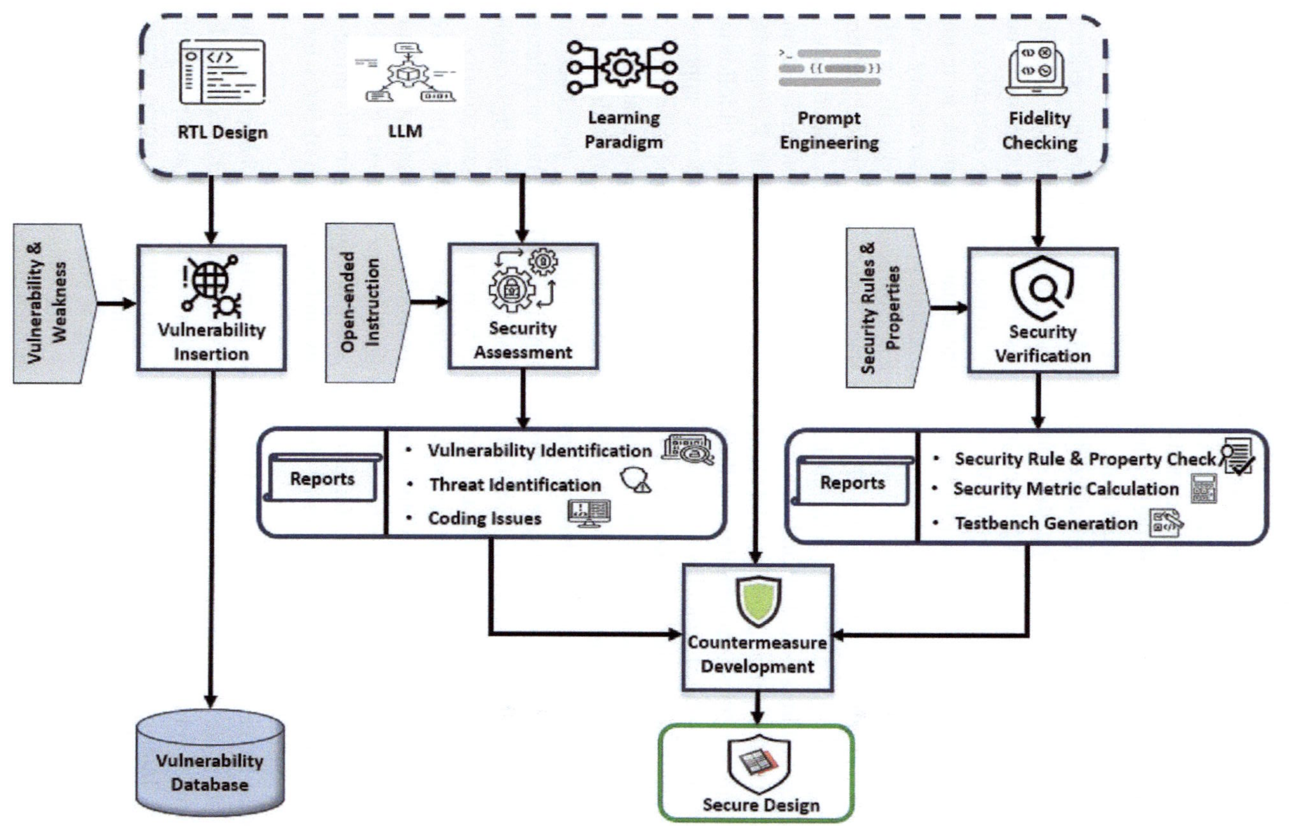

Fig. 6.1 Potential applications of LLM in SoC security

6.2 SoC Security

In the realm of hardware security, vulnerabilities represent potential points of exploitation that attackers could leverage to compromise the system. These vulnerabilities can manifest in various forms throughout the design process. Designers, despite their expertise, might inadvertently introduce security gaps due to oversights or a lack of understanding of the security implications of their design choices. Additionally, as designs transition from high-level abstractions to concrete implementations, the suite of computer-aided design (CAD) tools used may inadvertently introduce vulnerabilities during the synthesis and optimization phases [3]. The threat landscape is further complicated by the possibility of intentional malicious insertions by insiders or third-party IP providers, aiming to create backdoors for unauthorized access or control [3]. Moreover, the enhancements made to testing and debugging infrastructure to improve controllability and observability can, paradoxically, become conduits for security breaches, particularly post-silicon [2].

The RTL (Register-Transfer Level) design stage stands as a critical juncture in the design flow of modern integrated circuits. At this stage, the opportunity to detect and address potential vulnerabilities is the most strategic, potentially saving substantial time, effort, and financial resources in security verification. Recognizing and mitigating these vulnerabilities at the RTL stage is not just a matter of following a set of prescribed guidelines but requires a vigilant and proactive approach to design that prioritizes security as a foundational principle. The schematic of a typical SoC depicted in Fig. 6.2 illustrates the myriad potential locations for hardware vulnerabilities. These vulnerabilities can be preemptively addressed or mitigated during the design phase by adhering to a robust set of guidelines that govern the structures of hardware description language (HDL) code and design practices. Analyzing previously identified vulnerabilities to understand their origins and devise mitigation strategies is a critical step in bolstering hardware security.

The verification stage is increasingly recognized as a critical yet resource-intensive phase in chip design, consuming a disproportionate amount of time and resources [8]. Recent trends indicate that verification can dominate the project timeline, underscoring the need for efficiency and thoroughness in this process, especially when it comes to ensuring the security and compliance of SoCs for a wide range of applications.

Security verification in the pre-silicon phase is particularly challenging due to several factors. The global nature of supply chains and development cycles introduces complexities in ensuring consistent security measures across different stages. Interactions between hardware, software, and firmware can create security gaps that are difficult to predict and protect against. The sheer complexity of modern designs thinned the verification coverage, making it challenging to ensure that every aspect of the SoC is thoroughly vetted for vulnerabilities. Furthermore, the absence of standardized benchmarks makes it difficult to assess and compare the effectiveness of new verification techniques.

The current approaches to security verification and validation are often limited in their effectiveness. This can be attributed to a general lack of prioritization of

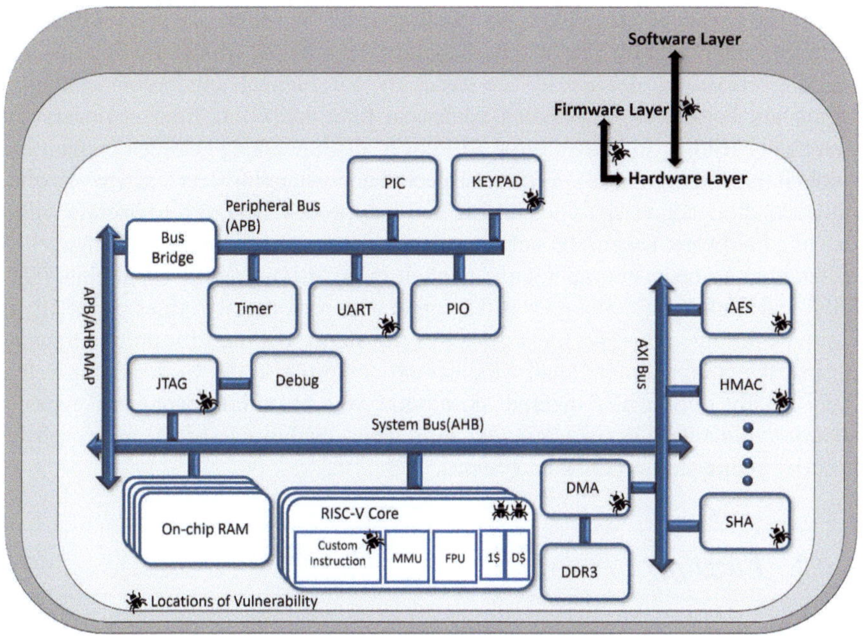

Fig. 6.2 A typical SoC with probable locations of hardware bugs

security concerns early in the design process, insufficiently robust threat models, and a scarcity of comprehensive vulnerability databases. The methods that are typically employed rely heavily on manual inspection and are ad hoc in nature, which is neither scalable nor reliable in the long term. The state-of-the-art verification methods encompass a range of strategies, from assertion-based security property verification [6], which checks for specific security conditions, to information flow tracking [33], which aims to ensure that sensitive information is not leaked or improperly accessed. Fuzz testing [11, 34–36] and Concolic testing [17] are dynamic techniques that simulate attacks or unusual conditions to discover vulnerabilities. Lastly, AI-based hardware verification represents an emerging frontier that leverages the power of machine learning and other AI techniques to automate and enhance the verification process.

6.2.1 Information Flow Tracking

IFT is a critical methodology in hardware verification, aimed at overseeing the propagation of data within a system to safeguard sensitive information and prevent security breaches. While IFT is instrumental in detecting vulnerabilities such as timing side channels and hardware Trojans [37, 38], its implementation is not without challenges. The utilization of IFT techniques like SecVerilog [37], Sapper [39], Caisson [40], and VeriCoqIFT [38] has proven effective in pinpointing information leakage bugs. However, these methods introduce several hurdles. For instance, some require the adoption of new languages, which presents a steep learning curve for designers. The necessity for manual annotation is another significant bottleneck, demanding additional time and effort from engineers and potentially leading to human error. Moreover, distinguishing between explicit and implicit information flows—a crucial aspect in understanding data leakage—is often not adequately addressed. The RTLLIFT [33] approach, designed to operate within existing hardware description languages, attempts to overcome these challenges by differentiating between implicit and explicit information flows. Despite this, RTL-LIFT and similar methodologies still struggle with performance issues, especially in complex designs. This can force designers to balance the need for precise tracking against the computational load, which could compromise the verification process. Additionally, current IFT methods do not have the breadth to capture all types of security vulnerabilities, indicating a gap in their capability to offer comprehensive security assurance for hardware systems.

6.2.2   Fuzzing

Fuzz testing, commonly referred to as "fuzzing," is an established practice in software testing and has gained significant traction in hardware security verification [41]. This method employs stochastic testing techniques on software to detect anomalies and vulnerabilities. Fuzzing usually encompasses an automated or partially automated procedure that aims to provoke faults in a program by testing it against a spectrum of invalid inputs, based on pre-established (instrumented) scenarios. Within the realm of hardware, fuzzing has been adopted to mitigate the issues of scale encountered in formal verification methods [42]. The approach of fuzzing is categorized into black box, gray box, and white box, depending on the level of access to the system information during the verification stage. Black-box fuzzing, which does not rely on internal signal knowledge, is applied based on design specifications alone [43]. White-box fuzzing is employed when there is full access to the design details [44], gray-box fuzzing combines elements from both white-box and black-box approaches to leverage their respective strengths [45]. Nonetheless, these techniques face several challenges, such as incomplete detection of vulnerabilities, diminished precision stemming from restricted visibility into the system, and the lack of well-defined metrics for measuring coverage [46].

6.2.3   Penetration Testing

Penetration testing is a proactive technique used to emulate possible attack scenarios on hardware systems or devices to identify potential security breaches [47]. The main objective is to determine the resilience of hardware components and systems against threats and unauthorized intrusions. Similar to fuzz testing, penetration testing may utilize black-box, white-box, or gray-box methodologies, contingent upon the nature of the threat and the resources at hand [48]. The process encompasses multiple phases, including the analysis of hardware design and the exploitation of identified vulnerabilities that require rectification. Distinguished from arbitrary test patterns, penetration testing employs specific knowledge regarding security attributes, existing vulnerabilities, and well-defined threat models. Nonetheless, its implementation in hardware is notably challenging; the diversity of hardware vulnerabilities necessitates tailored strategies for each distinct type of vulnerability within every penetration testing scenario [16].

6.2.4 Concolic Testing

Concolic testing, an innovative approach for automated generation of test vectors, merges actual program execution with symbolic analysis to examine and confirm system behavior [17, 18]. This technique fuses running a program or hardware design with real input values alongside symbolic abstract input representations. The aim of Concolic testing is to methodically traverse various execution pathways within a program or hardware design. Leveraging symbolic inputs, this method is capable of evaluating multiple paths in tandem, particularly those that might be elusive for conventional testing techniques. The application of Concolic testing, especially to large and intricate hardware designs or software, can demand substantial computational resources. It has been increasingly employed in the realm of hardware security verification, utilized in tasks such as uncovering hardware Trojans [18], detecting CPU core bugs in systems-on-chip (SoCs) [19], and affirming firmware integrity [9]. Nevertheless, these techniques are either confined to specific sections of the SoC due to scalability constraints or are only capable of identifying a subset of hardware vulnerabilities.

6.2.5 AI-Based Verification

The incorporation of Machine Learning (ML) and Deep Learning (DL) into the domain of hardware verification is a burgeoning area of interest, as these technologies have seen a rise in their application across various facets of the verification process. The use of ML techniques is proving to be invaluable for generating intricate test cases that challenge traditional methods and for verifying test results to ensure comprehensive coverage. The current trajectory of research is exploring several innovative avenues, including the production of constraint-random test vectors via supervised and reinforcement learning [49], optimizing the decision-making capabilities of SAT solvers [50], detecting hardware Trojans [51], and conducting thorough system failure analysis [52]. Despite the encouraging potential that these methods offer, the task of integrating AI into hardware verification processes comes with a unique set of challenges that need to be addressed.

One significant hurdle is the design-specific nature of current AI-based methods, which often lack the flexibility to generalize across diverse hardware architectures due to the scarcity of datasets encompassing a wide range of design scenarios and corner cases. This specialization limits the adaptability of AI solutions, constraining their applicability. The shortfall of comprehensive datasets also hampers the performance of ML models, undermining the verification process's efficacy and complicating the objective evaluation of AI-driven security methods due to a lack of standardized benchmarks. Moreover, the resource-intensive nature of AI methodologies raises concerns about their scalability and efficiency. Training and validating AI models require considerable computational power and time, which could delay the verification cycle and pose challenges in scaling to meet the demands of complex hardware designs. Additionally, the intricacies of feature selection and the crafting of objective functions for security verification tasks are non-trivial; they necessitate a keen understanding of the most relevant features for the model's learning process and require objective functions that precisely capture the criteria for successful verification.

6.3   Large Language Model

6.3.1   Evolution of GPT

The transformer model [53], introduced in 2017, revolutionized NLP by leveraging attention mechanisms, thus avoiding the need for traditional recurrent or convolutional layers. This innovation laid the groundwork for OpenAI's GPT series, which has since marked significant milestones in the evolution of language models. GPT-1 [20] emerged in 2018 as a pioneering language model featuring 117 million parameters, a substantial advance in language modeling. Its successor, GPT-2 [21], was released with 1.5 billion parameters, which not only brought improvements in text generation coherence but also raised concerns due to its potential to generate misleading information. In 2020, GPT-3 [22] was introduced with an unprecedented 175 billion parameters, pushing the boundaries of language models with its ability to generate near-human-like text and demonstrating impressive few-shot learning capabilities. However, despite its advances, GPT-3 still exhibited limitations.

The subsequent iterations, including GPT-3.5 and GPT-4, have sought to refine these capabilities further. GPT-3.5, with models such as "gpt-3.5-turbo," provided chat-optimized capabilities and extended token limits, while variants like "text-davinci-003" and "text-davinci-002" showed enhanced performance in language tasks. The Codex series, particularly "code-davinci-002," was tailored for coding applications. The GPT-3 series also includes a range of models such as "davinci," "curie," "babbage," and "ada," each with a token limit of 2,049, offering a spectrum

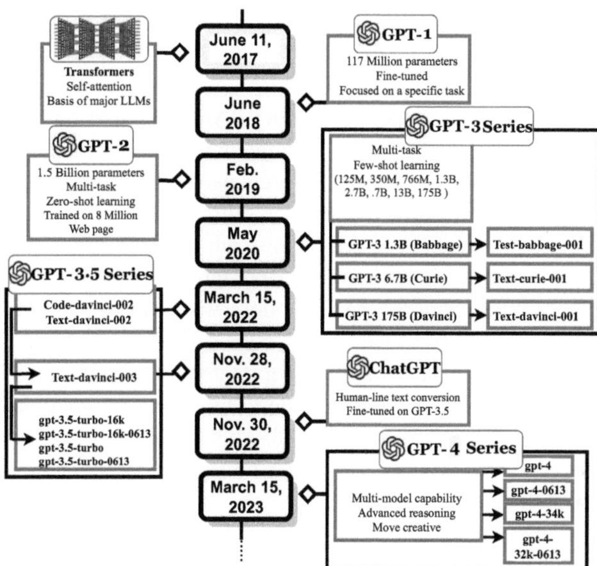

Fig. 6.3 Timeline for the evolution of GPTs

of efficiencies, speeds, and costs, with some available for fine-tuning. GPT-4's models, such as "gpt-4," "gpt-4-0613," "gpt-4-32k," and "gpt-4-32k-0613," extend the token limit even further, ranging from 8,192 to 32,768 tokens. These models represent the cutting-edge of LLMs, with certain versions optimized for chat and others capable of handling extended contexts.

The development trajectory of the GPT series is encapsulated in a timeline (shown in Fig. 6.3) that charts the key milestones from GPT-1 to GPT-4, reflecting the rapid advancements in language modeling capabilities and applications. This timeline serves as a testament to the relentless pace of innovation in the field of AI and NLP.

6.3.2 Model Architecture

LLMs have branched into a variety of sophisticated architectures, each designed to tackle specific challenges within the realm of natural language processing. Decoder-only models, such as the celebrated GPT-3 [22], utilize an autoregressive framework that excels in generating text sequentially, making it adept at producing contextually rich, coherent narratives and responses. Their strength lies in their ability to construct text that flows logically and engagingly, capturing the subtleties of human language with remarkable proficiency. On the other side of the spectrum, encoder-only architectures like BERT [24] focus on deeply analyzing and understanding input data. They transform input text into rich embeddings that capture a wealth of linguistic details, enabling the model to perform tasks that require nuanced language comprehension, such as sentiment analysis, question answering, and language inference. Encoder–decoder models, including BART [54] and T5 [55], combine the strengths of both encoder and decoder architectures, making them highly versatile. They first encode the input text into a comprehensive latent representation and then decode this information to generate a precise output. This makes them particularly suitable for complex tasks that require an intricate understanding of context before generating a response, such as summarizing lengthy documents or translating between languages. Furthermore, sparse models, which incorporate the mixture-of-experts (MoE) strategy, stand at the cutting-edge of LLM efficiency. By selectively engaging only relevant subsets of their parameters, they manage computational resources effectively, allowing them to scale to large and intricate datasets without a proportional increase in computation time. This makes them especially useful in scenarios where a vast amount of diverse data needs to be processed rapidly, without sacrificing the depth or accuracy of the analysis.

6.3.3 Learning Settings

The advent of LLMs such as GPT-3 has introduced a suite of advanced learning paradigms, each meticulously designed to refine the models' cognitive capability. Pre-training is a crucial phase, wherein the model digests a colossal corpus of text data, imbibing the essence of language structure, semantics, and the subtle nuances of context. This stage is pivotal in shaping the model's primary linguistic capabilities. Fine-tuning then meticulously hones these broad linguistic faculties, calibrating the model's parameters for heightened performance in niche tasks or domains. This process can take many forms, including supervised fine-tuning, which sharpens the model's skills on labeled datasets; instruction tuning, which aligns the model's outputs with specific user commands; and reinforcement learning with human feedback, which refines the model's responses based on human evaluation, thereby enhancing accuracy and mitigating biases. In-context learning (ICL) [56] offers LLM remarkable flexibility, enabling them to grasp and execute new tasks with minimal examples, effectively sidestepping the need for exhaustive retraining. This paradigm harnesses the innate ability of models like GPT-3 to infer and apply knowledge in novel scenarios, demonstrating a form of artificial intuition. Retrieval-Augmented In-Context Learning (RA-ICL) [57, 58] is on the cutting-edge of LLM training strategies. It enriches the model's repertoire by fetching up-to-the-minute information from vast external databases, allowing the LLM to generate responses that are not only contextually appropriate but are also saturated with the latest knowledge. These various learning paradigms collectively forge a multifaceted toolkit, elevating LLMs like GPT-3 to unparalleled levels of linguistic and contextual understanding.

6.3.4 Prompt Engineering

Prompt engineering or prompting [59] in the context of LLMs refers to the craft of designing inputs that effectively communicate a user's request to the model. These prompts act as a guide during the model's inference process, providing context or examples that assist the model in comprehending and executing the task at hand. The concept of prompting is integral to the interaction between users and LLMs, as it dictates the quality of the model's in-context learning and directly influences its output. The importance of prompt engineering cannot be overstated. A well-constructed prompt encapsulates the task requirements succinctly and steers the LLM toward producing the desired outcome. The effectiveness of an LLM's response hinges on the precision of the prompt, making prompt engineering a pivotal element in leveraging the full capabilities of these sophisticated models.

Prompting strategies [60–73] have evolved to optimize the interaction with LLMs. Techniques such as task decomposition [60, 66] break down complex problems into smaller, manageable tasks, enabling a stepwise approach to problem-solving. Sequential reasoning [68, 69], strategies guide the model through a logical progression of thoughts, enhancing its ability to tackle complex reasoning tasks. Self-evaluation and refinement techniques [61] allow the model to critique and improve its own outputs, fostering a more iterative and introspective problem-solving process. Visualization methods like the use of "scratchpads" [63] provide transparency into the model's intermediate steps, crucial for tasks that involve multi-step computations.

In the domain of SoC security, traditional questioning may not suffice due to the intricate nature of the tasks. Hence, the nuanced application of prompting strategies can be critical. Through meticulous prompt design and adjustment, one can better tap into the LLM's extensive knowledge base and sophisticated reasoning capabilities, thereby advancing the field of SoC security.

6.3.5 LLM for Coding Task

LLMs have shown impressive capabilities in coding tasks, bridging the gap between natural language processing and software development. Over the past few years, they have emerged as invaluable tools for developers, streamlining various stages of the software lifecycle. From code generation, defect detection, and auto-documentation, to assisting in debugging and even predicting potential software vulnerabilities, LLMs have transformed traditional coding paradigms. Various fine-tuned LLMs specially dedicated to coding tasks have been released by fine-tuning pre-trained models. After surveying the existing code-centric LLMs, we arrive at a few key observations. First, the LLMs have predominantly been developed with a focus on mainstream programming languages like Python. This emphasis on Python and similar languages is understandable given their widespread use in software development and data science. However, this has inadvertently led to a gap in the landscape of LLMs specifically fine-tuned for HDL such as Verilog and VHDL. Second, a significant number of these models have undergone fine-tuning, underscoring the importance of domain-specific training for coding tasks.

There is a noticeable scarcity of fine-tuned LLMs in Verilog, indicating a potential gap in the handling of HDL. However, given that these models already possess a foundational understanding of coding constructs and logic, there is a promising avenue to further fine-tune them for HDLs, leveraging their inherent coding expertise. This potential fine-tuning becomes even more significant. HDLs play a pivotal role in designing and verifying hardware components, making them crucial in ensuring the security and reliability of integrated circuits and systems. By adapting code-centric LLMs to better understand and generate HDL code, we can harness their capabilities to detect vulnerabilities, suggest mitigations, and even aid in the design of secure hardware components, thus bolstering the overall security posture of SoC designs.

- **Model Type and Architecture:** The dominant architectural choice among the existing code LLMs is the "Decoder-Only" configuration. Models like CodeGeex [74], CodeGen-Mono [75], Code LLama [76], and Codex have showcased a preference for this approach, suggesting its prominence in the field. However, there is a significant representation of the "Encoder–Decoder" structure, with models like AlphaCode [77] and CodeT5 [78] embodying this design. Unique and advanced architectures are also present: the MoE architecture seen in Incoder [79] and the Long-Range Transformer in LongCoder [80] provide indications of ongoing experimentation.
- **Model Size:** The range of model sizes is vast, from more compact models like CodeBERT [81], which has 125M parameters, to behemoths like StarCoder [82] and WizardCoder [83], boasting a staggering 15.5B. This spectrum suggests that while there is an exploration of the benefits of larger models, there is an acknowledgment of the value and utility that smaller models bring, especially in constrained computational environments.
- **Language:** Python stands out as the predominant language of choice for many models, including CodeGen-Mono [75], Codex [84], and PanGu-Coder [85]. This preference might stem from Python's widespread use in the software development realm. However, the list also features models with specialized language capabilities, such as VeriGen [86] which is tailored for Verilog, and some with multi-language functionalities.
- **Training Set Size:** There exists considerable variation in the volume of training data. While models like CodeBERT [81] use datasets as small as 6.4M codes, others like CodeGen2.5-Multi [87] rely on extensive data pools, up to 1.4T tokens. Larger datasets might offer better generalization, but they inherently demand more computational power and resources.
- **Abilities:** The capabilities of these LLMs span a wide spectrum. While models like AlphaCode [77] and CodeGeex [74] primarily focus on code generation, others such as CodeT5+ [88] encompass both understanding and summarization

tasks. There is also a trend toward advanced feature adoption, with models like Code-LLama [76] offering long-context handling and SantaCoder emphasizing infilling capabilities.

- **Availability:** A commendable fraction of LLMs is open source, indicative of the growing community-driven ethos in AI research. This openness fosters further research and diverse real-world applications. However, some models remain proprietary, likely due to business or strategic considerations.

6.4 Interaction Between LLM and SoC Security

6.4.1 Capability of LLM in SoC Security

LLMs possess an array of advanced capabilities that make them well suited to address the intricate security challenges of SoC designs:

- **Pattern Recognition:** LLMs have the ability to recognize patterns in data, a skill that is crucial for understanding the intricate designs of SoCs. They can analyze vast amounts of code and documentation to detect anomalies or patterns indicative of security vulnerabilities.
- **Natural Language Understanding (NLU):** With NLU, LLMs can interpret complex technical documentation written in human language. This allows them to understand the context and functionality described in SoC design documents, providing them with the knowledge required to identify potential security loopholes.
- **Advanced Reasoning:** LLMs can perform advanced reasoning tasks that go beyond simple pattern recognition. They can infer the implications of certain design choices, predict potential security issues, and propose solutions based on logical deductions.
- **Knowledge Transfer:** One of the most powerful capabilities of LLMs is their ability to transfer knowledge from one domain to another. In the context of SoC security, this means that an LLM trained in one set of designs can apply its learned knowledge to new, unseen designs, making it possible to uncover vulnerabilities that have not been explicitly programmed into its training data.
- **Automation:** LLMs have the potential to automate aspects of the verification process that are traditionally manual and labor-intensive. This includes tasks such as parsing design specifications, generating test cases, and scanning code for known vulnerability patterns.

These qualities enable LLMs to serve as sophisticated tools in the SoC design process, enhancing security protocols and significantly reducing the burden on human analysts. By leveraging their pattern recognition and reasoning, they can bring to light security weaknesses that might otherwise go unnoticed. Their capacity for knowledge transfer means that they can become increasingly effective over time as they are exposed to more designs, and their automation capabilities can speed up the security verification process, making it more cost-effective and scalable.

6.4.2 Role of Learning Setting in SoC Security

In the evolving landscape of SoC security, LLMs are emerging as an impressive force due to their sophisticated learning paradigms. These models undergo pre-training on extensive text data, which lays the groundwork for their language and semantic processing abilities, although this stage does not inherently focus on domain-specific expertise. The pre-training is crucial for LLMs to develop a foundational understanding of language, yet it often lacks the depth needed for specialized SoC security tasks, particularly due to the scarcity of hardware design language (HDL) codes in training datasets. To bridge this gap, fine-tuning is employed, allowing LLMs to refine their broad knowledge base toward particular SoC security applications. This process, however, faces challenges such as significant resource requirements and the "knowledge cut-off" issue, where models become outdated as new security threats emerge.

ICL and RA-ICL present more dynamic learning approaches within LLMs, offering adaptability without extensive retraining. ICL enables models to tackle new tasks by generating responses based on current instructions and examples, though it may struggle with maintaining long-term context for more complex security analyses. RA-ICL enhances ICL by incorporating current, external information, allowing LLMs to provide up-to-date security solutions and adapt to new threats continuously. This real-time adaptability is crucial in SoC security, where staying abreast of the latest vulnerabilities is key to maintaining robust defense mechanisms. Together, these learning paradigms position LLMs as powerful allies in the quest for advanced SoC security, promising enhanced adaptability, efficiency, and a continuous learning cycle that aligns with the rapidly advancing security landscape.

6.4.3 Choice of Model Architecture in SoC Security

In the specialized field of SoC security, model architectures can play pivotal roles. Encoder-only models are particularly suited for tasks that require a profound understanding of the hardware architecture and potential vulnerabilities, such as security verification and assessment. Decoder-only models are more inclined towards generative tasks, such as creating vulnerability test cases or generating security policies. Encoder–decoder models, with their dual capabilities, are well-positioned for complex tasks like vulnerability mitigation, where both analyzing the issue and producing a solution are required. Sparse models offer a balance of efficiency and precision, potentially serving as a scalable solution for analyzing extensive SoC designs and security datasets. Each architecture's application to SoC security tasks must be carefully considered to leverage its strengths while mitigating its limitations.

6.4.4   Fidelity Check in SoC Security

Fidelity checks play a critical role in the deployment of LLMs for SOC security tasks, serving as a vital step in ensuring that the models' outputs are not only theoretically correct but also practically applicable. When an LLM generates responses such as design files, testbenches, or SystemVerilog assertions, the accuracy and reliability of these outputs are paramount. In scenarios where LLMs are employed to introduce vulnerabilities into a design for testing purposes or, conversely, to mitigate them, fidelity checks are essential to verify the successful execution of these tasks. These checks confirm that vulnerabilities have been integrated or eradicated as intended and that no new security gaps have been inadvertently created in the process. Beyond vulnerability management, fidelity checks scrutinize the code quality, ensuring that the syntactical and functional aspects of the HDL code are precise and that the design adheres to best practices, ultimately representing a valid and operational hardware configuration.

Implementing robust fidelity checks within SoC security involves a multifaceted approach that encompasses both automated and manual verification techniques. Automated methods, such as static code analysis, leverage linting tools for standard checks and security-aware tools for more specialized vulnerability assessments, providing broad coverage and efficiency. However, due to the limitations of current automated tools in capturing the full spectrum of potential security issues, manual reviews still hold significant value, offering the nuanced analysis that only human expertise can provide. This dual strategy ensures a comprehensive review of the LLM's outputs, from the basic syntax to complex security implications, balancing the thoroughness of human analysis with the speed and repeatability of automated systems. Together, these fidelity checks form a rigorous validation framework that not only enhances the trustworthiness of the LLM-generated outputs but also supports the iterative improvement of the models, leading to more secure and reliable SoC designs.

6.4.5   LLM-Based Works on Hardware Design and Security

The disparity in the openness of software code and hardware designs has a pronounced impact on training datasets for LLM. With an abundance of software code available in the public domain, LLMs like GPT are well-versed in software-related queries and tasks, demonstrating a high level of proficiency. This extensive exposure enables these models to effectively comprehend, create, and modify software code. Conversely, the relative paucity of publicly shared hardware designs means that LLMs are less adept in the field of hardware description languages (HDLs), which reflects not an intrinsic shortcoming of the models but the data on which they have been trained. While there is an increasing volume of research exploring the competencies of LLMs within the software domain, their application in hardware remains less charted territory. The literature on LLM effectiveness in hardware design and security tasks is limited, highlighting an area that is ripe for investigation.

In the domain of hardware design generation, ChipChat [89] utilized iVerilog-compatible testbeds, chosen for their ease of simulation and testing, to carry out design verification through testbench prompting. This investigation was confined to a small scale, evaluating just eight benchmarks and comparing the performance of four LLMs—GPT-3.5, GPT-4, Bard, and HuggingChat. In particular, the study did not consider security implications and was not automated, which could present challenges in scaling up due to reliance on manual prompts. In contrast, the ChipGPT [90] research introduced an automated system that leverages an LLM to translate natural language specifications into hardware logic designs, effectively fitting into existing electronic design automation (EDA) workflow. To overcome the limitations often encountered with LLMs, such as manual code editing, this study applied prompt engineering specifically for HDL. By modifying the prompts in response to unmet power, performance, or area specifications, the system could iteratively regenerate the designs. Despite these advances, the ChipGPT framework was tested in a limited scope of only eight simple benchmarks and did not emphasize security in its priorities.

In the domain of security property and assertion generation, several studies have explored the application of LLMs to automate this critical aspect of hardware design verification. The research in [32] put forward an NLP-based Security Property Generator (NSPG) that leverages hardware documentation to autonomously extract sentences related to security properties. Utilizing the general BERT model [24], which was fine-tuned with SoC documentation, the study evaluated the model's performance on unseen OpenTitan design documents. However, the framework fell short of generating specific security properties in any particular language, such as SystemVerilog assertions, that could be directly applied to hardware designs. The evaluation's scope was also quite narrow, with only five previously unseen documents comprising the test set.

Kande et al. [31] developed a system for generating SystemVerilog assertions for hardware, using a combination of two manually crafted designs and modules from Hack@DAC [91] and OpenTitan [92]. Their approach was comprehensive, involving diverse parameter variations within the framework, and although their experiments centered on OpenAI's code-davinci-002, they also demonstrated scalability by testing three additional LLMs. The experiments indicated a foundational understanding of security assertions within LLMs, suggesting that accuracy could be improved with well-crafted prompts and judicious parameter selection, despite an initial average accuracy rate of 9.2%. Another study [93] presented a methodical approach that employed FPV and GPT-4 for the generation of syntactically and semantically correct SystemVerilog Assertions (SVAs) from RTL modules. When integrated with the AutoSVA framework [12], this enhanced AutoSVA2 framework proved capable of identifying previously undetected bugs in complex systems like the RISC-V CVA6 Ariane core. However, this process was not entirely automated, still requiring manual input.

Furthermore, Paria et al. [30] introduced an automated end-to-end framework designed to develop security policies with the aid of LLMs, producing relevant CWEs and SVAs from SoC specifications. Despite this, many of the assertions generated were syntactically incorrect, and a substantial number of the CWEs recommended by the LLMs were skewed toward software vulnerabilities, limiting their effectiveness for hardware security enhancement. Lastly, the study in [94] explored the use of LLMs in writing formal properties for functional verification, though it did not diversify its experiments and overlooked security aspects. It revealed that while ChatGPT could begin formulating correctness statements, there remained significant challenges, with a need for expert review and revisions to ensure their accuracy.

A particular study [29] highlighted the potential risks associated with prompt structures in ChatGPT, indicating that poorly structured prompts could unintentionally weave security vulnerabilities into hardware designs. The researchers in this study aimed to devise prompt design methods to foster the generation of secure designs. However, their analysis was narrow, examining only a small selection of examples across just ten common weakness enumerations (CWEs). The methods they proposed, which essentially involved appending additional sentences to existing prompts, were quite basic. These initial attempts at enhancing security through prompting should not be mistaken for fully fledged, robust prompting strategies. Comprehensive exploration into these preliminary methods is necessary, especially when one considers the broader spectrum of hardware-related CWEs, which surpass a hundred in count. Further scrutiny and development of these techniques are essential to ensure the security of LLM-generated hardware designs.

In the area of hardware bug repair, a study cited as [95] presented an innovative framework that utilizes Large Language Models (LLMs) to mend bugs in Verilog code. By creating benchmarks from open-source code and applying a blend of LLMs, the authors achieved a notable 31.9% success rate in repairing unseen code, surpassing previous approaches. However, the study also recognized certain limitations, including the necessity for human intervention in the bug localization process and the subjective nature of instruction variation. Additionally, the research did not conduct a thorough evaluation of the functional and security aspects, nor did it ascertain whether the applied fixes could potentially lead to new issues. The scope of the study was also constrained to a mere five CWEs, thereby leaving a wide array of security vulnerabilities unaddressed. To achieve a more in-depth understanding and remediation of hardware security risks, the study suggests that further research with expanded examples is essential.

6.5 Case Studies

We conducted an investigation of the capabilities of LLMs within the SoC security domain. To provide a basic understanding of how LLMs can be used in these tasks and to elucidate the optimal prompt structure, we selected four exemplary case studies for a detailed analysis. In this section, we describe these case studies in detail. These studies not only underline the applicability of LLMs in addressing SoC security challenges but also offer specific prompt guidelines that can significantly elevate the effectiveness of solutions in this realm.

6.5.1 Case Study I: Vulnerability Insertion

In this study, we explore the proficiency of GPT-3.5 in embedding vulnerabilities into hardware designs. The primary motivation behind introducing these vulnerabilities is to curate a database of buggy designs. Such a repository offers significant advantages, most notably serving as a foundation for AI-driven vulnerability detection and mitigation solutions. Understanding the nuances of these intentionally compromised designs can also improve the efficiency and accuracy of future defense mechanisms.

> **Prompt 1.1**
> Your task is to perform the following actions:
> Now, read the following Verilog code delimited by <>.
>   Code: <Input Design>.
> Modify the code by introducing/adding a static deadlock state to the existing state transitions in the case statement.
> Static deadlock refers to a situation when the FSM enters to that state from another state; it will not be able to come out from that state.
> To do this,
>
> *Step 1*:   From the parameter list, first select a state from the state transition graph.
> *Step 2*:   Change its state transition in the combinational block so that it connects to a new state called `deadlock_state`.
> *Step 3*:   Add new `deadlock_state` state in the case statement that has a self-connecting loop.

In this case, we focus on inserting vulnerabilities into FSM design. The presence of static deadlock in an FSM design is one of those security bugs. A static deadlock in an FSM can lead to denial of service (DoS), unintended data leakage, and

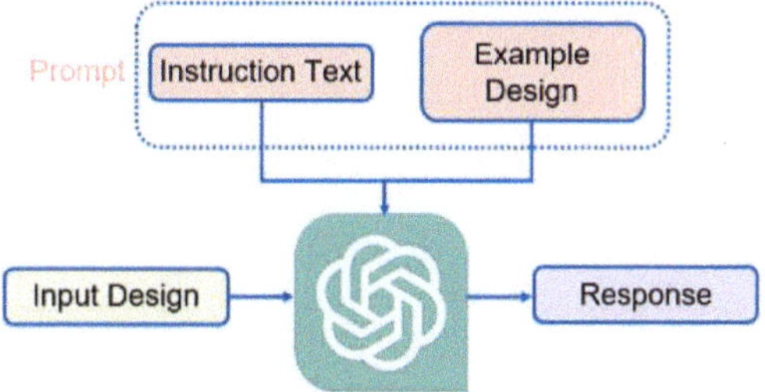

Fig. 6.4 Concept of one-hot prompting

exploitable system inconsistencies, compromising overall security and reliability of the system. Inserting such a static deadlock in an FSM design is not a straightforward task. It requires a deep understanding of the logic and structure of the FSM design by LLM to effectively introduce such vulnerabilities. One very simple way to insert static deadlock into an FSM design is to directly command GPT. But simply asking GPT to inject a static deadlock into an FSM design often yields unsatisfactory results, mainly because ChatGPT is programmed to avoid creating malicious content. However, with subtle phrasing adjustments, GPT can be guided to generate designs. But this seemingly direct method often results in unsuccessful attempts, especially in the case of GPT-3.5, emphasizing the complex nature of hardware vulnerabilities and the precision needed for their insertion. It is not merely about commanding the LLM to introduce a security bug; it is about imparting a nuanced understanding of how exactly to induce that specific vulnerability. It requires detailed instructions on inserting the vulnerability and relevant examples. The idea of including reference examples in the prompt is reminiscent of one-shot or few-shot learning in the context learning paradigm. Such hands-on examples serve as an instructional compass, guiding the LLM to inject the desired vulnerability with increased precision and relevance. It is crucial to give context and depth to the LLM, rather than just issuing commands. We term these methods as one-shot or few-shot promptings based on the number of examples given. The concept of one-shot prompting is depicted in Fig. 6.4.

> **Prompt 1.2**
> For example,
> Before deadlock:
>
> ```verilog
> parameter X=3'b000, Y=3'b001, Z=3'b011;
> case(current_state)
>     X: begin
>         next_state=Y;
>     end
>
>     Y: begin
>       next_state=Z;
>     end
> ```

(continued)

```verilog
10
11          Z: begin
12            next_state=X;
13          end
14
15      endcase
```

After deadlock:

```verilog
16      case(current_state)
17          X: begin
18            if (start)
19            next_state=Y;
20            else
21               next_state=deadlock_state;
22           end
23          Y: begin
24            next_state=Z;
25          end
26          Z: begin
27            next_state=X;
28          end
29          deadlock_state: begin
30            next_state=deadlock_state;
31          end
32      endcase
```

Here, when X transits to `deadlock_state`, FSM cannot get out of it.

Prompt 1.3

Now implement the deadlock in the provided code. Always implement deadlock state in the case statement. Do not modify the sequential block. Take care of the following steps:

1. Do not use semicolon(;) after "end" keyword.
2. For multiple statements, always use `begin..end`.
3. Use parameter instead of local parameter.
4. Put semicolon after the declaration of reg.
5. Put semicolon at the end of each statement.
6. Mark clock signal as "`clk`" and reset signal as "`rst`."
7. Make the module name "`fsm_module`."

Prompt 1.4

When giving a response, only write in the following format delimited by [ ]. Make sure that all three steps are followed.

(continued)

> Explanation: Where and how have Step 1, Step 2, and Step 3 been followed
> in the code? Tell me the line no. also where Steps 1, 2, and 3 have been
> implemented.
> Review 1: Have you implemented Step 2 in the case statement block? If
> not, rewrite the whole code by the case statement block as shown in the
> provided code.
> Review 2: Is there any issue regarding syntax, coding style, and synthesis? If
> so, correct the problems.
> [code: < modified code >]

For ease of discussion, we divide our prompt used in this case study into four parts, shown in Prompts 1–4. Here, we discuss the functions of these prompts:

- *Prompt 1.1*: Starting with *Prompt 1.1*, we present the input design and outline the scope of the task to GPT. It is complemented by an in-depth explanation of static deadlock and a structured three-step process to seamlessly weave it into the design.
- Prompt 1.2: Next, in Prompt 1.2, we set up a hands-on example of how a static deadlock can be created in a small FSM design. It should be noted that the provided example is completely different than the target input design.
- Prompt 1.3: Prompt 1.3 is dedicated to refining the model-generated design. Many of these post-processing steps mentioned in Prompt 1.3 are not necessary, in general. Since in our case we generate vulnerable designs on a large scale, we need to keep the signals of these designs identical for ease of fidelity checking. Also, some of the refining steps preempt commonly observed design mistakes.
- Prompt 1.4: Prompt 1.4 performs a self-review and prints out the output design. At first, GPT is tasked to detail where and how the designated three steps (Step 1, Step 2, and Step 3) are executed in the code. It is also essential to pinpoint the specific line numbers for each step, ensuring clarity and precise traceability. Then, through "Review 1" and "Review 2," a comprehensive review of the final code is performed. Prompt 1.4 not only seeks a response but also demands a holistic evaluation, ensuring that the logic of the code aligns with the steps and that the quality of the generated design is up to the mark.

The generated design through GPT-3.5 using such prompting is outlined in Listing 6.1. Upon analysis, it is evident that static deadlock has been inserted successfully by GPT-3.5. To simplify cross-referencing and traceability, the resultant design has incorporated commentary, marking the specific lines where each of the three steps from *Prompt 1.1* has been executed. This feature, in itself, showcases the attention to detail of GPT-3.5 and its ability to provide both the solution and its explanation concurrently. In a similar fashion, using simple natural language description through detailed prompting, we successfully inserted security different vulnerabilities into FSM designs. Because of such success, we have formulated another prompting guideline for vulnerability insertion using LLM. Without detailed instructions and examples absent in the prompt, the success rate of task completion (vulnerability insertion) by GPT-3.5 drops significantly. However, GPT-4 demonstrated a significantly superior ability to inject this vulnerability, even without an overly prescriptive prompt. But for complex cases, such as creating dynamic deadlock, the performance of GPT-4 is not notable.

It should be mentioned that during this experiment, we intentionally kept the temperature parameter 0, which makes the process very deterministic. While detailed prompts often simplify the vulnerability insertion process and increase the precision of the task to a great extent, there is a noticeable scope of the output becoming merely a replication of the provided example, rather than a genuine incorporation of the desired vulnerability. Furthermore, there is a significant challenge in guaranteeing the fidelity of the generated design. In other words, post-generation, we are faced with the task of verifying whether the intended vulnerability has indeed been seamlessly and accurately integrated into the design. This introduces an additional layer of complexity and reinforces the need for rigorous validation mechanisms.

As we mentioned before, the objective of this task is to form a vulnerable design dataset, which can be helpful for developing future AI solutions for vulnerability detection and mitigation. Formation of such a dataset manually can be very tiresome and time-consuming. But such quality of GPT can become a double-edged sword. With this capability, a malicious entity would not require profound knowledge about intricate hardware design nuances, breaking a fundamental assumption in many security threat models. Instead, they can utilize LLMs to simplify the complexity of embedding harmful vulnerabilities. This democratization of vulnerability insertion could drastically shift the landscape of SoC security, making it imperative for stakeholders to establish stringent ethical guidelines and prevent the LLM tools from generating malicious designs.

6.5.2 *Case Study II: Vulnerability Detection in RISC-V SoCs*

To examine vulnerabilities at the SoC level and evaluate their detection and security implications using GPT-4, we utilized two SoCs based on the RISC-V architecture: PULPissimo [96] and CVA6 [97]. PULPissimo employs a four-stage, in-order, single-issue core, while CVA6 features a six-stage, in-order, single-issue CPU with a 64-bit RISC-V instruction set. A commonality between them is the integrated debug module (JTAG). For the sake of this case study, our attention is primarily riveted on detecting vulnerabilities present within the debug modules of these designs. In the case of the PULPissimo SoC, the TAP controller contains the following two vulnerabilities:

- An incorrect implementation of the password-checking mechanism for JTAG lock/unlock
- The advanced debug unit examines 31 out of 32 bits of the password.

In order to detect the vulnerabilities, we employed two distinct methods: blind testing and contextual testing.

1. Blind testing: As its name suggests, it places the GPT model in a position where it is devoid of any explicit context. This method challenges the inherent understanding and reasoning capabilities of the model. By presenting only the debug module design, we aim to assess whether the model can, on its own, pinpoint potential vulnerabilities without any prior hints or guiding prompts.
2. Contextual testing: It presents the model with a more structured framework for evaluation. Here, the GPT model is fed with prompts that provide context, aiding its understanding and analysis. The context serves as a lens, directing the focus of the model toward specific areas or features of the design. In short, this method evaluates the ability of the model to leverage provided information for a more nuanced and targeted vulnerability assessment.

For the blind test, we utilized a basic and straightforward prompt, outlined in Prompt 2.1. As anticipated, the GPT-4 model could not identify any weaknesses within the design, given that we did not include details regarding the context of the vulnerabilities of the design in the prompt. On the other hand, in the contextual testing using Prompt 2.2, the model's response, as showcased in Response to Prompt 2.2, was more discerning. As evident in Response to Prompt 2.2, the model exhibited an enhanced aptitude for vulnerability detection, managed to identify the second vulnerability related to the bit-checking of the password. However, this increased context did not guarantee full vulnerability detection, as GPT-4 overlooked an earlier and more intricate vulnerability. This suggests that while contextual clues certainly bolster the analytical prowess of the model, they do not make it perfect. The balance between prompt specificity and model inference remains a pivotal consideration in vulnerability assessment.

```verilog
module fsm_module (
  input clk,
  input reset,
  input coin,
  input productSelected,
  output reg dispenseItem
);

  parameter IDLE = 3'b000;
  parameter ACCEPTING_COINS = 3'b001;
  parameter PRODUCT_SELECTED = 3'b010;
  parameter DISPENSING_ITEM = 3'b011;
  parameter DEADLOCK_STATE = 3'b100;
  //adding new state

  reg [2:0] current_state;
  reg [2:0] next_state;

  always @(posedge clk or posedge reset) begin
    if (reset) begin
      current_state <= IDLE;
    end else begin
```

```verilog
24        current_state <= next_state;
25      end
26    end
27
28    always @(current_state) begin
29      case (current_state)
30        IDLE:
31          begin
32            dispenseItem  = 0;
33            if (productSelected) begin
34              next_state = PRODUCT_SELECTED;
35            end
36            else begin
37              next_state = DEADLOCK_STATE;
38              //step 2: connecting IDLE to deadlock_state
39            end
40          end
41
42        ... ... ...
43
44        DISPENSING_ITEM:
45          begin
46            dispenseItem  = 1;
47            next_state    = IDLE;
48          end
49        DEADLOCK_STATE: //step 3: adding deadlock state
50          begin
51            dispenseItem = 0;
52            next_state = DEADLOCK_STATE;
53          end
54        default:
55          begin
56            dispenseItem  = 0;
57            next_state    = IDLE;
58          end
59      endcase
60    end
61  endmodule
```

Listing 6.1 GPT-3.5 generated design with static deadlock

Prompt 2.1
$Input Design$
Does this module have any security issues? Describe where and why?

Prompt 2.2
$Input Design$
Policy 1. Password-checking logic in debug unit should be correct.
Policy 2. It should check all bits of the password.
Policy 3. Debug register should not be accessed in halt mode.
Check if any of these policies are violated in the module. If violated, why and where? Give a response in the following pattern:

(continued)

Policy #: violated or not, explanation: if violated, why and where? line no.:
violated line

Response to Prompt 2.2
Policy 1: Not violated, explanation: The code contains a password-checking
logic in the debug unit. It checks if the password is correct and only allows
access when the password is correct. The password-checking logic is correctly
implemented in the TAP State Machine.
Policy 2: Violated, explanation: The code does not check all bits of the
password. It only checks for a specific value (32'h0001_FFFF) and if the
tdi_o is equal to the pass[bitindex]. This means that not all bits of the
password are being checked, line no.: 193–198

Prompt 2.3
$Input Design$
Policy 1. Password-checking logic in debug unit should be correct.
Check if this policy is violated in the module. If violated, why and where?

Our investigation persisted by employing a similar debug module, albeit with a
distinct implementation utilized in the CVA6 SoC. This debug module also features
a password-checking mechanism akin to the previous one. If valid, pass_check
signal of the design signifies the successful completion of the password-checking
process. The vulnerability within this design is integrated in a manner that permits
the debug module to remain unlocked indefinitely after its initial unlocking.
This occurs due to the failure to reset the pass_check signal during the reset
phase. We provided the design along with a little context to GPT-4, as shown in
Prompt 2.3.

```
 1 PassChkValid: begin
 2     if(hashValid) begin
 3         if(hash == pass_hash) begin
 4             pass_check = 1'b1;
 5         end else begin
 6             pass_check = 1'b0;
 7         end
 8         state_d = Idle;
 9     end else begin
10         state_d = PassChkValid;
11     end
12 end
```

Listing 6.2 Response from ChatGPT for prompt 2.3

The response depicted in Response to Prompt 2.3 initiates an intriguing discussion. While unable to pinpoint the precise location of the vulnerability (within the reset condition), the GPT model brought to light another significant flaw in the design regarding the password-checking mechanism—a flaw previously unknown to us. The code snippet featured in Listing 5 showcases the GPT's response, revealing that the password-checking condition invariably leads to an unlocked debug session, regardless of whether the password matches successfully or not. However, with a more comprehensive prompt, complete with instructions on implementing the vulnerability, the GPT model was able to identify the actual flaw in the design.

Response to Prompt 2.3

The DMI JTAG module described in your code seems to implement a kind of password-checking via the `PassChk` and `PassChkValid` states and the HMAC module, but there may be a potential violation of policy 1— *the password-checking logic in the debug unit should be correct*. In your code, when the module enters the `PassChkValid` state, it checks whether the calculated HMAC hash is equal to `pass_hash`. If they are equal, it sets `pass_check` to 1 (true), else it sets it to 0 (false). This procedure is secure in itself. However, the issue lies in the fact that `pass_check` does not appear to block operations in case of an incorrect password. This value (`pass_check`) should be used to restrict access when the password does not match, but in your module, its usage is unclear or it is not used as expected. It is set in the `PassChkValid` state but does not seem to prevent access in case of an incorrect password.

Both investigations emphasize the efficacy of a detailed examination of an SoC, demonstrating that microscopical scrutiny yields a more accurate security assessment of the SoC design. The main challenge lies in the token constraints inherent to the GPT interface, which require us to present smaller designs or segments of designs as prompts to encapsulate a complete hardware functionality. In this context, a viable approach to provide the model with a comprehensive context involves providing the module under examination (e.g., debug module). Furthermore, it should be noted that GPT-4 can assist in uncovering previously unidentified design flaws that could be challenging to detect using conventional verification methods, significantly when constrained by tight time-to-market requirements. Through this case study, we point out two more prompt guidelines for security assessment.

6.5.3   *Case Study III: Security Evaluation in FSM Design*

In this group of case studies, we primarily perform investigations on the competence of GPT in two scenarios: (1) calculation of security metric and (2) security assessment.

6.5.3.1   Case Study III-A: Security Metric Calculation

In this case study, we primarily perform investigations on the competence of GPT in the detection of security rule violations. The security metric plays an important role in the generation of secure designs and vulnerability detection. Calculating a security metric meticulously at RTL-level, it demands a proper understanding of design and the ability of arithmetic calculation. In this light, we check the competence of GPT in the calculation of a security metric named Fault Injection Feasibility (FIF) [1, 98] in this case. Drawing from the study [1], Kibria et al. set forth a security guideline: When a state transition occurs between two consecutive unprotected states, the Fault Injection Feasibility (FIF) metric should be "0." A design with an FIF metric of 1 for such states is deemed susceptible to fault injection. FIF metric can be calculated by

$$FIF = \prod_{i=0}^{n-1} \left((b_{x_i} \oplus b_{y_i}) + (b_{x_i} \cdot b_{p_i}) \right) \tag{6.1}$$

where b_{x_i}, b_{y_i}, and b_{p_i} denote bits of the present, next, and protected states at position i, respectively; the symbols $\oplus$, $+$, and $\cdot$ represent the operations XOR, OR, and AND, respectively.

Prompt 3.1
Your task is to perform the following actions:
First, read the following Verilog code delimited by <>.
Code: <input design>
Next, consider the "Result" state as a protected state, and the rest of the states are unprotected.
Next, analyze the state transitions that occurred in the design between unprotected states, and list all the state transitions. Consider the if, else if, and else conditions.
Then, remove the state transitions where the protected state is present.
While giving a response, only write down the modified state transition list (after removing the protected state) in the following format:
Modified state transition list: state transition 1: state_name (encoding) $\rightarrow$ state_name(encoding), protected_state: protected state (encoding state)

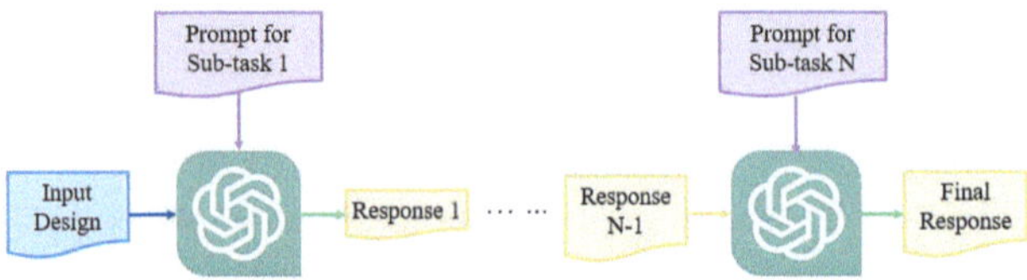

Fig. 6.5 Concept of multi-step prompting

From Eq. 6.1, we observe that the calculation of FIF metric is not a straightforward task. Given its complexity, we evaluated the proficiency of GPT-3.5 in deriving this particular security metric. To do this, our approach to guiding GPT-3.5 mirrored the principles of multi-step reasoning, a concept discussed earlier in Sect. 6.3.4. Without making the decision in a single step, we methodically segmented the task into sequential phases. Each phase is intrinsically linked, with the result from one acting as the foundation for the subsequent. This systematic breakdown not only simplifies the process for the model, but also accounts for the intricate nature of the FIF metric, and it has the potential to enhance the accuracy and efficiency of the computation performed by the model. By affording GPT-3.5 structured pathway, we enhance its chances of producing a more reliable and accurate output. The concept of design evaluation by multi-step prompting is illustrated in Fig. 6.5.

To compute the FIF metric, we employ the subsequent three-step methodology:

1. List all transitions that occur between unprotected states.
2. For every such transition, identify the values of bx_i, by_i, and bp_i for each respective bit position.
3. Use Eq. 6.1 to calculate the FIF metric for each of these listed state transitions.

We formulate three sequential prompts to execute the above-mentioned steps. The functionality of each prompt is discussed below:

- *Prompt 3.1*: In the initial phase of *Prompt 3.1*, we feed GPT-3.5 with the input design, identifying the protected state. We then guide the model to enumerate all state transitions and subsequently filter out those involving the protected state. Through this approach, we generate a list of state transitions with state encoding exclusive to unprotected states via prompting. The prompt and its response are presented in this document.
- *Prompt 3.2*: In the subsequent phase, the list of unprotected state transitions, along with their state encodings from the previous step, is introduced to *Prompt 3.2*. Initially, we explain the concept of FIF metric by presenting its corresponding equation and a brief description. This ensures that GPT-3.5 comprehends the exact information needed to compute this metric. To further aid understanding, we illustrate with a concise example, demonstrating how to derive the values of b_x, b_y, and b_p from a given state transition. Concluding the prompt, we task GPT-4 with computing the values for every state transition, instructing it to display the

Fig. 6.6 State transition
graph for the input design
used in Case Study III-A.

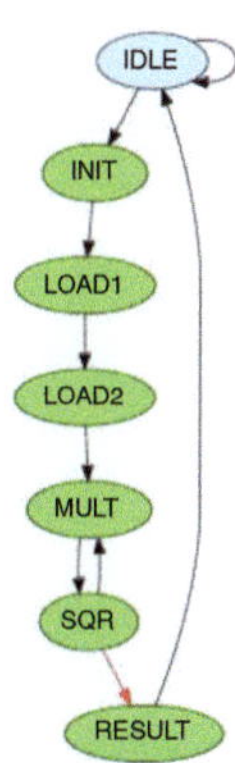

results in the prescribed format. Here, a tabular approach is used to specify the values for each position of the bit, primarily because GPT-4 processes the data more efficiently in this format.

- Prompt 3.3: In this final step, we initialize the prompt with the definition of the FIF metric. This provides the model with the requisite background knowledge. To ensure that GPT-4 grasps the computational intricacies involved, we illustrate with a detailed example explaining the process of calculating the FIF metric using values of b_x, b_y, and b_p from a representative state transition. Building on this understanding, we then direct GPT-4 to process the information generated in the preceding steps. Tasked with calculating the FIF metric in alignment with the provided instructions and example, we underscore the importance of adhering to the specified tabular format. This structured representation not only facilitates clearer visual comprehension but also ensures that GPT-4 can process and produce results with enhanced efficiency and accuracy

We meticulously devised this prompting strategy to compute the FIF metric across a large amount of FSM designs. For illustrative purposes, our chosen FSM design contains seven states. The state transition graph (STG) of this design mirrors the one depicted in Fig. 6.6. "Result" is considered as the protected state. The outcomes generated by *Prompt 3.1*, *Prompt 3.2*, and *Prompt 3.3* elucidate the step-wise systematic approach through which GPT-3.5 adeptly tackles the computationally intensive task of determining the FIF metric. To ensure clarity in our demonstration, we opted to abstain from displaying outputs for each individual state transition. A critical observation from this case study is the indispensable role of multi-stage prompting coupled with a structured tabular data representation in meticulously calculating such security metrics. This led to two more prompt guidelines for the calculation of extensive SoC security tasks.

Prompt 3.2

Your task is to perform the following actions:

1. Let us first know about the definition of the FIF metric.
   FIF = Product from $i = 0$ to n-1 of $[(bx_i$ XOR $by_i)$ OR $(bx_i$ AND $bp_i)]$
   where:

   - bx_i represents the bit of present state at position i.
   - by_i represents the bit of next state at position i.
   - bp_i represents the bits of protected state at position i.
   - n is the width of the state register (the total number of bits).
   - XOR is the bitwise exclusive-OR operation.
   - OR is the bitwise OR operation.
   - AND is the bitwise AND operation.

2. In this task, we want to identify the bits of b_x, b_y, and b_p.
   For example:
   state transition 1: A (11001) $\rightarrow$ B (01011), protected state (01100)
   $bx = 11001, by = 01011, bp = 01100$

i	0 (MSB)	1	2	3	4
bx_i	1	1	0	0	1
by_i	0	1	0	1	1
bp_i	0	1	1	0	0

3. Now read the following text delimited by <> State transitions: < state transition list >.
   For each of the state transitions, identify bx_i, by_i, and bp_i for all values of i. Put the information in tabular format.
   Review the TABLE again. Ensure that bx, by, and bp are listed in the same order.
   While giving the response, only write it down in the following format:
   < state transition 1: state1_name (encoding=bx) $\rightarrow$ state2_name(encoding=by), protected_state (bp)
   $bx =, by =, bp =, n =$

i	0 (MSB)	1	2	...	$n - 1$ (LSB)
bx_i					
by_i					
bp_i					

Prompt 3.3
Your task is to perform the following actions:

1. Let us first know about the definition of the FIF metric.
   FIF = Product from i = 0 to n-1 of [$(bx_i$ XOR $by_i)$ OR $(bx_i$ AND $bp_i)$]
   where:

   - bx_i represents the bit of present state at position i.
   - by_i represents the bit of next state at position i.
   - bp_i represents the bit of protected state at position i.
   - n is the width of the state register (the total number of bits).
   - XOR is the bitwise exclusive-OR operation.
   - OR is the bitwise OR operation.
   - AND is the bitwise AND operation.

2. Steps to calculate the FIF metric:
   Step 1:   Start from i = 0 and calculate.
   $$FIF_i = ((bx_i \text{ XOR } by_i) \text{ OR } (bx_i \text{ AND } bp_i))$$
   Step 2:   Repeat the process for other values of i up to n-1.
   Step 3:   Calculate overall FIF metric which is the product of all FIF_i
   values.
   For example, if bx (present state) = 010, by (next state) = 011, bp (protected
   state) = 000
   Step 1:   For i=0, $bx_i = 0, by_i = 0, bp_i = 0$
   FIF_0 = (0 XOR 0) OR (0 AND 0) = 0 OR 0 = 0
   Step 2:   For i= 1, $bx_i = 1, by_i = 1, bp_i = 0$
   FIF_1 = (1 XOR 1) OR (1 AND 0) = 0 OR 0 = 0
   For i= 2, $bx_i = 0, by_i = 1, bp_i = 0$
   FIF_2 = (0 XOR 1) OR (0 AND 0) = 1 OR 0 = 1
   Step 3:   Overall $FIF = FIF_0 \times FIF_1 \times FIF_2 = 1 \times 0 \times 0 = 0$
3. Now 1. read the following text delimited by <> Input information :
   <Previously generated table>
   For each state transition, follow step 1 to step 3 to calculate FIF.
   While giving a response, only write down in the following format:
   $FIF_i = ((bx_i \text{ XOR } by_i) \text{ OR } (bx_i \text{ AND } bp_i))$
   State transition 1: state1 (encoding) $\rightarrow$ state1 (encoding), protected
   (encoding)

(continued)

i	0	1	2	3
bx_i				
by_i				
bp_i				
bx_i XOR by_i				
bx_i AND bp_i				
Calculated FIF_i				

Overall $FIF = FIF_0 \times FIF_1 \times \ldots \times FIF_{(n-1)}$

Response to Prompt 3.1
modified state transition list:
state transition 1: IDLE (1000) $\to$ INIT (1100)
state transition 2: INIT (1100) $\to$ LOAD1 (0000)
state transition 3: LOAD1 (0000) $\to$ LOAD2 (0100)
state transition 4: LOAD2 (0100) $\to$ MULT (0010)
state transition 5: MULT (0010) $\to$ SQR (1010)
state transition 6: SQR (1010) $\to$ MULT (0010)
state transition 7: MULT (0010) $\to$ IDLE (1000)
protected_state: RESULT (1110)

Response to Prompt 3.2
State transition 1: IDLE (1000) $\to$ INIT (1100), protected_state (encoding=RESULT)
$bx = 1000, by = 1100, bp = 1110, n = 4$

i	0 (MSB)	1	2	3
bx_i	1	0	0	0
by_i	1	1	0	0
bp_i	1	1	1	0

State transition 2: INIT (1100) $\to$ LOAD1 (0000), protected_state (encoding=RESULT)

(continued)

$$bx = 1100, by = 0000, bp = 1110, n = 4$$

i	0 (MSB)	1	2	3
bx_i	1	1	0	0
by_i	0	0	0	0
bp_i	1	1	1	0

… … …

6.5.3.2   Case Study III-C: Security Assessment through Open-Ended Question

In the preceding investigations, namely Case Study III-A, our emphasis was on providing detailed prompts to GPT-3.5, targeting security metric computations. This approach was necessitated due to the relative limitations of GPT-3.5 in executing advanced reasoning tasks. Contrarily, in the present study, we prefer conducting a security assessment on GPT-4 using open-ended questions, deliberately avoiding any predefined prompting strategies. The FSM design under scrutiny in this study is presented in Listing 6. A thorough analysis of this design reveals that the state "s3" within the FSM stands isolated, devoid of any incoming transitions, making it an unreachable state. This configuration is not merely a design flaw; it poses a significant security risk. An attacker might leverage these states to communicate or transfer information without detection, thus compromising confidentiality. With this design as input, we prompted GPT-4, succinctly asking, "Is there any issue with this design?". GPT-4's feedback is visualized in Fig. 6.7. Impressively, GPT-4 identified a multitude of design issues, including the critical observation of the unreachable state. Beyond this primary concern, GPT-4 further pinpointed other design elements needing improvement, showcasing its comprehensive design evaluation capabilities.

In our evaluations spanning various scenarios, including static deadlock, dynamic deadlock, absence of default statement, and race condition, we consistently employed open-ended questions with GPT-4. Remarkably, GPT-4 demonstrated a significantly higher success rate in precisely assessing security concerns, with the exception of dynamic deadlock situations. Through these case studies, it became evident that GPT-4 possesses the capability to deeply dissect and comprehend FSM designs, promptly identifying simple and not-so-complicated design imperfections. Furthermore, contrasting its predecessors, GPT-4 accentuates its advanced reasoning skills by adeptly responding to open-ended inquiries without the necessity of iterative prompting.

The provided code describes a Finite State Machine (FSM) using two always blocks. The first one is used for sequential logic, updating the current state (**'curr_state'**) based on the clock and reset, and the second one is used for combinational logic, updating the next state (**'next_state'**) based on the current state.

Let's discuss some potential issues and improvements in this design:

1. **Active High Reset:** The reset in this design (**'resetn'**) is active low, which is evident from **if (resetn) curr_state <= S0**; line. As previously mentioned, active high resets are more common and intuitive in most designs, but if your design requirement or environment requires an active low reset, this would be fine.

2. **Latch Inference:** The case statement for **'S3'** state does not specify what happens to `sbit when in this state, this might cause latch inference for **'sbit'**.

3. **Unused State:** The state **'S3'** does not appear to be reached from any other state in your current FSM design. This makes it an unreachable state, and it would typically be removed to avoid confusion. If it's needed for future expansion or some other reason, that's fine, but it should be noted.

4. **State Encoding:** The state encoding (**'S0'** to **'S4'**) is not one-hot which might cause issues during synthesis for certain architectures. This might not be an issue depending upon the synthesis tool or target technology but it's good to note.

5. **Coding Style and Readability:** The **'next_state'** is always assigned to **'S0'** at the start of the second **'always'** block and then conditionally reassigned within the **'case'** statement. This could be confusing, and it might be clearer to just assign **'next_state'** within the **'case'** statement. This would make the code easier to understand and prevent potential mistakes.

Fig. 6.7 Security assessment by GPT-4 through open-ended question

Response to Prompt 3.3

$FIF_i = ((bx_i \text{ XOR } by_i) \text{ OR } (bx_i \text{ AND } bp_i))$

State transition 1: IDLE (1000) $\rightarrow$ INIT (1100), protected_state (encoding=RESULT)

i	0	1	2	3
bx_i	1	0	0	0
by_i	1	1	0	0
bp_i	1	1	1	0
bx_i XOR by_i	0	1	0	0
bx_i AND bp_i	1	0	0	0
Calculated FIF_i	1	1	0	0

Overall $FIF = FIF_0 \times FIF_1 \times FIF_2 \times FIF_3$

(continued)

$$= 1 \times 1 \times 0 \times 0 = 0$$

State transition 2: INIT (1100) $\rightarrow$ LOAD1 (0000), protected_state (encoding=RESULT)

i	0	1	2	3
bx_i	1	1	0	0
by_i	0	0	0	0
bp_i	1	1	1	0
bx_i XOR by_i	1	1	0	0
bx_i AND bp_i	1	1	0	0
Calculated FIF_i	1	1	0	0

Overall $FIF = FIF_0 \times FIF_1 \times FIF_2 \times FIF_3$
$$= 1 \times 1 \times 0 \times 0 = 0$$

. . . .

```
always @(*) begin
    next_state = s0;
    case (curr_state)
        so: begin
            sbit = 0;
            if(start) next_state = s1;
            else next_state = s0;
        end
        s1: begin
            sbit = ctrl;
            if (ctrl) next_state = s4;
            else next_state = s2;
        end
        s2: begin
            next_state = s0;
        end
        (*@\color{myred}\textbf{s3: begin}@*)
        (*@\color{myred}\textbf{    \hspace{3pt} if(finish) next\_state = s0;}@
            *)
        (*@\color{myred}\textbf{    \hspace{4pt} else next\_state=s3;}@*)
        end}@*)
        s4: begin
            if (sbit) next_state = s1;
            else next_state = s2;
        end
    endcase
end
endmodule
```

Listing 6.3 Input design used in security assessment by GPT-4 through open-ended question

6.5.4   Case Study IV: Countermeasure Development

In this exploration, we aim to assess how adept GPT-4 is at identifying and fixing vulnerabilities in FSM designs. We provided GPT-4 with an FSM design that had clear breaches of two important security rules related to state transitions and the use of default statements, as highlighted in [98]. Our approach in querying GPT-4 was strategic: We briefed the model about these security rules and the security assessment where we mentioned how the rules are violated in the provided design. Subsequently, we tasked GPT-4 with amending the design to ensure compliance with these rules. Additionally, we asked GPT-4 to review its own revised design, drawing from our approach in Case Study I. This method not only checks GPT-4's skill in addressing problems but also its ability to double-check its work, ensuring that corrections are both safe and aligned with recommended practices. Importantly, we refrained from providing GPT-4 with any specific mitigation strategies, requiring the model to derive solutions autonomously. A sample of the prompt is outlined in Prompt 4.1.

Listing 6.4 shows the input design used in this case study. Here, "WAIT_KEY" is considered the protected state. From careful observation, it becomes evident that the input design has two violations of security rules. At first, there is no default statement to handle the unspecified states. This can lead to unpredictable behavior when the system encounters an unspecified state. This unpredictability can be exploited by attackers. Second, the hamming distance between unprotected states becomes greater than 1 in the following transitions: like "WAIT_DATA" to "INITIAL_ROUND" and "DO_ROUND" to "FINAL_ROUND"'

```verilog
module fsm_module(KR, DS, clk, reset);
  input clk; // clock signal
  input reset; // reset input
  input KR, DS;
  (*@\color{myred}\textbf{parameter WAIT\_KEY = 3'b000, WAIT\_DATA = 3'b001,}@
      *)
      (*@\color{myred}\textbf{INITIAL\_ROUND = 3'b010, DO\_ROUND = 3'b011,}@
          *)
          (*@\color{myred}\textbf{FINAL\_ROUND =  3'b100;}@*)
  reg [2:0] current_state, next_state;
  always @(current_state, KR, DS)
  begin
   case(current_state)
     WAIT_KEY: begin
      if(KR == 1) next_state = WAIT_DATA;
      else next_state = WAIT_KEY;
     end

         ... ... ...

     FINAL_ROUND:
     begin
      if(KR == 0) next_state = WAIT_KEY;
      else next_state = WAIT_DATA;
     end

   endcase
  end
  // sequential memory of the Moore FSM
  always @(posedge clk, posedge reset)
```

```
29  begin
30    if(reset == 1)
31      current_state <= WAIT_KEY;
32    else
33      current_state <= next_state;
34  end
35 endmodule
```

Listing 6.4 Input design used in Case Study VI

- WAIT_DATA - INITIAL_ROUND : Hamming Distance = 2
- DO_ROUND - FINAL_ROUND : Hamming Distance = 3
- FINAL_ROUND - WAIT_DATA : Hamming Distance = 2

This scenario is particularly vulnerable to fault injection attacks. The generated design by GPT-4 is shown in Listing 6.5. It can be seen that the default statement has been successfully added to mitigate the security issue. For the second security breach, GPT-4 used gray encoding to restructure the current state encodings. This strategy successfully addressed two of the three highlighted vulnerabilities. However, the transition from FINAL_ROUND to WAIT_DATA remains a concern, with a Hamming distance of 3. While GPT-4 did not entirely eliminate this vulnerability, its approach did reduce the associated risk. The application of gray encoding, as suggested by GPT-4, is indeed a viable solution to rectify such design flaws. This demonstrates the ability of GPT-4 to autonomously derive effective solutions without external guidance. Ideally, to fully address the security concern, gray encoding should be applied to unprotected states initially, followed by the assignment of protected states. The solution provided by GPT-4 closely aligns with this optimal mitigation strategy. With more explicit context on the mitigation strategy, it is plausible that GPT-4 could have devised a complete countermeasure. It is worth noting that the mitigation of the second security violation is complex. In some cases, it may be challenging to find an encoding arrangement that fully adheres to this security rule.

```
1  module fsm_module(KR, DS, clk, reset);
2      input clk; // clock signal
3      input reset; // reset input
4      input KR, DS;
5      (*@\color{olive}\textbf{parameter WAIT\_KEY = 3'b000, WAIT\_DATA = 3'b001,}
            @*)
6      (*@\color{olive}\textbf{INITIAL\_ROUND =3'b011, DO\_ROUND = 3'b010,}@*)
7      (*@\color{olive}\textbf{FINAL\_ROUND = 3'b110;}@*)
8      reg [2:0] current_state, next_state;
9
10     always @(current_state, KR, DS)
11     begin
12         case(current_state)
13             WAIT_KEY: begin
14                 if(KR == 1) next_state = WAIT_DATA;
15                 else next_state = WAIT_KEY;
16             end
17
18             ... ... ...
19
20             end
21             FINAL_ROUND:
```

```verilog
22            begin
23                if(KR == 0) next_state = WAIT_KEY;
24                else next_state = WAIT_DATA;
25            end
26             (*@\color{olive}\textbf{default: next\_state = WAIT\_KEY; }@*)
27             (*@\color{olive}\textbf{// default to protected state}@*)
28        endcase
29    end
30
31    // sequential memory of the Moore FSM
32    always @(posedge clk, posedge reset)
33    begin
34        if(reset == 1)
35            current_state <= WAIT_KEY;
36        else
37            current_state <= next_state;
38    end
39 endmodule
```

Listing 6.5 Design generated by GPT-4 in case study VI

Prompt 4.1

<Input Design>

In this case, assume WAIT_KEY is the protected state, and other states are unprotected.

There are two security rules:

1. All unused states of a control FSM should be handled through the "default" statement in the RTL description.
2. When state transition takes place between two consecutive unprotected states, the hamming distance between the states should be 1.

Security Assessment:

These two rules are violated in this design in the following way:

1. There is no "default" statement through which unused states "101," "110," and "111" are handled.
2. There are following two state transitions between unprotected states where the hamming distance is not 1:
   WAIT_DATA - INDIAL_ROUND : 001 - 010 : HD=2
   DO_ROUND - FINAL_ROUND: 011 - 100: HD=3

Violation Mitigation Instructions:

Modify the FSM design so that the rules are followed. While modification the STG graph remains the same.

For the modified design, check if there is any rule violation in the provided design. If yes, continue modifying until two rules are followed in the modified design.

6.6 Prospects and Challenges of LLM-Based Security Verification

6.6.1 Prospects in Employing LLM in SoC Security

The integration of LLMs into SoC security heralds a significant shift toward more intelligent and adaptable security measures. The ability to fine-tune these models for specific security-related tasks suggests a future where AI can offer bespoke solutions to complex security challenges. With models such as GPT-4 and GPT-3.5 showing promise in identifying vulnerabilities and enhancing testbench creation, the application of LLMs in SoC security is poised for substantial growth.

The versatility of LLMs can be harnessed to refine the precision and efficiency of SoC security tasks. Advanced prompting techniques that facilitate multi-step reasoning and self-debugging are set to refine the performance in identifying and mitigating security risks. Moreover, success in software security can inspire focused research in leveraging LLMs for hardware design, exploring the potential of these models in creating extensive datasets for vulnerable designs, which are crucial for advancing AI-driven security solutions. Moreover, the ability of GPT-4 to uncover subtle design flaws, combined with the adaptability of models through real-time RA-ICL, can ensure solutions remain at the cutting-edge of current knowledge. The integration of automated tools and the development of self-scrutinizing LLMs could further streamline the creation of optimized hardware designs, free from syntactical errors.

The success of LLMs in inserting and mitigating vulnerabilities suggests an untapped potential for these models to automate the development of databases for vulnerable designs, aiding the creation of robust security tools. The prospect of enhancing the consistency and accuracy of GPT models in vulnerability detection by fine-tuning parameters and employing strategic prompting is particularly exciting. The promising results in vulnerability mitigation hint at a future where LLMs can revolutionize hardware security, making them indispensable tools for developers. With continued research and development, LLMs are on track to become a central component of SoC security, offering new levels of efficiency and protection in hardware design.

6.6.2 Challenges in Employing LLM in SoC Security

Applying LLMs to the security of SoC presents a nuanced set of challenges that can be both technical and conceptual in nature. A primary hurdle is the scarcity of domain-specific data for training LLMs in the nuances of hardware security. SoC designs often remain proprietary, limiting the models' exposure to a diverse range of security scenarios and hindering their ability to provide specialized solutions.

The models' limitations in maintaining long-term context also pose a challenge, particularly for complex security issues that require a comprehensive understanding of extended interactions. This can lead to fragmented solutions that do not address security vulnerabilities effectively.

Economic and accessibility factors also play a role; the high cost and proprietary nature of models like GPT-3 and GPT-4 restrict their widespread use and the community's ability to conduct in-depth research or customization for SoC security. The specialization gap is another concern. Most LLMs excel with mainstream programming languages but lack proficiency in HDLs, which are essential for embedding security into hardware.

Token constraints in LLMs introduce further complications, restricting the size of the designs that can be processed and thus challenging comprehensive security assessments of complete SoC designs. Although LLMs can swiftly generate test-benches, the verification of their correctness and the accuracy of the vulnerabilities they detect remain complex issues that require sophisticated automated workflows for validation. Moreover, the fast-evolving landscape of security threats means LLMs need continuous updates to remain effective, a requirement that current models may not always meet. Achieving consistent results with LLMs is also problematic due to their inherent non-determinism, underscoring the need for advanced prompting strategies and parameter tuning to mitigate variability.

Finally, there is a risk associated with the potential misuse of LLMs. Without proper safeguards, these models could be exploited to introduce vulnerabilities into hardware designs, raising the stakes for ensuring that LLMs are trained in secure data and that their use is governed by strong ethical guidelines and security protocols.

6.7 Conclusion

The rise of LLMs represents a paradigm shift in natural language processing, coding, and complex reasoning tasks. The sophisticated linguistic and cognitive abilities of LLMs present a novel approach to tackling the intricate security requirements of SoCs. This chapter explores the intersection of LLMs and SoC security, seeking to unlock the capabilities of these powerful models for enhancing SoC security protocols. We conducted an in-depth analysis of the current landscape of LLMs, charting their progress, exploring their potential, and identifying the avenues where they can contribute to SoC security measures. This journey through the capabilities and potential applications of LLMs has revealed a spectrum of challenges and promising opportunities within the realm of SoC security validation. By synthesizing the capabilities of LLMs with the security demands of SoCs, our work provides a foundational framework for future exploration. This synthesis offers a springboard for further research and development, paving the way for industry and academia to leverage LLMs in creating more secure and resilient SoC designs.

References

1. A. Nahiyan, K. Xiao, K. Yang, Y. Jin, D. Forte, M. Tehranipoor, AVFSM: a framework for identifying and mitigating vulnerabilities in FSMs, in *2016 53nd ACM/EDAC/IEEE Design Automation Conference (DAC)* (2016), pp. 1–6. https://doi.org/10.1145/2897937.2897992
2. G.K. Contreras, A. Nahiyan, S. Bhunia, D. Forte, M. Tehranipoor, Security vulnerability analysis of design-for-test exploits for asset protection in SoCs, in *2017 22nd Asia and South Pacific Design Automation Conference (ASP-DAC)* (IEEE, 2017), pp. 617–622
3. P. Mishra, M. Tehranipoor, S. Bhunia, Security and trust vulnerabilities in third-party IPs, in *Hardware IP Security and Trust* (Springer, Berlin, 2017), pp. 3–14
4. J. Lee, M. Tebranipoor, J. Plusquellic, A low-cost solution for protecting IPs against scan-based side-channel attacks, in *24th IEEE VLSI Test Symposium* (IEEE, 2006), 6 pp.
5. N. Pundir, J. Park, F. Farahmandi, M. Tehranipoor, Power side-channel leakage assessment framework at register-transfer level. IEEE Trans. Very Large Scale Integr. Syst. **30**(9), 1207–1218 (2022)
6. N. Farzana, F. Rahman, M. Tehranipoor, F. Farahmandi, Soc security verification using property checking, in *2019 IEEE International Test Conference (ITC)* (IEEE, 2019), pp. 1–10
7. M. Tehranipoor, F. Koushanfar, A survey of hardware trojan taxonomy and detection. IEEE Design Test Comput. **27**(1), 10–25 (2010). https://doi.org/10.1109/MDT.2010.7
8. W. Chen, S. Ray, J. Bhadra, M. Abadir, L.-C. Wang, Challenges and trends in modern SoC design verification. IEEE Design Test **34**(5), 7–22 (2017). https://doi.org/10.1109/MDAT.2017.2735383
9. N. Corteggiani, G. Camurati, A. Francillon, Inception: system-wide security testing of real-world embedded systems software, in *Proceedings of the 27th USENIX Conference on Security Symposium*. SEC'18 (USENIX Association, Baltimore, 2018), pp. 309–326. ISBN: 9781931971461
10. F. Wijitrisnanto, S. Sutikno, S.D. Putra, Efficient machine learning model for hardware trojan detection on register transfer level, in *2021 4th International Conference on Signal Processing and Information Security (ICSPIS)* (IEEE, 2021), pp. 37–40
11. M.M. Hossain, A. Vafaei, K.Z. Azar, F. Rahman, F. Farahmandi, M. Tehranipoor, SoCFuzzer: SoC vulnerability detection using cost function enabled fuzz testing, in *2023 Design, Automation & Test in Europe Conference & Exhibition (DATE)* (IEEE, 2023), pp. 1–6
12. M. Orenes-Vera et al., AutoSVA: democratizing formal verification of RTL module interactions, in *ACM/IEEE Design Automation Conference (DAC)* (2021), pp. 535–540
13. B. Ahmad et al., Don't CWEAT it: toward CWE analysis techniques in early stages of hardware design, in *Proceedings of the 41st IEEE/ACM International Conference on Computer-Aided Design* (2022), pp. 1–9
14. H. Witharana et al., Automated generation of security assertions for RTL models. J. Emerg. Technol. Comput. Syst **19**, 1–27 (2023)
15. A. Ardeshiricham, W. Hu, J. Marxen, R. Kastner, Register transfer level information flow tracking for provably secure hardware design, in *Design, Automation & Test in Europe Conference & Exhibition (DATE), 2017* (IEEE, 2017), pp. 1691–1696
16. H. Al-Shaikh, A. Vafaei, M. Md Mashahedur Rahman, K.Z. Azar, F. Rahman, F. Farahmandi, M. Tehranipoor, Sharpen: SoC security verification by hardware penetration test, in *Proceedings of the 28th Asia and South Pacific Design Automation Conference* (2023), pp. 579–584
17. K. Sen, Concolic testing, in *Proceedings of the 22nd IEEE/ACM International Conference on Automated Software Engineering, ASE '07* (Association for Computing Machinery, Atlanta, 2007), pp. 571–572. ISBN: 9781595938824. https://doi.org/10.1145/1321631.1321746
18. A. Ahmed, F. Farahmandi, P. Mishra, Directed test generation using concolic testing on RTL models, in *2018 Design, Automation & Test in Europe Conference & Exhibition (DATE)* (IEEE, 2018), pp. 1538–1543

19. L. Shen, D. Mu, G. Cao, M. Qin, J. Blackstone, R. Kastner, Symbolic execution based test-patterns generation algorithm for hardware trojan detection. Comput. Secur. **78** (2018). https://doi.org/10.1016/j.cose.2018.07.006
20. A. Radford, K. Narasimhan, T. Salimans, I. Sutskever, et al., Improving language understanding by generative pre-training (2018)
21. A. Radford, J. Wu, R. Child, D. Luan, D. Amodei, I. Sutskever, et al., Language models are unsupervised multitask learners. OpenAI Blog **1**(8), 9 (2019)
22. T. Brown, B. Mann, N. Ryder, M. Subbiah, J.D. Kaplan, P. Dhariwal, A. Neelakantan, P. Shyam, G. Sastry, A. Askell, et al., Language models are few-shot learners. Adv. Neural Inform. Process. Syst. **33**, 1877–1901 (2020)
23. GPT-4 Technical Report. https://arxiv.org/pdf/2303.08774.pdf
24. J. Devlin, M.-W. Chang, K. Lee, K. Toutanova, Bert: pre-training of deep bidirectional transformers for language understanding (2018). arXiv preprint arXiv:1810.04805
25. A. Chowdhery, S.J. Devlin, M. Bosma, G. Mishra, A. Roberts, P. Barham, H.W. Chung, C. Sutton, S. Gehrmann, et al., Palm: Scaling language modeling with pathways (2022). arXiv preprint arXiv:2204.02311
26. Q. Zhong, L. Ding, J. Liu, B. Du, D. Tao, Can ChatGPT understand too? A comparative study on ChatGPT and fine-tuned bert (2023). arXiv preprint arXiv:2302.10198
27. C. Qin, A. Zhang, Z. Zhang, J. Chen, M. Yasunaga, D. Yang, Is ChatGPT a general-purpose natural language processing task solver? (2023). arXiv preprint arXiv:2302.06476
28. T. Webb, K.J. Holyoak, H. Lu, Emergent analogical reasoning in large language models. Nat. Hum. Behav. **7**(9), 1526–1541 (2023)
29. M. Nair, R. Sadhukhan, D. Mukhopadhyay, How hardened is your hardware? Guiding ChatGPT to generate secure hardware resistant to CWEs, in *International Symposium on Cyber Security, Cryptology, and Machine Learning* (Springer, Berlin, 2023), pp. 320–336
30. S. Bhunia, S. Paria, A. Dasgupta, *DIVAS: An LLM-based End-to-End Framework for SoC Security Analysis and Policy-based Protection* (2023). eprint: 2308.06932. https://arxiv.org/abs/2308.06932.
31. R. Kande, H. Pearce, B. Tan, B. Dolan-Gavitt, S. Thakur, R. Karri, J. Rajendran, LLM-assisted generation of hardware assertions (2023). arXiv preprint arXiv:2306.14027
32. X. Meng, A. Srivastava, A. Arunachalam, A. Ray, P. Henrique Silva, R. Psiakis, Y. Makris, K. Basu, *Unlocking Hardware Security Assurance: The Potential of LLMs* (2023). arXiv: `2308.11042 [cs.CR]`
33. A. Ardeshiricham, W. Hu, J. Marxen, R. Kastner, Register transfer level information flow tracking for provably secure hardware design, in *Design, Automation & Test in Europe Conference & Exhibition (DATE), 2017* (2017), pp. 1691–1696. https://doi.org/10.23919/DATE.2017.7927266
34. B.P. Miller, L. Fredriksen, B. So, An empirical study of the reliability of UNIX utilities. Commun. ACM **33**(12), 32–44 (1990). ISSN: 0001-0782. https://doi.org/10.1145/96267.96279
35. N. Farzana, M.M. Hossain, K.Z. Azar, F. Farahmandi, M. Tehranipoor, FormalFuzzer: formal verification assisted fuzz testing for SoC vulnerability detection, in *Asia and South Pacific Design Automation Conference (ASP-DAC)* (IEEE, 2024), pp. 1–6
36. M.M. Hossain, K.Z. Azar, F. Farahmandi, M. Tehranipoor, TaintFuzzer: SoC security verification using taint inference-enabled fuzzing, in *International Conference On Computer Aided Design (ICCAD)* (IEEE, 2023), pp. 1–9
37. D. Zhang, Y. Wang, G. Edward Suh, A.C. Myers, A hardware design language for timing-sensitive information-flow security. SIGARCH Comput. Archit. News **43**(1), 503–516 (2015). ISSN: 0163-5964. https://doi.org/10.1145/2786763.2694372
38. M.-M. Bidmeshki, Y. Makris, Toward automatic proof generation for information flow policies in third-party hardware IP, in *2015 IEEE International Symposium on Hardware Oriented Security and Trust (HOST)* (2015), pp. 163–168. https://doi.org/10.1109/HST.2015.7140256
39. X. Li, V. Kashyap, J.K. Oberg, M. Tiwari, V.R. Rajarathinam, R. Kastner, T. Sherwood, B. Hardekopf, F.T. Chong, Sapper: a language for hardware-level security policy enforcement, in

Proceedings of the 19th International Conference on Architectural Support for Programming Languages and Operating Systems. ASPLOS '14 (Salt Lake City, 2014), pp. 97–112. ISBN: 9781450323055. https://doi.org/10.1145/2541940.2541947

40. X. Li, M. Tiwari, J.K. Oberg, V. Kashyap, F.T. Chong, T. Sherwood, B. Hardekopf, Caisson: a hardware description language for secure information flow. SIGPLAN Not **46**(6), 109–120 (2011). ISSN: 0362-1340. https://doi.org/10.1145/1993316.1993512

41. K.Z. Azar, M.M. Hossain, A. Vafaei, H. Al Shaikh, N.N. Mondol, F. Rahman, M. Tehranipoor, F. Farahmandi, Fuzz, penetration, and ai testing for soc security verification: challenges and solutions (2022). Cryptology ePrint Archive

42. T. Trippel, K.G. Shin, A. Chernyakhovsky, G. Kelly, D. Rizzo, M. Hicks, *Fuzzing Hardware Like Software* (2021). arXiv: 2102.02308 [cs.AR]

43. J. De Ruiter, E. Poll, Protocol state fuzzing of TLS implementations, in *Proceedings of the 24th USENIX Conference on Security Symposium*. SEC'15 (USENIX Association, Washington, 2015), pp. 193–206. ISBN: 9781931971232

44. V. Ganesh, T. Leek, M. Rinard, Taint-based directed whitebox fuzzing, in *2009 IEEE 31st International Conference on Software Engineering* (2009), pp. 474–484. https://doi.org/10.1109/ICSE.2009.5070546

45. M.M. Hossain, F. Farahmandi, M. Tehranipoor, F. Rahman, BOFT: exploitable buffer overflow detection by information flow tracking, in *2021 Design, Automation & Test in Europe Conference & Exhibition (DATE)* (2021), pp. 1126–1129. https://doi.org/10.23919/DATE51398.2021.9474045

46. K.Z. Azar, et al., Fuzz, penetration, and AI testing for SoC security verification: challenges and solutions (2022). Cryptology ePrint Archive

47. S. Shah, B. Mehtre, An overview of vulnerability assessment and penetration testing techniques. J. Comput. Virol. Hacking Techn. **11**, 27–49 (2014). https://doi.org/10.1007/s11416-014-0231-x

48. M. Fischer, F. Langer, J. Mono, C. Nasenberg, N. Albartus, Hardware penetration testing knocks your SoCs off. IEEE Design Test **38**(1), 14–21 (2021). https://doi.org/10.1109/MDAT.2020.3013730

49. W. Hughes, S. Srinivasan, R. Suvarna, M. Kulkarni, *Optimizing Design Verification Using Machine Learning: Doing Better Than Random* (2019)

50. F. Hutter, D. Babic, H.H. Hoos, A.J. Hu, Boosting verification by automatic tuning of decision procedures, in *Formal Methods in Computer Aided Design (FMCAD'07)* (2007), pp. 27–34. https://doi.org/10.1109/FAMCAD.2007.9

51. Z. Huang, Q. Wang, Y. Chen, X. Jiang, A survey on machine learning against hardware trojan attacks: recent advances and challenges. IEEE Access **8**, 10796–10826 (2020). https://doi.org/10.1109/ACCESS.2020.2965016

52. P. Gaur, S.S. Rout, S. Deb, Efficient hardware verification using machine learning approach, in *2019 IEEE International Symposium on Smart Electronic Systems (iSES) (Formerly iNiS)* (2019), pp. 168–171. https://doi.org/10.1109/iSES47678.2019.00045

53. A. Vaswani, N. Shazeer, N. Parmar, J. Uszkoreit, L. Jones, A.N. Gomez, L. Kaiser, I. Polosukhin, Attention is all you need, in *Advances in Neural Information Processing Systems*, vol. 30 (2017)

54. M. Lewis, Y. Liu, N. Goyal, M. Ghazvininejad, A. Mohamed, O. Levy, V. Stoyanov, L. Zettlemoyer, Bart: denoising sequence-to-sequence pre-training for natural language generation, translation, and comprehension (2019). arXiv preprint arXiv:1910.13461

55. C. Raffel, N. Shazeer, A. Roberts, K. Lee, S. Narang, M. Matena, Y. Zhou, W. Li, P.J. Liu, Exploring the limits of transfer learning with a unified text-to-text transformer. J. Mach. Learn. Res. **21**(1), 5485–5551 (2020)

56. E. Akyürek, D. Schuurmans, J. Andreas, T. Ma, D. Zhou, What learning algorithm is in-context learning? Investigations with linear models (2022). arXiv preprint arXiv:2211.15661

57. O. Ram, Y. Levine, I. Dalmedigos, D. Muhlgay, A. Shashua, K. Leyton-Brown, Y. Shoham, In-context retrieval-augmented language models (2023). arXiv preprint arXiv:2302.00083
58. P. Lewis, E. Perez, A. Piktus, F. Petroni, V. Karpukhin, N. Goyal, H. Küttler, M. Lewis, W.-T. Yih, T. Rocktäschel, et al., Retrieval-augmented generation for knowledge-intensive NLP tasks. Adv. Neural Inform. Process. Syst. **33**, 9459–9474 (2020)
59. L. Reynolds, K. McDonell, Prompt programming for large language models: beyond the few-shot paradigm, in *Extended Abstracts of the 2021 CHI Conference on Human Factors in Computing Systems* (2021), pp. 1–7
60. D. Zhou, N. Schärli, L. Hou, J. Wei, N. Scales, X. Wang, D. Schuurmans, C. Cui, O. Bousquet, Q. Le, et al., Least-to-most prompting enables complex reasoning in large language models (2022). arXiv preprint arXiv:2205.10625
61. X. Chen, M. Lin, N. Schärli, D. Zhou, Teaching large language models to self-debug (2023). arXiv preprint arXiv:2304.05128
62. G. Kim, P. Baldi, S. McAleer, Language models can solve computer tasks (2023). arXiv preprint arXiv:2303.17491
63. M. Nye, A.J. Andreassen, G. Gur-Ari, H. Michalewski, J. Austin, D. Bieber, D. Dohan, A. Lewkowycz, M. Bosma, D. Luan, et al., Show your work: scratchpads for intermediate computation with language models (2021). arXiv preprint arXiv:2112.00114
64. A. Creswell, M. Shanahan, Faithful reasoning using large language models (2022). arXiv preprint arXiv:2208.14271
65. O. Rubin, J. Herzig, J. Berant, Learning to retrieve prompts for in-context learning (2021). arXiv preprint arXiv:2112.08633
66. D. Dua, S. Gupta, S. Singh, M. Gardner, Successive prompting for decomposing complex questions (2022). arXiv preprint arXiv:2212.04092
67. T. Wu, M. Terry, C.J. Cai., Ai chains: transparent and controllable human-ai interaction by chaining large language model prompts, in *Proceedings of the 2022 CHI Conference on Human Factors in Computing Systems* (2022), pp. 1–22
68. J. Wei, X. Wang, D. Schuurmans, M. Bosma, F. Xia, E. Chi, Q.V. Le, D. Zhou, et al., Chain-of-thought prompting elicits reasoning in large language models. Adv. Neural Inform. Process. Syst. **35**, 24824–24837 (2022)
69. T. Kojima, S.S. Gu, M. Reid, Y. Matsuo, Y. Iwasawa, Large language models are zero-shot reasoners. Adv. Neural Inform. Process. Syst. **35**, 22199–22213 (2022)
70. S. Yao, D. Yu, J. Zhao, I. Shafran, T.L Griffiths, Y. Cao, K. Narasimhan, Tree of thoughts: deliberate problem solving with large language models (2023). arXiv preprint arXiv:2305.10601
71. Y. Fu, H. Peng, A. Sabharwal, P. Clark, T. Khot, Complexity-based prompting for multi-step reasoning (2022). arXiv preprint arXiv:2210.00720
72. Z. Gou, Z. Shao, Y. Gong, Y. Shen, Y. Yang, N. Duan, W. Chen, Critic: large language models can self-correct with tool-interactive critiquing (2023). arXiv preprint arXiv:2305.11738
73. A. Madaan, N. Tandon, P. Gupta, S. Hallinan, L. Gao, S. Wiegreffe, U. Alon, N. Dziri, S. Prabhumoye, Y. Yang, et al., Self-refine: iterative refinement with self-feedback (2023). arXiv preprint arXiv:2303.17651
74. Q. Zheng, X. Xia, X. Zou, Y. Dong, S. Wang, Y. Xue, Z. Wang, L. Shen, A. Wang, Y. Li, et al., CodeGeex: a pre-trained model for code generation with multilingual evaluations on HumanEval-x (2023). arXiv preprint arXiv:2303.17568
75. E. Nijkamp, B. Pang, H. Hayashi, L. Tu, H. Wang, Y. Zhou, S. Savarese, C. Xiong, Codegen: an open large language model for code with multi-turn program synthesis (2022). arXiv preprint arXiv:2203.13474
76. Introducing Code Llama, Software. Meta AI, 2023. https://github.com/facebookresearch/codellama
77. Y. Li, D. Choi, J. Chung, N. Kushman, J. Schrittwieser, R. Leblond, T. Eccles, J. Keeling, F. Gimeno, A.D. Lago, et al., Competition-level code generation with alphacode. Science **378**(6624), 1092–1097 (2022)

78. Y. Wang, W. Wang, S. Joty, S.C.H. Hoi, Codet5: identifier-aware unified pre-trained encoder-decoder models for code understanding and generation (2021). arXiv preprint arXiv:2109.00859

79. D. Fried, A. Aghajanyan, J. Lin, S. Wang, E. Wallace, F. Shi, R. Zhong, W.-T. Yih, L. Zettlemoyer, M. Lewis, Incoder: a generative model for code infilling and synthesis (2022). arXiv preprint arXiv:2204.05999

80. D. Guo, C. Xu, N. Duan, J. Yin, J. McAuley, Long-coder: a long-range pre-trained language model for code completion (2023). arXiv preprint arXiv:2306.14893

81. Z. Feng, D. Guo, D. Tang, N. Duan, X. Feng, M. Gong, L. Shou, B. Qin, T. Liu, D. Jiang, et al., CodeBERT: a pretrained model for programming and natural languages (2020). arXiv preprint arXiv:2002.08155

82. R. Li, L.B. Allal, Y. Zi, N. Muennighoff, D. Kocetkov, C. Mou, M. Marone, C. Akiki, J. Li, J. Chim, et al., StarCoder: may the source be with you! arXiv preprint arXiv:2305.06161 (2023)

83. Z. Luo, C. Xu, P. Zhao, Q. Sun, X. Geng, W. Hu, C. Tao, J. Ma, Q. Lin, D. Jiang, WizardCoder: empowering code large language models with evol-instruct (2023). arXiv preprint arXiv:2306.08568

84. M. Chen, J. Tworek, H. Jun, Q. Yuan, H.P. de Oliveira Pinto, J. Kaplan, H. Edwards, Y. Burda, N. Joseph, G. Brockman, et al., Evaluating large language models trained on code (2021). arXiv preprint arXiv:2107.03374

85. F. Christopoulou, G. Lampouras, M. Gritta, G. Zhang, Y. Guo, Z. Li, Q. Zhang, M. Xiao, B. Shen, L. Li, et al., Pangu-coder: program synthesis with function-level language modeling (2022). arXiv preprint arXiv:2207.11280

86. S. Thakur, B. Ahmad, H. Pearce, B. Tan, B. Dolan-Gavitt, R. Karri, S. Garg, VeriGen: a large language model for verilog code generation (2023). arXiv preprint arXiv:2308.00708

87. E. Nijkamp, H. Hayashi, C. Xiong, S. Savarese, Y. Zhou, CodeGen2: lessons for training LLMs on programming and natural languages, in *ICLR* (2023)

88. Y. Wang, H. Le, A.D. Gotmare, N.D.Q. Bui, J. Li, S.C.H. Hoi, Codet5+: open code large language models for code understanding and generation (2023). arXiv preprint arXiv:2305.07922

89. J. Blocklove, S. Garg, R. Karri, H. Pearce, Chip-chat: challenges and opportunities in conversational hardware design (2023)

90. K. Chang, Y. Wang, H. Ren, M. Wang, S. Liang, Y. Han, H. Li, X. Li, ChipGPT: how far are we from natural language hardware design (2023). arXiv: 2305.14019 [cs.AI]

91. HACK@DAC'23. HACK@DAC%E2%80%9923,%E2%80%9D%20 https://hackatevent.org/hackdac23/

92. R. Hansen, D. Rizzo, OpenTitan. *Open-Source Project* (2019). https://opentitan.org/

93. M. Orenes-Vera, M. Martonosi, D. Wentzlaff, *Using LLMs to Facilitate Formal Verification of RTL* (2023). arXiv: 2309.09437 [cs.AR]

94. P. Srikumar, Fast and wrong: the case for formally specifying hardware with LLMs, in *Proceedings of the International Conference on Architectural Support for Programming Languages and Operating Systems (ASPLOS)* (ACM. ACM Press, 2023). https://asplos-conference.org/wp-content/uploads/2023/waci/6-Fast-and-wrong-priyasrikumarpdf.pdf

95. B. Ahmad, S. Thakur, B. Tan, R. Karri, H. Pearce, *Fixing Hardware Security Bugs with Large Language Models* (2023)

96. P.D. Schiavone, D. Rossi, A. Pullini, A. Di Mauro, F. Conti, L. Benini, QuentIn: an ultra-low-power PULPissimo SoC in 22nm FDX, in *2018 IEEE SOI-3D-Subthreshold Microelectronics Technology Unified Conference (S3S)* (2018), pp. 1–3. https://doi.org/10.1109/S3S.2018.8640145

97. F. Zaruba, et al., The cost of application-class processing: energy and performance analysis of a Linux-ready 1.7-GHz 64-bit RISC-V core in 22-nm FDSOI technology. IEEE Trans. Very Large Scale Integr. Syst. **27**(11), 2629–2640 (2019)

98. R. Kibria, F. Farahmandi, M. Tehranipoor, ARC-FSMG: automatic security rule checking for finite state machine at the netlist abstraction (2023). Cryptology ePrint Archive

Chapter 7
Power Side-Channel Evaluation
in Post-quantum Cryptography

7.1 Introduction

IBM has released a new 127-quantum bit (qubit) processor, named "Eagle," in 2021, which is a step toward creating a 433-qubit chip called "Osprey" next year, followed by a 1121-qubit chip called "Condor" in 2023 [1, 2]. Other companies, including Google, Honeywell Quantum Solutions, and well-funded start-up companies, such as IonQ, have a similarly ambitious strategy to make a useful and error-corrected quantum computer [3]. Due to the fast progress in the development of quantum computers, existing public key algorithms like RSA and Elliptic Curve Cryptography (ECC) are needed to be replaced since they could be broken by practical quantum computing [4].

To address this need, the American National Institute of Standards and Technology (NIST) started the post-quantum cryptography (PQC) standardization process for key encapsulation mechanisms (KEMs)/public key encryption (PKE) and digital signature schemes in 2016 [5]. The competition began with 69 proper submissions in December 2017. As of July 22, 2020, the competition entered the third round with seven finalist algorithms (four KEM/PKE and three signature). In contrast to traditional cryptography, PQC relies on different mathematical hard problems, which are believed to be secure against quantum attacks. There are three general hard problem families to build these schemes: code-based cryptography, lattice-based cryptography, and multivariate-based cryptography. Among the different schemes of PQC, lattice-based cryptography is one of the most promising approaches due to its simplicity, performance, and small key sizes. Five out of seven third-round finalist algorithms of the NIST processes are based on lattice-based schemes. The two remaining non-lattice schemes are not suited for all applications, so the new standard will probably include lattice-based schemes, one of the three KEM/PKE (CRYSTALS-KYBER, Saber, or NTRU), and one of the two signature (FALCON or CRYSTALS-DILITHIUM) schemes. The components of lattice-based schemes are different compared to today's prevalent asymmetric cryptographic schemes.

M. Tehranipoor et al., *Hardware Security*,
https://doi.org/10.1007/978-3-031-58687-3_7

The recent advance in the PQC field has gradually shifted from the theory to the implementation of the cryptosystem, especially on the hardware platforms. During the standardization process, it is necessary to validate the candidates' implementation with secure countermeasures with regard to hardware vulnerability to side-channel attacks as well as mathematical cryptoanalysis.

Even though existing lattice-based KEM/PKE schemes are proven to be resistant to known mathematical cryptanalytic attacks, their hardware vulnerability has been studied recently. Especially, physical side-channel leakages, such as power and electromagnetic radiation, have been exploited to extract secret information [6, 7], and the primary threat to PQC implementations is caused by the physical side-channel leakages as well. In [8], various side-channel techniques, such as vertical correlation power analysis, horizontal in-depth correlation power analysis (HICPA), online template attacks, and chosen-input simple power analysis, are exploited to recover the entire private key from NTRU-Prime, which targets generic polynomial multiplications. Xu et al. proposed power side-channel attacks on lattice-based Kyber to extract the whole secret key with less than 960 traces, targeting message-recovery decoding functions [9]. To prevent side-channel attacks, Beirendonck et al. proposed a side-channel resistant Saber implementation on an Arm Cortex-M4 (Arm CM4) core using a first-order masking method with a factor of 2.5x overhead [10]. However, the first-order countermeasures can be broken by deep learning-based approaches in which masked values and random masks are trained to recover messages in a decoding step so that the secret key can be extracted in an IND-CCA secure Saber [11].

Based on many works of literature related to PQC side-channel attacks and countermeasures, there are challenging problems in this area as follows: (i) Further study is necessary to discover more vulnerabilities against both traditional power/EM side-channel attacks and powerful AI-based side-channel attacks such that required countermeasures can be developed to satisfy both security standards and consumer constraints. (ii) Existing side-channel evaluation methods are specific to a PQC implementation, so they are limited to various PQC HW or SW implementations. Therefore, a generic side-channel evaluation framework is required to verify the leakages of different PQC implementations efficiently. (iii) Most existing side-channel leakage assessments have been performed in post-silicon validation. If the assessment result does not satisfy the security standard, the implementation should be modified at the expense of cost and time. Side-channel assessments at early design phases such as RTL or gate level can make a secure PQC implementation in an SoC or a standalone IP more efficient.

In this chapter, we attempt to exploit *PQC-SEP*, a completely automated side-channel evaluation platform to evaluate side-channel leakages of NIST KEM/PKE schemes at both pre- and post-silicon levels. PQC-SEP will provide the following distinctive capabilities to a chip designer and a security evaluator:

- It automatically estimates the amount of side-channel leakage in the power profile of a PQC design at early design stages, i.e., RTL, gate level, and physical layout level.

- It identifies which modules or primitive cells of a design are mainly leaking secret information.
- The analysis is scalable to large complex systems-on-chip (SoCs) designs, which include a PQC module as an intellectual property (IP) block and master processors, hence accounting for the noise induced by other IP blocks in an SoC.
- It efficiently validates side-channel leakages at the post-silicon level against AI-based SCA models as well as traditional SCA models.

7.2 Preliminaries

7.2.1 Notation

We describe the ring of integers modulo q as $\mathbb{Z}_q$ and the polynomial ring as $R_q(X) = \mathbb{Z}_q[X]/(X^N+1)$. An $l_1 \times l_2$ matrix over R_q is denoted as $R_q^{l_1 \times l_2}$. Matrices will be written in uppercase bold letters, vectors in lowercase bold, and polynomials in normal letters, e.g., $\mathbf{A}$, $\mathbf{s}$, and v, respectively. For a vector v (or matrix $\mathbf{A}$), we denote by v^T (or $\mathbf{A}^T$) its transpose. The number theoretic transform (NTT) of any element a is represented as $\hat{a}$. We denote by $\circ$ the point multiplication.

We denote the sampling x according to a distribution $\mathcal{X}$ as $x \leftarrow \mathcal{X}$, e.g., $x \leftarrow \mathcal{U}$, where $\mathcal{U}$ is the uniform distribution. The binomial distribution with the parameter μ is denoted as β_μ. We can extend the sampling notation into a polynomial matrix sampled by a distribution such as $\mathbf{X} \leftarrow \mathcal{X}(R_q^{l_1 \times l_2})$, where the coefficient of $\mathbf{X}$ is sampled independently by the distribution $\mathcal{X}$.

Three hash functions, $\mathcal{F}$, $\mathcal{G}$, and $\mathcal{H}$, are instantiated with SHA3-256, SHA3-512, and SHA3-256, respectively. We denote the flooring operation and the rounding operation as $\lfloor \cdot \rfloor$ and $\lfloor \cdot \rceil$, respectively.

7.2.2 NIST Round 3 Candidates

In the third round of the PQC competition, the selected candidate algorithms are designated as either seven finalist algorithms (four KEM/PKE and three digital signatures) or eight alternate candidates (five KEM/PKE and three digital signatures). The finalists are the more likely schemes to be considered for standardization. At the same time, the alternates are schemes advanced into the third round with some, but the very low likelihood of being standardized [12]. The candidate algorithms represent multiple categories of cryptographic schemes by their underlying mathematical formulation: (1) code-based, (2) hash-based, (3) lattice-based, and (4) multivariate PKE-based cryptography. Table 7.1 shows the third-round candidates, their placement, and algorithm categories.

Table 7.1 NIST PQC competition round 3 candidates

	Placement	Algorithm	Candidates
Finalist	KEM/PKE	Code-based	Classic McEliece [13]
		Lattice-based	CRYSTALS-Kyber [14]
			NTRU [15]
			Saber [16]
	Digital signatures	Lattice-based	CRYSTALS-Dilithium [17]
			FALCON [18]
Alternate	KEM/PKE	Code-based	BIKE [19]
Candidates			HQC [20]
		Lattice-based	FrodoKEM [21]
			NTRU Prime [22]
	Digital signature	Hash-based	SPHINCS+ [23]
		Multivariate PKE	GeMSS [24]

7.2.3   Lattice-Based KEM/Encryption Schemes

A lattice is the set of intersection points in n-dimensional space with a periodic structure (i.e., $L = \left\{ \sum_{i=1}^{n} a_i \boldsymbol{b}_i : a_i \in \mathbb{Z} \right\}$, where $\boldsymbol{b}_i$ is a basis vector). Several complex mathematical problems are used to construct lattice-based schemes. The two most fundamental problems on lattices are the shortest vector problem (SVP) and the closest vector problem (CVP). Given a basis of a lattice L, SVP asks us to attempt to find the shortest nonzero vector of L. Given a basis of a lattice L and a target vector $\boldsymbol{t}$, CVP asks us to find an element of L closest to the target vector $\boldsymbol{t}$. The lattice-based schemes are based on the three problems such as standard learning with errors (SLWE) problem, ring learning with errors (RLWE) problem, and module learning with errors (MLWEs) problem. The RLWE and MLWE are potentially reducible to SVP. The lattice-based cryptographic implementation hierarchy is shown in Fig. 7.1. The SLWE, RLWE, and MLWE are built on the matrix/vector arithmetic, the arithmetic of polynomials, and the polynomial arithmetic and matric/vector arithmetic, respectively. Moreover, symmetric-key primitives (i.e., PRNG, hash functions) and error correction modules are required to build the schemes. The main components (bottom-level components) used in lattice-based methods are modular arithmetic, polynomial multiplication, polynomial inversion, error sampler, encoder, and decoder.

In the finalists, three KEM/Encryption algorithms (CRYSTAL-Kyber [14], NTRU [15], and Saber [16]) and two digital signatures (CRYSTALS-Dilithium [17] and FALCON [18]) are based on lattice algorithm (see Table 7.1). In this chapter, we mainly focus on lattice-based KEM/PKE schemes such as Saber and CRYSTALS-Kyber. These schemes are fulfilled by the NIST security level 5 (i.e., very high-security level), which is considered as hard to break as an exhaustive key search attack on AES-256. A public key encryption scheme (PKE) has private and public keys, where we can encrypt a message using the public key and decrypt using

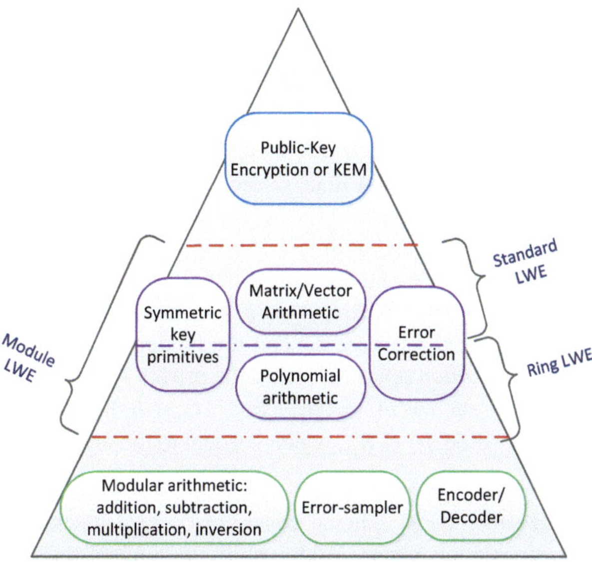

Fig. 7.1 Lattice-based cryptographic implementation hierarchy

the private key. A key encapsulation mechanism (KEM) is a scheme with public
and private keys, where we can use the public and private key pair to generate and
securely exchange session keys. Specifically, Alice first generates the key pair, keeps
the private key, and distributes only her public key. Bob can use Alice's public key
to generate a ciphertext c and common secret key K. The ciphertext can now be
sent to Alice. Alice uses her private key to decrypt the ciphertext c and generate the
common secret key K. Next, we briefly discuss the Saber and CRYSTALS-Kyber.

7.2.3.1   SABER

It is the third structured lattice-based KEM whose security relies on the hardness
of the module learning with rounding problem (MLWR), which is a variant of
the LWE problem [16]. There are three versions of Saber: LightSaber (NIST
security level 1: similar to AES-128), Saber (NIST security level 3: similar to
AES-192), and FireSaber (NIST security level 5: similar to AES-256). Saber [16]
uses the MLWR problem with both p and q power of two to construct an
IND-CPA (indistinguishability under chosen-plaintext attack) secure PKE scheme.
Following that, an IND-CCA (indistinguishability under chosen-ciphertext attack)
secure Saber KEM scheme uses the post-quantum variant of the Fujisaki–Okamoto
transformation [25].

Saber PKE The saber PKE scheme is also composed of three phases: key generation, encryption, and decryption. These operations are built on top of Saber PKE

operations. In a *key generation*, the first generates a public matrix of polynomials $\mathbf{A}$ from randomly generating a seed ($seed_\mathbf{A}$) and a secret vector of polynomials s from a centered binomial distribution with parameter μ, i.e., $\mathbf{A} \leftarrow \mathcal{U}(R_q^{l \times l})$, $s \leftarrow \beta_\mu(R_q^{l \times 1})$, where q is a power-of-two modulus. Then, it computes the vector b by scaling and rounding the product ($b = \lfloor \mathbf{A}^T s \rceil_p$, where p is a rounding modulus). At last, the public key consists of $\mathbf{A}$ matrix seed ($seed_\mathbf{A}$) and b, while the secret key consists of the secret vector s ($pk := (seed_\mathbf{A}, b), sk := (s)$). An *encryption* consists of generating a new secret s' and adding a 256-bit message m to the inner product v' between the public vector b and the new secret s', i.e., $c_m = (v' + h_1 - 2^{\epsilon_p - 1} m \mod p) \gg (\epsilon_p - \epsilon_T) \in R_T$, where $v' = b^T(s' \mod p) \in R_p$. The ciphertext c consists of the encrypted message c_m and the hidden secret $b' = \lfloor \mathbf{A}s' \rceil_p$, i.e., $c := (c_m, b')$. The *decryption operation* takes the private key s and ciphertext c and produces a message m' such as $m' = ((v - 2^{\epsilon_p - \epsilon_T} c_m + h_2) \mod p) \gg (\epsilon_p - 1) \in R_2$, where $v = b'^T(s \mod p) \in R_p$, which is equal to the original message m with high probability.

Saber KEM The saber KEM scheme consists of a key generation, an encapsulation, and a decapsulation phase. Additionally, it requires three hash functions that model random oracles: $\mathcal{F}, \mathcal{G}$, and $\mathcal{H}$, which are instantiated with SHA3-256, SHA3-512, and SHA3-256, respectively. The *KEM key generation* is a similar operation as the PKE key generation to generate the public key ($seed_\mathbf{A}, b$) and the secret vector s except for including a hashed public key, $pkh = \mathcal{F}(pk)$, i.e., $pk := (seed_\mathbf{A}, b), sk := (s, z, pkh)$, where $z = \mathcal{U}(\{0, 1\}^{256})$. The *KEM Encapsulation* is constructed from Saber PKE encryption operation. It takes the public key and produces a common secret key K and ciphertext c. The *Saber KEM decapsulation* algorithm is based on the Saber PKE encryption and decryption algorithms. It decrypts the ciphertext via Saber PKE decryption with the private key s and generates the shared secret key K. The decrypted message is re-encrypted to check the integrity of the ciphertext. For further details, one may refer to [16].

7.2.3.2 CRYSTALS-KYBER

It is a lattice-based PQC whose security is based on the hardness of solving the LWE problem over module lattices (MLWE) [14]. It follows the conventional construction method to build an IND-CPA PKE scheme firstly and then turns it into an IND-CCA KEM through the tweaked Fujisaki–Okamoto transform. It involves finding a vector s when given a matrix A and a vector $t = As + e$, where e is a small (random masking) error vector which is used to hide the private key. In this scheme, vector–vector and matrix–vector multiplication can be optimized with the fast number theoretic transform (NTT), which can reduce computational complexity from $O(n^2)$ to roughly $O(n \log n)$. Depending on the security level, Kyber comes in three versions such as Kyber-512 (security roughly equivalent to AES-128), Kyber-768 (security roughly equivalent to AES-192), and Kyber-1024 (security roughly equivalent to AES-256).

Kyber PKE It is composed of three phases: key generation, encryption, and decryption. In the *key generation phase*, using a random seed, it generates the public matrix parameter A from a uniform distribution and secret vectors s and e sampled from a centered binomial distribution. Then, the LWE instance can be calculated as $\hat{t} = \hat{A} \circ \hat{s} + \hat{e}$ in the NTT domain. In the *encryption phase*, three vectors, s', $e1$, and $e2$, are sampled from uniformly distributed random numbers and centered binomial distributions. The message m to be encrypted is first encoded to $m' = enc(m)$. Then, the ciphertext c_1 is calculated as $A^T s' + e1$, while the ciphertext c_2 is formed by embedding the message into an LWE instance as $c_2 = t^T s' + e2 + m'$. Then, both $(c1, c_2)$ are rounded and published as the ciphertext $ct = (c_1, c_2)$. In the *decryption phase*, decompress (c_1, c_2) and calculate $r = (c_2 - c_1 s)$, which when decoded as $dec(r)$ yields the message m.

Kyber KEM It is constructed from Kyber-PKE operations and consists of three phases: key generation, encapsulation, and decapsulation. First, Alice generates a matrix A and computes a vector t by following a similar operation as Kyber PKE key generation operation. Then, the seed used to generate a matrix A and the computed t encoded as public key is sent to Bob for encapsulation operation. Furthermore, secret vector s encoded as private key is kept for decapsulation operation later. In the encapsulation phase, Bob generates the ciphertext c by using Kyber-PKE encryption algorithm. He also computes the shared secret K by using Alice's public key, message, and the hashed value of the ciphertext. In the decapsulation phase, Alice takes the ciphertext c and the private key s and then generates the message m' by using Kyber-PKE decryption algorithm. Then, she verifies whether it can be encrypted to the same ciphertext sent by Bob by using Kyber-PKE encryption algorithm. If ciphertexts match ($c = c'$), Alice computes the shared secret K by using the ciphertext c, the message, and her public key. Otherwise, she computes the shared secret K by using a random value and the ciphertext c. For more details of the algorithm, one may refer to [14].

7.3   Prevailing Side-Channel Attacks

The existing lattice-based PKE and KEM schemes are resistant against known mathematical cryptanalytic attacks, such as chosen-plaintext attacks (CPAs) and chosen-ciphertext attacks (CCAs), respectively. However, they are vulnerable to physical attacks, such as power/EM side-channel attacks or fault injection attacks [9, 11, 26–33]. It can be categorized into two kinds of side-channel attacks depending on indirect exploitation of side-channel leakages or direct exploitation of them to extract secret keys: (1) algorithmic-level attacks [9, 11, 26–28] and (2) implementation-level attacks [29–33]. Both can exploit traditional side-channel attacks, such as simple power/EM side-channel attacks (SPAs), differential power/EM side-channel attacks (DPAs), correlation power/EM side-channel attacks (CPAs), and AI-based side-channel attacks to extract the secret asset. An adversary

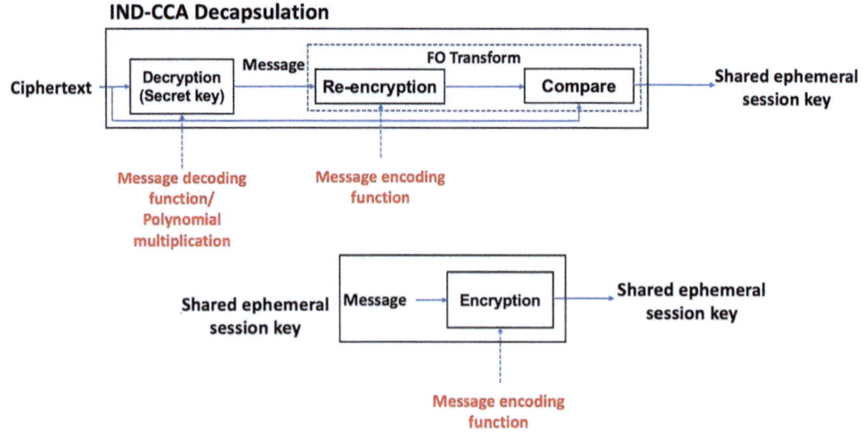

(a) Targets of PQC KEM schemes.

	Target function	Attack	Goal
Decapsulation	Message decoding function	Algorithmic-level SCA with CCA	Long-term secret key
	Message encoding function	Algorithmic-level SCA	Shared ephemeral session key
	Polynomial multiplication	Implementation-level SCA	Long-term secret key
Encapsulation	Message encoding function	Algorithmic-level SCA	Shared ephemeral session key

(b) SCA attacks and goals.

Fig. 7.2 Side-channel attacks on PQC KEM schemes

can target both the decapsulation and encapsulation parties to extract the long-term secret key or the shared ephemeral session key as shown in Fig. 7.2.

7.3.1 Algorithmic-Level Attacks

Power/EM side-channel leakages can assist mathematical cryptanalyses, such as CCAs, even though lattice-based KEMs have indistinguishability under chosen-ciphertext attack (IND-CCA) security by using well-known Fujisaki–Okamoto (FO) transform theoretically. In the FO transform scheme of the decapsulation part, an adversary does not have information of decrypted messages without physical attacks so that she/he cannot succeed CCAs. However, the message decoding function with invalidated ciphertexts in the decryption phase can be targeted by power/EM side-channel attacks to extract messages used to reveal the *long-term secret key* by a

CCA. A single-bit message converted from each coefficient of a polynomial in the message decoding function can be distinguishable by simple power analysis [34]. Ravi et al. recovered the secret key using EM side-channel-assisted CCAs with templates to classify a single-bit/byte message on NewHope KEM, Kyber KEM, Saber KEM, LAC KEM, and Round5 PKE implemented on an Arm Cortex-M4 core [26].

In addition, the message encoding function in the encapsulation phase can be targeted to reveal a randomly generated secret message. A *shared ephemeral session key* can be generated by the recovered message and public values. In the CRYSTALS-Kyber algorithm, a bit message is converted into 0×0000 or 0xFFFF by the message encoding function, i.e., when a bit message is 0, the coefficient of the polynomial is encoded into 0×0000, and when a bit message is 1, the coefficient of the polynomial is encoded into 0xFFFF. Each message can be observed through power/EM signatures since power consumption depends on the Hamming weight of the encoded coefficient. This attack scenario can use only a single trace since the target message is randomly generated every time. Sim et al. achieved a 100% single-trace attack to recover the message on CRYSTALS-Kyber and Saber SW implementations on an Arm Cortex-M4 core (ST32F3) [28].

7.3.2 *Implementation-Level Attacks*

Adversaries can target some functions directly related to the secret key in lattice-based KEM/PKE algorithms such as polynomial/matrix-vector multiplication, error/secret sampling, or extendable output function using SHAKE to extract the secret key without the assistance of cryptanalytic methods [29–33]. For example, an input of the polynomial multiplication is the secret key in the lattice-based KEM/PKE algorithm. Power/EM signatures depend on the intermediate values of the polynomial multiplication. This side-channel information can be used to reveal the secret key for horizontal differential power analysis (DPA) attacks, template attacks, or AI-based side-channel attacks. Since a single trace includes several multiplications of a target subkey, single-trace key recovery is possible by exploiting sub-traces corresponding to target multiplications in a single trace. Primas et al. [30] performed successful single-trace template attacks on number theoretic transform (NTT) in the ring lattice learning-with-error (LWE) decryption phase to extract the entire private key. Recently, Pessl and Primas [31] successfully demonstrated an advanced single-trace attack on NTT of CRYSTALS-Kyber software implementation running on an Arm Cortex-M4 microcontroller. Table 7.2 summarizes existing side-channel attacks on various PQC implementations with the target operations, leakage sources, the type of SCAs, and the required number of traces to satisfy a success rate.

Table 7.2 Side-channel attacks on lattice-based PQC

Work	PQC	Imp.	Target	Leakage	Method	# Query	Success rate
[26]	NewHope, Kyber Saber, Round5 LAC	ARM CM4	Message Decoding	EM	SPA	490	100 %
[35]	Saber	ARM CM4	Polynomial multiplication	Power	TVLA	–	$|t| > 4.5$
[10]	Saber	ARM CM4	A2A, `Keccak-f`, `SecBitSliced Sampler`	Power	TVLA	10,000	$|t| > 4.5$
[36]	KYBER	ARM CM4	Barrett reduction	Power	CPA	12	100%
[9]	KYBER	ARM CM4	Inverse NTT	EM	SPA	960	100%
[37]	KYBER	ARM CM4	Message encoding	Power	SPA	525	68.6%
[38]	KYBER	ARM CM4	Polynomial multiplication	Power	CPA	200	100%
[39]	NTRU	XMEGA 128D4	Modular addition	Power	SPA	1	100%
[40]	NTRU Prime	ARM CM4	Polynomial multiplication	Power	HICPA	1	100%
[41]	NTRU	ARM CM4	Modular reduction (mod3())	EM	SPA	1	75%
[28]	KYBER	ARM CM4	Message encoding	Power	MLP	500	100%
[28]	Saber	ARM CM4	Message encoding	Power	MLP	10,000	100%
[28]	FrodoKEM	ARM CM4	Message encoding	Power	MLP	10,000	79%
[42]	Saber	ARM CM4	`POLY2MSG`	Power	MLP	100,000	97.4%
[43]	Saber	ARM CM4	`Poly_A2A`	Power	Ensembled MLP	7,800	100%
[44]	Frodo & NewHope	Xilinx Spartan-6	Polynomial Multiplication	Power	CNN	–	100%
[45]	Frodo	Xilinx Spartan-6	Polynomial Multiplication	Power	2-D CNN on images	2,200	100%
[45]	NewHope	Xilinx Spartan-6	Polynomial Multiplication	Power	2-D CNN on images	3,300	100%

7.3.3   AI-Based Side-Channel Attacks

Researchers have recently become interested in an AI-based side-channel attack because of its potential to compromise secure hardware implementation. Few studies use AI-based methodologies to undertake side-channel attacks against software and/or hardware implementations of various PQC algorithms in the literature.

Using the deep learning model, Bo-Yeaon et al. [28] attempted side-channel attacks on Saber, Kyber, and FrodoKEM. In this study, points of interest (POI) were first detected using a sum of squared pairwise t-differences (SOST) value of collected power traces, and then the message was recovered using a multilayer perceptron model (MLP). One of the work's shortcomings is that the technique was evaluated on unprotected software implementations. Ngo et al. executed a message recovery attack on Saber using an MLP architecture. The authors successfully recovered both the session key and the long-term secret key from few traces by performing experiments on three separate devices [42]. In a separate study, Ngo et al. applied ensembled deep learning networks to attack the software version of CCA secure Saber KEM secured by first-order masking and shuffle. Aydin et al. [44] demonstrated the capability of the CNN by attacking hardware implementations of the Frodo and NewHope protocols with a single trace, as well as demonstrating that traditional attacks such horizontal TA and DPA were outperformed by AI-based attacks by up to 25% and 900%, respectively. In a separate study, 1D time-series power measurement data was converted into 2D pictures, and DL techniques were performed to the SCA images [45]. The results of such attacks on the hardware implementation of NewHope and Frodo demonstrated their superiority over traditional tactics. Table 7.2 summarizes prevailing side-channel attacks on Lattice-based PQC protocols, including the target implementation, data set size, SCA method, and success rate.

7.4   PQC Side-Channel Evaluation Platform (SEP)

In recent years, the area of PQC hardware security and trust has seen vastly increasing research activity. A large population of academic and industry researchers has been working on various aspects of securing a cryptographic module from power/EM side-channel analysis attacks. Most research in this area is still carried out in an *ad hoc* fashion without the help of well-established metrics, test methods, and EDA tools. Although semiconductor companies are becoming increasingly aware of the requirement of automatic SCA resistance analysis and protection against it, a systematic framework to accomplish these goals is notably lacking in the industry. In this section, for the first time, we propose to develop a systematic framework for comprehensive side-channel evaluation of NIST PQC implementations during design phases (RTL, gate level, and layout level) that can be seamlessly integrated into the existing design flow as well as post-silicon validation shown in Fig. 7.3.

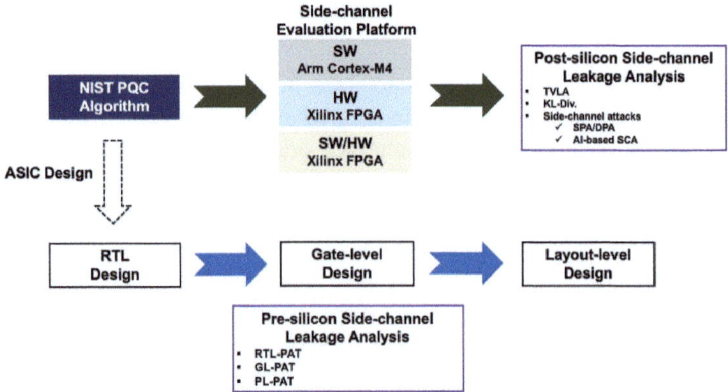

Fig. 7.3 PQC side-channel evaluation platform

7.4.1 Pre-silicon Side-Channel Leakage Assessment

Power side-channel assessment of NIST PQC designs starts from early design phase and moves coherently forward to different levels of design abstractions (RTL $\rightarrow$ gate level $\rightarrow$ layout level) using our power side-channel analysis tools (PAT).

7.4.1.1 RTL-PAT

At the RTL phase, vulnerable modules in the top-level design of the NIST PQC can be searched by estimating a statistical distance between two sets of switching activities in each module. For example, randomly generated input vectors given a secret key are used to make a set of switching activities. A statistical distance of two groups corresponding to two different secret keys is calculated based on KL divergence [46]. If the KL divergence of any module is larger than a predefined threshold value, the module is determined as a vulnerable module. The selected vulnerable modules can be replaced into secure enhanced modules at the early design phase. The secure RTL design will be evaluated repeatedly until passing the test, and then it will move forward to gate-level design abstraction.

7.4.1.2 GL-PAT

After logic synthesis, a gate-level netlist includes timing delay of gates and flip-flops so that switching activities of all nodes of gates and flip-flops will be simulated in fine-grained timescale, which causes infeasible analysis of the top-level design due to tremendous simulation time or a huge amount of memory requirements. To address this issue, gate-level assessments focus on only vulnerable modules selected at the previous design phase, and vulnerable gates and flip-flops in the vulnerable modules will be ranked. By continuously reducing target areas for the assessments according to design abstractions levels, the analysis can be manageable and efficient. The vulnerable cells will be replaced to secure primitives, such as masked logic gates, wave dynamic differential logic (WDDL) [47], or t-private logic circuits [48]. The secure gate-level modules will be evaluated to verify side-channel robustness.

7.4.1.3 PL-PAT

After designing a power distribution network, synthesizing clock trees, placing standard cells, and routing wires, dynamic power estimation can be estimated accurately due to the physical information of each cell by which capacitance and resistance of all nodes in the target cells can be calculated. Dynamic power or current of vulnerable cells will be simulated using commercial power estimation tools such as Cadence Voltus [49] to assess side-channel leakages. Even if secure cells or primitives in the previous design phase can mitigate side-channel leakages, it may not be enough to pass the layout-level side-channel leakage assessment due to broken power balance after placement and routing. In this case, inserting decaps, isolating the power network of vulnerable cells, or random frequency clock may be helpful to make data-dependent dynamic power balanced or randomized. Finally, a physically enhanced secure layout design will be evaluated. Figure 7.4 shows RTL-PAT, GL-PAT, and PL-PAT tools which can be incorporated in Cadence/Synopsys ASIC design flow. For simulation, randomly generated input vectors or specific input vectors by PSC-TG [50, 51] which can significantly reduce the required

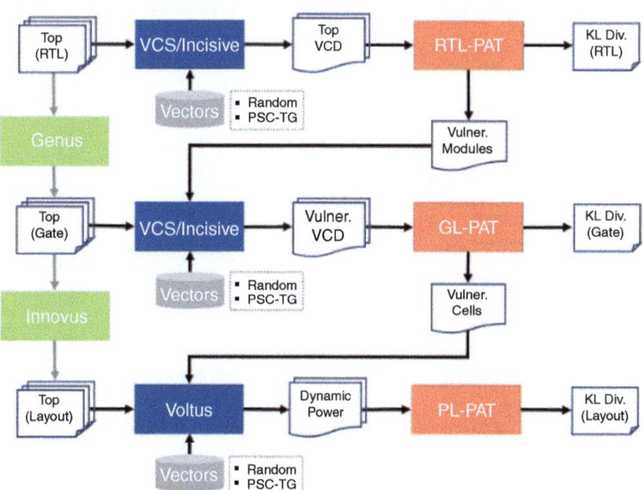

Fig. 7.4 Pre-silicon side-channel leakage CAD tool with ASIC design flow: RTL-PAT, GL-PAT, and PL-PAT

number of input vectors for accurate assessment are used. All switching activities are stored in VCD files during the evaluations. Results of the evaluation at each design phase will be described in terms of KL divergence or new AI-based metrics.

This pre-silicon side-channel leakage approach has some features as follows:

- **Various design-level analysis**: Early design estimation gives flexibility to modify the design, and all design-level analysis can measure contribution of various countermeasures suitable to each design abstraction to mitigate side-channel leakages.
- **Scalability**: Due to significant performance improvement, the PAT tools can analyze an SoC design, including PQC IPs, small standalone crypto IPs, and microcontrollers, or large scale designs such as PQC implementations which take thousands of clock cycles to finish.
- **Accuracy**: Powerful AI-based approaches as well as traditional statistical approaches can improve the accuracy of side-channel leakage estimation.
- **Efficiency**: By reducing target area according to the level of design abstraction and the number of input vectors based on PSC-TG method, efficient assessments are possible.
- **Generality**: This framework can be used to analyze different cryptographic algorithms without any customization to analyze for the leakage.

7.4.2　Post-silicon Side-Channel Leakage Assessment

For measuring power or EM side-channel leakages of NIST PQC implementations, our side-channel evaluation FPGA platform in Fig. 7.5 is used. All SW, HW, and SW/HW co-design of NIST PQC algorithm can be implemented on our customized

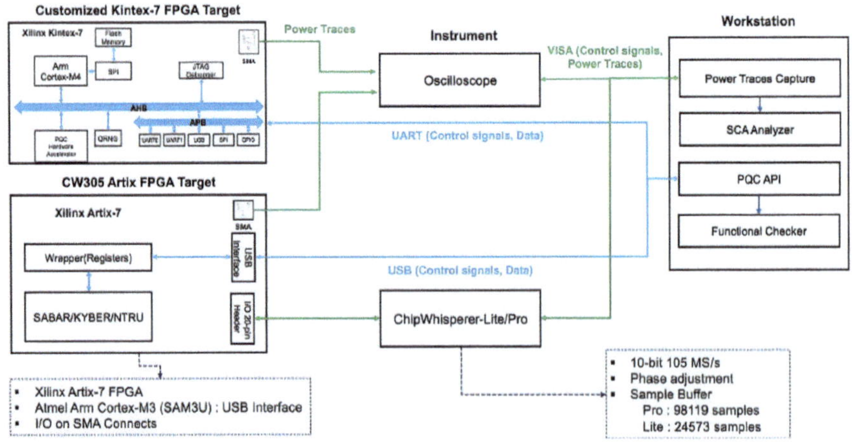

Fig. 7.5　Side-channel evaluation FPGA platform

Xilinx Kintex-7 FPGA/Xilinx Artix-7 FPGA board and collecting power traces, and statistical t-statistic or KL divergence tests and SCA attacks are executed automatically in an assessment flow without additional tools. These assessment results are compared to pre-silicon side-channel leakages so that the accuracy of pre-silicon side-channel estimation can be analyzed, and it can be scrutinized how much contribution countermeasures at each level of design abstraction have to improve side-channel mitigation of the final implementation. Our SCA FPGA evaluation platform consists of three major parts as follows:

- **FPGA boards:** We can use our customized Xilinx Kintex-7 board or ChipWhisperer Artix-7 board to implement PQC designs. Since an Arm Cortex-M4-based SoC design is feasible on the customized FPGA board, we can evaluate SW, HW, and SW/HW PQC implementations using the board. The ChipWhisperer Artix-7 FPGA board [52] is generally used in academic and industry research areas. To compare with other research group works, we can use the ChipWhisperer board. However, it is not suitable for a large scale SoC design due to the limited resource of the Xilinx Artix-7 FPGA.[1]

  The power analysis attack is mounted by measuring the voltage across the shunt resistor placed in between the wire connecting the power supply to the voltage pin of the target FPGA. Both FPGA boards also have an operational amplifier that will amplify the value of the power signal so that it is readable by the capture board or the oscilloscope. Figure 7.6 shows both FPGA target boards. Specially, a quantum random number generator (QRNG) [53] is installed to generate high-entropy random numbers in the customized FPGA board.

- **Measurement instruments:** For collecting power traces, we use either a Tektronix oscilloscope with 1 GHz bandwidth and 5Gs/s sampling rate (MDO3102) or the NewAE ChipWhisperer-Pro/Lite [54, 55] designed to conduct power side-channel analysis. The NewAE ChipWhisperer-Pro and Lite capture boards have a 10-bit analog-to-digital converter (ADC) with 105 MS/s but have different sizes of buffers to store 98119 samples and 24573 samples, respectively. Both can be connected via an SMA connector on the target board and a USB port to communicate with the workstation shown in Fig. 7.5.

- **Workstation:** The workstation sends commands to control target boards such as start or stop operations and data such as plaintexts or random numbers and receives data such as ciphertexts or generated keys though a UART or USB channels. PQC APIs are developed to execute the PQC algorithm in an ARM Cortex-M4 core or an FPGA-based hardware design. A functional checker is installed to verify the functionality of the PQC implementations by comparing simulation results of the verified reference design in the workstation (see Fig. 7.7). In addition, a power trace capture tool collects received data from the instruments, and a power side-channel analyzer performs power side-channel leakage assessment and side-channel attacks.

---

[1] Xilinx Artix-7 XC7A100T has 103K logic cells, 4.8Mb BRAMs, and 240 DSP slices.

(a) Customized FPGA board with a QRNG.

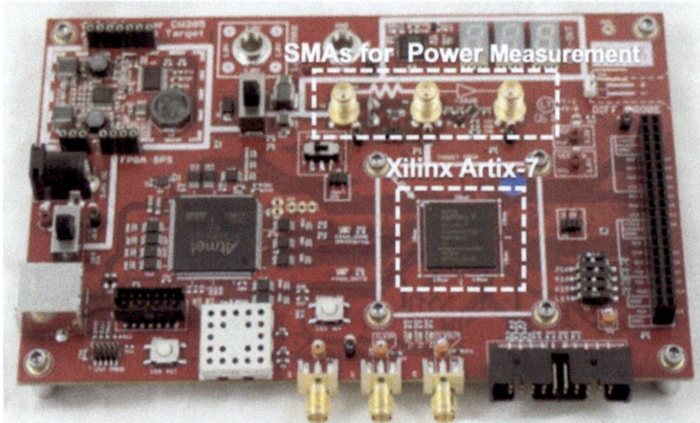

(b) NewAE ChipWhisperer CW305 FPGA board.

Fig. 7.6 Target FPGA boards used for PQC evaluation

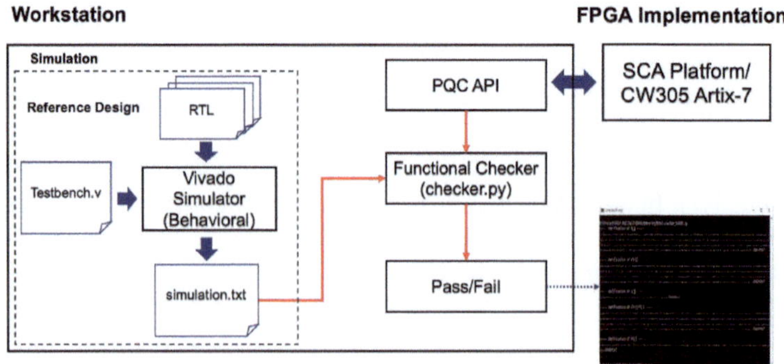

Fig. 7.7 Functional verification of PQC implementations

7.4.2.1 Leakage Assessment

To evaluate side-channel leakage of PQC implementations, generally used statistical methods such as test vector leakage assessment (TVLA) [56] or KL divergence test are significantly exploited.

TVLA Methodology It is based on Welch's t-test which is used to test the hypothesis that two populations have equal means when two samples have unequal variance and unequal sample size. In the side-channel evaluation process, n power side-channel measurements are collected, while the device under the test operates with an asset. The n measurements, $\boldsymbol{p}^i = [p_0^i, \ldots, p_{m-1}^i]$ for $i = 1, \ldots, n$, where m is the number of the sampling points, are classified into two sets by the determinant function D: $S_0 = \{\boldsymbol{p}^i | D = 0\}$ and $S_1 = \{\boldsymbol{p}^i | D = 1\}$. If the t-test statistic,

$$t = \frac{\mu_0 - \mu_1}{\sqrt{\frac{\sigma_0^2}{N_0} + \frac{\sigma_1^2}{N_1}}} \tag{7.1}$$

is out of the confidence interval, $|t| > C$, the null hypothesis, $H_o : \mu_0 = \mu_1$ is rejected, which means that two groups are distinguishable, and the implementation has a high probability of leaking information. Thus, it does not pass the leakage assessment test. In our experiment, the threshold value C is chosen as 4.5, which leads to a confidence of > 0.99999 to reject the null hypothesis.

KL Divergence Even though we can determine if the PQC implementation is vulnerable based on the TVLA test, the t-test statistic cannot estimate how much resistance or vulnerability the implement has against side-channel attacks. In order to quantify side-channel leakage, KL divergence between two sets can be exploited. Let $f_0(p)$ and $f_1(p)$ be the probability density functions of two measurement sets S_0 and S_1, respectively. KL divergence is defined as the following equation [46]:

$$D_{KL}(S_0||S_1) = \int f_0(p) \log \frac{f_0(p)}{f_1(p)} dp \tag{7.2}$$

If S_0 and S_1 are the normal distribution with μ_0, σ_0^2 and μ_1, σ_1^2, respectively, then

$$D_{KL}(S_0||S_1) = \left\{ (\mu_0 - \mu_1)^2 + \sigma_0^2 - \sigma_1^2 \right\} / (2\sigma_1^2) + \ln(\sigma_1/\sigma_0) \tag{7.3}$$

Also, KL divergence is related to the number of traces N necessary to assert with a confidence of $(1 - \alpha)$ that the two normal distributions S_0 and S_1 are different. The number of traces N is a significant contributor in quantifying a lower bound on the attack complexity. The smallest number of traces to satisfy that $\Pr\left[|\bar{S}_0 - \bar{S}_1 - (\mu_0 - \mu_1)| < \epsilon\right] = (1 - \alpha)$ is

$$N \geq \frac{(\sigma_0 + \sigma_1)^2}{\epsilon^2(\mu_0 - \mu_1)^2} \cdot z_{1-\alpha/2}^2 \tag{7.4}$$

where the quantile $z_{1-\alpha/2}$ of the standard normal distribution has the property that $\Pr\left[Z \leq z_{1-\alpha/2}\right] = 1 - \alpha/2$. Comparing to Eq. (7.3), as the KL divergence of two random variables increases, the number of traces N decreases.

7.4.3 AI-Based SCA Attacks

Side-channel analysis has seen remarkable growth in the last 5 years because of the inclusion of machine learning and deep learning-based approaches for attacks and vulnerability assessment. In this section, we first discuss the road map for AI-based side-channel analysis. Later, we describe a framework for AI-based side-channel attacks using signal decomposition. Finally, we narrate the training scheme and model setup we used in the experiments for the framework.

7.4.3.1　Roadmap for AI-based Side-Channel Analysis

The overview of an AI-based profiling side-channel attack is shown in Fig. 7.8. The collected power trace is first passed through the *data preprocessing* step. In this stage, the collected data passes through one or many algorithms from a pool of data preprocessing techniques such as data augmentation, data standardization, noise reduction, etc. Data normalization through normalization or standardization is one of the most common practices for data preprocessing. Noise cancellation from data using signal processing techniques or auto-encoder neural networks can also be applied for preprocessing the data [57, 58], particularly power traces. In addition, data augmentation can increase the quality of an attack by introducing new data observations in the training set and reducing overfitting.

Feature engineering is the next step that extracts features from the preprocessed raw power traces. In several AI-based side-channel attacks, feature engineering is considered an optional step. In these cases, features are extracted from the neural network end-to-end. Additionally, signal decomposition techniques, such as empirical mode decomposition (EMD) [59], Hilbert vibration decomposition (HVD) [60], variational mode decomposition (VMD) [61], and other methods, can be applied. Such decomposition techniques depict the intrinsic features unseen in

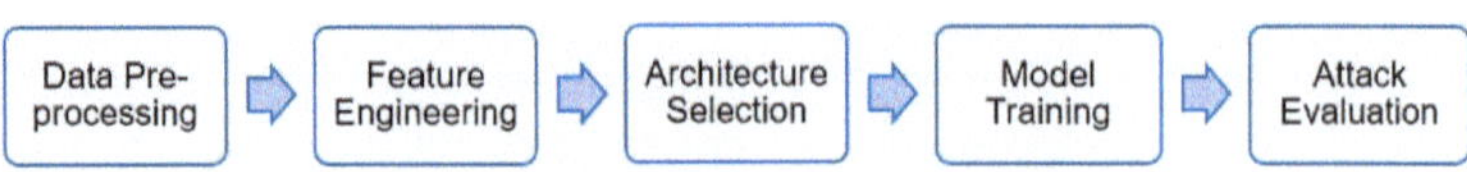

Fig. 7.8 Major steps involved to perform an AI-based side-channel attack

raw signals. Notably, EMD can be used as a denoising step to remove unnecessary information from the power traces. 2D transformation of 1-D data can be an effective way of adding new discriminative features. SCA images generated through 2-D transformation algorithms such as Gramian angular difference field (GADF), Gramian angular difference field (GADF), Markov transition field (MTF) [62], recurrence plots [63], spectrogram, and scalogram can improve the quality of side-channel attack at the cost of computational complexity.

After feature engineering, *selecting an effective algorithm* is a crucial step. Deep neural networks such as convolution neural networks (CNNs) or multilayer perceptron (MLP) are used in most AI-based approaches. The performance of an attack largely depends on challenging tasks such as the model training and selection of appropriate hyperparameters. Automated hyperparameter tuning through Bayesian optimization or reinforcement learning (RL) is a viable way to adjust these hyperparameters properly. At the *attack* phase, the quality of attack is measured through different SCA evaluation metrics that differ from traditional machine learning evaluation metrics.

7.4.3.2 DL-Based Framework Using Signal Decomposition

In this work, we propose a generic framework of deep learning-assisted side-channel attack as shown in Fig. 7.9. The framework includes signal decomposition technique, enabling of neural networks, and automated hyperparameter tuning. From Fig. 7.9, it can be seen that the raw power signal and decomposed signals generated from raw traces are processed from a step named "Partition and Prediction." Figure 7.9 shows the partition and prediction approach of the methodology when the input is the raw trace. In this stage, at first, each trace is preprocessed and then partitioned into multiple sub-traces, and for each sub-trace, a separate neural network is trained at the profiling stage. The choice of these neural networks is decided through automated hyperparameter tuning. In the attack phase, key probabilities of each sub-trace of both raw and decomposed signals are summed up. Finally, the secret key is recovered from this aggregated key probability. As a signal decomposition technique, empirical mode decomposition is used. EMD [59] is a recursive observation-based approach for decomposing a signal into numerous intrinsic mode functions (IMFs). EMD uses the signal's rhythmic activity at the local level to expose intrinsic properties, unlike Fourier transforms and wavelet decomposition. Each IMF's envelopes function as a fundamental oscillatory mode, with an equal number of extrema and zero crossings. This work sets the maximum number of iterations per single sifting at 1000 while deconstructing a trace using EMD. The energy ratio and standard deviation threshold values are set at 0.2 as a termination condition in each IMF examination. The total power per EMD decomposition threshold value is set at 0.005.

Like other deep learning-based side-channel attack methodologies, the approach is performed in two stages: profiling and attack. Algorithms 13 and 14 outline the profiling and attack stage of the method, respectively. This method targets

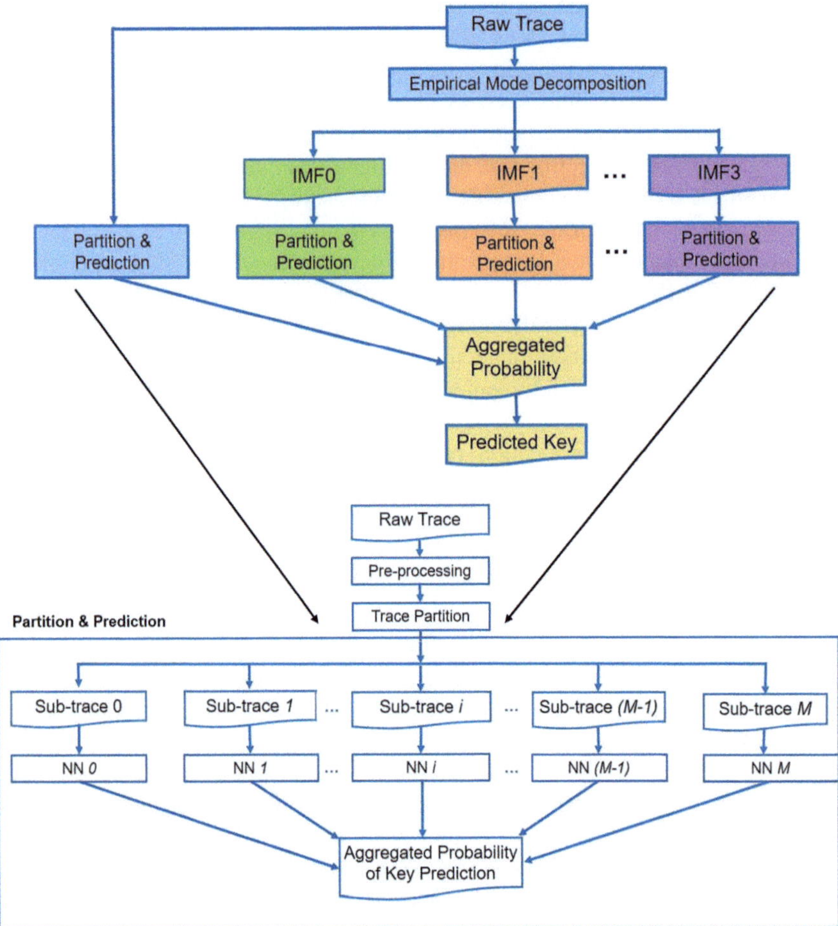

Fig. 7.9 Overview of the generic framework of deep learning-based power side-channel attack using signal decomposition

the matrix multiplication operation of Saber. As seen in Algorithm 13, in the profiling stage, post-silicon traces and corresponding key values are available to train neural network architecture. The profiling stage starts with decomposing the raw traces into $N + 1$ distinct modes using EMD. The algorithm takes the raw trace and splits it into $M + 1$ sub-traces in the next step. Each sub-trace contains an equal number of samples. Afterward, sub-traces of the same order are selected to construct a dataset, where the corresponding labels are the key values. Normalization is used to preprocess the data. Next, the entire dataset of the sub-traces is split into the train, validation, and test sets. Before training a neural network, the most suitable hyperparameters are selected through Bayesian optimization (BO)-based automated hyperparameter tuning [64]. After choosing the

---

Algorithm 13 Profiling stage

1: **Input:** Raw $(t_r^i \in \mathcal{T}_r, i = 1, 2, 3, \ldots, n)$, key $(k^i \in \mathcal{K}, i = 1, 2, 3, \ldots, n)$
2: **Output:** Trained $\mathcal{NN}$
3: Apply signal decomposition on raw trace $\mathcal{T}_r$ and derive $\mathcal{T}_r^j$, $j = 0, 1, 2, \ldots, N$
4: Take raw trace
5: Apply partition so that $t_r^i = \{t_{r,0}^i, t_{r,1}^i, \ldots\ldots\ldots, t_{r,M}^i\}$
6: **for** $j=0:M$ **do**
7:      Construct dataset, $\mathcal{D}_j = (\mathcal{T}_{r,j}, \mathcal{L})$
8:      Apply preprocessing on $\mathcal{T}_{r,j}$
9:      Split the dataset $\mathcal{D}_j$ into $\mathcal{D}_{j,train}, \mathcal{D}_{j,val}, \mathcal{D}_{j,test}$
10:     Select hyperparameters using automated hyperparameter tuning
11:     Train $\mathcal{NN}_j$ on $\mathcal{D}_j$
12: return Trained $\mathcal{NN}_j$
13: **for** $l=0:N$ **do**
14:     Take$(\mathcal{T}_l, \mathcal{L})$
15:     Repeat lines 5-12 on dataset $(\mathcal{T}_l, \mathcal{L})$
16: **end for**

---

hyperparameters, a separate neural network architecture is trained for each dataset of sub-trace coming from raw traces. Later, the same process is repeated for all orders of sub-traces. Consequently, $M + 1$ trained neural networks are obtained. Such steps of trace partition, dataset construction, automated hyperparameter tuning, and neural network training, described in Lines 5–12, are repeated for other decomposed traces. In this way, the profiling stage returns $(N + 2) \times (M + 1)$ neural networks altogether.

Algorithm 14 Attack stage

1: **Input:** Raw trace t_r^{test}, Trained $\mathcal{NN}$
2: **Output:** Predicted key, k_p
3: **for** $j = 0 : M$ **do**
4:     Calculate key probability $p_{k,j}$ for sub-trace $t_{r,j}^{test}$
5: Calculate key probability $p_{r,k}$ for trace t_r^{test}
6: $p_{r,k} = \sum_{j=0}^{M} p_{k,j}$
7: **for** $l=0:N$ **do**
8:     determine t_l^{test}
9:     Repeat Lines 3-6 and calculate $p_{l,k}$
10: **end for**
11: Calculate aggregated key probability S_k for key k
12: $S_k = p_{r,k} + \sum_{l=0}^{N} p_{l,k}$
13: Determine the predicted key, $k_p = \underset{k}{\mathrm{argmax}}\, S(k)$

The attack algorithm of the methodology SEY is described in Algorithm 14. In the first step, the raw trace is partitioned into $M + 1$ number of sub-traces and preprocessed. Using the previously trained neural networks, the key probability for each sub-trace is calculated, described in Lines 3 and 4. The key probability for the whole raw trace is calculated by summing up the individual probabilities. Next, the decomposed signals are generated from the raw trace using the selected decomposition technique named EMD. The key probability for each sub-trace of each decomposed signal is measured following the steps described in Lines 7–10. Finally, the aggregated key probability is calculated, and the key is predicted from the probability.

7.4.3.3   Training Scheme

In this work, MLP is used as the network architecture. Each MLP architecture contains multiple fully connected layers with SeLU activation functions to solve nonlinear complexity. Each hidden layer is followed by batch normalization to fasten the learning and work as a regularizer. Dropout is used between these hidden layers to prevent overfitting. The number of layers and number of neurons in each layer is decided through the implementation of Bayesian optimization (BO)-based automated hyperparameter tuning [64]. This automated process searches for the best hyperparameter by maximizing the specified objective function in a minimum number of iterations through exploration and exploitation approach. In order to find the best model architecture, we set the percentage of neurons at the initial layer and the percentage of neuron shrink in the subsequent layers as the hyperparameters. In this study, ten steps of random exploration are performed with 100 iterations.

The full dataset is split into training, validation, and test sets for the experiments using 80%, 10%, and 20%, respectively. We use mini-batch optimization. RMSprop algorithm is used for optimization, and categorical cross-entropy is used as an objective function. We find the batch size and learning rate values through automated hyperparameter searching by defining a search space.

7.5   Experimental Results

7.5.1   Pre-silicon Side-Channel Analysis Results

In this section, we present the efficacy of RTL-PAT to assess a Saber design [65] at the RTL, which can evaluate the individual modules of the Saber design for PSC leakage. Saber is a full coprocessor hardware implementation of PQC based on learning with rounding (LWR). The regular submodules, i.e., `polynomial multiplier` (using quadratic-complexity schoolbook algorithm), `SHA3/SHAKE128`, `binomial sampler`, `Addpacks`, `AddRound`, `Verify`, `CMOV`, and `CopyWords`, are connected with the data memory (block memory) and program memory. Program memory controls the instruction execution, whereas data memory provides the needed data to the modules. In the RTL simulation, the random seed required by the Saber is provided through the SystemVerilog's internal *urandom* command. Saber uses the power of two for the polynomial multiplication, removing the need for modular reduction. In this implementation, the 256-bit polynomial multiplication is done in parallel. For more details on each subprocess of algorithm and the hardware implementation, refer to [65].

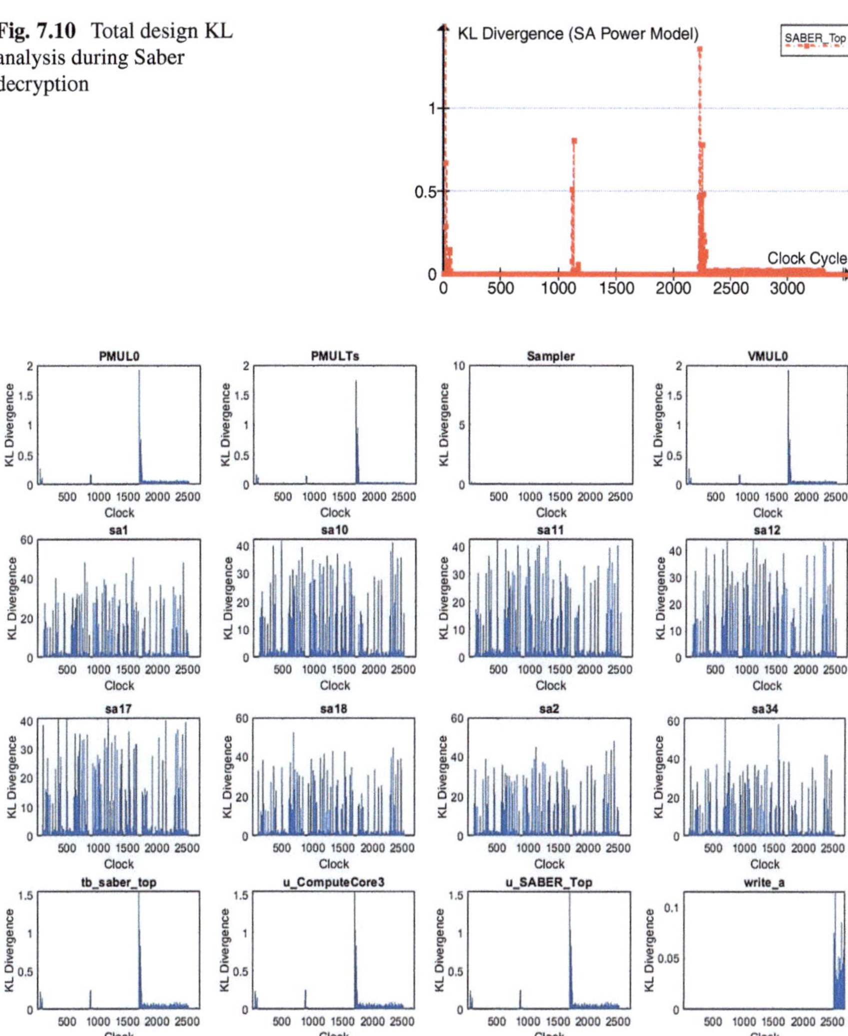

Fig. 7.10 Total design KL analysis during Saber decryption

Fig. 7.11 Modular KL analysis during Saber decryption

During encryption, a secret key is only used once during "Key Generation" from the random seed, making it very difficult for an attacker to extract the key from a single power trace. Therefore, for power side-channel analysis in Saber, we target the decryption process of the implementation. We decrypt 1000 random ciphertexts for two different keys, and it takes more than 3K clock cycles to finish one decryption. Figure 7.10 shows the KL divergence analysis of the top module. It shows that the design is leaky at various places. The high peaks in the figure correspond to reading/writing data from/to the RAM. Figure 7.11 shows the modular KL divergence analysis results of the leaky modules. It is clear that modules related to multiplication, i.e., vector-multiplication (VMUL) and polynomial-256-multiplication (PMUL) consisting of schoolbook-based multipliers (sa), are the leakiest. Their KL divergence graphs are very similar to the total design leakage. Besides multiplication, the binomial sampler module also shows leakiness used to sample secret coefficients from a centered binomial distribution. The U_computeCore3 module is just a wrapper to handle the calling of different modules throughout the decryption. Based on RTL-PAT, we can identify vulnerable modules, such as sa, VMUL, and PMUL, in the Saber HW design. The lower-level PAT tools, i.e., GL-PAT and PL-PAT, will focus on the vulnerable modules activated by specific input vectors using PSC-TG [50]. These lower-level analyses and other design's analyses are out of scope in this chapter.

7.5.2 Post-silicon Side-Channel Analysis Results

In this section, we describe the experimental platforms and implementations used for our experiments. We are currently evaluating both software and hardware implementations of target PQC KEM candidates such as Saber and KYBER.

7.5.2.1 SW Implementations

We use our side-channel evaluation FPGA platform to access side-channel leakages of SW implementations on an ARM Cortex-M4 core as mentioned in Sect. 7.4.2. The SEGGER J-link EDU [66] and embedded studio are used for debugging/probing and programming the ARM-based PQC SW implementations. The system clock of the ARM core has 60 MHz frequency, and the sampling rate to collect power traces is set to 10 MS/s. Generating trigger signals at the start and the end of a target function, e.g., a vector multiplication during decapsulation, can help collect aligned power traces corresponding to the target operation. Figure 7.12 shows a C code to generate trigger signals and a collected power trace via a GPIO port at the start and the end time of a vector multiplication in a Saber KEM.Dec function.

  We analyze three versions of a Saber SW implementation,[2] i.e., LightSaber, Saber, and FireSaber, and three versions of a Kyber SW implementation in pqm4 testbench [67], i.e., Kyber512, Kyber768, and Kyber1024. We focus on the vector multiplication, which is the most vulnerable operation based on the pre-silicon analysis in Sect. 7.5.1. The Saber and Kyber are implemented with 4-way Toom-Cook-based multipliers and NTT-based multipliers, respectively. Figures 7.13 and 7.14 show TVLA and KL divergence testing results of Saber

---

[2] The Saber SW implementation is open source and can be downloaded at https://github.com/ KULeuven-COSIC/Saber.

```c
void indcpa_kem_dec(const unsigned char *sk, const unsigned char *ciphertext, unsigned char message_dec[])
{

        uint32_t i,j;
        uint16_t sksv[SABER_K][SABER_N]; //secret key of the server
        uint16_t v[SABER_N];

        gpio_init_pin(GPIOB, GPIO_PIN_ALL, IO_DIR_OUTPUT, IO_OUTPUT_TYPE_NORMAL, IO_PULLUP_NONE);

        BS2POLVECq(sk, sksv); //sksv is the secret-key

#if Saber_type == 1
            // SABER_un_pack3bit(scale_ar, op);
            SABER_un_pack3bit(&ciphertext[SABER_POLYVECCOMPRESSEDBYTES], v);
#elif Saber_type == 2
            // SABER_un_pack4bit(scale_ar, op);
            SABER_un_pack4bit(&ciphertext[SABER_POLYVECCOMPRESSEDBYTES], v);
#elif Saber_type == 3
            // SABER_un_pack6bit(scale_ar, op);
            SABER_un_pack6bit(&ciphertext[SABER_POLYVECCOMPRESSEDBYTES], v);
#endif

        for (i = 0; i < SABER_N; ++i) {
                v[i] = h2 - (v[i] << (SABER_EP-SABER_ET));
        }
        gpio_write_pin(GPIOB, GPIO_PIN_0, IO_STATE_HIGH);    // Trigger signal
        VectorMul(ciphertext, sksv, v);
        gpio_write_pin(GPIOB, GPIO_PIN_0, IO_STATE_LOW);     // Trigger signal

        //Extraction
        // //addition of h1
        for (i = 0; i < SABER_N; ++i) {
                v[i] = (v[i] & (SABER_P-1)) >> (SABER_EP-1);
        }

        POL2MSG(v, message_dec);

}
```

(a) C Code to Generate Trigger Signal

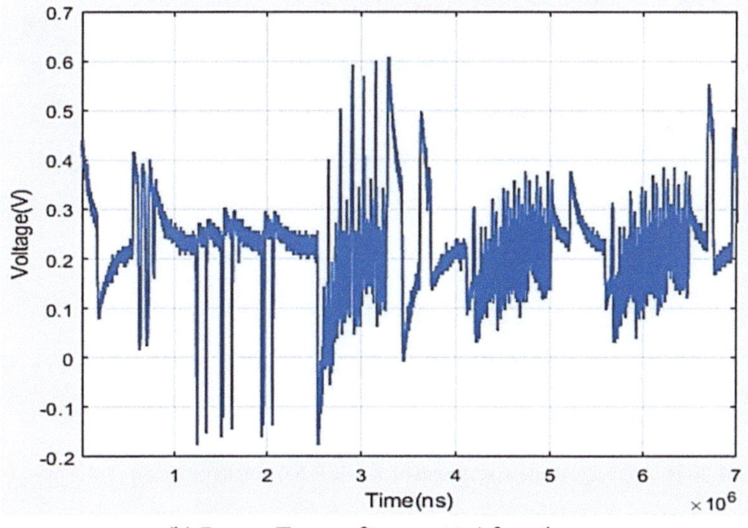

(b) Power Trace of VectorMul function.

Fig. 7.12   C code snippet for collecting power traces and a measured power trace

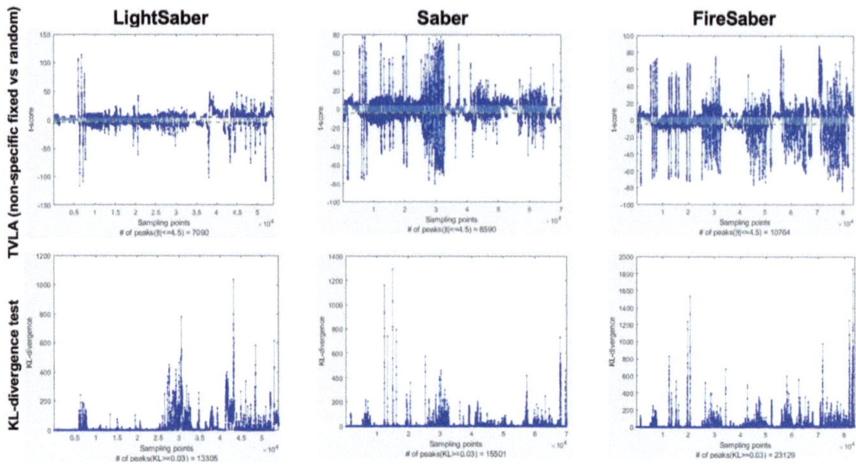

Fig. 7.13 Side-channel leakage assessment of Saber SW implementation

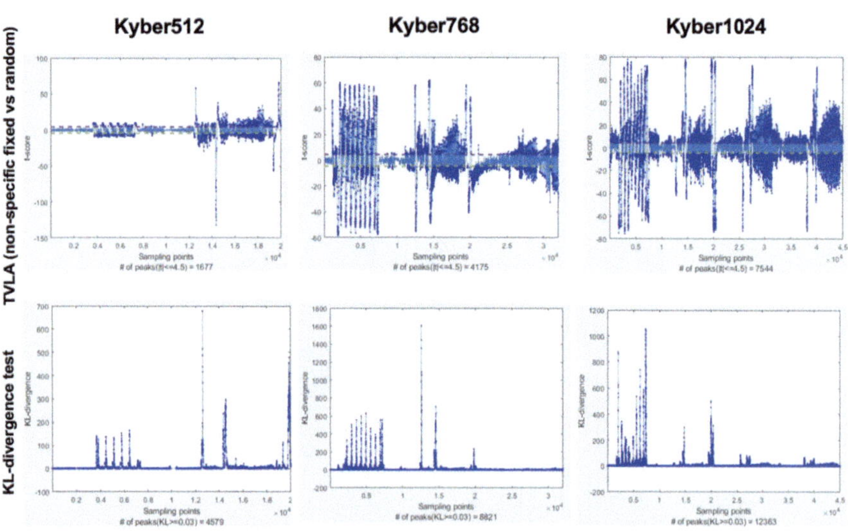

Fig. 7.14 Side-channel leakage assessment of KYBER SW implementation

SW and Kaber SW implementations, respectively. Based on these results, both SW implementations have many vulnerable leakage points during vector multiplications which occupy from 8% to 16% based on TVLA t-statistics (>4.5 or < -4.5). Table 7.3 summarizes the target function, the processing time, and the results of TVLA and KL divergent tests of Saber and Kyber SW implementations.

Table 7.3 Post-silicon side-channel leakage assessment of Saber and Kyber SW implementations

	Target	Time	TVLA (# of peaks $(|t| \leq 4.5)$/ # of samples)	KL (# of peaks $(D_{KL} \geq 0.3)$/ # of samples)	max (D_{KL})
Light Sabe	`VectorMul`	5.4 ms (324K cycles)	7090/54000 = 13.13%	13305/54000 = 24.64%	1035.3
Saber	(4-way Toom-	7 ms (420K cycles)	8590/70000 = 12.27%	15501/70000 = 22.14%	1293.7
Fire Saber	Cook)	8.4 ms (504K cycles)	10764/84000 = 12.81%	23129/84000 = 27.53 %	1853.6
Kyber 512	`Poly_ frombyte_ mul`	2 ms (120K cycles)	1677/20000 = 8.38%	4579/20000 = 22.90 %	677.1
Kyber 768	(NTT)	3.2 ms (192K cycles)	4275/32000 = 13.35%	8821/32000 = 27.56%	1602.8
Kyber 1024		4.5 ms (270K cycles)	7544/45000 = 16.76%	12363/45000 = 27.47%	1053.9

7.5.2.2 HW Implementations

Xilinx Vivado 2020.2 is used for design synthesis, placement, and routing of PQC hardware implementations on Xilinx Kintex-7 or Artix-7 FPGA. To capture power traces from the target board, we used NewAE ChipWhisperer-Lite (CW-Lite) capture board [55] that has a 10-bit analog-to-digital converter (ADC) with a 105 MS/s sampling rate and a high-performance oscilloscope with the maximum sampling rate of 5 GS/s, which is used to convert the analog power trace into discrete values. Then, these discrete values are stored in files on the connected computer, which are later used for TVLA and Kullback–Leibler (KL) divergence. A script on the host PC written in Python communicates with a control program running on the FPGAs via UART. The control program is responsible for communicating inputs and outputs with the target PQC IP cores.

CRYSTALS-KYBER We implemented the CRYSTALS-Kyber512 KEM variant based on Xing et al. [68] on the Xilinx Virtex-7 FPGA platform. CRYSTALS-Kyber512 is based on a polynomial ring $R_q = Z_q[X]/(X^n + 1)$ of the dimension $n = 256$ and modulus $q = 3329$. Other public parameters for Kyber512 are $\eta = 2$ (as the n-th primitive root of unity in Z_q), $k = 2$ (which represents module dimension in MLWE), and $\eta = 2$ (binomial distribution with parameter). The Kyber512-KEM is constructed from three top-level Kyber-PKE block modules: key generation, encryption and decryption operations. We briefly explained the Kyber-KEM algorithms in Sect. 7.2.3.2 and give the block diagram of the Kyber-KEM in Fig. 7.15. The block diagram of the Kyber-KEM performs key generation, encapsulation, and decapsulation steps one by one (as explained in Sect. 7.2.3.2). In the key generation step, two random seed values (ρ, σ) are fed to the key generation block to generate the sampled public matrix parameter $\hat{A}$ and secret vectors $\hat{s}$ and $\hat{e}$

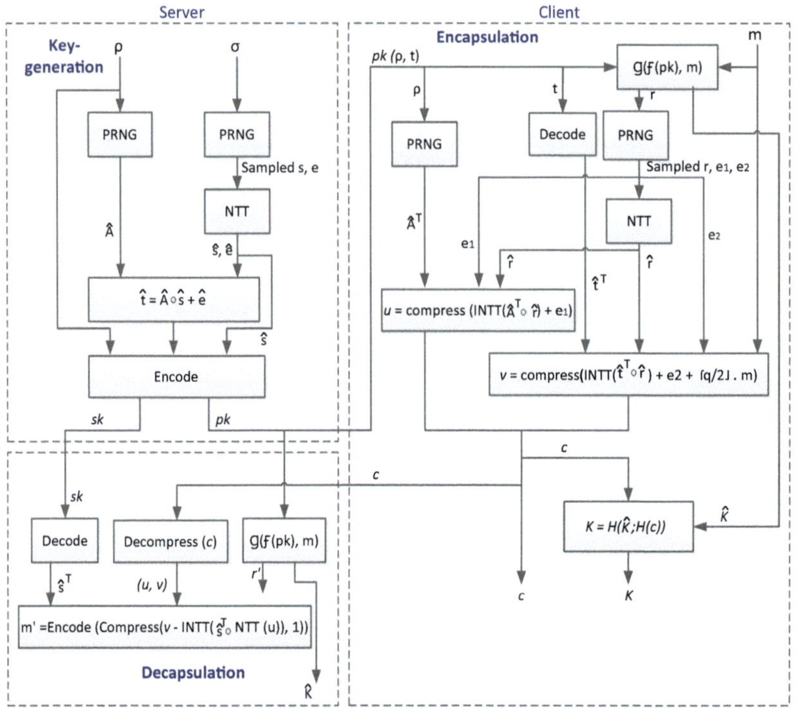

Fig. 7.15 Block diagram of the Kyber-KEM

then perform the computation $\hat{t} = \hat{A} \circ \hat{s} + \hat{e}$ in the NTT domain, and the results of encoded secret key are $\hat{s}$ and the public key pk (encoded t with ρ) is stored in the corresponding registers. In the encapsulation step, encoded public key pk, random message m, and random seed value r are fed to the encryption block, first decode pk to generate the matrix A and $\hat{t}^T$, vectors r, e_1, and e_2 then perform the computation $u = \text{INTT}(\hat{A}^T \circ \hat{r}) + e_1$ and $v = \text{INTT}(\hat{t}^T \circ \hat{r}) + e_2 + Decompress(Decode(m))$, and the results are stored in the corresponding registers. Next, the u, v values are fed decompressed and decoded units to compute the ciphertext $c = (c_1 || c_2)$, and their results are stored in the corresponding registers. In the decapsulation step, the secret $\hat{s}$, c_1 values are fed decryption block to calculate the message m^1. Then, the m^1, encoded public key pk, and random seed value r' are fed again to the encryption block to compute the another ciphertext c'. If the ciphertexts match ($c = c'$), the shared secret K is computed using the ciphertext c, the value m, and the public key. Otherwise, computes the shared secret K by using a random value and the ciphertext c.

For leakage detection, we rely on the widely used Welch t-test-based TVLA comparing fixed with random input measurements, and the resulting t-value is compared to a set threshold of ± 4.5, representing $\alpha = 0.0001$. Informally, if the threshold is exceeded, it is assumed to be possible to distinguish between fixed

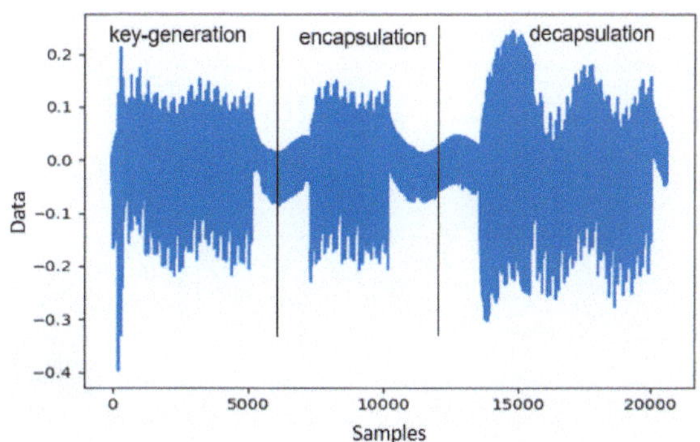

Fig. 7.16 Power traces of hardware implementation of Kyber512 KEM algorithm collected on CW-lite board

and random inputs, which confirms the existence of exploitable leakage. For more details on TVLA, the readers are referred to [56]. Also, we use another vulnerability assessment such as KL divergence to evaluate vulnerability of the targeted design. The KL divergence is calculated based on power leakage distribution of the design. If KL divergence of any design is greater than predefined threshold value, the design is considered to be the vulnerable one [69]. Once Kyber512 KEM has been implemented on CW305 Artix-7 FPGA board, then we can start to capture the power traces based on feeding different input sets to the Kyber512 KEM core, which is target on the CW305 Artix-7 FPGA board. Our Python script is responsible for configuring bit file, arming CW-lite capture board, and capturing traces from the CW-lite board. The trigger signal is send to CW-lite board from the target FPGA enable trace capture when the CW-lite capture board is armed. The single power traces of the non-masked Kyber512 KEM hardware implementation are illustrated in Fig. 7.16.

In our TVLA experiments, we use a non-specific fix-vs.-random test. We measure the power consumption of the Artix-7 FPGA board executing 10,000 capturing traces of the full Kyber512 KEM (i.e., including key generation, encapsulation, and decapsulation steps) on fixed values (ρ, σ, m) for each execution and another set of 10,000 capturing traces on random values (ρ, σ, m). The results of the t-test and KL divergence for these sets are shown in Fig. 7.17a and b. As shown in Fig. 7.17a and b, the results of the t-test and KL divergence tests for full Kyber512 KEM show the presence of lot of leakages in many locations on the traces such as 5656 leakage points in t-test analysis and 4537 vulnerability points in KL divergence analysis.

Figure 7.18a and b shows the result of the t-test and KL divergence test for only decapsulation step of Kyber512 KEM on the Artix-7 FPGA. In this experiment, we measure 10,000 capturing traces of the decapsulation step only on a fixed secret key (sk), fixed cybertext (c) for each execution at first set, and another set of 10,000 capturing traces on a fixed secret key (sk), random public cybertexts (c). As shown in Fig. 7.18a and b, the results of the t-test and KL divergence tests are 3283 leakage points in t-test analysis and 1128 vulnerability points, respectively.

Figure 7.19a and b shows another experiment result of the t-test and KL divergence test for key generation step of Kyber-512 on the FPGA. In this experiment,

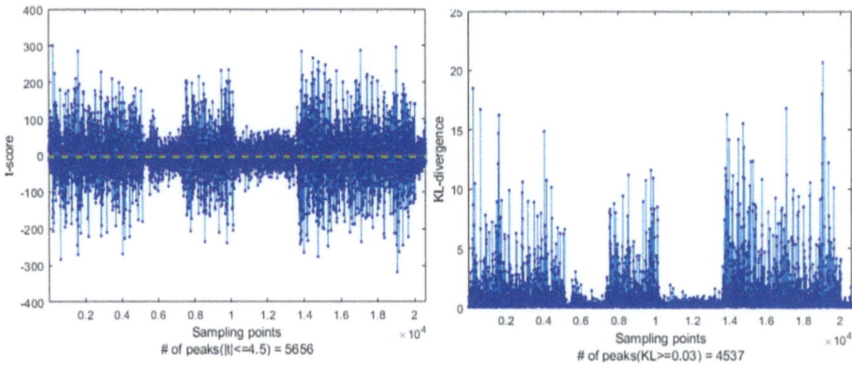

(a) TVLA for full Kyber512 KEM. (b) KL for full Kyber512 KEM.

Fig. 7.17 Side-channel leakage assessment of full Kyber512 KEM

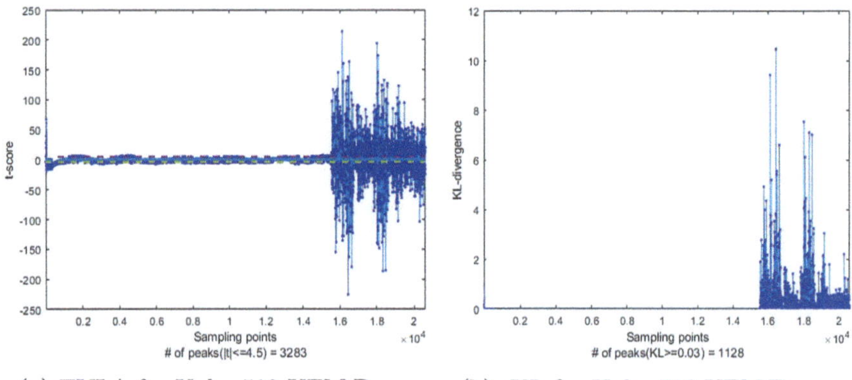

(a) TVLA for Kyber512 KEM.Decap. (b) KL for Kyber512 KEM.Decap.

Fig. 7.18 Side-channel leakage assessment of decapsulation step of Kyber512 KEM

we measure 10,000 capturing traces of the key generation step of Kyber512 KEM on a fixed secret seed (σ), fixed public seed (ρ) for each execution, and another set of 10,000 capturing traces on a fixed secret seed (σ), random public seed (ρ). As shown in Fig. 7.19a and b, the results of the t-test and KL divergence tests are 1640 leakage points in t-test analysis and 759 vulnerability points, respectively.

SABER We implemented the Saber instruction-set coprocessor [65] integrated into an ARM Cortex-M4 architecture via an AHB interface on a Xilinx Kintex-7 FPGA and a stand-alone version of the Saber instruction-set architecture on a Xilinx Artix-7 FPGA. Figure 7.20 shows the block diagram of the Saber architecture that has a 35-bit wide instruction set consisting of a 5-bit OP code, two 10-bit input operand addresses, and a 10-bit result address, and a data memory with 64-bit words. Similar to the Saber SW analysis in Sect. 7.5.2.1, the polynomial multiplier in the Saber

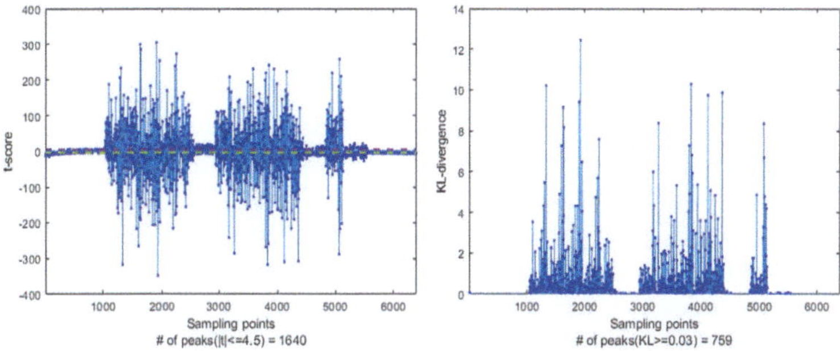

(a) TVLA for Kyber512 KEM.Keygen.   (b) KL for Kyber512 KEM.Keygen.

Fig. 7.19 Side-channel leakage assessment of key generation step of Kyber512 KEM

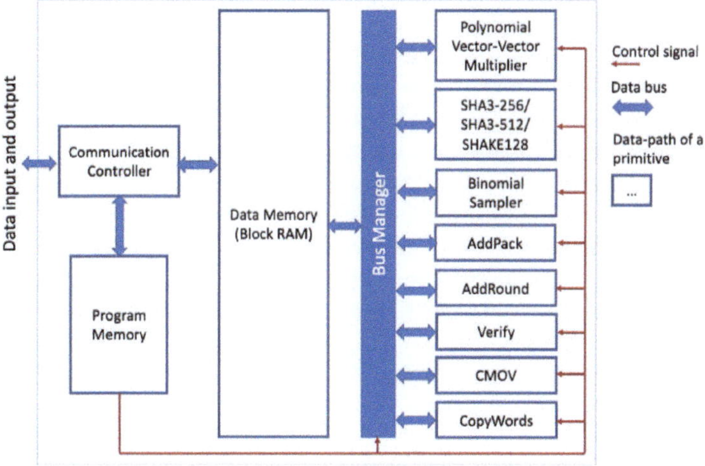

Fig. 7.20 Block diagram of Saber instruction-set architecture [65]

instruction-set architecture is scrutinized. It is based on optimized schoolbook-based multiplication and consists of parallel 256 multiplier-and-accumulate (MAC) units. For computing a matrix–vector multiplication, i.e., $\mathbf{As} = \begin{bmatrix} A_{0,0} & A_{0,1} & A_{0,1} \\ A_{1,0} & A_{1,1} & A_{1,2} \\ A_{2,0} & A_{2,1} & A_{2,2} \end{bmatrix} \begin{bmatrix} s_0 \\ s_1 \\ s_2 \end{bmatrix}$, polynomial multiplications, $A_{i,j}s_k$, are repeatedly executed. In this work, we only focus on the polynomial multiplication. For various TVLA and KL divergence tests, multiple sets are made as follows:

- Set 1: a fixed key, s_1 with n random $\mathbf{A}$
- Set 2: a fixed key, s_2 with n random $\mathbf{A}$
- Set 3: a fixed key, s_1 with a fixed $\mathbf{A}$
- Set 4: a fixed key, s_2 with a fixed $\mathbf{A}$

During executing multiplications in each set, we can collect n power traces and then perform both TVLA and KL divergence tests in such a way that (i) *random-vs.-random* with Set 1 and Set 2, (ii) *random-vs.-fixed* with Set 1 and Set 3, and (iii) *fixed-vs.-fixed* with Set 3 and Set 4. Two FPGAs are working at different clock frequencies; ChipWhisperer and our customized boards operate at 12 MHz and 60 MHz, respectively. Figure 7.21 shows a collected power trace during a polynomial multiplication in each FPGA. Measured power traces from the customized FPGA are more distorted for the high-frequency clock. Each set consists of 1000 power traces on the ChipWhisperer CW305 board. On the other hand, 10,000 power traces per set are collected from our Kintex-7 FPGA board.

Figures 7.22 and 7.23 show the results based on TVLA and KL divergence metrics from two different boards, which identify many vulnerable points during polynomial multiplications. Especially, Test2, i.e., fixed-vs.-random test, detects the most vulnerable points on both FPGA boards. Even if the Saber HW designs on two different boards are the same, vulnerable sampling points and quantitative leakage based on KL divergence are entirely different for the other clock frequency and integration style, a stand-alone Saber HW with a 12 MHz system clock on ChipWhisperer CW board and an ARM Cortex-M4-based SoC including the Saber HW with a 60 MHz system clock on our customized FPGA board. This brings us a new challenging issue, such as how to build a reference FPGA platform and measurement setups to evaluate side-channel leakages of any kind of PQC designs efficiently and accurately. We will deal with this issue again in Sect. 7.6.

Finally, Table 7.4 summarizes the target function, the processing time, and the results of TVLA and KL divergent tests of two Saber HW implementations.

7.5.2.3 AI-Based SCA Attacks

In this work, extensive experiments are performed to evaluate the efficacy of the DL-based framework of side-channel attack using signal decomposition technique.

Dataset There is no publicly available dataset on the Saber algorithm's hardware implementation. A hardware implementation of Saber is set up utilizing the CW305 Artix-7 FPGA board for testing. In all, 16000 power traces were captured from a CW305 Artix-7 FPGA board at 12 MHz clock frequency using a Tektronix MDO3102 in Fig. 7.5 for all possible subkeys from 0×0 to 0xF, i.e., 1000 power traces per subkey. The sampling rate and bandwidth of the oscilloscope are set at 100 MS/s and 250 MHz, respectively. Each power trace corresponds to a vector multiplication in this scenario, with a total of 9000 sample points. Each sub-trace corresponds to a different polynomial key s_i for $i = 0, 1$, and 2. This work focuses on extracting a coefficient of the polynomial key, i.e., a 4-bit subkey among the 256×4-bit entire key. In the Saber HW implementation, 256 multipliers with a 13-bit operand (a coefficient of A) and a 4-bit subkey (a coefficient of s) are executed simultaneously for a clock cycle. It takes 256 cycles to finish a polynomial multiplication. Figure 7.24 shows the structure of the polynomial

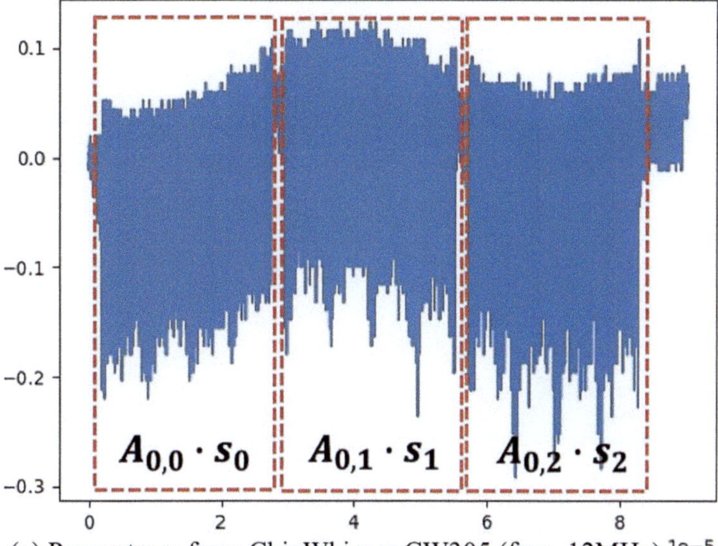

(a) Power trace from ChipWhisper CW305 (freq. 12MHz)

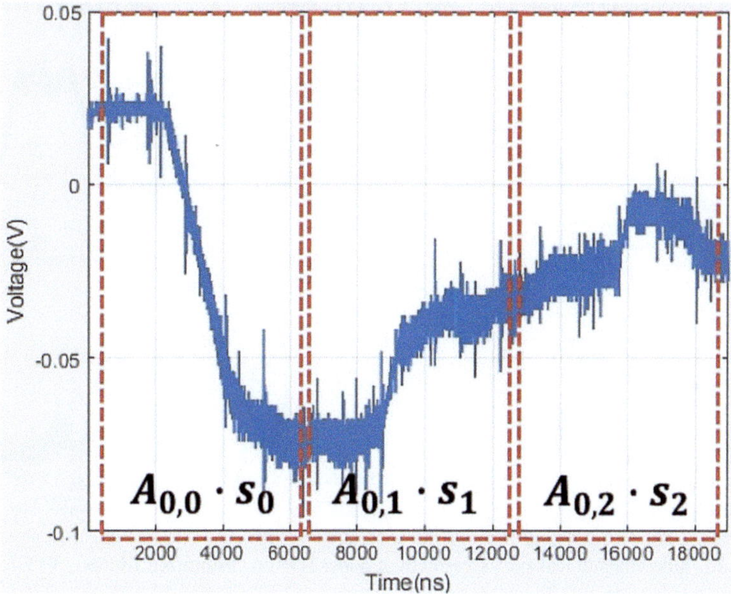

(b) Power trace from custom Kintex-7 (freq. 60MHz)

Fig. 7.21 Power trace during a polynomial multiplication of Saber HW

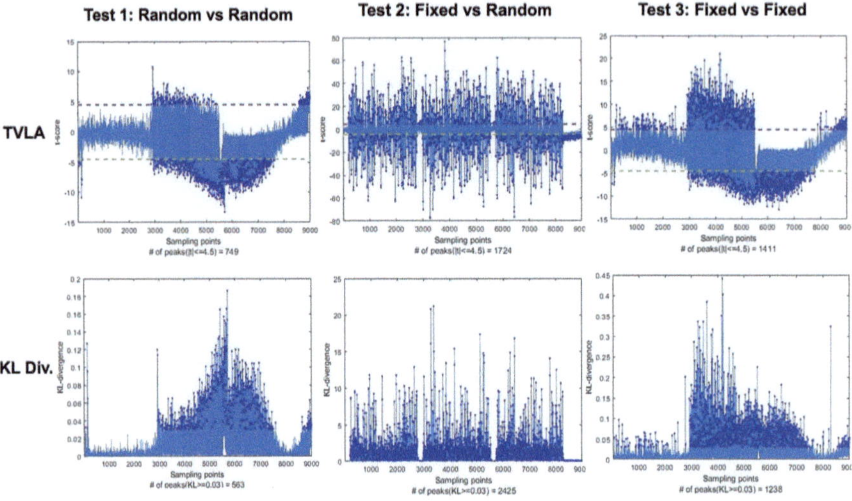

Fig. 7.22 Side-channel leakage assessment of Saber HW on ChipWhisperer CW305 FPGA board

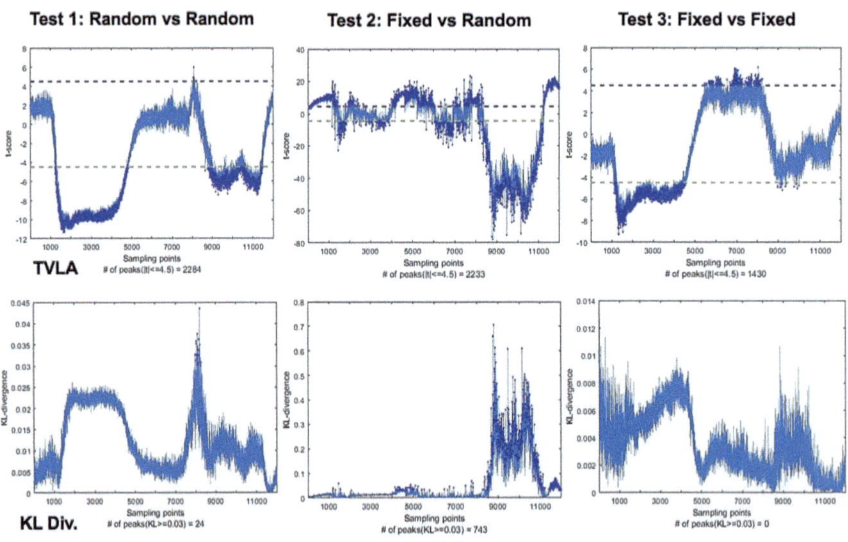

Fig. 7.23 Side-channel leakage assessment of Saber HW on customized Kintex-7 FPGA board

multiplier consisting of 256 multiplier-and-accumulate (MAC) units. At the $(i+1)$st clock cycle, the ith coefficient of $A_{0,0}$, denoted by $a_{0,0}[i]$, is multiplied with all coefficients of a polynomial key, $s_0[j]$, $\forall j = 0, 1, \ldots, 255$. The first part of the power traces in Fig. 7.21a includes 256 multiplications with the target of the 4-bit subkey.

Table 7.4 Post-silicon side-channel leakage assessment of Saber HW implementations

	Target	Time	TVLA (# of peaks ($|t| \leq 4.5$)/ # of samples)	KL (# of peaks ($D_{KL} \geq 0.3$)/ # of samples)	max (D_{KL})
Saber (CW305)	Polynomial Multiplier	360 us	1724/9000 = 19.15 %	2425/9000 = 26.90%	21.23
Saber (Our custom)		60 us	2233/12000 = 18.6%	743/12000 = 6.19%	0.7

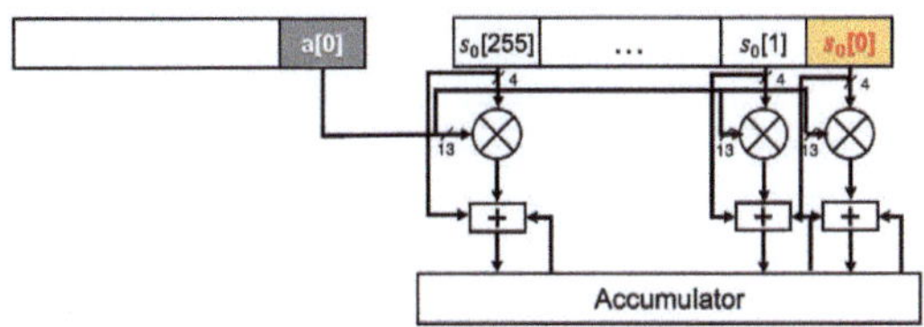

Fig. 7.24 Polynomial multiplier Architecture in Saber HW; $a[0]$ is multiplied at the first clock cycle and $s_0[0]$ is the target subkey

Evaluation Metrics The performance of the methods used in experiments is measured in terms of success rate and guessing entropy (GE) in this study. The success rate is defined as the proportion of successful key recovery attempts to total key recovery attempts. When the predicted key matches the correct secret key, the attempt is successful. The GE, on the other hand, represents the average rank of the right key overall potential key–value combinations. For the GE calculation, we repeat the assault 200 times using randomly selected sub-samples of the test to determine the average number of traces necessary for key extraction. We also use a confusion matrix to visualize the key classification result properly.

Results Figure 7.25a depicts the performance of the proposed technique. When the input features are varied, the performance of the method described in Sect. 7.4.3.2 is compared. When just raw traces are employed, the success percentages are the lowest. The inclusion of IMF0 to raw traces improves the performance of multiple-trace attacks by 76.59 %. EMD's first decomposed signal, IMF0, adds unique observations to the profiled attacks. In both metrics, the addition of IMF1 enhances the attack performance. The addition of decomposed traces as input features in the proposed framework ensures that the attack's unique intrinsic features are extended, which improves the attack's quality. When raw trace and four additional decomposed signals from IMF0 to IMF3 are combined as input features, the overall attack performance improves. A roughly 75% success rate is attained using several traces using this framework.[3] Although the additional complexity comes at the expense of enhanced performance, the computation cost is still acceptable and not prohibitive compared to the gains in performance.

---

[3] For 256 simultaneous multiplications, the SNR is noticeably low compared to other implementations in Table 7.2.

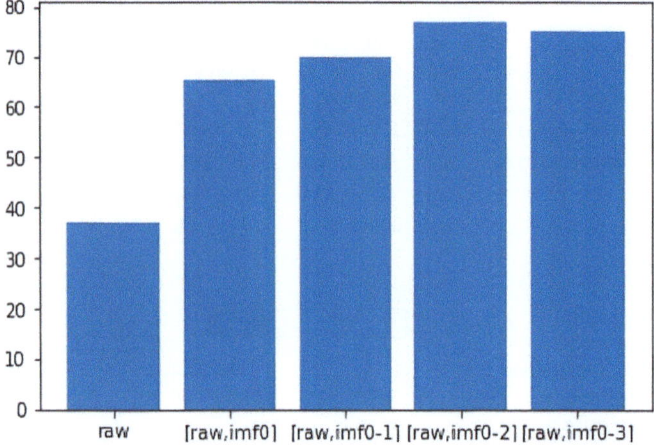

(a) AI-based using Raw Traces and Different Empirical modes of EMD

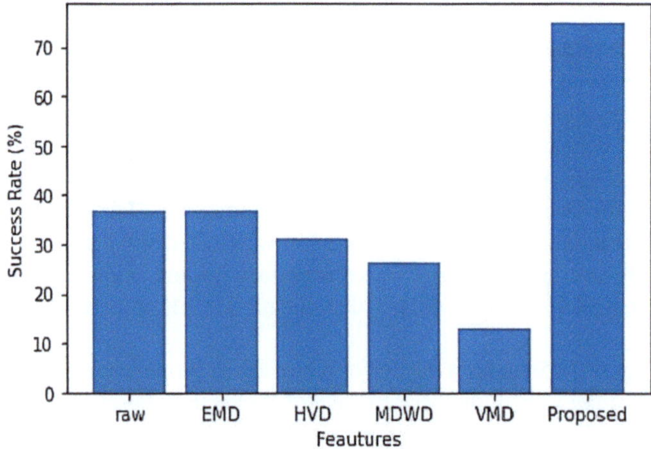

(b) Performance of AI-based using Different Signal Decomposition Techniques

Fig. 7.25 Performance of AI-based side-channel attacks on Saber HW

The confusion matrix for a single trace attack with the framework is shown in Fig. 7.26a. From the figure, it can be seen that for keys 5, 6, and 14, the key detection accuracy is higher than other key classes. Guessing entropy for key 5 is plotted in Fig. 7.26b. The figure shows that the attack phase requires less than 20 traces to find the secret key when the correct key is key 5. The average guessing entropy calculated for all secret keys is also shown in Fig. 7.26b. From the figure, it can be noted that the overall guessing entropy is less than 2 when around 20 traces are used. That means the correct key is within the top three suggestions of the proposed method.

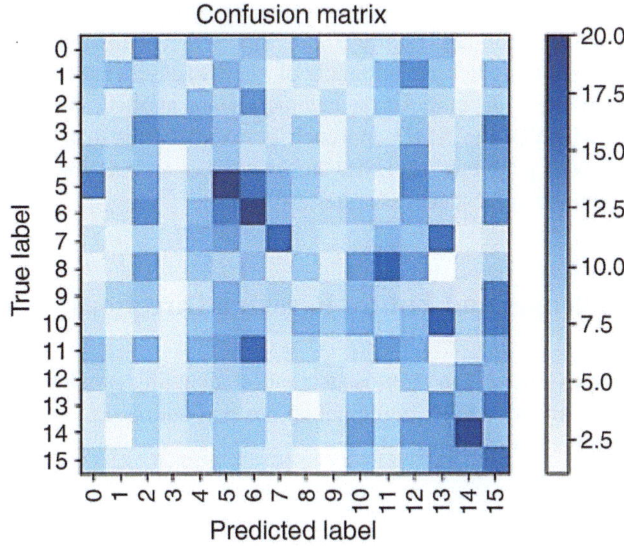

(a) Confusion Matrix for DL-based Side-channel Attack using EMD.

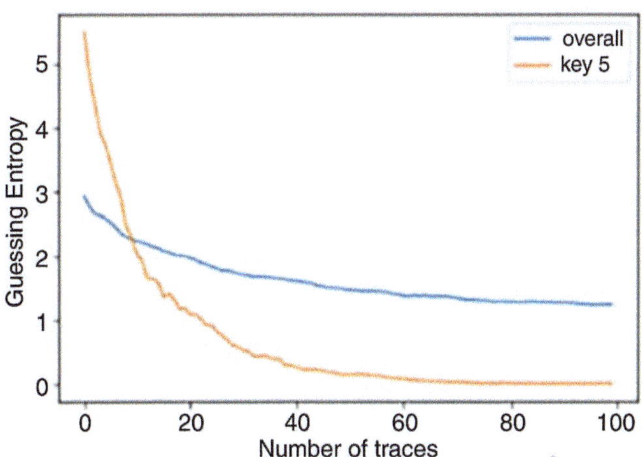

(b) Single-trace Side-channel using EMD in terms of Guessing Entropy.

Fig. 7.26 Performance of single-trace side-channel attack using EMD

As mentioned before, the proposed approach uses EMD as the signal decomposition technique. The performance of this decomposition technique is compared to other methods when different decomposition techniques are used. In comparing methods, four different signal decomposition techniques have been used in this work. The comparing decomposition techniques are Hilbert vibration decomposition (HVD), multilevel discrete wavelet decomposition (MDWD), and variational mode decomposition (VMD). Each of these techniques focuses on different signal properties when splitting the raw traces into multiple decomposed signals and generates unique features. Figure 7.25b shows the comparison. From the table, it can be seen that the performance of the attack is increased in all metrics when any of these decomposition techniques are applied. Among EMD, VMD, HVD, and MDWD, the EMD approach performs better than others. When both empirical modes generated by EMD and raw traces are ensembled in the proposed methodology, the highest success rates are attained.

7.6 Challenges and Future Research Directions

This section discusses the open issues and challenging problems of side-channel leakage assessment of PQC implementations and addresses high-level approaches for future research directions.

7.6.1 Side-Channel Leakage Assessment

Although our PQC-SEP can analyze PQC IPs and a SOC, including the PQC IPs, automatically at pre- and post-silicon levels, we need to address the challenging problems as follows:

- **Need for reference evaluation platform:** In Sect. 7.5.2.2, even if the Saber HW designs on two different boards are the same, vulnerable sampling points and quantitative leakage based on KL divergence are entirely different. The side-channel leakage analysis will depend on the system clock frequency, the method to integrate the PQC design into any architecture, and measurement setups. A reference FPGA platform and the measurement setup are required to evaluate the side-channel leakages of PQC designs efficiently and accurately.
- **Need for reference testing methodology:** We need a guideline to perform a side-channel leakage assessment of PQC designs, such as TVLA for AES designs. PQC algorithms have an extended length of keys and tremendous computational complexity, making the evaluation harder. For example, we chose only a pair of two keys to calculate the statistical distance between two different sets in this work. There is a considerable sample space to select a key pair, and we cannot guarantee which one is the best to estimate side-channel leakages accurately. Maybe it will be infeasible to search for a key pair as a reference for the PQC analysis.

To build the reference evaluation platform and testing methodology, tremendous experiments under various setups, e.g., different clock frequencies, other FPGA boards, various measurement instruments, and many input test patterns, should be accomplished. Since these experiments will need significant time, we plan to optimize PQC-PAT with additional FPGA evaluation boards to reduce experimental processing time.

7.6.2 AI-Based Side-Channel Attack

Although a good amount of work on side-channel attacks in AES uses machine learning and deep learning, in the case of PQC algorithms, this field has not been explored thoroughly. Nevertheless, the current works show the potential of AI in side-channel attacks on PQC protocols. The existing open challenges for AI-based side-channel attacks are described below:

- **Lack of data:** Since AI-based side-channel attack is a data-driven approach, the availability of data is crucial. Previously, in Sect. 7.4.3, existing works of AI-based side-channel attacks on PQC algorithms are described. But none of these studies make the data publicly available. Such unavailability of data makes it difficult to check the efficacy of varieties of AI algorithms to perform a successful side-channel attack on different PQC algorithms. It also makes it hard to compare the side-channel vulnerability of different PQC algorithms for a specific AI-based attack approach.
- **Lack of hardware implementation:** Among the existing studies, only in the case of AI-based side-channel attacks on Frodo and NewHope, the attack is performed on hardware implementation.
- **Lack of explainable approach:** The lack of explainability is another open challenge for AI-based side-channel attacks. Undoubtedly, neural networks have demonstrated tremendous effectiveness even in the face of advanced counter-measures. However, it is uncertain how these networks actually work. It is tough to tell if a failed attack was due to a bad countermeasure or a less effective AI strategy. Similarly, none of the existing works has been able to determine the design's specific weakness in the case of a successful attack. Without any explanation, it is difficult to devise effective countermeasures.
- **Lack of clear guidance**: Any profiled DL-based side-channel attack requires several steps to be followed sequentially: preprocessing, feature engineering, algorithm selection, and attack evaluation. There is still no proper guideline for a post-silicon or a pre-silicon side-channel attack on PQC protocols for any of these steps. It is unclear what type of preprocessing and feature extraction approaches would work efficiently.

7.6.3 Secure PQC Implementation

Masking is commonly used to prevent higher-order side-channel attacks. Target functions, such as a polynomial multiplication in an LWE scheme, can be protected by an arithmetic masking or Boolean masking according to nth order side-channel attack models. However, many random bits, n linear masked modules, nonlinear masked modules, and conversion modules between Boolean domain and the arithmetic domain will result in tremendous area/power overhead and lower

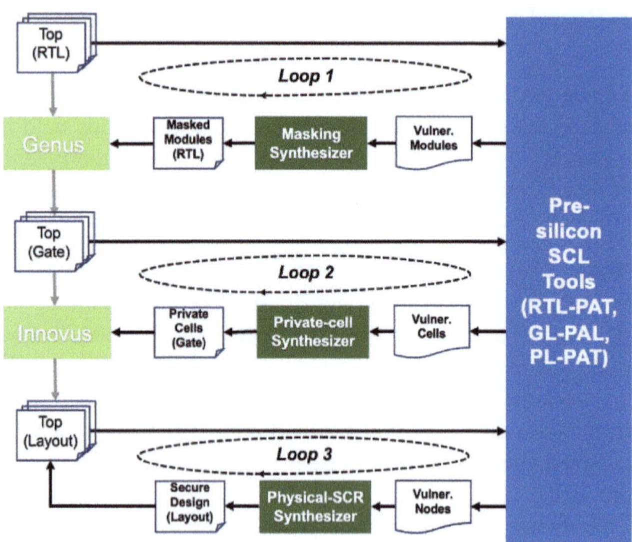

Fig. 7.27   A framework of secure PQC IP design

performance. In addition, after placing and routing, different analysis results can
be generated by low-level side-channel leakage analysis, i.e., PL-PAT, without
additional countermeasures, e.g., hiding by inserting random noise. Only the
masking technique cannot guarantee power/EM side-channel protection.

In this regard, we need a generic framework of secure PQC IP designs against
power/EM side-channel attacks combined with pre-silicon side-channel leakage
analysis shown in Fig. 7.4. It consists of the following synthesizers as shown in
Fig. 7.27:

1. **Masking synthesizer (RTL)**: Based on the result of the pre-silicon side-channel
   leakage assessment in Sect. 7.4.1, vulnerable modules will be replaced into
   masked modules depending on nth order attack models automatically by the
   masking synthesizer. The new top design with the generated masked modules
   will be analyzed in terms of side-channel leakages. If not satisfying side-channel
   resistance, it will be re-masked until passing the side-channel leakage test
   (Loop 1).
2. **Private-cell synthesizer (gate level):** After logic synthesis of the masked design,
   vulnerable cells will be identified by the GL-PAT in Fig. 7.4. The vulnerable
   cells will be converted into private cells, such as WDDL or t-private cells,
   automatically by the private-cell synthesizer. Repeatedly, the above gate-level
   processes continue until satisfying any SCL standard and constraints, such as
   area overhead, power consumption, and performance (Loop 2).
3. **Physical-SCR synthesizer (layout level):** Although the gate-level design has
   side-channel resistance, its layout design can be vulnerable to side-channel
   attacks. The physical side-channel resistant (SCR) synthesizer will automatically

generate a secure physical layout design by connecting physical countermeasure circuits, such as a randomizer, into vulnerable nodes based on a PL-PAT or isolating power networks of vulnerable nodes. It will be finished if the regenerated layout design passes layout-level SCL tests; otherwise, the secure physical design will continue (Loop 3).

References

1. J. Chow, O. Dial, J. Gambetta, *IBM Quantum Breaks the 100-qubit Processor Barrier* (2021). https://research.ibm.com/blog/127-qubit-quantum-processor-eagle
2. J. Gambetta, *IBM's Roadmap For Scaling Quantum Technology.* https://research.ibm.com/blog/ibm-quantum-roadmap
3. P. Ball, *First Quantum Computer to Pack 100 Qubits Enters Crowded Race* (2021). https://www.nature.com/articles/d41586-021-03476-5
4. P.W. Shor, Polynomial-time algorithms for prime factorization and discrete logarithms on a quantum computer. SIAM J. Comput. **26**(5), 1484–1509 (1997). issn: 1095-7111. https://doi.org/10.1137/s0097539795293172. http://dx.doi.org/10.1137/S0097539795293172
5. NIST, *NIST Post-Quantum Cryptography Standardization.* https://csrc.nist.gov/projects/post-quantum-cryptography/post-quantum-cryptography-standardization
6. Y. Bai, A. Stern, J. Park, M. Tehranipoor, D. Forte, RASCv2: enabling remote access to side-channels for mission critical and IoT systems. ACM Trans. Des. Autom. Electron. Syst. (2022). Just Accepted. issn: 1084-4309. https://doi.org/10.1145/3524123
7. J. Park, X. Xu, Y. Jin, D. Forte, M. Tehranipoor, Power-based side-channel instruction-level disassembler, in *2018 55th ACM/ESDA/IEEE Design Automation Conference (DAC)* (2018), pp. 1–6. https://doi.org/10.1109/DAC.2018.8465848
8. W.-L. Huang, J.-P. Chen, B.-Y. Yang, Power analysis on NTRU Prime. IACR Transactions on Cryptographic Hardware and Embedded Systems **2020**(1), 123–151 (2019). https://doi.org/10.13154/tches.v2020.i1.123-151. https://tches.iacr.org/index.php/TCHES/article/view/8395
9. Z. Xu, O.M. Pemberton, S.S. Roy, D. Oswald, W. Yao, Z. Zheng, Magnifying side-channel leakage of lattice-based cryptosystems with chosen ciphertexts: the case study of Kyber. IEEE Trans. Comput. **71**(9), 2163–2176 (2021). https://doi.org/10.1109/TC.2021.3122997
10. M. Van Beirendonck, J.-P. D'anvers, A. Karmakar, J. Balasch, I. Verbauwhede, A side-channel-resistant implementation of SABER. J. Emerg. Technol. Comput. Syst. **17**(2), 1–26 (2021). issn: 1550-4832
11. K. Ngo, E. Dubrova, Q. Guo, T. Johansson, A side-channel attack on a masked IND-CCA secure saber KEM implementation. IACR Transactions on Cryptographic Hardware and Embedded Systems **2021**(4), 676–707 (2021). https://doi.org/10.46586/tches.v2021.i4.676-707. https://tches.iacr.org/index.php/TCHES/article/view/9079
12. NIST, *PQC Standardization Process: Third Round Candidate Announcement.* https://csrc.nist.gov/news/2020/pqc-third-round-candidate-announcement
13. R.J. McEliece, *A Public Key Cryptosystem Based on Algebraic Coding Theory* (1978)
14. J. Bos, L. Ducas, E. Kiltz, T. Lepoint, V. Lyubashevsky, J.M. Schanck, P. Schwabe, G. Seiler, D. Stehle, CRYSTALS-Kyber: a CCA-secure module-lattice-based KEM, in *2018 IEEE European Symposium on Security and Privacy (EuroS P)* (2018), pp. 353–367. https://doi.org/10.1109/EuroSP.2018.00032
15. J. Hoffstein, J. Pipher, J.H. Silverman, NTRU: a ring-based public key cryptosystem, in *Algorithmic Number Theory*, ed. by J.P. Buhler (Springer, Berlin, 1998), pp. 267–288
16. J.-P. D'Anvers, A. Karmakar, S.S. Roy, F. Vercauteren, Saber: module-LWR based key exchange, CPA-secure encryption and CCA-secure KEM, in *Progress in Cryptology—AFRICACRYPT 2018—10th International Conference on Cryptology in Africa, Marrakesh,*

Morocco, May 7–9, 2018, Proceedings, vol. 10831, ed. by A. Joux, A. Nitaj, T. Rachidi. Lecture Notes in Computer Science. (Springer, Berlin, 2018), pp. 282–305. https://doi.org/10.1007/978-3-319-89339-6_16. https://doi.org/10.1007/978-3-319-89339-6%5C_16

17. L. Ducas, E. Kiltz, T. Lepoint, V. Lyubashevsky, P. Schwabe, G. Seiler, D. Stehlé, CRYSTALS-Dilithium: a lattice-based digital signature scheme. IACR Transactions on Cryptographic Hardware and Embedded Systems **2018**(1), 238–268 (2018). https://doi.org/10.13154/tches.v2018.i1.238-268. https://tches.iacr.org/index.php/TCHES/article/view/839

18. P.-A. Fouque, J. Hoffstein, P. Kirchner, V. Lyubashevsky, T. Pornin, T. Prest, T. Ricosset, G. Seiler, W. Whyte, Z. Zhang, Falcon: Fast-Fourier lattice-based compact signatures over NTRU. in *Submission to the NIST's Post-quantum Cryptography Standardization Process* , vol. 36 (2018)

19. N. Aragon, P. Barreto, S. Bettaieb, L. Bidoux, O. Blazy, J.-C. Deneuville, P. Gaborit, S. Gueron, T. Guneysu, C.A. Melchor, R. Misoczki, E. Persichetti, N. Sendrier, J.-P. Tillich, G. Zé mor, *BIKE: Bit Flipping Key Encapsulation* (2017)

20. C. Aguilar-Melchor, O. Blazy, J.-C. Deneuville, P. Gaborit, G. Zémor, Efficient encryption from random quasi-cyclic codes. IEEE Trans. Inf. Theory **64**(5), 3927–3943 (2018). https://doi.org/10.1109/TIT.2018.2804444

21. J. Bos, C. Costello, L. Ducas, I. Mironov, M. Naehrig, V. Nikolaenko, A. Raghunathan, D. Stebila, Frodo: take off the Ring! Practical, Quantum-Secure Key Exchange from LWE, in *Proceedings of the 2016 ACM SIGSAC Conference on Computer and Communications Security*. CCS '16 (Association for Computing Machinery, Vienna, 2016), 1006–1018. isbn: 9781450341394. https://doi.org/10.1145/2976749.2978425

22. D.J. Bernstein, C. Chuengsatiansup, T. Lange, C. van Vredendaal, NTRU Prime: reducing attack surface at low cost, in *Selected Areas in Cryptography—SAC 2017—24th International Conference, Ottawa, ON, Canada, August 16–18, 2017, Revised Selected Papers*, vol. 10719, ed. by C. Adams, J. Camenisch. Lecture Notes in Computer Science (Springer, Berlin, 2017), pp. 235–260. https://doi.org/10.1007/978-3-319-72565-9_12. https://doi.org/10.1007/978-3-319-72565-9%5C_12

23. D.J. Bernstein, A. Hülsing, S. Kölbl, R. Niederhagen, J. Rijneveld, P. Schwabe, The SPHINCS+ Signature Frame-work, in *Proceedings of the 2019 ACM SIGSAC Conference on Computer and Communications Security*. CCS '19 (Association for Computing Machinery, London, 2019), pp. 2129–2146. isbn: 9781450367479. https://doi.org/10.1145/3319535.3363229

24. A. Casanova, J.-C. Fauge're, G. Macario-Rat, J. Patarin, L. Perret, J. Ryckeghem, *GeMSS: A Great Multivariate Short Signature* (2017)

25. D. Hofheinz, K. Hövelmanns, E. Kiltz, A modular analysis of the Fujisaki-Okamoto transformation, in *Theory of Cryptography*, ed. by Y. Kalai, L. Reyzin (Springer International Publishing, Cham, 2017), pp. 341–371

26. P. Ravi, S. Bhasin, S.S. Roy, A. Chattopadhyay, Drop by Drop you break the rock—Exploiting generic vulnerabilities in Lattice-based PKE/KEMs using EM-based Physical Attacks, in *IACR Cryptol. ePrint Arch.* (2020), p. 549. https://eprint.iacr.org/2020/549

27. S. Bhasin, J.-P. D'Anvers, D. Heinz, T. Pöppelmann, M. Van Beirendonck. *Attacking and Defending Masked Polynomial Comparison for Lattice-Based Cryptography*. Cryptology ePrint Archive, Report 2021/104. https://ia.cr/2021/104.2021

28. B.-Y. Sim, J. Kwon, J. Lee, I.-J. Kim, T.-H. Lee, J. Han, H. Yoon, J. Cho, D.-G. Han, Single-trace attacks on message encoding in lattice-based KEMs. IEEE Access **8**, 183175–183191 (2020)

29. F. Aydin, A. Aysu, M. Tiwari, A. Gerstlauer, M. Orshansky, Horizontal side-channel vulnerabilities of post-quantum key exchange and encapsulation protocols. ACM Trans. Embed. Comput. Syst. **20**(6), 1–22 (2021). issn: 1539-9087. https://doi.org/10.1145/3476799

30. R. Primas, P. Pessl, S. Mangard, *Single-Trace Side-Channel Attacks on Masked Lattice-Based Encryption*. Cryptology ePrint Archive, Report 2017/594 (2017). https://ia.cr/2017/594

31. P. Pessl, R. Primas, *More Practical Single-Trace Attacks on the Number Theoretic Transform*. Cryptology ePrint Archive, Report 2019/795 (2019). https://ia.cr/2019/795

32. M.J. Kannwischer, P. Pessl, R. Primas, Single-trace attacks on Keccak. IACR Transactions on Cryptographic Hardware and Embedded Systems **2020**(3), 243–268 (2020). https://doi.org/10.13154/tches.v2020.i3.243-268. https://tches.iacr.org/index.php/TCHES/article/view/8590

33. J.W. Bos, S. Friedberger, M. Martinoli, E. Oswald, M. Stam, *Assessing the Feasibility of Single Trace Power Analysis of Frodo*. Cryptology ePrint Archive, Report 2018/687 (2018). https://ia.cr/2018/687

34. N.N. Anandakumar, M.P.L. Das, S.K. Sanadhya, M.S. Hashmi, Reconfigurable hardware architecture for authenticated key agreement protocol over binary Edwards curve. ACM Trans. Reconfigurable Technol. Syst. **11**(2), 12:1–12:19 (2018)

35. A. Abdulrahman, J.-P. Chen, Y.-J. Chen, V. Hwang, M.J. Kannwischer, B.-Y. Yang, Multi-moduli NTTs for Saber on Cortex-M3 and Cortex-M4, in *IACR Cryptol. ePrint Arch.* (2021), p. 995

36. B.-Y. Sim, A. Park, D.-G. Han, Chosen-ciphertext clustering attack on CRYSTALS-KYBER using the side-channel leakage of Barrett reduction, in *IACR Cryptol. ePrint Arch.* (2021), p. 874. https://eprint.iacr.org/2021/874

37. H.M. Steffen, L.J. Kogelheide, T. Bartkewitz, In-depth analysis of side-channel countermeasures for CRYSTALS-Kyber message encoding on ARM Cortex-M4, in *IACR Cryptol. ePrint Arch.* (2021), p. 1307

38. A. Karlov, N.L. de Guertechin, Power analysis attack on Kyber, in *IACR Cryptol. ePrint Arch.* (2021), p. 1311

39. A. Park, D.-G. Han, Chosen ciphertext simple power analysis on software 8-bit implementation of ring-LWE encryption, in *2016 IEEE Asian Hardware-Oriented Security and Trust (AsianHOST)* (2016), pp. 1–6. https://doi.org/10.1109/AsianHOST.2016.7835555

40. W.-L. Huang, J.-P. Chen, B.-Y. Yang, Power analysis on NTRU prime. IACR Trans. Cryptogr. Hardw. Embed. Syst. **2020**(1), 123–151 (2020). https://doi.org/10.13154/tches.v2020.i1.123-151

41. A. Askeland, S. Rønjom, *A Side-Channel Assisted Attack on NTRU*. Cryptology ePrint Archive, Report 2021/790. https://ia.cr/2021/790.2021

42. K. Ngo, E. Dubrova, Q. Guo, T. Johansson, A side-channel attack on a masked IND-CCA secure saber KEM, in *IACR Cryptol. ePrint Arch.*, vol. 2021 (2021), p. 79

43. K. Ngo, E. Dubrova, T. Johansson, Breaking masked and shuffled CCA secure saber KEM by power analysis, in *Cryptology ePrint Archive* (2021)

44. F. Aydin, P. Kashyap, S. Potluri, P. Franzon, A. Aysu, DeePar-SCA: Breaking parallel architectures of lattice cryptography via learning based side-channel attacks, in *International Conference on Embedded Computer Systems* (Springer, Berlin, 2020), pp. 262–280

45. P. Kashyap, F. Aydin, S. Potluri, P.D. Franzon, A.-d. Aysu, 2Deep: enhancing side-channel attacks on lattice-based key-exchange via 2-D deep learning. IEEE Trans. Comput. Aided Des. Integr. Circuits Syst. **40**(6), 1217–1229 (2020)

46. S. Kullback, R.A. Leibler, On Information and Sufficiency. Ann. Math. Stat. **22**(1), 79–86 (1951)

47. K. Tiri, I. Verbauwhede, A logic level design methodology for a secure DPA resistant ASIC or FPGA implementation, in *Proceedings of the Conference on Design, Automation and Test in Europe—Volume 1*. DATE '04 (IEEE Computer Society, USA, 2004), p. 10246. isbn: 0769520855

48. Y. Ishai, A. Sahai, D. Wagner, Private circuits: securing hardware against probing attacks, in *Advances in Cryptology—CRYPTO 2003*, ed. by D. Boneh (Springer, Berlin, 2003), pp. 463–481

49. Cadence, *Cadence Voltus Power Integrity Solution* (2021). https://www.cadence.com/en_US/home/tools/digital-design-and-signoff/silicon-signoff/voltus-ic-power-integrity-solution.html

50. T. Zhang, J. Park, M. Tehranipoor, F. Farahmandi, PSC-TG: RTL power side-channel leakage assessment with test pattern generation, in *2021 58th ACM/IEEE Design Automation Conference (DAC)* (2021), pp. 709–714

51. A. Nahiyan, M. He, J. Park, M. Tehranipoor, *Chapter 7. CAD for Side-channel Leakage Assessment, Emerging Topics in Hardware Security* (2021). https://doi.org/10.1007/978-3-030-64448-2

52. NewAE Technology Inc, *CW305 Artix FPGA Target*. https://www.newae.com/products/NAE-CW305

53. J. Park, S. Cho, T. Lim, S. Bhunia, M. Tehranipoor, SCR-QRNG: side-channel resistant design using quantum random number generator, in *2019 IEEE/ACM International Conference on Computer-Aided Design (ICCAD)* (2019), pp. 1–8. https://doi.org/10.1109/ICCAD45719.2019.8942152

54. NewAE Technology Inc, *CW1200 ChipWhisperer-Pro*. https://rtfm.newae.com/Capture/ChipWhisperer-Pro/

55. NewAE Technology Inc, *ChipWhisperer-Lite (CW1173) Two-Part Version*. https://store.newae.com/chipwhisperer-lite-cw1173-two-part-version/

56. J. Cooper, E. DeMulder, G. Goodwill, J. Jaffe, G. Kenworthy, P. Rohatgi, et al., Test vector leakage assessment (TVLA) methodology in practice, in *International Cryptographic Module Conference*, vol. 20 (2013)

57. N.N. Mondol, A. Vafaei, K.Z. Azar, F. Farahmandi, M. Tehranipoor, RL-TPG: automated pre-silicon security verification through reinforcement learning-based test pattern generation, in *Design, Automation and Test in Europe (DATE)* (IEEE, New York, 2024), pp. 1–6

58. K.Z. Azar, H.M. Kamali, H. Homayoun, A. Sasan, NNgSAT: neural network guided SAT attack on logic locked complex structures, in *IEEE/ACM International Conference On Computer Aided Design (ICCAD)* (2020), pp. 1–9

59. N.E. Huang, Z. Shen, S.R. Long, M.C. Wu, H.H. Shih, Q. Zheng, N.-C. Yen, C.C. Tung, H.H. Liu, The empirical mode decomposition and the Hilbert spectrum for nonlinear and non-stationary time series analysis. Proc. R. Soc. London, Ser. A Math. Phys. Eng. Sci. **454**(1971), 903–995 (1998)

60. M. Feldman, Time-varying vibration decomposition and analysis based on the Hilbert transform. J. Sound Vib. **295**(3), 518–530 (2006). issn: 0022-460X. https://doi.org/10.1016/j.jsv.2005.12.058. https://www.sciencedirect.com/science/article/pii/S0022460X06001556

61. K. Dragomiretskiy, D. Zosso, Variational mode decomposition. IEEE Trans. Signal Process. **62**(3), 531–544 (2014). https://doi.org/10.1109/TSP.2013.2288675

62. Z. Wang, T. Oates, *Imaging Time-Series to Improve Classification and Imputation* (2015). arXiv: 1506.00327 [cs.LG]

63. J.-P. Eckmann, S. Oliffson Kamphorst, D. Ruelle, Recurrence plots of dynamical systems. Europhys. Lett. (EPL) **4**(9), 973–977 (1987). https://doi.org/10.1209/0295-5075/4/9/004

64. F. Nogueira, *Bayesian Optimization: Open Source Constrained Global Optimization Tool for Python* (2014). https://github.com/fmfn/BayesianOptimization

65. S.S. Roy, A. Basso, High-speed instruction-set coprocessor for lattice-based key encapsulation mechanism: Saber in hardware, in *IACR Transactions on Cryptographic Hardware and Embedded Systems* (2020), pp. 443–466

66. SEGGER, *J-Link EDU—The Educational J-Link*. https://www.segger.com/products/debug-probes/j-link/models/j-link-edu/

67. M.J. Kannwischer, J. Rijneveld, P. Schwabe, K. Stoffelen, *pqm4: Testing and Benchmarking NIST PQC on ARM Cortex-M4*. Cryptology ePrint Archive, Report 2019/844. https://ia.cr/2019/844.2019

68. Y. Xing, S. Li, A compact hardware implementation of CCA-secure key exchange mechanism CRYSTALS-KYBER on FPGA. IACR Transactions on Cryptographic Hardware and Embedded Systems **2021**(2), 328–356 (2021). https://doi.org/10.46586/tches.v2021.i2.328-356

69. M. He, J. Park, A. Nahiyan, A. Vassilev, Y. Jin, M. Tehranipoor, RTL-PSC: automated power side-channel leakage assessment at register-transfer level, in *2019 IEEE 37th VLSI Test Symposium (VTS)* (2019), pp. 1–6

Chapter 8
Digital Twin for Secure Semiconductor Lifecycle Management

8.1 Introduction

Amidst rising threats in the supply chain and the ever expanding attack surface, ensuring the security of semiconductor devices across their entire lifecycle has become a challenging and complex endeavor. Although established practice puts security at the forefront of each stage of the software development lifecycle [1], such efforts to secure the hardware lifecycle is in its infancy due to the unique challenges associated with it [2, 3]. The traditional hardware verification and testing methodologies, which focus on functional verification as their primary objective, are often ineffective in detecting security vulnerabilities, which may be introduced through malicious 3PIPs or security unaware design practices [4–6]. If security vulnerabilities evade detection and verification efforts, they can later be exploited by malicious entities in the supply chain [7, 8]. Unlike software, however, hardware cannot be easily patched, which makes identifying the root origin of the vulnerability in the silicon lifecycle paramount to ensure generational improvement in security assurance.

Digital twins (DTs) have experienced exponential growth in academia as frameworks to monitor, maintain, and control quality and reliability of different products across their different stages of the lifecycle [9]. Although originally conceived in [10] as a high-fidelity digital representation of aerospace vehicles, the concept and definition of the digital twin has evolved to encompass any virtual representation of a physical object, process, or operation which is continually updated by data that is collected across the lifecycle using which it provides optimization feedback on the functionality and control of the physical counterpart (as shown in Fig. 8.1 [11–13]. Although DTs have received much recognition as tools of managing product maintenance, fault diagnosis, and monitoring in the aerospace and manufacturing industry [14–17], DTs that consider the full lifecycle are very rare. In fact, Liu et al. report that only 5% of their reviewed papers on DTs considered the whole lifecycle [18]. Furthermore, addressing cyber and hardware security issues that

M. Tehranipoor et al., *Hardware Security*,
https://doi.org/10.1007/978-3-031-58687-3_8

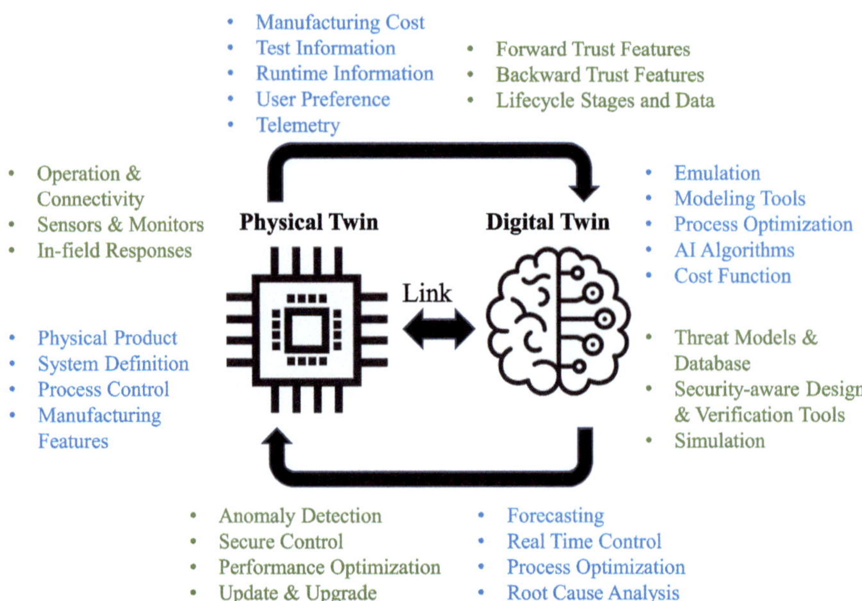

Fig. 8.1 Digital twin overview for secure semiconductor lifecycle management

are intertwined with cyber-physical systems utilizing DTs have also been rarely considered [19].

Existing solutions proposed in academia and industry to address hardware security concerns also suffer from major limitations when applied in the context of end-to-end secure lifecycle management. Firstly, almost all proposed detection and prevention mechanisms are concentrated on very specific threats at specific parts of the lifecycle [20, 21] with little to no scalability when considering the entire lifecycle and other threat models.

Secondly, although there are multiple detection and prevention methods against hardware attacks including hardware Trojans [22–25], counterfeits [26–31], information leakage [32, 33], fault injection [34–36], and side-channel attacks [37–39], the analysis of these methods starts with the *a priori* assumption that the defender knows what attack vector is principally responsible for an anomalous behavior. It is far more likely that the designer or defender would only get to *observe* the anomalous behavior, rather than *knowing* what precise attack vector is causing said behavior. For example, from a hardware security perspective, a chip in a system may experience accelerated aging due to many possible reasons. It can fail before its intended lifespan because (i) it is a recycled chip that was unknowingly used by the system designer, (ii) it is a defective chip that was shipped without authorization by an untrusted foundry or a rogue employee working in a trusted foundry [40], or (iii) it experienced accelerated aging due to being taken over by a parametric hardware Trojan [41].

Thirdly, once an attack or vulnerability has been detected, thus far none of the proposed solutions have the capability to trace the lifecycle stage where it originated from. For example, if an information leakage is detected through formal verification, existing methods cannot infer whether the problem was introduced during high-level architecture specification (also known as electronic system-level specification) or during the formulation of the logic design (through hardware description languages) of the circuit. The ability to track down the origin is absolutely vital if we want to facilitate generational improvement in security assurance of the design.

Lastly, there have been suggestions in literature and the semiconductor industry to embed different types of sensors on the chip so that it has a defense against certain attack vectors [42, 43]. In addition to more area, power, and performance overhead, these approaches are not scalable in the context of emerging future threats. New threats and attack vectors are always being developed by researchers and malicious actors. As an example, in the initial years of hardware Trojan research, it was frequently assumed that hardware Trojans need to be activated by rare signals and node to avoid detection. However, researchers have since demonstrated that it is possible to design hardware Trojans that do not need to be triggered by rare events yet easily escape traditional testing and verification efforts [44]. It is not feasible to keep continually adding new sensors to a design to tackle new threats as they emerge.

Therefore, we argue that without a comprehensive framework, such as the one we are proposing in this chapter, security assurance in the semiconductor lifecycle would only be limited to partial effectiveness with severely limited scalability (no matter how robust individual detection algorithms or protection mechanisms are) [45]. A DT with bidirectional data flow and feedback, as shown in Fig. 8.1, between the real world and the virtual presents a suitable concept around which data collection and analysis tools and algorithms can be leveraged to build a comprehensive framework to provide security assurance across the whole lifecycle by addressing each of the aforementioned challenges. Accordingly, this chapter will cover the following:

- We propose a digital twin framework that can provide security assurance across the entire lifecycle by considering the potentially malicious supply chain entities and vulnerable cycle phases. DT deconstructs the problem by analyzing causal relationships between available data and hardware security vulnerabilities. Thus, instead of addressing one or two attack vectors, DT provides a scalable methodology to combat potentially all possible hardware attack vectors.
- Our proposed methodology theorizes the use of data that is already being gathered by the traditional process flows in the silicon lifecycle. Consequently, adoption of our framework incurs no hardware overhead and offers a promising prospect of being seamlessly integrated into existing flows.
- We define the feedback from the DT to the physical world in terms of two functionalities: namely, *Backward* and *Forward Trust Analyses*, respectively. *Backward Trust Analysis* provides traceability through root cause analysis of observed anomalous behavior in device security policies at any stage of the

lifecycle. We demonstrate how artificial intelligence (AI) algorithms can be used to perform reliable root cause analysis in the hardware security domain. To perform this root cause analysis, we explore three different statistical relational learning algorithms, namely Bayesian Networks, Hidden Markov Models, and Markov Logic Networks, by each of which causal inference can be performed. Additionally, we demonstrate how they can be adapted to the problems of silicon lifecycle security.

- The dichotomy of security assurance is that on the one hand, as time passes, novel threats emerge that circumvent existing protection and detection measures [46]. On the other hand, the collective understanding of these newly emerging threats calcify, which gives rise to better performing prevention and detection methodologies. Through our proposed framework, we demonstrate how it can be made scalable and continually updatable, which in turn can preserve applicability of its ability for root cause analysis (even against unforeseen threats). This scalability and adaptability is what we refer to as *Forward Trust Analysis* [47].

8.2 Preliminaries

8.2.1 *Digital Twin at a Glance*

The concept of DT has evolved to encompass many different definitions [11]. Some authors have put strong emphasis on the simulation aspect of DTs, whereas others have argued for a clear definition of three aspects (physical, virtual, and connection parts) as the criterion for a framework to be called a digital twin [48]. We use the definition provided by Madni [13] in context of the lifecycle management of products to illustrate the different components of a DT system in Fig. 8.1. At the core of a DT is the collection of sensor, simulation, emulation, and preliminary analytics data that are gathered across a physical device's lifecycle. The physical process, or device, is also referred to as *Physical* twin. The twins are housed within environments that are referred to as physical and virtual environments, respectively. The *Digital* counterpart is formed by continually updating the database hosted in the virtual environment. The DT is capable of providing intelligent feedback (e.g., forecasting, optimization of parameters, root cause analysis, real-time control) to the physical world through a combination of simulation, emulation, data analytics, and AI modeling. The communication links between the physical and virtual environments are also essential components of the DT. It is imperative to note here that a digital twin is not merely a single algorithm or a single technology [18], but rather a framework around which a systematic methodology can be built to combat product lifecycle issues. For security assurance across the whole lifecycle of a semiconductor device, it should be noted that having only the traditional components and transactions is not sufficient since they do not necessarily offer security-aware features. Hence, additional transactions and functionalities are required as indicated in green boxes and circles in Fig. 8.1. It also calls for advanced machine learning, statistical relational learning, and other data analytic-related algorithms to gleam insight from gathered data. The methods, algorithms, structure, and contingencies required to realize these additional components and transactions are discussed throughout the rest of the chapter.

8.2.2 Digital Twin for Lifecycle Management and Cybersecurity

Fault diagnosis or root cause analysis as a core functionality of digital twins in context of product lifecycle management and industrial production has been explored in several works [49–51]. DTs have been demonstrated to be applicable for lifecycle management in agricultural [52] and Building Information Modeling (BIM) systems [53]. The existing literature on DT for cybersecurity focuses mainly on network and software security [54]. The focus has been on identifying intrusion [55] or false data injection attacks in an industrial setting. Bitton et al. proposed the use of a DT specified *automatically* from a rule set derived from tests and a so-called problem builder derived the constraints by solving a nonlinear maximization problem [56]. In a similar setting of an ICS, DT has been used to resolve security issues associated with a refueling machine [57]. Saad et al. addressed attacks from potentially multiple coordinated sources on a networked microgrid [58]. The reader should note that these approaches only consider a specific type of control systems, not the security issues associated with the entire lifecycle. Lifecycle management of products, especially security management, requires additional capabilities, considerations, and bidirectional transactions.

A high-level formulation of digital twins for semiconductor reliability can be found in [59]. Reliability concerns are inherently limited to considering a subset of the lifecycle as a vast majority of semiconductor reliability concerns originate from fabrication and packaging processes. Another discussion of digital twins in context of the semiconductor fabrication process can be found in [60]. Again, this discussion is limited to only one phase of the lifecycle in context of smart manufacturing and not related to security concerns. The current dominant trend in academia, which is evident in this brief literature review section as well, is to utilize digital twin for systems which are almost exclusively manufacturing systems or processes. We buck that trend in our chapter by showcasing how digital twins can contribute significantly in secure lifecycle management as well.

An overview of the papers discussed in the preceding can be found in Table 8.1.

Table 8.1 Digital twin applications for lifecycle management and cybersecurity

Paper	Application area	Comments
Bitton et al. [56]	Cybersecurity of Industrial Control Systems (ICS)	Proposed the use of DT to overcome the limitations of existing network penetration testing when applied to industrial SCADA systems
Lou et al. [57]	Cybersecurity of ICS	Demonstrated the use of DT to address security issues of a refueling machine
Balta et al. [61]	Process management	Proposed a DT for anomaly detection and process monitoring of the fused deposition modeling AM process
Eckhart et al. [55]	Network and CPS security	Proposed a CPS twinning system where states of the physical systems are mirrored through the DT that can incorporate security enhancing features, such as intrusion detection
Saad et al. [58]	Network and grid security	Illustrated a DT's capability in providing security against false data injection, Distributed Denial-of-Service (DDoS) and network delay attacks in microgrids
Li et al. [62]	Product lifecycle management	Proposed fault diagnosis and prognosis technique in aircraft wings through a dynamic Bayesian network-driven DT
Sleuters et al. [51]	System management	Proposed a DT to capture the operational behavior of a distributed IoT system
Wang et al. [60]	Smart manufacturing	Discussed how a DT may be used for intelligent semiconductor manufacturing
Jain et al. [50]	System management	Proposed a DT to offer real-time analysis and control of a photovoltaic system
Xu et al. [49]	Process management	Demonstrated a DT that offers real-time diagnosis and predictive maintenance of a car body-side production line
Kaewunruen et al. [63]	Operational lifecycle management	Proposed a DT for sustainable management of railway turnout systems
Heterogeneous Integration Roadmap 2021 Ed. [59]	Reliability management for Semiconductor	Briefly discussed possible DT prospects for reliability management of semiconductor devices
Alves et al. [52]	System management	Developed a DT to monitor and control water management in agricultural farms
Tchana et al. [52]	Operational lifecycle management	Developed a DT to address operational issues in linear construction sector

8.2.3   Hardware Security and Trust

Over past several decades, hardware of a computer system has traditionally been considered as the emphroot of trust to guard against attacks on the software running on the system. The underlying assumption here is that since hardware is less easily malleable than software, it is likely to be robust and secure against different types of attacks [24]. However, emerging hardware attacks that exploit intralayer and cross-layer vulnerabilities have propelled hardware security as a widely researched topic. The recent proliferation of reported attacks on hardware is not surprising given how the business model of the semiconductor industry has evolved over the course of past few decades. Previously, all stages associated with bringing a semiconductor chip to the market (namely design, fabrication, test, and debug) were handled by a single entity. To address aggressive time-to-market demands and profitability concerns, the global semiconductor industry has gradually adopted a horizontal model of business wherein each previously mentioned stage may be handled by completely different entities often situated in different parts of the globe.

Several of the IPs used in a typical SoC are procured from third parties across the globe [64].

Today, most SoC design houses are "fabless," meaning that they do not own a foundry to physically fabricate the chips that they design. They rely on a foundry and a third party to, respectively, fabricate and distribute the chips for them. The consequence of this distributed manufacturing and supply process is that the security of the designed SoC may become compromised through malicious modification by multiple entities across the entire supply chain. This raises the issue of trust between multiple entities referred to as hardware trust. Hardware attacks can take place in the form of a malicious modification of circuit (hardware Trojan), stealing of the IP by the foundry, recycling and remarking of chips, physical tampering [65, 66], reverse engineering [67, 68], and side-channel attacks by end-users. These attacks might be carried out by different actors in the supply chain who may have different goals. In addition to these attacks, various vulnerabilities might be introduced unintentionally in the design, such as the leakage of a security-critical asset through an output port or to an unauthorized IP. The possible hardware attacks and the stages in which they might occur are highlighted in red on the right side of Fig. 8.2. These attack vectors have highly varied associated threat models, characteristic symptoms, and detection methodologies.

8.2.4 Data-Driven Assurance in Semiconductor

The basic building blocks of a digital twin (i.e., data collection and analytics) are already an indispensable part of existing flows in traditional semiconductor lifecycle; however, data-driven approaches that leverage this sizeable amount of data to manage the whole lifecycle have rarely been reported.

Data obtained from these steps can be analyzed to provide assurance to broadly three aspects of the semiconductor lifecycle, namely quality, reliability, and security. As such, a digital twin framework can be constructed to enhance each of these aspects without drastic modification of its existing design and process flows.

It should be noted that there is a fundamental difference between product lifecycle management and security assurance through lifecycle management. While

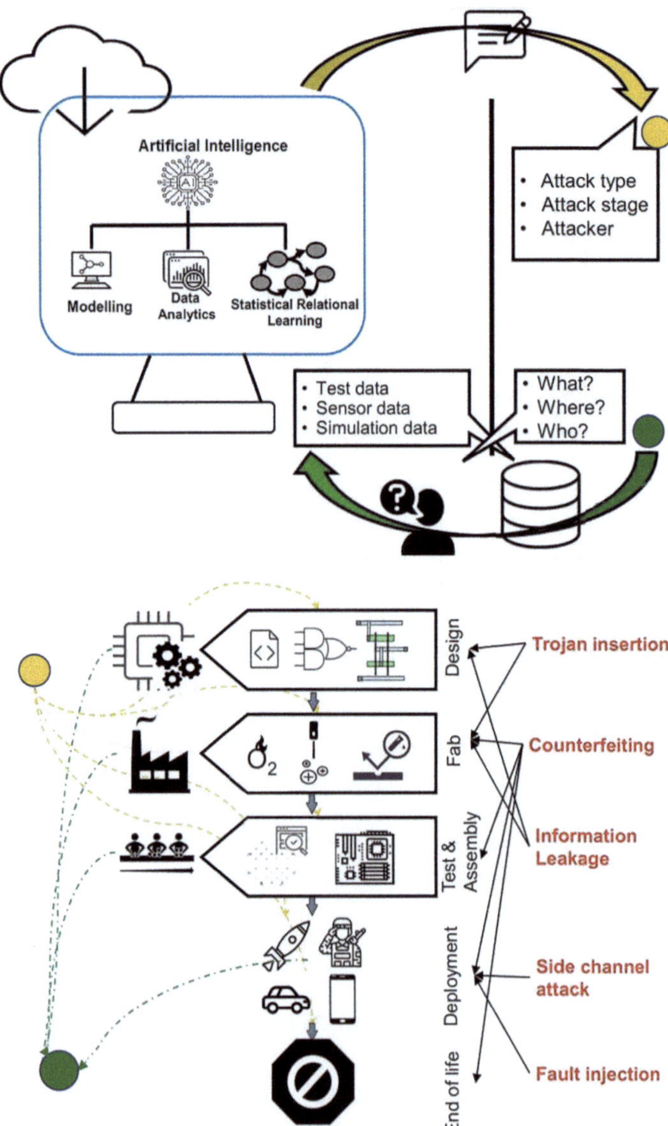

Fig. 8.2 A high-level overview of AI-driven DT framework security assurance in semiconductor lifecycle against various hardware attacks

the former is concerned with satisfying the functional requirements of a product and diagnosis of the underlying causes upon failure to do so, the latter is concerned with preserving the desired security properties of a system against attacks or unintentional mistakes of the designer.

In the last two decades, malicious modifications, vulnerabilities, and attacks on hardware have been extensively reported in the literature and the press [69–73].

To the best of our knowledge, there have only been two approaches in the silicon lifecycle management in the literature that attempt to provide security assurance to the lifecycle. In [42], the authors present their Synopsys SLM platform to assure quality, reliability, and security across the lifecycle. The proposed platform uses proprietary data engines to gain *actionable insights* to address various design and manufacturing issues. Although the authors claim that the analytics engines can be used for bolstering security defenses, there is no clear guideline provided on what data items are related to security vulnerabilities and how these relationships can be leveraged to defend against different types of security threats. Inspired by similar practices in software domain, the authors present a hardware secure development lifecycle (HSDL) [74] composed of five phases to identify and mitigate security issues as early as possible in the lifecycle. However, the proposed approach is a general pointer on *what* steps to follow for secure hardware development without specifics on *how* to achieve them through a singular framework. In [21], traceability for hardware Trojans is provided through a unified framework; however, it only does so for a specific hardware threat vector.

8.3   DT Motivating Examples for Secure Lifecycle Management

In the hardware security domain, academia has proposed many different algorithms and testing methodologies to detect different types of hardware attacks. Also, many proposals called design for security (DfS) approaches have been inspired by established design for testing (DfT) practices, which advocate for embedding different sensors into a chip or leveraging data from existing chips to better prevent attacks. However, the challenge is that the device is more likely to exhibit an anomalous behavior during its operation or when subjected to a test; thus it is up to the defender to understand why this behavior is occurring. As semiconductor industry has gradually shifted from a vertically integrated business model to a globally distributed one, there can be multiple possible explanations for a single anomalous behavior as there are many untrusted entities in the supply chain. The vast majority of existing literature on defense against hardware attacks has the underlying assumption that the attack vector is already known and detection or prevention methods against that attack vector need to be developed. This assumption makes sense if the threat model under consideration makes appropriate assumptions. In context of the whole lifecycle though, such restrictive threat models do not apply. A naive solution might be to put preventive measures in place on the chip to address all possible attacks; however, as the sensors and circuitry required are different, the performance penalty and hardware overhead for doing so would be unacceptable.

As a motivating example, let us consider three different scenarios. In the first two cases, the semiconductor device is a chip designed by a fabless design house.

Scenario 1

The design house receives customer feedback that a certain number of chips designed by them are experiencing accelerated aging. For the sake of focused discussion on security assurance, let us also assume that the accelerated failure is not a reliability issue as the design passed through all reliability checks during the design phase. In this scenario, the designers and the CAD tools used by the design house are considered trusted. Now, the design house has to consider at least three possible explanations behind this behavior:

   i. The failing chips are recycled or remarked chips that got resold as after they had reached their end of life.
  ii. The failing chips are out-of-spec or defective chips that should have failed the burn-in or wafer probe test during test and assembly. A rogue employee in the foundry or potentially the untrusted foundry itself is shipping some of the chips that failed these tests.
 iii. The failing chips are infested with a process reliability-based Trojan inserted by a rogue employee or the untrusted foundry.

Scenario 2

Infield testing such as JTAG testing and Built-In Self-Test (BIST) are often carried out after deployment to debug performance anomalies. In this scenario, let us assume that during such testing it is found that a confidential asset such as a secret key can be observed through the debug ports. Similar to Scenario 1, there might be multiple possible explanations, each of which arises from either security vulnerabilities introduced or attacks performed earlier in the lifecycle. We assume that the CAD tools can extensively verify information leakage flows. The possible explanations for this behavior are as follows:

   i. Designer overlooked the proper implementation of security policies while writing the hardware description code or even earlier at high-level design specification stage.
  ii. A malicious information leaking hardware Trojan was introduced in the circuit in the design phase through 3PIPs or inserted by the untrusted foundry.

Given these two scenarios, backward trust functionality of our proposed DT functionality will assign a probable provenance to the observed anomalous behavior through root cause analysis. Backward trust also entails identifying the possible causes of an observed anomalous behavior in the first place. This functionality is illustrated in Fig. 8.2 where the queries driving backward trust analysis are highlighted: What type of attack it is, where in the lifecycle it originated from, and who was responsible. The DT framework will also facilitate forward trust by ensuring that it is adaptable to future threats insofar as their successful identification and application of root cause analysis are concerned.

8.4   Semiconductor Lifecycle Data for Digital Twin Mapping

At the heart of every digital twin, there is bidirectional data flow between the physical and virtual environments. Data can be exchanged either as it is collected or after preliminary analytics has been performed on it. As a semiconductor device moves through various phases of its lifespan, the tools and software that are used to design it, the machinery that are used to manufacture it as well as the tests that are carried out to ensure its proper operation generate a huge volume of data. The reader can consult Fig. 8.3 to get a glance of the numerous steps a chip has to go through before it is ready for use for in-field applications. The life of a semiconductor chip ends at the recycle facility. In between, the chip is fabricated by a foundry, assembled and tested, and distributed to the market. Additional design features have to be added to each designed chip to reduce effort and complexity of testing and debugging. These design for test (DfT) and design for debug (DfD) infrastructures are sometimes outsourced to third parties. Each of these stages of the lifecycle consists of several sub-phases, a high-level overview showcasing every sub-phase for the design, fabrication, test, and assembly stages in Fig. 8.3.

A discussion on DT for secure semiconductor lifecycle is impossible without an understanding of the lifecycle stages and available data therein. Therefore, this section presents a brief description of each of the lifecycle stages along with an emphasis on the gathered data. At each stage of the lifecycle, industrial practice dictates the extensive collection and analysis of data to ensure reliability as well as satisfactory performance. Security is an afterthought in most cases although the rising threats in the global supply chain necessitate that the collected data be used and analyzed for security assurance as well. Academia has suggested various secondary analysis on the available data that may be used for that purpose. For a detailed reference to how the collected data may be used to perform security assurance evaluations, the reader is advised to consult Sect. 8.5.2.

8.4.1   Pre-silicon Design Stage

In the design phase of a semiconductor chip, a blueprint of the chip to be fabricated is prepared and delivered to the foundry in the form of a GDSII file [75]. The design must satisfy all specifications and perform desirably under all operating conditions and constraints (in terms of power, area, timing, etc.) of interest. The design phase itself can be further subdivided into multiple sub-phases, all of which form a sequential flow starting from architecture specification and ending at tape out [76]. Today, most of the following sub-phases in the process are automated using a combination of commercially available and open-source software.

(a) **Architecture specification:** At this phase, a high-level description of the circuit to be designed is prepared taking various trade-offs and customer feedback into consideration. The vast majority of circuits ship with a specification sheet

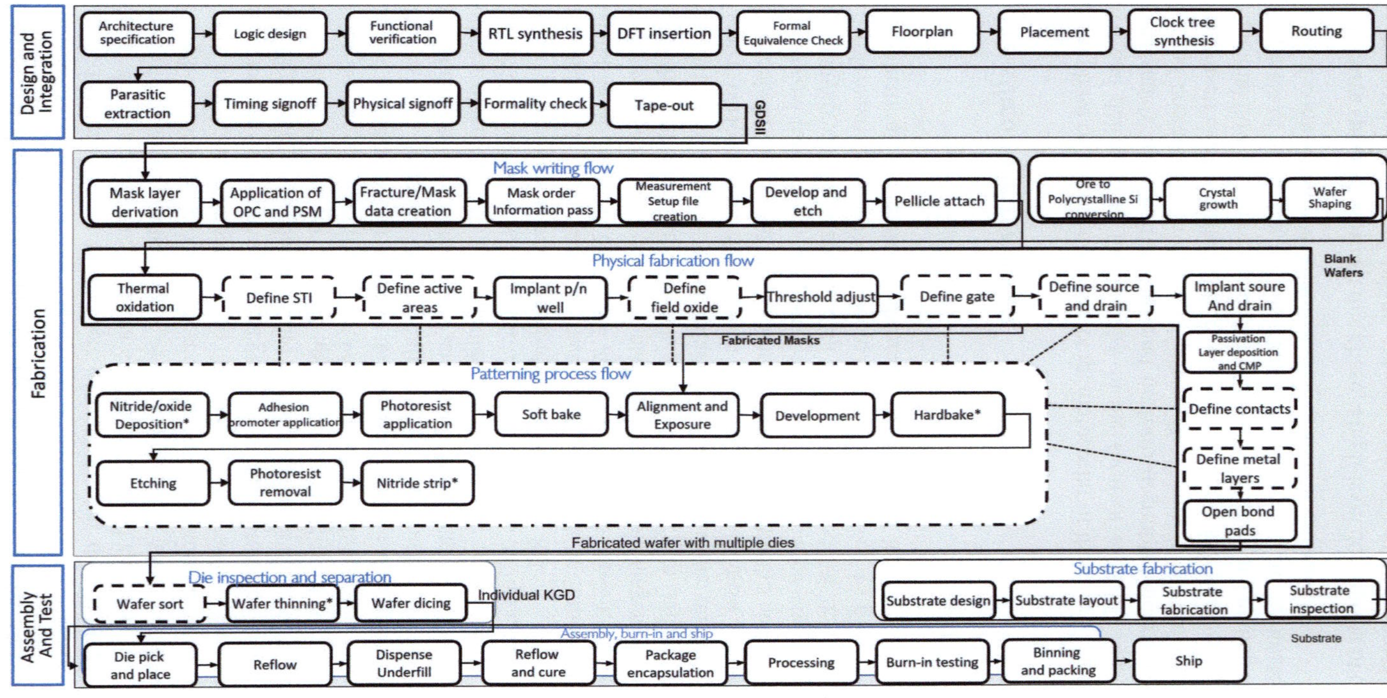

Fig. 8.3 A high-level overview of the first three stages of semiconductor lifecycle: design, fabrication, and assembly and test

that has detailed data on intended functionality, user guide on how to use the setup and use the hardware debugging features, data on important electrical, physical and architectural properties, such as operating temperature, clock speed, memory size, interface protocols, etc.

(b) **Logic design:** Hardware Description Languages (HDLs) such as Verilog, VHDL, SystemVerilog, etc. are used to describe and capture the specifications determined in the previous step. The code written at this stage is also known as Register-Transfer Level (RTL) code. The code itself may describe the behavior and/or the structure of the circuit in a specialized language.

(c) **Functional verification:** Next, the written RTL code is tested against the specification to verify whether it has successfully captured the intended behavior and functionality. The results of functional verification, also known as logic simulation, are often stored in a Value Change Dump (VCD) file which contains information on the sequence of value changes in different signal variables with respect to time along with the file metadata, definition of signal variables, and timescale [77].

(d) **RTL to gate-level synthesis:** The RTL code is then synthesized to produce a schematic of the circuit in terms of constituent logical gates. This process is entirely automated with the help of commercially available tools. The synthesized netlist is a description of the nodes in the circuit along with the interconnection between these nodes.

(e) **DfT insertion:** As mentioned previously, the complexity of modern VLSI circuits necessitates the inclusion of additional features in the design for increased testability of designs. DfT insertion step has similar outputs to RTL to gate-level synthesis step.

(f) **Formal equivalence check:** The design is verified by formal assertions in the form of logical and mathematical properties at this step. It provides a mathematical proof of functional equivalence between the intended design and the synthesized netlist. At the end of the verification, the designer is informed of the no. of points in the design that are equivalent to the intended functionality of the design.

(g) **Floorplanning and placement:** Floorplanning refers to the organization of circuit blocks within small rectangular spaces of the available space. The precise location of the I/O pins, power, and clock distributions are determined in the placement step.

(h) **Clock tree synthesis:** Clock tree synthesis step ensures the even distribution of the clock to all sequential elements in the design, satisfaction of all timing constraints, as well as minimization of skew and latency by clock tree building and balancing.

(i) **Routing:** In the routing step, the myriad of interconnects that connect different cells with each other as well as the individual gates within each cell get outlined.

(j) **Power, timing, and physical sign-off:** Physical sign-off involves the verification of the physical design performed in the last four steps against technology node defined design rules. Timing and power sign-off verifies the physical design against timing and power requirements. At the end of verification, the designer has detailed information on whether important circuit parameters such as hold and setup time, dynamic, static and leakage power, and interconnect and pat delay meet design requirements.

(k) **Tape out:** The verified design is then shipped out to the foundry in the form of a GDSII file.

As mentioned previously, automation effort in the design process is achieved by the use of commercially available software. This software collects and analyzes data at each step to optimize performance and reliability of the design. The data available from each stage are often stored in the form of different software file formats. In the literature, various types of secondary analysis have been proposed which can be performed on each of these files to derive secondary data of interest. For example, the RTL file describing the behavioral specification of the circuit can be analyzed to get information about branching probability, control flow graph (CFG), and data flow graph (DFG) which in turn may be used for security and performance optimization purposes [78, 79].

A summary of available data from these sub-stages of design phase can be found in Table 8.2. The readers are advised to note that this table is not exhaustive; particularly, the data file formats and available secondary data may change depending on the software being used and the type of analysis being performed, respectively.

8.4.2   Fabrication and Manufacturing Stage

CMOS fabrication is an extremely sophisticated process in which the exact steps followed depend on a large variety of factors including the technology node, the operating conditions that the chip is expected to experience, performance as well as cost considerations, the device or application in which it is to be used on, and many more.

The most important technique during the fabrication process is photolithography which refers to a process to transfer an etched pattern from a chromium or quartz plate—called a photomask—onto a silicon wafer using light or electromagnetic radiation [93].

The middle box in Fig. 8.3 shows sub-stages of an example fabrication process [94, 95]. This section describes the CMOS VLSI fabrication process in brief and also provides a summary of data collected from various equipment on the manufacturing floor.

8.4.2.1   Mask Writing Flow

The process of manufacturing photomasks is known as mask writing. The goal of the mask writing process is to transform the GDSII file, which is a binary file format, to a format which is understood by mask writing tools [96] as well as to break the

Table 8.2 Data available at design stage

Design sub-stage	Output data files	Primary information/data	Secondary analysis data
Arch. specification	Specification datasheet	Description of the intended functionality, operating limits, specific protocols and technologies used	–
Logic design	RTL	Behavioral and/or structural description of the circuit	Branching probability, relative branching probability [79], side-channel leakage assessment score [80]
Functional verification	SAIF, VCD	File metadata, timescale, variable type and bit-length, identifier codes	Coverage metrics [81, 82]
RTL to gate-level synthesis	Netlist, SDF	List of nodes and their interconnections, delay information	Controllability and observability of nets [83], rare nets [84, 85]
DfT insertion	DfT inserted netlist, SDF	Same as RTL to gate-level synthesis	Same as RTL to gate-level synthesis
Formal equivalence and model checking	Equivalency report	No. of primary outputs, and points in the circuit that are equivalent	Test pattern generation for hardware Trojan detection [86], secure information flow verified 3PIP cores [32], code, functional, and toggle coverage [87]
Floorplan and placement	Floorplan and placement db	Size and coordinates of the IO pads, IP blocks, and power network	Chip temperature [88], power supply noise [89], overall electromagnetic radiation [90]
Clock tree synthesis	Clock tree db	Clock skew and latency	Unused spaces in the layout [91]
Routing	Routed db (e.g., DEF), post place, and route netlist	Detailed location and geometry of interconnects	Wire length adjustments required for suppressing electromagnetic leakage [90]
Parasitic extraction	SPEF	Parasitic resistance and capacitance of nets and interconnects	–
Power, timing, and physical sign-off	Timing, power, and physical sign-off reports	Static, dynamic, and leakage power, delay information, setup and hold time, and no. of DRC and LVS violations	Path delay fingerprinting for hardware Trojan detection [92]
Tape out	GDSII	Binary representation of layout geometries	–

complex shapes present in the GDSII into simpler polygons. A data preparation step starts the mask writing process by performing graphical operations such as using Boolean mask operations to derive mask layers that were not part of the original input data.

Next, to facilitate printing of features that are much smaller than what would be possible for a particular wavelength of incident light, the geometrical shapes that are present in the GDSII need to be augmented by applying different resolution enhancement techniques (RETs) such as optical proximity error correction (OPC) for nearby features, phase shifting features, scattering bars, etc.

For verification and metrology, the masks contain barcodes and mask IDs. The RET applied mask data along with these additional data is then "fractured" into simpler polygons that can be handled by the mask writing tools. The consequence of addition of all these data to the original design data is that the fractured file size is often several times more than the original GDSII [97, 98]. The mask data then needs to be passed to a mask shop, typically outside the foundry, which will manufacture these masks.

Often instructions regarding how and where to carry out these measurements are contained in a specific file, known as measurement setup file, which are loaded into the tools. Once this file has been created, the masks are then physically fabricated using similar processes to those that are used in fabrication of the chip itself. For the sake of brevity, the discussion of these processes is only presented once, later in this section. The manufactured masks are extremely expensive costing up to millions of dollars. As such, they are encased in a protective membrane called pellicle to protect against erosion, dust particle adherence, and other mechanical damages.

8.4.2.2 Physical Fabrication

The polycrystalline Electron Grade Silicon (EGS) is processed through an apparatus called crystal puller to create the silicon ingot which is then mechanically shaped to a closed disc-shaped wafer [94]. These blank wafer preparation steps are shown in the top right corner of the middle box in Fig. 8.3. Using the masks prepared in mask writing flow and the blank wafers, the physical fabrication of the circuit commences. Again, the reader should be advised that the process flow that is presented in Fig. 8.3 or 8.4 and described in a greater detail in this section is representative, not exhaustive. For circuits that are fabricated using a different, for example, silicon-on-insulator (SOI) technology, the precise steps may differ.

The blank wafer is first thermally oxidized to form a layer of SiO_2 on top of it. This process is known as oxidation. The oxidation step is carried out in oxidation ovens or furnaces which can tightly control the temperature and gas flow rate at which the oxidation reaction occurs. A real-time monitoring of the oxide thickness and oxidation rate is also possible in modern systems [95]. Then, different regions in the circuit such as active area, isolation, the gate, the p/n well, metal layers, and contacts are patterned using photolithography through a series of chemical and mechanical process shown in Fig. 8.4.

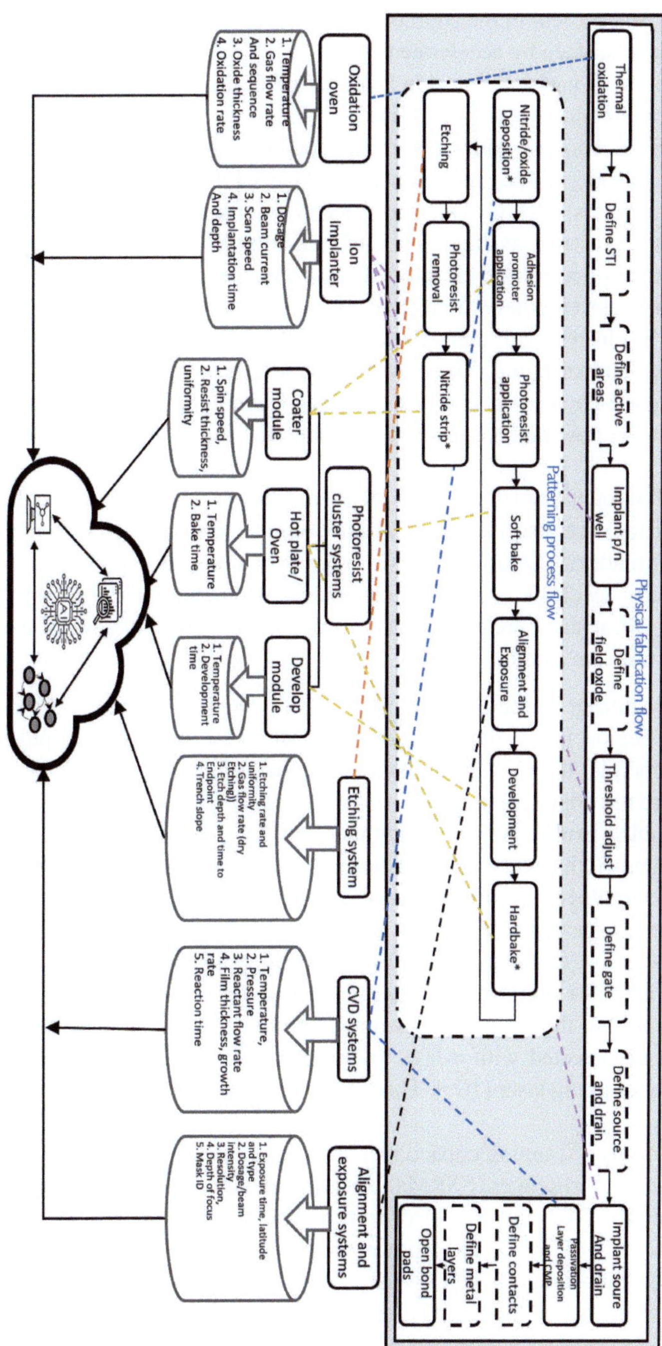

Fig. 8.4 Representative fabrication steps and data obtained therefrom

The dopant present in the source and drain areas is introduced through bombarding the wafer surface by accelerated ions.

Deposition of metal and nitride layer on top of the wafer happens through either physical or chemical vapor deposition.

In newer technology nodes, chemical vapor deposition (CVD) is used extensively to deposit nitride, oxide, and even metal layers [99]. The CVD systems, similar to oxide furnaces, also provide the foundry with extensive real-time data on the temperature, pressure, reactant flow rate, reaction time, and growth diagnostics. The spin speed, applied torque, developed resist thickness and uniformity data, baking temperature, development time, and spray pressure data are all available from the modern cluster equipment that can combinedly perform photoresist coating, developing, and stripping. The etching system carries out the etching of materials underlying the photoresist in either dry or wet etching process. Etching rate, trench slope, and gas flow rate are examples of data that can be obtained from an etching system.

The aforementioned data from these machinery is collected through built-in sensors. These are also known as in-line, online, or *in situ* data. In addition to *in situ* data, foundries also utilize a wide range of *ex situ* tests to provide a stronger and more precise feedback of fabrication process parameters. *Ex situ* tests, otherwise known as offline tests, are also generators of a large volume of data. Some examples of offline tests are as follows:

- **C–V profiling:** Through the application of a DC voltage, the width of the space charge region in the junction area of a MOSFET may be manipulated. Using this principle, the C–V profiling test can determine the type of the dopant and measure the doping density.
- **Four-point probe:** A typical methodology for measuring semiconductor resistivity, linewidths, and sheet resistance is the four-point probe method. It is most usually employed as an absolute measurement that does not rely on calibrated standards. Four-point probe tests can be used to construct sheet resistance contour maps which in turn can be used to infer the doping density profile of the wafer.
- **Thermal wave imaging:** The wafer is subjected to periodic heating stimuli. In the vicinity of the surface, the heating stimuli produces minute volume changes. These are detected with a laser by measuring the change in reflectance of the incident or pump laser [100]. The test data is represented in the form of a contour map.
- **Microscopy:** Scanning capacitance microscopes (SCMs) or scanning spreading resistance microscopes (SSRMs) can be used to build a later doping profile of the sample [101, 102]. Atomic force microscopes (AFMs) and transmission electron microscopes (TEMs) can capture images up to nanometer resolution. These high resolution images can be used for a variety of purposes including critical dimension control, topography analysis, electrical potential measurement, etc.

8.4.3   Post-silicon Packaging, Assembly, and Test

8.4.3.1   Wafer Test

After the completion of the BBEOL process, wafers containing multiple dies are placed onto an automatic test equipment (ATE) known as a wafer probe station.

A wafer prober is a highly sophisticated equipment that applies test patterns to check whether a given die on a wafer meets functional and parametric requirements based on which a chip is either accepted or rejected for packaging. The dies that are accepted and rejected together form a color-coded wafer map which can be viewed by an operator in a computer. The entire wafer sort process is illustrated in Fig. 8.5 along with the data obtained from these steps.

The wafer probing test itself applies test patterns in response to which the following data are gathered [103]:

(a) **Functional test data:** No. of stuck-at, transistor open and short faults, the nets they occur in, fault and test coverage, no. of untestable of faults, and automatic test pattern generation (ATPG) algorithm effectiveness.

(b) **Electrical parametric test data:** Various AC and DC parameters such as output drive and short current, contact resistance, input high and low voltages, terminal impedance, and reactance.

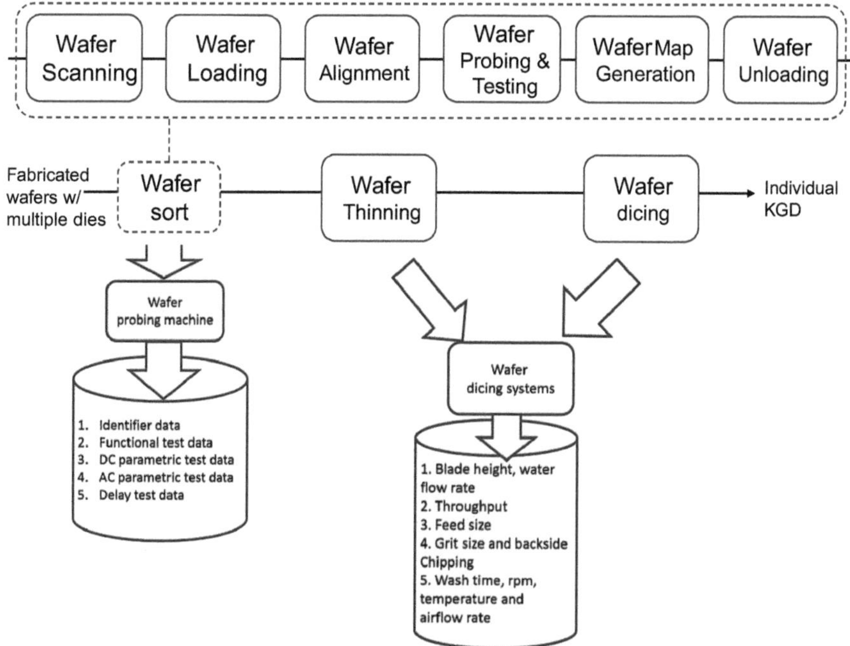

Fig. 8.5 Wafer testing process and possible data obtained

(c) **Delay test data:** The rise and fall times of transition and the setup and hold times of sequential circuits are some of the data available from delay tests performed at wafer test step.
(d) **Test identifier data:** These data include device, lot and wafer ID, wafer flat position, which is used for aligning, and no. of wafers discarded after wafer probing.

After the wafers have been probed, the dies that pass performance and functional requirements are sliced in wafer dicing systems.

The data available from wafer dicing systems include the blade rpm, wash time, temperature, water flow rate, etc., which are also shown in Fig. 8.5.

8.4.3.2   Packaging and Assembly

Packaging refers to the process of encapsulating the known good dies (KGDs) in protective insulating material and attaching metal balls or pins to them so that they can be accessed from the outside.

Assembly refers to the process of binding all of these different ICs and electronic components to a printed circuit board (PCB). Rapid device scaling, growth in the number of I/O pins, necessity of access to DfT and DfD features, thermal, mechanical, and economic considerations have meant that packaging technology has continually evolved over the past 60 years.

Through the years, packaging technologies such as surface-mount technology (SMT), quad flat packaging (QFP), pin grid array (PGA), and ball grid array (BGA) have been used. Recently, 2.5D and 3D packaging technologies have also been proposed. Depending on the particular technologies being used, the steps followed in the packaging process would be different. Figure 8.3 shows the steps for BGA packaging, more specifically flip chip BGA (FCBGA) packaging.

After the bond pads have been opened, a metal "bump" or ball is deposited on top of these pads. This process is known as bumping. These bumps will form the bond between the substrate of the PCB and the die when the dies are "flipped" to be conjoined.

The wafer is then diced, and KGDs are picked and placed by an automatic machine to its appropriate place on the substrate ball side down.

An epoxy type material is deposited by capillary action underneath to fill the space between the balls and the package. This step is known as underfilling.

Underfill flow rate, chemical composition, and fluid temperature are some examples of the data available at this stage.

8.4.3.3   Package and Burn-In Testing

Once the packaging and assembly steps are completed, the fabricated ICs are subjected to elaborate stress testing comprised of package and burn-in testing to evaluate their longevity under real-world operating conditions. Combined they are

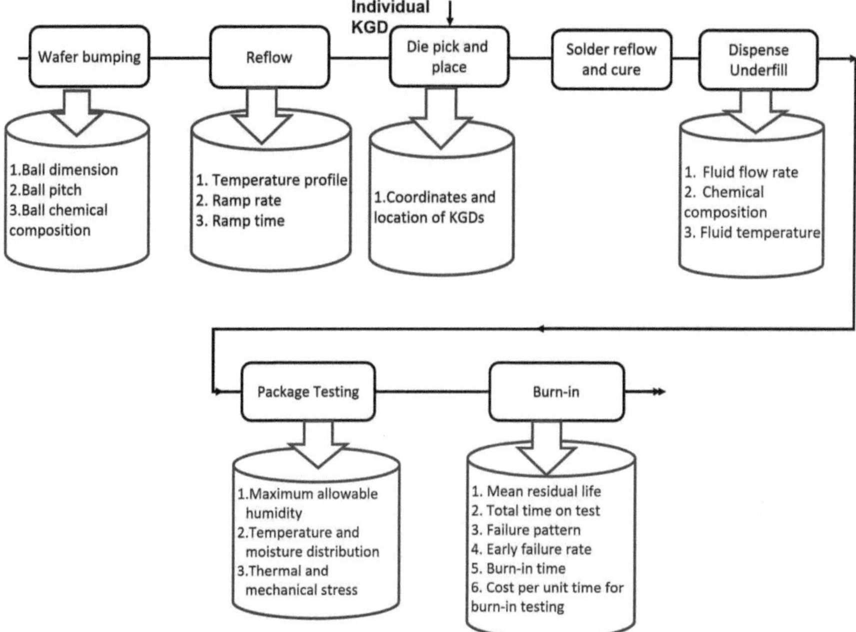

Fig. 8.6 Data available from assembly, packaging, and reliability test flows

also sometimes known as reliability tests. Preconditioning, temperature cycling, thermal shock, and temperature–humidity accelerated stress testing form a partial list of the series of stress tests that the chip is subjected to [104]. Some of the data items available from this series of stress tests are shown in Fig. 8.6.

8.4.4   In-Field Deployment Stage

On-chip performance, voltage, and temperature monitors monitor relevant circuit and software parameters and collectively form a report on the health and performance of the device. The availability of the tests listed in the following, therefore, depends largely on the on-chip sensors, DfT facilities, and the interfacing software for a particular chip. Different vendors also enhance the existing standards to offer additional debugging and testing features into their chip, and as such, the data available from these tests would largely depend on the specific vendor and type of the chip.

1. Built-In Self-Test (BIST): BIST is used to periodically test the circuit subsystems and their operation [105]. Its main purpose is to verify whether different components are working properly and in some cases apply appropriate coun-

Table 8.3 Data available from in-field

In-field source of data	Available data
BIST	Different subsystems in the device such as UART, memories, LED system, etc. working or not. Certain vendors and chips may offer additional functionality to show coverage of stuck-at faults and memory faults tested
JTAG debugging	Interconnect open and shorts and associated nets/ units, existence of stuck-at, crosstalk faults, device ID, non-testable nets and coverage statistics, real-time program counter
BSDL description	JTAG instructions and available registers, signal mapping, package information, type of boundary cell available for signals
Hardware performance counters	Cache references, branch misses, bus cycles, cycles, cache misses, CPU cycles, L1-dcache and L1-icache loads, stores and load-store misses, LLC load, stores, load-store misses

termeasures. Two types of BIST are widely used: Logic (LBIST) and Memory (MBIST). LBIST generates input patterns for internal scan chains using a pseudo-random pattern generator such as linear feedback shift register. MBIST is used for detecting memory defects and in some cases repair those defects.

2. Joint Test Action Group (JTAG) debugging: JTAG is a standard to access the boundary scan DfT features in a chip to verify its functionality [106]. Although originally conceived as a means of overcoming limitations of bed-of-nails fixtures of testing PCBs after manufacturing, today it is used for diagnosis of failing interconnects, memories, and testing functionality of ICs. Often, a boundary scan description language (BSDL) specification of existing JTAG features on a chip is provided by vendors to customers. This ensures that customers can have a useful manual on what test features are present in their device and how to use them.

3. Hardware Performance Counters (HPCs): HPCs are special purpose registers provided in a chip that stores various performance related activities in the device. These statistics can usually be accessed by an operating system (e.g., in Linux these may be accessed by the *perf* instruction) or special purpose software for the purpose. A list of data available from these tests may be found in Table 8.3.

8.5 Hardware Attacks: Data and Security Perspectives for DT

To keep our discussion focused, we describe three hardware security threats associated with the scenarios we mentioned earlier in Sect. 8.3, namely hardware Trojans, counterfeits, and information leakage. In this section, we also highlight the data items that have a correlation with these attack vectors.

8.5.1   Attack Vectors

8.5.1.1   Hardware Trojan (HT)

A hardware Trojan (HT) is a term that refers to a malicious alteration of a circuit. The Trojan may take control of, alter, or obstruct the underlying computing device's components and communications. Trojan trigger refers to the specific condition or the circuit that leads to the dormant Trojan being activated. Trojan payload refers to the functionality that a Trojan achieves once it has been activated. Trojan payload may either leak sensitive information to an externally observable port, cause denial of service, degrade performance, cause accelerated aging or change the functionality entirely. There are a variety of ways in which hardware Trojan may be categorized. A detailed Trojan taxonomy is provided in [107]. One important distinction we want to note here is between functional and parametric Trojans. Functional Trojans are Trojans that inherently change the functionality of the circuit, whereas parametric Trojans manipulate certain electrical or physical parameters of the device to cause performance degradation.

In Fig. 8.7, the red marked boxes show the lifecycle stages where either a functional or parametric Trojans may be inserted. In the design phase, functional hardware Trojans may be inserted through acquired 3PIPs. During the logic design phase, RTL to gate-level synthesis, and placement stages, these IPs are categorized as soft IPs, firm IPs, and hard IPs, respectively. DfT features, when outsourced to a third-party test vendor, may also be a source of Trojans. The Trojans inserted in the design phase are almost all exclusively functional Trojans. Functional Trojans may also be introduced by the untrusted foundry. In that case, functional Trojans would require the modification of the GDSII file provided by the design house which translates to a manipulation of the mask writing data in the mask layer derivation and mask data creation phases. Due to the need of adding additional logic into the circuit, for this type of Trojan additional active areas need to be included.

Parametric HTs may be inserted by manipulating fabrication recipes across a wide range of steps in the physical fabrication flow shown in Fig. 8.7. For example, doping dosage, oxide thickness [41], and threshold voltage [108] may be manipulated to cause accelerated aging or parametric failure of the device. Critical dimension change of the gate and channel lengths by changing the etching depth can also cause such aging.

Trojan detection is challenging even for state-of-the-art detection techniques. To begin with, the intrinsic opaqueness of integrated circuit internals makes it difficult to identify manipulated components; typical parametric IC testing procedures are often ineffective due to limited coverage during testing. Even if a testing method could be devised that reaches extremely wide coverage, hardware Trojans can be sequentially activated meaning only a very specific sequence events can trigger it. Formal verification methods often fail due to state explosion problems in such cases. Destructive tests and IC reverse engineering techniques are time-consuming and costly. When technology scales to the boundaries of device physics and mask

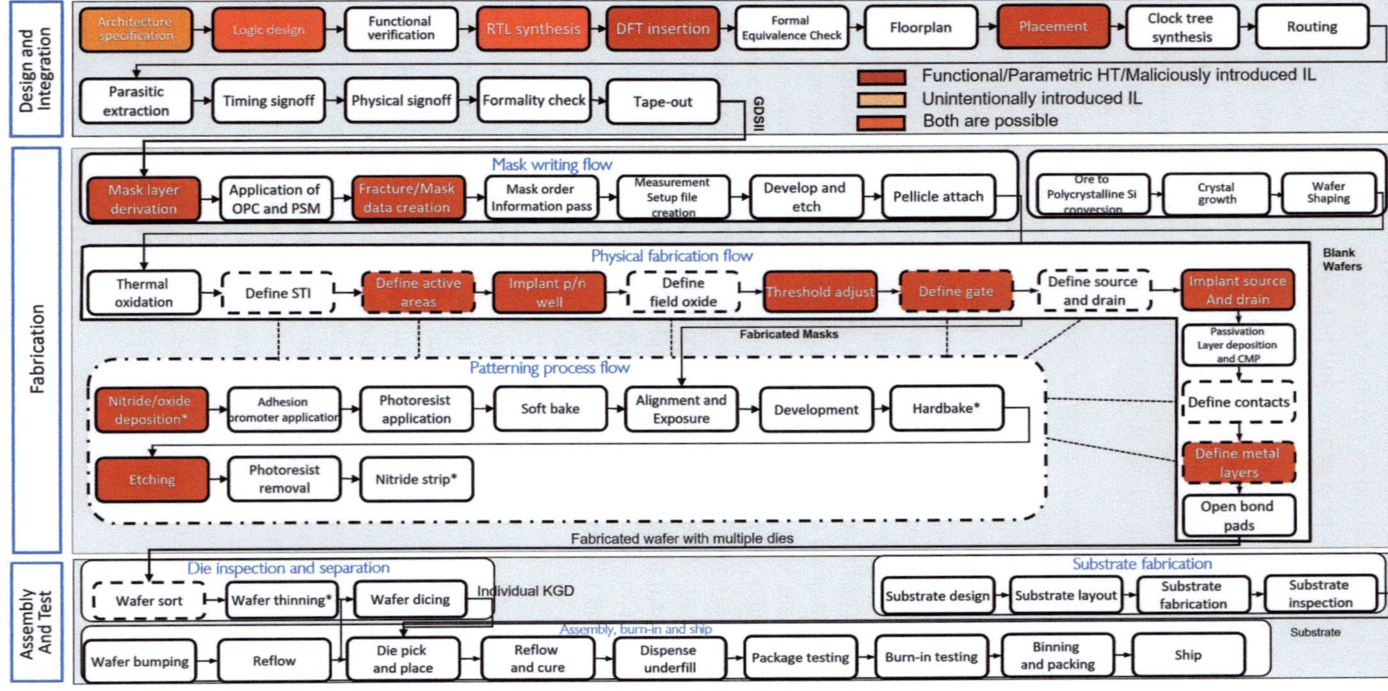

Fig. 8.7 Example lifecycle sub-stages where a hardware Trojan inserted or information leakage vulnerability may be introduced unintentionally

imprecisions, a chip's properties become nondeterministic, making the difference between what is a device affected with merely process variation and a device infested with Trojans difficult to surmise. Finally, the layout of the design may have "empty" spaces that serve as HT insertion spots for a malicious foundry.

8.5.1.2   Counterfeits

Due to the complex globally distributed horizontal nature of the electronics supply chain, it is difficult to trace the authenticity of each component that goes into an electronic system. The most frequent hazard associated with an untrustworthy electronics supply chain is the availability of various forms of counterfeit devices. Counterfeits can be of the following types:

- **Recycled:** The recycled electronic components are recovered from used PCBs that are disposed of as e-waste, repackaged, and resold in the market as brand new components. Despite the fact that such devices and systems may still be functional, there are performance and life expectancy difficulties associated with them because of aging process and various adverse effects resulting from exposure to chemicals during the recycling process. Recycling, therefore, is an end-of-life issue from the viewpoint of provenance.
- **Remarked:** Electronic components that have had the labeling on their package or the die replaced with falsified information are known as remarked chips. New electronic equipment might also be intentionally mislabeled with a higher standard by the untrusted foundry or other actors in the supply chain. For example, a chip may be designated as industrial or defense grade despite only meeting the requirements of a commercial grade one.
- **Overproduced:** Untrustworthy foundries, assembly plants, and test facilities that have access to the original design may be responsible for overproduction. These parties may be able to fabricate more chips or systems than the number specified in the contract and resell them without permission. Overproduction thus originates from the fabrication phase of the lifecycle.
- **Defective and/or Out-of-Specification:** Failure to comply with functional or parametric standards or grades (e.g., commercial, industrial, and military) results in the rejection of the device during fabrication and testing phases. However, the untrusted foundry or testing facility may ship defective or out-of-spec components into the market as genuine integrated circuits or systems without the knowledge of the design house.
- **IC Cloning and IP Theft:** A cloning attack can be carried out by any untrustworthy party in the electronics supply chain. Clones are direct copies of the original design that are created without the consent of the original component manufacturer (OCM). Cloning may be accomplished by reverse engineering an IC or a system that has been purchased from the market. IP theft refers to the stealing of intellectual property design components or tools and selling them to other parties without compensating the original owner. This may include things

like the HDL code of IP cores. Cloning may happen at any time during post silicon phases, whereas IP theft is mostly a pre-silicon issue.

Due to the overarching nature of the origin of these counterfeit types insofar as lifecycle sub-stages are concerned, they have not been explicitly shown in Fig. 8.7. Various approaches have been proposed in the literature to combat counterfeits. For preventing the shipping of defective, out-of-spec and overproduced chips hardware metering approaches such as Secure Split Test (SST) have been proposed [109]. Although these methods are effective, due to the requirement of new industrial practices for successful realization, these have not yet been fully integrated into the traditional lifecycle, and many design houses still have to almost blindly trust the foundry to get their chips into the market. Aging-based statistical analysis parametric fingerprints are used for the detection of recycled ICs [110, 111], but they may suffer from reduced accuracy due to process variations.

8.5.1.3   Information Leakage (IL)

Information leakage refers to the breach of integrity and/or confidentiality requirements in the security policies of a device. Confidentiality requirement violation results in unauthorized parties being privy to sensitive assets on the device, while integrity violation results in such parties being able to modify these assets. Information may be leaked through primary debug or test access ports due to unintentional mistakes made in the design phase or architecture specification stage by the designers or due to the insertion of a malicious hardware Trojan [112, 113]. They can also be leaked unintentionally through side channels such as timing, power, acoustic, etc. Inserted hardware Trojans can be responsible for leaking information through observable points in the circuit. In summary, there are two types of IL: maliciously introduced through HTs and unintentionally introduced. Maliciously introduced IL sources in lifecycle stages are highlighted in red, while unintentional ones are highlighted in orange. In the logic design and RTL to gate-level synthesis steps, both types of IL may be introduced. The IL vulnerability unintentionally introduced in RTL to gate-level synthesis step is mainly due to CAD tools that do not take security concerns into consideration.

The challenges in detecting intentional IL caused by the insertion of a hardware Trojan are the same as they are for hardware Trojans. Unintentional IL can be detected by formal verification methods [32, 114] and information flow tracking (IFT) methods [115, 116]. Formal verification methods are entirely reliant on the expertise of the verification engineer and the capability of the verification software. If formal assertions are not written properly, they might throw false positives. Model checking software used for formal verification also has the state explosion problem as the design can be "unrolled" only to a limited number of cycles.

In addition to the challenges outlined above, hardware attacks are evolving and new threats are proposed in the literature that circumvent traditional and literature proposed detection schemes. For example, there are always new Trojans being designed by researchers and malicious actors alike that defeat existing traditional verification methods as well as the ones proposed in literature [117, 118]. So it is entirely feasible that a hardware Trojan may go unnoticed until it is triggered in the field or in a more advantageous situation, detected at a later testing stage than the one it was inserted in. However the case may be, once it has been detected, it is impossible for almost all existing hardware Trojan detection techniques to inform the defender on where the Trojan was inserted. For instance, if a Trojan is detected at the formal equivalence check step, the designer cannot infer, merely from the result of the detection test, whether it was a soft or firm IP or the test vendor that introduced the Trojan. Similar arguments may be made for all hardware attack vectors including IL and counterfeits discussed in this section. This is where the proposed DT architecture adds new dimensionality to the hardware security threat analysis and defense.

8.5.2 Data and Security Perspectives

8.5.2.1 Hardware Trojan and Information Leakage

Maliciously introduced information leakage may be caused by Trojan insertion, and as such any data item that is related to hardware Trojan insertion is also related to maliciously introduced information leakage. We list some data items that are related to HT and IL threats which are already available from traditional flows of the semiconductor lifecycle.

- Pattern density refers to the number and width of features that need be transferred from a mask to the wafer in unit area of the mask [119, 120]. If functional Trojans are inserted, the pattern area density available from mask writing tools might be different from what was nominal for the layout delivered with the GDSII file as heretofore absent features need to be added to the mask so that it translates to the new malicious logic added to the circuit.
- Electron Beam Lithography (EBL) systems use electron beams to etch pattern onto the mask according to the output of mask layer creation software. The shot time of incident electron beams is proportional to pattern density, which means that if hardware Trojans are inserted, shot time may be changed as well. The governing equation is $T = \frac{T_0}{1+2\alpha\eta}$, where T_0 is the shot time at zero pattern density and T is the shot time at α pattern density.
- Electron Beam Proximity errors are encountered in mask writing when neighboring features are too adjacent. They need to be compensated for by dosage correction. This dosage correction is inversely proportional to pattern density which in turn is related to hardware Trojan insertion.

- If functional hardware Trojans are inserted, it can be reasonably deduced that to add the necessary logic into the circuit, more materials will need to be etched at most of the fabrication steps. The more material that needs to be etched, the more the etching plasma will be depleted and hence reduce the etch rate. This is known as the etch-loading effect, and it is typically more pronounced for dry etching. This data is usually available from the etching system used in photolithography process.
- When hardware Trojans are inserted by a malicious foundry, they are typically inserted in the empty spaces in the layout. This increases the pattern density as mentioned before. Additionally, as the features increase in adjacency, we can reasonably expect that the foundry must compensate for these additional features by adding more OPC features such as vertices and line segments. This should increase the file size and runtime of the OPC program.
- Many hardware Trojans are triggered by rare signals present in the circuit. The controllability and observability of nodes can be measured through SCOAP measure which is available with most modern commercial synthesis tools.
- The slope of an etched slope is inversely proportional to the density of features. So, the addition of functional Trojan features may impact the slope of the features which might be visually inspected by high resolution microscopy images.
- The intensity profile of diffracted rays through the mask is effected if patterns are too close by. This data is available from lithography simulation software.
- Critical dimension manipulation can be checked for through offline test methodologies or through high resolution microscopy imaging. Critical dimension data such as oxide thickness and gate dimension are related to parametric Trojans. These data are also available from the respective production equipment in which these steps are performed.
- Recipe changes to insert parametric Trojans such as doping dosage change and etching depth change can be traced from different offline tests as well as inline sensors in the production equipment. The contour map generated from offline tests such as thermal wave imaging and the four-point probe are indicative of the doping ion used and doping dosage. So any change in doping dosage may be traced back to the results of these tests.
- Branching probability metrics proposed in [79], which measure the likelihood of certain branches in RTL code being taken, can be extracted from HDL description of the circuit. These features are correlated with hardware Trojans activated on rare condition triggers.
- The authors identified 11 most important features extracted from the netlist of a circuit related to hardware Trojans in [121]. These include features such as how many logic levels away a net are from the primary output/input, the number of FFs a certain logic level away from that net, etc.
- For unintentional information leakage verification, formal verification methods can be useful. Although care must be taken to consider the possibility of false positives and the limited capability of the verification tool.

8.5.2.2   Counterfeits

The following are some of the data that can be analyzed to relate to different types of counterfeits:

- Package markings are carefully placed by OCMs. Package markings may include things like country of origin, lot identification number, pin orientation marking, etc. A deviation from OCM provided specifications indicates remarking of the chip.
- The pitch, alignment of pins, balls, and leads are carefully selected by OCM. Deviation indicates potential recycling. The dimensions of leads are also carefully selected and monitored, and as such deviation may indicate a recycled/remarked chip.
- Leads, pins, or balls of a chip may be reworked during recycling resulting in a different material than that of the authentic chip. A cloned chip's lead might not also have the same material composition as the original. This material composition can be verified from tests like energy dispersive microscopy.
- Crudely recycled or defective chips may have damaged bond wires. The length, shape, and tolerable stress of these wires may be found from the wire bonding stage. The wire bonding stage is common for DIP and QFP packaging technologies. This stage does not exist for BGA packages which is why it is not shown in Fig. 8.3.
- The standard test data format (STDF) is a widely used specification in the semiconductor industry [122]. Many ATEs used in the industry upload their test data to the database maintained by the foundry or the test facility in this format. It has also evolved as the *de facto* standard for organizing and collecting other test data from the fabrication and test facility. The STDF test specification lists 25 different types of record that are cataloged in an STDF file. Among these wafer result and master result records are of interest in dealing with overproduced, defective, and out-of-spec chips. Combined these keep a track record of the no. of parts that were tested to be of acceptable quality in each wafer. As such, these might be verified to trace whether yields were falsified. Furthermore, hardware bin records are kept to list which chips were placed in which bin in the testing facility. So, if a rogue employee is shipping defective chips out of the foundry, this record may be analyzed to find evidence of unaccounted dies and chips. STDF also comes with audit trail record which keeps a detailed history of each modification made to the file. This can help in dealing with insiders that might be falsifying the test data.
- Curve tracing and early failure rate, available from parametric testing and burn-in testing, respectively, can be performed to detect counterfeit ICs [123].

A summary of these relationships is listed in Table 8.4. These relationships establish that if anomalies in relevant data items are found, then it could be reasoned

Table 8.4 Relationship between available data and hardware security threats

Data item	Related security vulnerability	Available from (stage)	Available from (equipment/test/software)
Diffracted intensity	Hardware Trojan, maliciously introduced information leakage	Mask writing	Lithography simulation software
Branching probability, relative branching probability, controllability Index	Hardware Trojan, maliciously introduced information leakage	Logic design	RTL parser/static HDL code analyzer
Code, functional, and toggle coverage	Hardware Trojan, maliciously introduced information leakage	Formal verification	Cadence JasperGold®
Pattern area density, shot time, dosage of radiation	Hardware Trojan, maliciously introduced information leakage	Mask writing	E-beam mask writer systems
Out_flipflop_x, in_flipflop_x, in_nearest_pin etc.	Hardware Trojan, maliciously introduced information leakage	RTL to gate-level synthesis	GLN parser
OPC program runtime and file size	Hardware Trojan, information leakage	OPC correction	OPC software such as Cadence® Process Proximity Compensation (PPC)
Doping density	Hardware Trojan, maliciously introduced information leakage	Ion implantation	Ion implantation systems, Four-point probe, thermal wave imaging, C–V profiling
Gate and oxide dimension	Hardware Trojan, maliciously introduced information leakage	Gate definition, oxidation	Four-point probe, AFM/TEM, oxidation furnace
Etching depth	Hardware Trojan, maliciously introduced information leakage	Etching, post-fabrication	Etching systems, AFM/TEM
SCOAP controllability and observability	Hardware Trojan	RTL to gate-level synthesis	Synopsys TetraMAX®
Contour map	Hardware Trojan, maliciously introduced information leakage	Any time after ion implantation	Modulated photoreflectance, four-point probe, C–V profiling
Lead plating, ball chemical composition	Recycled, remarked, cloned	Wire bonding	Energy dispersive microscopy

Texture of package	Recycled, cloned	Package test	Any photograph taken of golden chip available in the market or after fabrication
Bond wire, ball, pin dimension, and count	Recycled, defective	Wire bonding, package test	Visual inspection or microscope imaging
Invalid markings on the package (e.g., lot identification code, CAGE code, pin orientation marking)	Remarked	Any time after burn-in test	Any photograph taken
Lead, pin, or ball straightness, pitch, alignment	Recycled	Wire bonding, flip chip attach	Visual inspection, microscope image
No. of good and functional parts tested	Out-of-spec/defective, overproduced	Wafer sort	STDF database (master results record)
Bin no., number of parts in the bin	Out-of-spec/defective	Wafer sort, various offline and inline test performed in the manufacturing floor	STDF database (hardware bin record)
Wafer ID, no. of good parts tested per wafer	Defective/out-of-spec, overproduced	Wafer sort	STDF database (wafer results record)
Early failure rate	Recycled	Burn-in testing	Wafer prober or burn-in tester
Curve trace	Different types of counterfeits	Wafer sort	Wafer prober

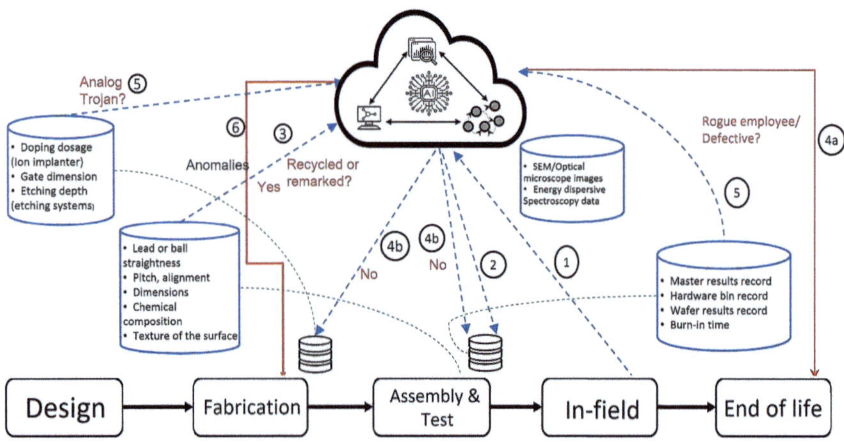

Fig. 8.8 An example of using lifecycle data for anomaly tracing and reasoning

automatically *where* an attack took place. Figure 8.8 illustrates how some of these datasets may be used to trace the root cause of the observed accelerated failure of chips described previously. For example, we start with some in-field data or report collected for the chip for an operation of interest, i.e., accelerated failure in this example. After a report of the observed accelerated failure is uploaded to the DT, DT would reason that one of the three hardware vulnerabilities mentioned earlier is the probable source of it. Then, the DT would require images obtained from microscopy of the device and/or a simple high resolution photograph (shown with ① in the figure). Data from more involved tests such as energy dispersive spectroscopy may be uploaded to the DT as well. At this point, from historical database of assembly and packaging stage, the DT cross-checks the uploaded data with data items that have a relationship with recycled and remarked chips (e.g., lead, ball dimensions, count, chemical composition, and texture of the surface of the chip) (② and ③). If the uploaded data do not match with previously stored specifications, then it is likely a recycled or remarked chip (④a). However, if anomaly cannot be found, then other explanations must be explored. As such, the DT would look at data associated with process variation-based Trojan-related data (e.g., gate and oxide dimension, etching depth, doping dosage, etc.) and shipping of defective/out-of-spec chip-related data (e.g., master results and wafer results record in the STDF database) to determine if there is an observed anomaly (④b –⑥). There are some pertinent issues to be mindful of. Anomaly detection algorithms for any of these data items may not be robust, and thus there is a significant element of uncertainty in assigning a deterministic cause to any of these underlying possible explanations. Uncertainty may also arise from the fact that collected data may be inadequate or too noisy to reach a conclusive decision. Therefore, the DT for secure semiconductor lifecycle would require an AI algorithm that can model this uncertainty and reason accordingly. The AI algorithms that may be used for this purpose are explained in more detail in Sect. 8.6.

8.6 DT Structure and Modeling for Secure Semiconductor Lifecycle

We illustrated in Sect. 8.5.2 how different data collected across different stages of the lifecycle may be utilized by a DT to provide traceability to observed anomalous behavior in the lifecycle. However, the view of the framework in Fig. 8.8 is not complete since the full functionality and underlying algorithms were opaque. We present the complete DT structure in Fig. 8.9. The DT is driven by a collection of AI and data analytics algorithms. There are four distinct components of the proposed framework:

  i. The initial anomaly discovery is driven by a data analytics algorithm that uses verification, testing, or sensor data to find violations in the specified security policies of the device. This component is shown at the top right side of Fig. 8.9.
 ii. Feature extractors are simulation tools, scripts, emulators, EDA tools, parsers, or scripts that extract and normalize features in context of the threat model and core AI algorithm.
iii. Anomaly detectors find the threshold of relevant features to label instances within the database as anomalous. We argue that if a hardware attack has taken place, one or possibly more of the features that have a correlation with the threats at hand would contain traces of anomaly. The anomaly detectors are data analytics algorithm (e.g., statistical models, time-series analyzers) or even ML or DL models that can find said anomalies in the extracted features.

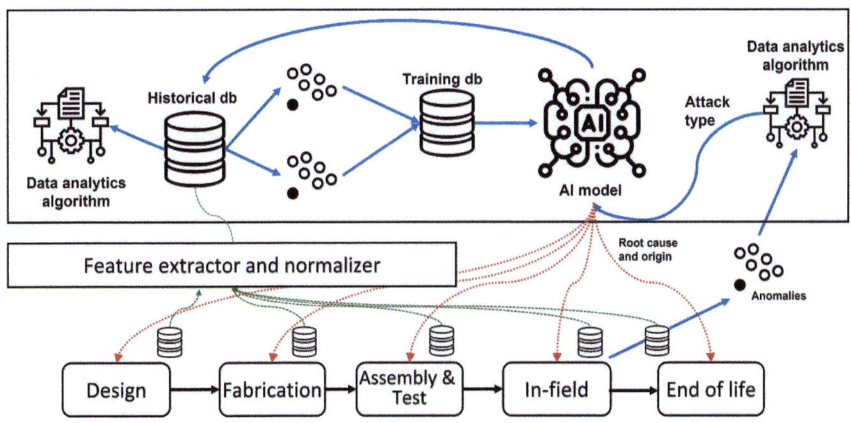

Fig. 8.9 An in-depth view of proposed AI-based DT framework

iv. Using these evidences of anomaly, the core AI model will infer the lifecycle stage where the problem originated from. We note here that not all relevant features will contain conclusive evidence of anomaly in them. Consequently, we propose to use an AI algorithm to infer the probable cause in the absence of complete consensus of extracted anomalies.

For scenario 1 described earlier, the analysis commences with the observation of an anomalous behavior. In this case, the observed anomaly is a simple one, i.e., accelerated failure of the chip. Once the possible causes have been identified, the DT would consult its historical database. For scenario 1, the relevant databases are that of fabrication and assembly and test stages. The DT would now require another set of auxiliary data analytics algorithms to find traces of anomaly in these databases. Once these anomalies have been found, these can be incorporated as evidence or knowledge base for training the central AI algorithm which is responsible for root cause analysis. For the core AI algorithm, we are proposing statistical relational learning in this chapter although other approaches with similar capabilities would also suffice.

8.6.1 Backward Trust Analysis

We define the first functionality of the proposed DT for security assurance in semiconductor lifecycle as Backward Trust Analysis. It has the following three components:

- When an anomalous behavior is suspected in the performance of a chip, the DT should be able to analyze the uploaded data to confirm or deny that suspicion.
- If the suspicion is confirmed, then the DT should also be able to identify what type of attack might the device be under. This step would require incorporating domain knowledge as well as data-driven processes.
- Once possible causes have been identified, the DT should be able to analyze historical data to find traces of anomaly in relevant data items. Using these evidences of anomaly, the DT should be able to assign a *most probable* cause to the observed behavior and determine where the attack took place and who the attacker was.

It is important to note that the data available from different stages of the lifecycle are not i.i.d. (independent and identically distributed). Traditional machine learning (ML) and deep learning (DL) algorithms are not suited for causal reasoning that is required for carrying out backward trust analysis as all of them implicitly carry the i.i.d. assumption [124]. Consequently, we propose to use statistical relational learning (SRL) models to reason by combining probability theory, machine learning, and mathematical logic. We discuss three possible SRL models, namely Bayesian Networks (BNs), Hidden Markov Models (HMMs), and Markov Logic Networks (MLNs).

8.6.2 *Statistical Relational Learning Models for DT*

8.6.2.1 Bayesian Networks

As a probabilistic graphic model (PGM), BNs make use of Bayesian inference for probability computations. They describe conditional dependency between random variables as edges in a directed acyclic graph (DAG). Bayesian networks attempt to model conditional reliance and, by extension, causality between random variables through describing the joint probability distribution of constituent random variables. More specifically, a Bayesian Network $B(G, \Theta)$ over the set of random variables $\mathbf{X} = (X_1, X_2, \ldots, X_N)$ has two constituents:

1. A graph G consisting of nodes representing the set of random variables and edges representing causal relationships among these variables
2. A set of conditional probability distributions Θ associated with each node of the graph

Figure 8.10a shows a simple Bayesian Network which signifies that there is a causal relationship between events A and B or C. Event A can cause either event B or event C to happen which is signified by an "edge" pointing from A to B and C. In BN terminology, A is the parent node, whereas B and C are child nodes. Starting

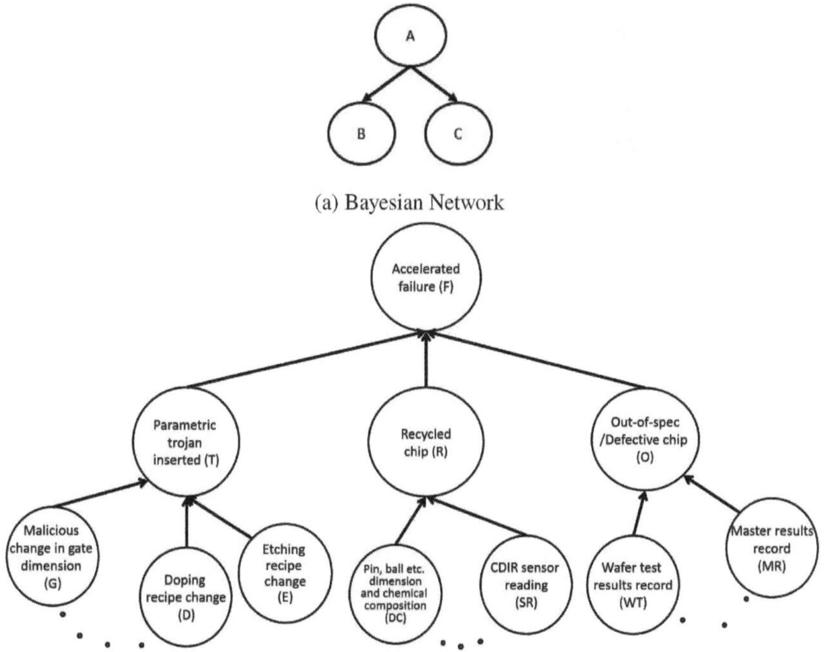

Fig. 8.10 Bayesian networks for root cause analysis

from any node in the graph and following along the edges, one cannot end up in the same node that they started from. This is why it is called an acyclic graph. The joint probability density function for any node in a BN is given by the formula:

$$P(X) = \prod_{i=1}^{N} P(X_i | parents(X_i)) \tag{8.1}$$

Inference in Bayesian Networks

Inference in BN refers to the process of finding the probability of any given node when conditional probability distributions of all other nodes are known. Inference in BNs can take two forms: The first is a straightforward calculation of the joint probability of a certain value assignment for each variable (or subset) in the network [125]. Because we already have a factorized representation of the joint distribution, we can simply evaluate it using the specified conditional probabilities. If we are only interested in a subset of variables, we must eliminate those that are irrelevant. This task is known as *belief revision*. The second one is a calculation of the probability of a node X given some observed values of evidence nodes E, i.e., $P(X|E)$. This type of inference is known as *belief updating* or *belief propagation*. More formally, the task of belief updating may be represented by the following formula:

$$P(X|E) = \sum_{\forall y \in Y} P(x, e, y) \tag{8.2}$$

where Y is the set of random variables that do not appear in either x or e.

Learning in Bayesian Networks

The inference in BNs requires the prior knowledge of the conditional probability table (CPT) of each node. CPTs of each node can be learnt in one of the three ways: (i) expert elicitation, (ii) applying learning algorithms on historical observed values [126], or (iii) assigning an initial CPT and having it be updated on new observed data [127].

In the presence of availability of a fully labeled dataset, method (ii) is more appropriate as CPT of the nodes can be learnt from historical values. One of the popular algorithms to learn parameters of BNs from historical dataset is the Maximum Likelihood Estimation (MLE) algorithm. In the absence of a large dataset, domain knowledge in form of prior beliefs can be incorporated into the learning process. This is done through the Maximum *A Posteriori* (MAP) algorithm.

BN for DT

A possible formulation of a BN for the causal relationships in scenario 1 described in Sect. 8.3 is shown in Fig. 8.10b. For our purposes, we propose to construct BNs in three "tiers": (i) observed anomaly node (e.g., accelerated failure), (ii) primary possible explanations of observed failure; the nodes denoting parametric Trojan inserted, recycled chip, and defective chip—all with common child nodes are this type of nodes, and (iii) the set of data items or features that have a correlation with the second tier of nodes. As an example, gate dimension, doping recipe, and etching depth have correlation with parametric Trojans. Thus they form the third tier in Fig. 8.10b. Due to space constraints, we do not show the entire set of features as discussed in Sect. 8.5.2. These nodes do not have any parent nodes.

BNs are suited for capturing relationships inherent in backward trust assurance problem for the DT since we would have evidence of some type of anomaly which can be explained by many underlying causes. These causal relationships are easily represented by a BN. Furthermore, the posterior probabilities $P(T|F)$, $P(R|F)$ and $P(O|F)$ obtained through inference for the BN shown in Fig. 8.10b would assign a probable cause to the query "Why are chips experiencing accelerated aging in the field?" Furthermore, if the probabilities such as $P(G|F)$, $P(E|F)$ can be calculated, we would know, within the bounds of a confidence level, that a certain lifecycle stage is where the problem originated from. For the illustrative BN shown in Fig. 8.10b, the CPT of each of the nodes would be best learnt by encoding the prior belief that recycling is much more likely to be the problem. The probabilities associated with each node can be updated by incorporating new data points.

8.6.2.2   Hidden Markov Models

An HMM is an augmented Markov Chain that models the situation where a sequence of hidden processes influences the outcome of a set of observable events. Like Markov Chains, HMMs postulate that the future state of the system can be predicted by knowledge of only the present state of the system. Formally, an HMM is characterized by the following set of parameters:

- A set of N states $Q = q_1, q_2, \ldots, q_N$
- A transition probability matrix $A = a_{11}, \ldots, a_{ij}, \ldots, a_{NN}$ with each a_{ij} representing the probability of the system transitioning from state i to state j
- A sequence of observations $O = o_1, o_2, \ldots, o_T$
- The emission probability matrix B which is the probability of a state i generating the observation o_T
- An initial probability distribution of states $\pi = \pi_1, \pi_2, \ldots, \pi_N$

Fundamental Problems of HMMs

According to Rabiner [128, 129], there are three fundamental problems that can be answered through HMM modeling:

i. **Likelihood:** Given an HMM (A, B), the likelihood problem is to determine how likely an observation sequence is to be obtained from that HMM.
ii. **Decoding:** Given an HMM and an obtained observation sequence, the decoding problem is to find the best sequence of hidden states.
iii. **Learning:** The learning problem is to calculate the parameter matrices A and B when presented with an observation sequence.

HMM for DT

Let us consider scenario 2 and assume that a hardware Trojan was inserted. In an ideal case, one of the pre-silicon verification tests would be able to detect that a Trojan was inserted. This is shown in Fig. 8.11a where we show how the lifecycle process in a case where formal verification was able to detect a Trojan insertion may be modeled as an HMM. The actual Trojan was inserted in the logic design phase through a 3P soft IP. This caused the circuit to transition from a Trojan free state to Trojan infested state. In this particular threat scenario, the transition of the circuit from Trojan free to Trojan infested also represents it transitioning from

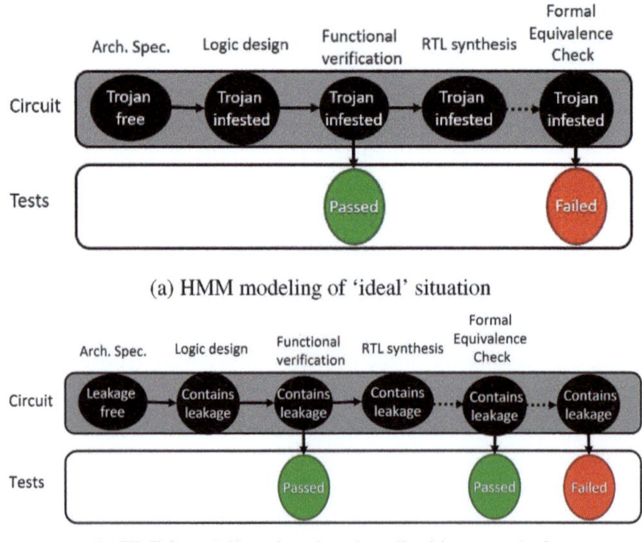

(a) HMM modeling of 'ideal' situation

(b) HMM modeling situation described in scenario 2

Fig. 8.11 HMM modeling for ensuring backward trust in semiconductor lifecycle

an information leakage free circuit to a leaky circuit. The argument for modeling hardware security threat scenarios as HMM is that Trojan insertion (or a circuit becoming information leakage prone) is a clandestine event and the state of circuit is not observable to us unless a sequence of tests is performed. The progression of the lifecycle stages constitutes the hidden Markov process, whereas the sequence of tests performed constitutes the observable process which is influenced by the hidden states. Figure 8.11b shows the more likely scenario. Here, we model the fact that none of the pre-silicon verification tests were successful (indicated by a "passed" observation). The anomalous behavior, i.e., the leakage of the key, was detected at deployment stage through JTAG test.

For security assurance through the proposed DT, the root cause problem we are interested in solving can be framed as the decoding problem of HMM. In Fig. 8.11, the Viterbi algorithm [130] may be used to find the most probable sequence of events. If the modeling is successful, we can reasonably expect the constructed HMM model to infer that the circuit changed state in the logic design phase. From the description of the decoding problem given earlier, to apply Viterbi algorithm prior knowledge of the transition and emission matrices is needed. Obviously, it is challenging to know what is the emission probability from a Trojan free state to a functional test giving a "pass" result. For this reason, initially from historical observation data analysis, matrices A and B must be learned. Here, certain facts should be noted. The probability that a Trojan infested circuit would transition to a Trojan free circuit is 0. Therefore, certain transition probabilities can be assigned from domain knowledge, whereas others would need to be learned. The learning algorithm will also need to be adapted for such that only the parameters that are not set by domain knowledge are learnt. The entire process flow required for HMM realization is shown in Fig. 8.12b. The logged historical data acquired from previous progressions of a device through lifecycle steps can be used to learn the transition and emission probabilities.

8.6.2.3   Markov Logic Networks

An MLN is an SRL model that attributes a weight to a set of first-order logic formulae known as the knowledge base (KB) [131]. For any domain, a KB written in first-order formulae can succinctly represent prior domain knowledge. Constants, variables, functions, and predicates are four types of symbols that are used to build formulas in first-order logic. Constant symbols are used to represent items within a certain scope of interest. Variable symbols span the domain's objects. Function symbols denote mappings between tuples of objects and individual objects. Predicate symbols describe relationships between items or properties of objects in the domain. An interpretation describes which symbols are used to represent which objects, functions, and relations in the domain. Certain logical connectives and quantifiers (e.g., $\wedge$, $\vee$, $\implies$, $\exists$) are used to construct and qualify these formulae.

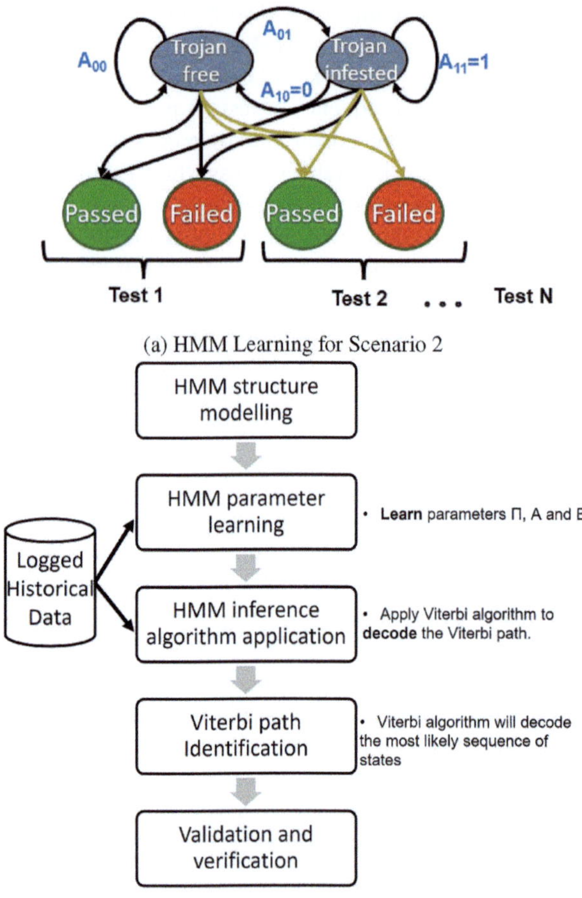

(a) HMM Learning for Scenario 2

(b) HMM Learning and Inference Process

Fig. 8.12 HMM learning and decoding for ensuring backward trust

However, first-order logic by itself is not suitable to deal with uncertainty inherent in semiconductor lifecycle. MLNs add the capacity to deal rationally with uncertainty, and to tolerate unsure and conflicting knowledge, by constructing a Markov network utilizing each grounding of a formula in the KB. More specifically, MLN is an instantiation of a Markov Random Field (MRF) where groundings of the predicate (atom) constitute the nodes and groundings of the formulae (clauses) represent the features. The atoms in the field are then described by the joint probability distribution:

$$P(X = x) = \frac{1}{Z} exp \left(\sum_i w_i n_i(x) \right) \tag{8.3}$$

where w_i is the weight associated with the ith formula f_i, Z is a constant, and $n_i(x)$ is the number of true groundings of the formula f_i in the world x. A world corresponds to a small subset of all possible groundings of atoms.

The formulae in the KB are usually designed by experts in the field [132]. The weight associated with each formula brings about the notion of "hard" and "soft" formulae. A hard formula has a weight of $+\infty$ or $-\infty$ signifying always true and false, respectively. A soft formula on the other hand may have weights in between signifying the "degree" of truth in them. This incorporation of weights is what would allow us to deal with uncertainty inherent in root cause analysis problem of ensuring backward trust.

Additionally, two types of statistical inference are possible in MLNs. Given some evidence, maximum *a posteriori* inference (MAP) finds most probable world. The second type of inference, namely marginal inference, calculates the posterior probability distribution over the values of all variables. We are interested in MAP since in backward trust the DT would be presented with some evidence of anomaly and it would have to find the most probable world in which the atoms are satisfied.

8.6.3   Anomaly Detection

The DT has another important component to its digital analytics engine and that is the anomaly detection methods collectively referred to as *data analytics algorithm* in Fig. 8.9. In scenario 1, the initial observed anomaly is a report on accelerated failure of the chip which may be available from the customer's feedback. However, in the more general case, the suspected behavior may not be so easy to confirm. For example, the authors in [133] had to devise their own Pipeline Emission Analytics to discover that a military grade commercially available FPGA had a possible backdoor in it. Depending on the suspected anomalous behavior, the analysis technique would have to be different. As a result, the DT requires a preliminary data analytics algorithm that would find the anomaly in the suspected data stream and subsequently infer the list of possible causes. This data analytics algorithm can be something as simple as visual inspection of a high resolution microscopy image to more involved analytics like time series analysis methods, statistical signal processing algorithms, or machine learning and deep learning models.

Furthermore, it is the evidence of anomaly gathered from various different algorithms that drives the training database of the SRL model. The data collected from various stages of the lifecycle are extremely varied in nature. Data can be textual such as the RTL code from design phase or the Jobdeck file from mask writing phase. The fabrication floor data such as gate dimension, etching depth, and doping dosage are numerical data. As the type of data is different, the analytics algorithm would also be different. This is also the reason why we have included a "feature extractor" block that takes these disparate types of data and translates them into legible values understandable by core reasoning algorithms.

We discuss some of the data analytics that can be performed to find traces of anomaly in different types of data. Let us consider threat scenario 1. To find evidence of anomaly in data such as doping dosage or etching depth, a method such as the one proposed in [134] may be used. If an insider, for example, changes the recipe of the doping dosage, this would be detectable by the Kernel Density Estimation (KDE) and the Kullback–Liebler divergence-based change point anomaly detection. A mismatch between the OCM specification and the DUT's observed measurements represents anomaly in data like lead, ball, or pin dimensions, chemical composition, straightness, alignment, etc. To identify it, a simple visual inspection from an X-ray or SEM image may suffice. More complex techniques suggested in the literature can also be employed to enhance confidence in the obtained evidence. The 4D enhanced SEM imaging proposed in [135] and the machine vision and advanced image processing methods demonstrated in [136, 137] represent such more sophisticated methods. For data such as wafer and master results record, a cross match of the STDF file with expected values would reveal anomalies. This cross match may be automated through natural language processing algorithms. For scenario 2, the data to inspect would be the RTL and netlist files in the design phase in addition to the fabrication stage data discussed for scenario 1.

In case an information leaking hardware Trojan was inserted, a plot of relative branching probability (R_p) vs effective branching probability (P_{eff}), as defined in [79], may demonstrate an anomaly. For example, in Fig. 8.13, we show the plot of R_p vs P_{eff} for the Trojan-free version of the circuit AES-T1300 (an example of an information leaking Trojan available from the TrustHub database [138, 139]) as compared to the Trojan infested version. There is a very distant cluster of branches corresponding to the Trojan trigger branches as measured by cluster centers derived from the Mean Shift algorithm. The structural features of the circuit were extracted by an RTL parser which is fulfilling the role of "Feature extractor or normalizer" in Fig. 8.9. We have discussed previously in Sect. 8.5 how traditional methods for testing for unintentional information leakage and information leaking hardware

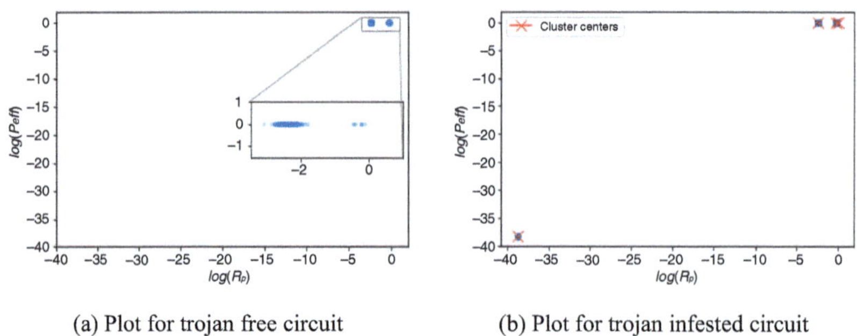

(a) Plot for trojan free circuit   (b) Plot for trojan infested circuit

Fig. 8.13 Anomaly can be detected in a plot of relative branching probability vs effective branching probability when comparing Trojan infested and Trojan free versions of AES-T1300. The cluster centers were determined by Mean Shift algorithm

Trojan provide low coverage and low confidence. New testing methods such as concolic testing [140] and RTL IFT [141] have been proposed that offer a greater deal of certainty in finding implicit and explicit information flows in the design. Any one of these testing methods may be used to find evidences of information leakage in the design which would serve as evidence of anomaly.

We answer this question in 8.6.4 through the explanation of the second proposed functionality of the DT framework, i.e., forward trust analysis.

8.6.4   Forward Trust Analysis

We have discussed how existing solutions in hardware security do not scale very well either because they address only one or two specific vulnerability or because they require the embedding of overhead incurring sensors and/or modification of existing process flows. Enabling forward trust refers to the scalability and adaptability of the DT framework. Scalability means that the proposed framework would be usable to address all possible attack vectors instead of a subset of the attack vectors. Adaptability refers to the ability of DT to address emerging threats by effectively carrying out root cause analysis, even unforeseen emerging threats. As ease of accessibility to computation resources keeps increasing and the semiconductor industry becomes more horizontally globalized, new threats would continue to emerge. Adaptability addresses this issue by leveraging the fact that as hardware security threats evolve over time, so do our collective understanding of their underlying causes and the corresponding anomaly detection algorithms.

Scalability is possible in the proposed DT framework by simply translating the domain knowledge to an equivalent KB for MLN or to a DAG for characterizing the BN and HMM. The same fundamental principles would apply, as demonstrated by the scenarios discussed in this chapter, when considering other attack vectors like side-channel and fault injection attacks. Adaptability is also an inherent feature of the SRL models discussed previously. Let us consider an example. Researchers in [118] demonstrated a novel Trojan that circumvents existing verification efforts and introduces capacitive crosstalk in between metal layers to realize its payload. Now, to enable root cause analysis of this scenario, all we are required to do is to add a parent node to the child node "parametric Trojan inserted" called "rerouting the metal layers." Similarly for MLNs, this would amount to a new grounding of the formulae in the KB. Without changing the core algorithms drastically, we can still model new threats so long as the underlying causes are known. Furthermore, as the individual anomaly detection algorithms are "detached" from the core functionality of the root cause analysis AI model, they can be updated as better anomaly detection algorithms emerge. One of the examples in the context of information flow tracking was already discussed in 8.6.3.

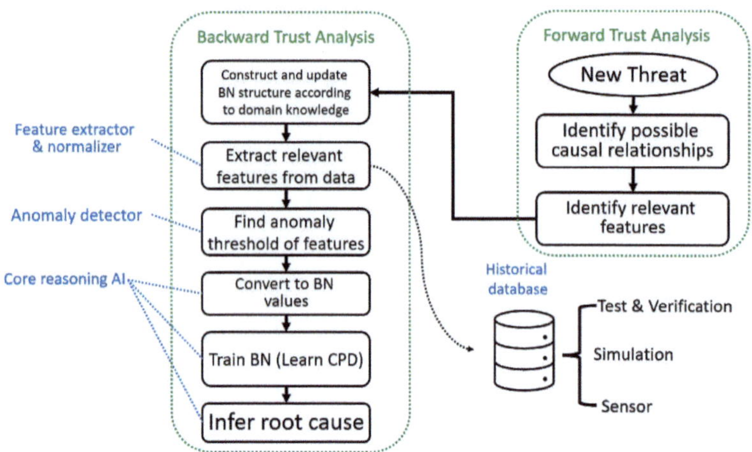

Fig. 8.14 Backward and forward trust analysis process flows

8.6.5 Process Flow for Trust and Security Analysis

We now summarize the entire process flow with the help of Fig. 8.14 and scenarios mentioned earlier. For the sake of simplicity, we constrain our discussion of the process flow with the assumption that BN is the core AI algorithm for root cause analysis. We have already shown the BN network structure for scenarios 1 and 3 (shown in Fig. 8.10b). Next, we have the knowledge of relevant features of the threat model, which the reader can consult from Sect. 8.5.2. For example, we can extract relevant features from RTL and GLN, an RTL, and netlist parser or using commercial EDA tools in scenario 2. The anomaly detector would then find the associated threshold of the features that signify an anomaly. A BN with binary outcomes for each node can be constructed by determining whether corresponding features exceed the threshold. For each entry in the database, the BN can then be trained to learn the CPT parameters. This builds the initial CPT of the nodes in the BN. Next, the features of design under test can be extracted in the same process, and CPT parameters can be updated according to a learning rate. Lastly, inference from BN can point us to the probable cause. The process thus far described constitutes the Backward Trust Analysis. For adaptability to new threats, the domain knowledge needs to be updated. This amounts to updating the BN structure by adding the proper child or parent nodes and identifying relevant features. We have already discussed one example of this earlier in 8.6.4. Once this has been updated, we can proceed as before in Backward Trust Analysis to infer the origin of the threat.

8.7 Challenges and Future Research Avenues

We have laid a clear outline of the algorithms and components required to realize a digital twin for end-to-end semiconductor lifecycle management in the preceding sections. A full implementation of the proposed framework when working on real datasets will present several challenges pertaining to optimization, computational complexity, logistics of data transfer, and adopted technologies. We discuss these challenges in this section and provide some indications on how to overcome them.

8.7.1 Defining the Virtual Environment

According to the definition of digital twins used for lifecycle management, there are five principal dimensions: physical part, digital part, data, service, and connection [48]. For secure semiconductor lifecycle, the physical part is a process instead of a physical object. More specifically, we are modeling the lifecycle of a physical object as it progresses through different stages. We have also defined the data that the DT will leverage and the services it will provide as well, namely forward and backward trust analysis. In the DT literature, virtual environments are also defined as the containers of the digitized twin. It is most often a cloud platform or data warehouse to host the relevant database and run AI models on said database although it need not be [9]. In the preceding, we have not defined the cloud platform to host the database and the digitized twin. Platforms such as AWS or Microsoft Azure can be used for this purpose.

8.7.2 Data Acquisition and Security

For the first scenario mentioned in Sect. 8.3, we illustrated which data can be used to assign a probable provenance to a hardware attack in Sect. 8.5.2. However, this ignores the fact that the foundry may not be willing to share its manufacturing data. This does not automatically diminish the usefulness of our proposed framework. In such a case, destructive reverse engineering of a population of the used available chips will be required to find evidences of CD change in devices which indicates a parametric Trojan [142]. Similarly, other innovative data mining and acquisition techniques will need to be employed in case certain data items are not available to the entity with the access to the DT.

Furthermore, secure acquisition of relevant data items is an avenue we have not discussed. In creating a cyber-physical system, there is always the risk of false data injection attacks which may be carried out by an insider through a network. Falsification of test data at the fabrication step, for example, may be traced through the Audit Trail Record of STDF database. At other steps, or through the network, tracing data injection attacks would require a completely separate approach which is not explored in this chapter.

8.7.3   State Space Complexity

The Bayesian Network presented in Sect. 8.6.2 shows how a child may have three parent nodes. According to [143], this presents a significant challenge to keep the task of populating the associated CPTs tractable. Indeed, for many threat vectors, each child node may have quite a few more than three parent nodes. A similar performance speed optimization problem may be encountered for modeling cybersecurity threats through MLN KB. As the domain of discourse for predicates may become unacceptably large for quick convergence and training, parallelization [144] of the inference process or preprocessing of the data [145] will be required in more unrestricted threat scenarios.

8.7.4   Model Optimization

Due to the complexity of the model and potentially multiple faceted causal relationships, it might become computationally expensive and time-consuming for an SRL model to converge to a solution. Additionally, since data-driven secure lifecycle management approaches in the semiconductor industry are still in their infancy, there might be limited availability of labeled datasets. Therefore, a multi-factor optimization of the network would be needed considering the large state space, learning parameter constraints, and limitations of the enabling technologies applied. Design space exploration and hyperparameter tuning thus becomes a very significant challenge for more involved and complex threat models.

8.7.5   Model Upgradeability

Forward trust analysis is an indispensable cornerstone of the proposed DT functionality. For many novel threats reported in literature, the underlying causal relationships are known to an extent as soon as they are discovered. For example, in the first report of Spectre attacks, the authors clearly identified that they were exploiting speculative execution vulnerability in the hardware implementation of the processors. However, in case of many zero-day threats that are not discovered by researchers beforehand, the causal relationships may not be readily understood. In such a case, the core SRL model structure would need to be learnt from the data available. One example of how it can be done is the learning of Bayesian network structure through evolutionary algorithms [146].

8.7.6 Threat Mitigation

Although the proposed framework offers scalability to future threats through inherently expandable and updatable AI algorithms, it does not provide any provision for mitigation of threats by itself. Once the root cause has been identified, the appropriate countermeasure to be taken is left to the discretion of the defender. Countermeasures can be taken across both hardware and software layers of the computing stack. Implementing hardware countermeasures once it has been deployed is extremely difficult especially for ASIC platform. For FPGA platform, the uploaded design bitstream may be comparatively easily updated with proper measures although monetary cost of taking a system offline to do so is a great concern. Hardware patching [147] has been proposed recently although such efforts remain in infancy in terms of both adoption and usability. Embedded FPGA [148–151] and programmable hardware monitors [152] represent other alternatives which require further investigation.

Software patching remains the go to option for addressing hardware security concerns due to the relative ease of their implementation. For instance, even though both Meltdown and Spectre attacks exploit vulnerabilities present in hardware, the vast majority of defenses deployed infield have been software countermeasures. Effective implementation of threat mitigation of a wide range of hardware security concerns may have to be tackled through both hardware–software "co-upgrading" similar to hardware–software co-design and co-verification practices being proposed in literature. We plan to visit these questions in our future works on the proposed DT framework.

References

1. H. Khattri, N.K.V. Mangipudi, S. Mandujano, HSDL: a security development lifecycle for hardware technologies, in *2012 IEEE International Symposium on Hardware-Oriented Security and Trust* (IEEE, Piscataway, 2012), pp. 116–121
2. G. Dessouky, D. Gens, P. Haney, G. Persyn, A. Kanuparthi, H. Khattri, J.M. Fung, A.-R. Sadeghi, J. Rajendran, {HardFails}: insights into {Software-Exploitable} hardware bugs, in *28th USENIX Security Symposium (USENIX Security 19)* (2019), pp. 213–230
3. Md.S. Ul Islam Sami, F. Rahman, F. Farahmandi, A. Cron, M. Borza, M. Tehranipoor, End-to-end secure SoC lifecycle management, in *2021 58th ACM/IEEE Design Automation Conference (DAC)* (IEEE, Piscataway, 2021), pp. 1295–1298
4. S. Ray, E. Peeters, M.M. Tehranipoor, S. Bhunia, System-on-chip platform security assurance: architecture and validation. Proc. IEEE **106**(1), 21–37 (2017)
5. M.M. Hossain, A. Vafaei, K.Z. Azar, F. Rahman, F. Farahmandi, M. Tehranipoor, SoCFuzzer: SoC vulnerability detection using cost function enabled fuzz testing, in *2023 Design, Automation & Test in Europe Conference & Exhibition (DATE)* (IEEE, Piscataway, 2023), pp. 1–6
6. H. Al-Shaikh, A. Vafaei, M.Md. Mashahedur Rahman, K.Z. Azar, F. Rahman, F. Farahmandi, M. Tehranipoor, Sharpen: SoC security verification by hardware penetration test, in *Proceedings of the 28th Asia and South Pacific Design Automation Conference* (2023), pp. 579–584

7. X. Xu, F. Rahman, B. Shakya, A. Vassilev, D. Forte, M. Tehranipoor, Electronics supply chain integrity enabled by blockchain. ACM Trans. Des. Autom. Electron. Syst. **24**(3), 1–25 (2019)

8. D. Zhang, X. Wang, Md.T. Rahman, M. Tehranipoor, An on-chip dynamically obfuscated wrapper for protecting supply chain against IP and IC piracies. IEEE Trans. Very Large Scale Integr. VLSI Syst. **26**(11), 2456–2469 (2018)

9. D. Jones, C. Snider, A. Nassehi, J. Yon, B. Hicks, Characterising the Digital Twin: A systematic literature review. CIRP J. Manuf. Sci. Technol. **29**, 36–52 (2020)

10. E. Glaessgen, D. Stargel, The digital twin paradigm for future NASA and US Air Force vehicles, in *53rd AIAA/ASME/ASCE/AHS/ASC Structures, Structural Dynamics and Materials Conference 20th AIAA/ASME/AHS Adaptive Structures Conference 14th AIAA* (2012), p. 1818

11. A. Fuller, Z. Fan, C. Day, C. Barlow, Digital twin: enabling technologies, challenges and open research. IEEE Access **8**, 108952–108971 (2020)

12. J. Bao, D. Guo, J. Li, J. Zhang, The modelling and operations for the digital twin in the context of manufacturing. Enterp. Inf. Syst. **13**(4), 534–556 (2019)

13. A.M. Madni, C.C. Madni, S.D. Lucero, Leveraging digital twin technology in model-based systems engineering. Systems **7**(1), 7 (2019)

14. G. Bachelor, E. Brusa, D. Ferretto, A. Mitschke, Model-based design of complex aeronautical systems through digital twin and thread concepts. IEEE Syst. J. **14**(2), 1568–1579 (2019)

15. G.S. Martinez, S. Sierla, T. Karhela, V. Vyatkin, Automatic generation of a simulation-based digital twin of an industrial process plant, in *IECON 2018-44th Annual Conference of the IEEE Industrial Electronics Society* (IEEE, Piscataway, 2018), pp. 3084–3089

16. M. Kritzler, J. Hodges, D. Yu, K. Garcia, H. Shukla, F. Michahelles, Digital companion for industry, in *Companion Proceedings of The 2019 World Wide Web Conference* (2019), pp. 663–667

17. Q. Qi, F. Tao, Digital twin and big data towards smart manufacturing and industry 4.0: 360 degree comparison. IEEE Access **6**, 3585–3593 (2018)

18. M. Liu, S. Fang, H. Dong, C. Xu, Review of digital twin about concepts, technologies, and industrial applications. J. Manuf. Syst. **58**, 346–361 (2021)

19. J. Wurm, Y. Jin, Y. Liu, S. Hu, K. Heffner, F. Rahman, M. Tehranipoor, Introduction to cyber-physical system security: a cross-layer perspective. IEEE Trans. Multi-Scale Comput. Syst. **3**(3), 215–227 (2016)

20. U. Guin, Q. Shi, D. Forte, M. Tehranipoor, FORTIS: a comprehensive solution for establishing forward trust for protecting IPs and ICs. ACM Trans. Des. Autom. Electron. Syst. **21**(4), 1–20 (2016)

21. A. Dabrowski, H. Hobel, J. Ullrich, K. Krombholz, E. Weippl, Towards a hardware Trojan detection cycle, in *2014 Ninth International Conference on Availability, Reliability and Security* (IEEE, Piscataway, 2014), pp. 287–294

22. M. Tehranipoor, H. Salmani, X. Zhang, *Integrated Circuit Authentication*, vol. 10 (Springer, Cham, 2014), pp. 978–3

23. S. Bhunia, M. Abramovici, D. Agrawal, P. Bradley, M.S. Hsiao, J. Plusquellic, M. Tehranipoor, Protection against hardware trojan attacks: towards a comprehensive solution. IEEE Des. Test **30**(3), 6–17 (2013)

24. S. Bhunia, M. Tehranipoor, *Hardware Security: A Hands-on Learning Approach* (Morgan Kaufmann, Burlington, 2018)

25. K. Xiao, D. Forte, Y. Jin, R. Karri, S. Bhunia, M. Tehranipoor, Hardware trojans: lessons learned after one decade of research. ACM Trans. Des. Autom. Electron. Syst. **22**(1), 1–23 (2016)

26. U. Guin, X. Zhang, D. Forte, M. Tehranipoor, Low-cost on-chip structures for combating die and IC recycling, in *2014 51st ACM/EDAC/IEEE Design Automation Conference (DAC)* (IEEE, Piscataway, 2014), pp. 1–6

27. U. Guin, K. Huang, D. DiMase, J.M. Carulli, M. Tehranipoor, Y. Makris, Counterfeit integrated circuits: a rising threat in the global semiconductor supply chain. Proc. IEEE **102**(8), 1207–1228 (2014)

28. S. Wang, J. Chen, M. Tehranipoor, Representative critical reliability paths for low-cost and accurate on-chip aging evaluation, in *Proceedings of the International Conference on Computer-Aided Design* (2012), pp. 736–741

29. U. Guin, D. DiMase, M. Tehranipoor, A comprehensive framework for counterfeit defect coverage analysis and detection assessment. J. Electron. Test. **30**(1), 25–40 (2014)

30. Y. Alkabani, F. Koushanfar, Active hardware metering for intellectual property protection and security, in *USENIX Security Symposium* (2007), pp. 291–306

31. K. Huang, Y. Liu, N. Korolija, J.M. Carulli, Y. Makris, Recycled IC detection based on statistical methods. IEEE Trans. Comput. Aided Des. Integr. Circuits Syst. **34**(6), 947–960 (2015)

32. J. Rajendran, A.M. Dhandayuthapany, V. Vedula, R. Karri, Formal security verification of third party intellectual property cores for information leakage, in *2016 29th International Conference on VLSI Design and 2016 15th International Conference on Embedded Systems (VLSID)* (IEEE, Piscataway, 2016), pp. 547–552

33. X. Guo, R.G. Dutta, J. He, M.M. Tehranipoor, Y. Jin, QIF-Verilog: quantitative information-flow based hardware description languages for pre-silicon security assessment, in *2019 IEEE International Symposium on Hardware Oriented Security and Trust (HOST)* (IEEE, Piscataway, 2019), pp. 91–100

34. H. Wang, H. Li, F. Rahman, M.M. Tehranipoor, F. Farahmandi, SoFI: security property-driven vulnerability assessments of ICs against fault-injection attacks, in *IEEE Transactions on Computer-Aided Design of Integrated Circuits and Systems* (2021)

35. A. Nahiyan, K. Xiao, K. Yang, Y. Jin, D. Forte, M. Tehranipoor, AVFSM: a framework for identifying and mitigating vulnerabilities in FSMs, in *2016 53nd ACM/EDAC/IEEE Design Automation Conference (DAC)* (IEEE, Piscataway, 2016), pp. 1–6

36. A. Nahiyan, F. Farahmandi, P. Mishra, D. Forte, M. Tehranipoor, Security-aware FSM design flow for identifying and mitigating vulnerabilities to fault attacks. IEEE Trans. Comput. Aided Des. Integr. Circuits Syst. **38**(6), 1003–1016 (2018)

37. J. Park, F. Rahman, A. Vassilev, D. Forte, M. Tehranipoor, Leveraging side-channel information for disassembly and security. ACM J. Emerging Technol. Comput. Syst. **16**(1), 1–21 (2019)

38. W. Shan, S. Zhang, J. Xu, M. Lu, L. Shi, J. Yang, Machine learning assisted side-channel-attack countermeasure and its application on a 28-nm AES circuit. IEEE J. Solid-State Circuits **55**(3), 794–804 (2019)

39. A. Nahiyan, J. Park, M. He, Y. Iskander, F. Farahmandi, D. Forte, M. Tehranipoor, Script: a cad framework for power side-channel vulnerability assessment using information flow tracking and pattern generation. ACM Trans. Des. Autom. Electron. Syst. **25**(3), 1–27 (2020)

40. M.M. Tehranipoor, U. Guin, D. Forte, Counterfeit test coverage: an assessment of current counterfeit detection methods, in *Counterfeit Integrated Circuits* (Springer, Berlin, 2015), pp. 109–131

41. Y. Shiyanovskii, F. Wolff, A. Rajendran, C. Papachristou, D. Weyer, W. Clay, Process reliability based trojans through NBTI and HCI effects, in *2010 NASA/ESA Conference on Adaptive Hardware and Systems* (IEEE, Piscataway, 2010), pp. 215–222

42. R. Kashyap, Silicon lifecycle management (SLM) with in-chip monitoring, in *2021 IEEE International Reliability Physics Symposium (IRPS)* (IEEE, Piscataway, 2021), pp. 1–4

43. F. Khalid, S.R. Hasan, O. Hasan, M. Shafique, SIMCom: statistical sniffing of inter-module communications for runtime hardware trojan detection. Microprocess. Microsyst. **77**, 103122 (2020)

44. L. Lin, W. Burleson, C. Paar, MOLES: malicious offchip leakage enabled by side-channels, in *2009 IEEE/ACM International Conference on Computer-Aided Design-Digest of Technical Papers* (IEEE, Piscataway, 2009), pp. 117–122

45. K.Z. Azar, H.M. Kamali, F. Farahmandi, M.H. Tehranipoor, *Understanding Logic Locking*, (Springer, 2024)

46. H.M. Kamali, K.Z. Azar, F. Farahmandi, M. Tehranipoor, Advances in logic locking: past, present, and prospects, in *Cryptology ePrint Archive* (2022)

47. K.Z. Azar, F. Farahmand, H.M. Kamali, S. Roshanisefat, H. Homayoun, W. Diehl, K. Gaj, A. Sasan, COMA: communication and obfuscation management architecture, in *Int'l Symposium on Research in Attacks, Intrusions and Defenses (RAID)* (2019), pp. 181–195
48. F. Tao, H. Zhang, A. Liu, A.Y.C. Nee, Digital twin in industry: state-of-the-art. IEEE Trans. Ind. Inf. **15**(4), 2405–2415 (2018)
49. Y. Xu, Y. Sun, X. Liu, Y. Zheng, A digital-twin-assisted fault diagnosis using deep transfer learning. IEEE Access **7**, 19990–19999 (2019)
50. P. Jain, J. Poon, J.P. Singh, C. Spanos, S.R. Sanders, S.K. Panda, A digital twin approach for fault diagnosis in dis-tributed photovoltaic systems. IEEE Trans. Power Electron. **35**(1), 940–956 (2019)
51. J. Sleuters, Y. Li, J. Verriet, M. Velikova, R. Doornbos, A digital twin method for automated behavior analysis of largescale distributed IoT systems, in *2019 14th Annual Conference System of Systems Engineering (SoSE)* (IEEE, Piscataway, 2019), pp. 7–12
52. R.G. Alves, G. Souza, R.F. Maia, A.L.H. Tran, C. Kamienski, J.-P. Soininen, P.T. Aquino, F. Lima, A digital twin for smart farming, in *2019 IEEE Global Humanitarian Technology Conference (GHTC)* (IEEE, Piscataway, 2019), pp. 1–4
53. Y. Tchana, G. Ducellier, S. Remy, Designing a unique Digital Twin for linear infrastructures lifecycle management. Proc. CIRP **84**, 545–549 (2019)
54. A. Pokhrel, V. Katta, R. Colomo-Palacios, Digital twin for cybersecurity incident prediction: a multivocal literature review, in *Proceedings of the IEEE/ACM 42nd International Conference on Software Engineering Workshops* (2020), pp. 671–678
55. M. Eckhart, A. Ekelhart, A specification-based state replication approach for digital twins, in *CPS-SPC18. Toronto, Canada: Association for Computing Machinery* (2018), pp. 36–47. ISBN: 9781450359924. https://doi.org/10.1145/3264888.3264892
56. R. Bitton, T. Gluck, O. Stan, M. Inokuchi, Y. Ohta, Y. Yamada, T. Yagyu, Y. Elovici, A. Shabtai, Deriving a cost-effective digital twin of an ICS to facilitate security evaluation, in *European Symposium on Research in Computer Security* (Springer, Berlin, 2018), pp. 533–554
57. X. Lou, Y. Guo, Y. Gao, K. Waedt, M. Parekh, An idea of using Digital Twin to perform the functional safety and cybersecurity analysis, in *INFORMATIK 2019: 50 Jahre Gesellschaft für Informatik–Informatik für Gesellschaft (Workshop-Beiträge)* (Gesellschaft für Informatik eV., Berlin, 2019)
58. A. Saad, S. Faddel, T. Youssef, O.A. Mohammed, On the implementation of IoT-based digital twin for networked microgrids resiliency against cyber attacks. IEEE Trans. Smart Grid **11**(6), 5138–5150 (2020)
59. Reliability, in *Heterogeneous Integration Roadmap*, ed. by P. Wesling (IEEE Electronics Packaging Society, Piscataway, 2021), Chap. 24, pp. 1–30. https://eps.ieee.org/images/files/HIR%5C_2021/ch24%5C_rel.pdf
60. H. Wang, S. Chen, Md.S. Ul Islam Sami, F. Rahman, M. Tehranipoor, Digital twin with a perspective from manufacturing industry, in *Emerging Topics in Hardware Security* (2021), pp. 27–59
61. E.C. Balta, D.M. Tilbury, K. Barton, A digital twin framework for performance monitoring and anomaly detection in fused deposition modeling, in *2019 IEEE 15th International Conference on Automation Science and Engineering (CASE)* (IEEE, Piscataway, 2019), pp. 823–829
62. C. Li, S. Mahadevan, Y. Ling, S. Choze, L. Wang, Dynamic Bayesian network for aircraft wing health monitoring digital twin. AIAA J. **55**(3), 930–941 (2017)
63. S. Kaewunruen, Q. Lian, Digital twin aided sustainability-based lifecycle management for railway turnout systems. J. Cleaner Prod. **228**, 1537–1551 (2019)
64. M. Rostami, F. Koushanfar, R. Karri, A primer on hardware security: models, methods, and metrics. Proc. IEEE **102**(8), 1283–1295 (2014)
65. N. Asadizanjani, M.T. Rahman, M. Tehranipoor, M.H. Tehranipoor, *Physical Assurance: For Electronic Devices and Systems* (Springer, Berlin, 2021)

66. M.T. Rahman, Q. Shi, S. Tajik, H. Shen, D.L. Woodard, M. Tehranipoor, N. Asadizanjani, Physical inspection & attacks: new frontier in hardware security, in *2018 IEEE 3rd International Verification and Security Workshop (IVSW)* (IEEE, Piscataway, 2018), pp. 93–102

67. S.E. Quadir, J. Chen, D. Forte, N. Asadizanjani, S. Shahbazmohamadi, L. Wang, J. Chandy, M. Tehranipoor, A survey on chip to system reverse engineering. ACM J. Emerging Technol. Comput. Syst. **13**(1), 1–34 (2016)

68. U.J. Botero, R. Wilson, H. Lu, M.T. Rahman, M.A. Mallaiyan, F. Ganji, N. Asadizanjani, M.M. Tehranipoor, D.L. Woodard, D. Forte, Hardware trust and assurance through reverse engineering: a tutorial and outlook from image analysis and machine learning perspectives. ACM J. Emerging Technol. Comput. Syst. **17**(4), 1–53 (2021)

69. M. Sharma, *AMD Hardware Security Tricks Can Be Bypassed with a Shock of Electricity*, ed. by Techradar.com (2021). https://www.techradar.com/news/amd-hardware-security-tricks-can-be-bypassed-with-a-shock-of-electricity

70. D. Goodin, *Trusted Platform Module Security Defeated in 30 Minutes, No Soldering Required*, ed. by arstechnica.com (2021). https://arstechnica.com/gadgets/2021/08/how-to-go-from-stolen-pc-to-network-intrusion-in-30-minutes/

71. A.P. Fournaris, L.P. Fraile, O. Koufopavlou, Exploiting hardware vulnerabilities to attack embedded system devices: a survey of potent microarchitectural attacks. Electronics **6**(3), 52 (2017)

72. S.T. King, J. Tucek, A. Cozzie, C. Grier, W. Jiang, Y. Zhou, Designing and implementing malicious hardware. Leet **8**, 1–8 (2008)

73. D. Lee, D. Jung, I.T. Fang, C.-C. Tsai, R.A. Popa, An off-chip attack on hardware enclaves via the memory bus, in *29th {USENIX} Security Symposium ({USENIX} Security 20)* (2020)

74. J. Grand, Practical secure hardware design for embedded systems, in *Proceedings of the 2004 Embedded Systems Conference*, San Francisco (2004)

75. M.S. Rahman, A. Nahiyan, F. Rahman, S. Fazzari, K. Plaks, F. Farahmandi, D. Forte, M. Tehranipoor, Security assessment of dynamically obfuscated scan chain against oracle-guided attacks. ACM Trans. Des. Autom. Electron. Syst. **26**(4), 1–27 (2021)

76. B. Ahmed, Md.K. Bepary, N. Pundir, M. Borza, O. Raikhman, A. Garg, D. Donchin, A. Cron, M.A. Abdel-Moneum, F. Farahmandi, et al. Quantifiable assurance: from IPs to platforms, in *Cryptology ePrint Archive* (2021)

77. IEEE standard for verilog hardware description language, in *IEEE Std 1364-2005 (Revision of IEEE Std 1364-2001)* (2006), pp. 325–348. https://doi.org/10.1109/IEEESTD.2006.99495

78. D. Forte, S. Bhunia, M. Tehranipoor, *Hardware Protection through Obfuscation* (Springer, Berlin, 2017)

79. H.S. Choo, C.Y. Ooi, M. Inoue, N. Ismail, M. Moghbel, C.H. Kok, Register-transfer-level features for machine-learning-based hardware trojan detection. IEICE Trans. Fundam. Electron. Commun. Comput. Sci. **103**(2), 502–509 (2020)

80. M. He, J. Park, A. Nahiyan, A. Vassilev, Y. Jin, M. Tehranipoor, RTL-PSC: automated power side-channel leakage assessment at register-transfer level, in *2019 IEEE 37th VLSI Test Symposium (VTS)* (IEEE, Piscataway, 2019), pp. 1–6

81. F. Fallah, S. Devadas, K. Keutzer, OCCOM-efficient computation of observability-based code coverage metrics for functional verification. IEEE Trans. Comput. Aided Des. Integr. Circuits Syst. **20**(8), 1003–1015 (2001)

82. Y. Serrestou, V. Beroulle, C. Robach, Functional verification of RTL designs driven by mutation testing metrics, in *10th Euromicro Conference on Digital System Design Architectures, Methods and Tools (DSD 2007)* (IEEE, Piscataway, 2007), pp. 222–227

83. L.H. Goldstein, E.L. Thigpen, SCOAP: sandia controllability/observability analysis program, in *Proceedings of the 17th Design Automation Conference* (1980), pp. 190–196

84. A. Waksman, M. Suozzo, S. Sethumadhavan, FANCI: identification of stealthy malicious logic using boolean functional analysis, in *Proceedings of the 2013 ACM SIGSAC Conference on Computer & Communications Security* (2013), pp. 697–708

85. R.S. Chakraborty, F. Wolff, S. Paul, C. Papachristou, S. Bhunia, MERO: a statistical approach for hardware Trojan detection, in *International Workshop on Cryptographic Hardware and Embedded Systems* (Springer, Berlin, 2009), pp. 396–410

86. J. Cruz, F. Farahmandi, A. Ahmed, P. Mishra, Hardware Trojan detection using ATPG and model checking, in *2018 31st International Conference on VLSI Design and 2018 17th International Conference on Embedded Systems (VLSID)* (IEEE, Piscataway, 2018), pp. 91–96

87. A. Nahiyan, M. Tehranipoor, Code coverage analysis for IP trust verification, in *Hardware IP Security and Trust* (Springer, Piscataway, 2017), pp. 53–72

88. Q. Xu, S. Chen, Fast thermal analysis for fixed-outline 3D floorplanning. Integration **59**, 157–167 (2017)

89. J. Cong, G. Nataneli, M. Romesis, J.R. Shinnerl, An area-optimality study of floorplanning, in *Proceedings of the 2004 International Symposium on Physical Design* (2004), pp. 78–83

90. H. Ma, J. He, Y. Liu, L. Liu, Y. Zhao, Y. Jin, Security-driven placement and routing tools for electromagnetic side-channel protection. IEEE Trans. Comput. Aided Des. Integr. Circuits Syst. **40**(6), 1077–1089 (2020)

91. K. Xiao, M. Tehranipoor, BISA: built-in self-authentication for preventing hardware Trojan insertion, in *2013 IEEE International Symposium on Hardware-Oriented Security and Trust (HOST)* (IEEE, Piscataway, 2013), pp. 45–50

92. H. Salmani, M. Tehranipoor, J. Plusquellic, New design strategy for improving hardware Trojan detection and reducing Trojan activation time, in *2009 IEEE International Workshop on Hardware-Oriented Security and Trust* (IEEE, Piscataway, 2009), pp. 66–73

93. C. Mack, *Fundamental Principles of Optical Lithography: The Science of Microfabrication* (John Wiley & Sons, Hoboken, 2008)

94. G. May, S. Sze, *Fundamentals of Semiconductor Fabrication*, 1st edn. (Wiley, Hoboken, 2003). ISBN: 0-471-23279-3

95. S. Campbell, *Fabrication Engineering at the Micro and Nanoscale*, 3rd edn. (Oxford University Press, Oxford, 2008), pp. 74–106

96. N.B. Cobb, E.Y. Sahouria, Hierarchical GDSII-based fracturing and job deck system, in *21st Annual BACUS Symposium on Photomask Technology*, vol. 4562 (International Society for Optics and Photonics, Bellingham, 2002), pp. 734–742

97. F.E. Abboud, M. Asturias, M. Chandramouli, Mask data processing in the era of multibeam writers, in *BACUS News* (2015). https://spie.org/Documents/Membership/BacusNewsletters/BACUS-Newsletter-January-2015.pdf

98. S.F. Schulze, P. LaCour, P.D. Buck, GDS-based mask data preparation flow: data volume containment by hierarchical data processing, in *22nd Annual BACUS Symposium on Photomask Technology*, vol. 4889 (International Society for Optics and Photonics, Bellingham, 2002), pp. 104–114

99. M.J. Hampden-Smith, T.T. Kodas, Chemical vapor deposition of metals: Part 1. An overview of CVD processes. Chem. Vap. Deposition **1**(1), 8–23 (1995)

100. W.L. Smith, A. Rosencwaig, D.L. Willenborg, Ion implant monitoring with thermal wave technology. Appl. Phys. Lett. **47**(6), 584–586 (1985)

101. G. Neubauer, A. Erickson, C.C. Williams, J.J. Kopanski, M. Rodgers, D. Adderton, Two-dimensional scanning capacitance microscopy measurements of cross-sectioned very large scale integration test structures. J. Vac. Sci. Technol. B Microelectron. Nanometer Struct. Process. Measure. Phenom. **14**(1), 426–432 (1996)

102. W. Vandervorst, P. Eyben, S. Callewaert, T. Hantschel, N. Duhayon, M. Xu, T. Trenkler, T. Clarysse, Towards routine, quantitative two-dimensional carrier profiling with scanning spreading resistance microscopy, in *AIP Conference Proceedings*, vol. 550 (American Institute of Physics, College Park, 2001), pp. 613–619

103. M. Bushnell, V. Agrawal, *Essentials of Electronic Testing for Digital, Memory and Mixed-Signal VLSI Circuits*, vol. 17 (Springer Science & Business Media, Berlin, 2004)

104. A. Chen, R.H.-Y. Lo, *Semiconductor Packaging: Materials Interaction and Reliability* (CRC Press, Boca Raton, 2017)

105. M. Nourani, M. Tehranipoor, N. Ahmed, Low-transition test pattern generation for BIST-based applications. IEEE Trans. Comput. **57**(3), 303–315 (2008)

106. M.H. Tehranipour, N. Ahmed, M. Nourani, Testing SoC interconnects for signal integrity using boundary scan, in *Proceedings. 21st VLSI Test Symposium, 2003* (IEEE, Piscataway, 2003), pp. 158–163

107. M. Tehranipoor, F. Koushanfar, A survey of hardware trojan taxonomy and detection. IEEE Desig. Test Comput. **27**(1), 10–25 (2010)

108. V.C. Patil, A. Vijayakumar, S. Kundu, Manufacturer turned attacker: dangers of stealthy trojans via threshold voltage manipulation, in *2017 IEEE North Atlantic Test Workshop (NATW)* (IEEE, Piscataway, 2017), pp. 1–6

109. G.K. Contreras, Md.T. Rahman, M. Tehranipoor, Secure split-test for preventing IC piracy by untrusted foundry and assembly, in *2013 IEEE International Symposium on Defect and Fault Tolerance in VLSI and Nanotechnology Systems (DFTS)* (IEEE, Piscataway, 2013), pp. 196–203

110. K. Huang, J.M. Carulli, Y. Makris, Parametric counterfeit IC detection via support vector machines, in *2012 IEEE International Symposium on Defect and Fault Tolerance in VLSI and Nanotechnology Systems (DFT)* (IEEE, Piscataway, 2012), pp. 7–12

111. X. Zhang, K. Xiao, M. Tehranipoor, Path-delay fingerprinting for identification of recovered ICs, in *2012 IEEE International Symposium on Defect and Fault Tolerance in VLSI and Nanotechnology Systems (DFT)* (IEEE, Piscataway, 2012), pp. 13–18

112. A. Nahiyan, M. Sadi, R. Vittal, G. Contreras, D. Forte, M. Tehranipoor, Hardware trojan detection through information flow security verification, in *2017 IEEE International Test Conference (ITC)* (IEEE, Piscataway, 2017), pp. 1–10

113. H. Al-Shaikh, M.B. Monjil, K.Z. Azar, F. Farahmandi, M. Tehranipoor, F. Rahman, QuardTropy: detecting and quantifying unauthorized information leakage in hardware designs using g-entropy, in *2023 IEEE International Symposium on Defect and Fault Tolerance in VLSI and Nanotechnology Systems (DFT)* (IEEE, Piscataway, 2023), pp. 1–6

114. N. Farzana, M.M. Hossain, K.Z. Azar, F. Farahmandi, M. Tehranipoor, FormalFuzzer: formal verification assisted fuzz testing for SoC vulnerability detection, in *Asia and South Pacific Design Automation Conference (ASP-DAC)* (IEEE, Piscataway, 2024), pp. 1–6

115. W. Hu, D. Mu, J. Oberg, B. Mao, M. Tiwari, T. Sherwood, R. Kastner, Gate-level information flow tracking for security lattices. ACM Trans. Des. Autom. Electron. Syst. **20**(1), 1–25 (2014)

116. M.M. Hossain, K.Z. Azar, F. Farahmandi, M. Tehranipoor, TaintFuzzer: SoC security verification using taint inference-enabled fuzzing, in *International Conference On Computer Aided Design (ICCAD)* (IEEE, Piscataway, 2023), pp. 1–9

117. Y. Liu, Y. Jin, A. Nosratinia, Y. Makris, Silicon demonstration of hardware Trojan design and detection in wireless cryptographic ICs. IEEE Trans. Very Large Scale Integr. VLSI Syst. **25**(4), 1506–1519 (2016)

118. C. Kison, O.M. Awad, M. Fyrbiak, C. Paar, Security implications of intentional capacitive crosstalk. IEEE Trans. Inf. Forensics Secur. **14**(12), 3246–3258 (2019)

119. S. Rizvi, *Handbook of Photomask Manufacturing Technology* (CRC Press, Boca Raton, 2018)

120. K. Takahashi, M. Osawa, M. Sato, H. Arimoto, K. Ogino, H. Hoshino, Y. Machida, Proximity effect correction using pattern shape modification and area density map. J. Vac. Sci. Technol. B Microelectron. Nanometer Struct. Process. Measure. Phenom. **18**(6), 3150–3157 (2000)

121. K. Hasegawa, M. Yanagisawa, N. Togawa, Trojan-feature extraction at gate-level netlists and its application to hardware-Trojan detection using random forest classifier, in *2017 IEEE International Symposium on Circuits and Systems (ISCAS)* (IEEE, Piscataway, 2017), pp. 1–4

122. M. Sharma, Standard test data format specification (2021). http://www.kanwoda.com/wp-content/uploads/2015/05/std-spec.pdf

123. U. Guin, D. Forte, M. Tehranipoor, Anti-counterfeit techniques: from design to resign, in: *2013 14th International Workshop on Microprocessor Test and Verification* (IEEE, Piscataway, 2013), pp. 89–94

124. D. Koller, N. Friedman, S. Džeroski, C. Sutton, A. Mc-Callum, A. Pfeffer, P. Abbeel, M.-F. Wong, C. Meek, J. Neville, et al., *Introduction to Statistical Relational Learning* (MIT Press, Cambridge, 2007)

125. H. Guo, W. Hsu, A survey of algorithms for real-time Bayesian network inference, in *Join Workshop on Real Time Decision Support and Diagnosis Systems* (2002)
126. S.H. Chen, C.A. Pollino, Good practice in Bayesian network modelling. Environ. Model. Software **37**, 134–145 (2012)
127. D. Heckerman, D. Geiger, D.M. Chickering, Learning Bayesian networks: the combination of knowledge and statistical data. Mach. Learn. **20**(3), 197–243 (1995)
128. L.R. Rabiner, A tutorial on hidden Markov models and selected applications in speech recognition. Proc. IEEE **77**(2), 257–286 (1989)
129. L. Rabiner, B. Juang, An introduction to hidden Markov models. IEEE ASSP Mag. **3**(1), 4–16 (1986)
130. G.D. Forney, The viterbi algorithm. Proc. IEEE **61**(3), 268–278 (1973)
131. M. Richardson, P. Domingos, Markov logic networks. Mach. Learn. **62**(1–2), 107–136 (2006)
132. D. Lowd, P. Domingos, Efficient weight learning for Markov logic networks, in *European Conference on Principles of Data Mining and Knowledge Discovery* (Springer, Berlin, 2007), pp. 200–211
133. S. Skorobogatov, C. Woods, Breakthrough silicon scanning discovers backdoor in military chip, in *International Workshop on Cryptographic Hardware and Embedded Systems* (Springer, Berlin, 2012), pp. 23– 40
134. N. Laptev, S. Amizadeh, I. Flint, Generic and scalable framework for automated time-series anomaly detection, in *Proceedings of the 21th ACM SIGKDD International Conference on Knowledge Discovery and Data Mining* (2015), pp. 1939–1947
135. S. Shahbazmohamadi, D. Forte, M. Tehranipoor, Advanced physical inspection methods for counterfeit IC detection, in *ISTFA 2014: Conference Proceedings from the 40th International Symposium for Testing and Failure Analysis* (ASM International, Detroit, 2014), p. 55
136. M. Baygin, M. Karakose, A. Sarimaden, A. Erhan, Machine vision based defect detection approach using image processing, in *2017 International Artificial Intelligence and Data Processing Symposium (IDAP)* (IEEE, Piscataway, 2017), pp. 1–5
137. P. Ghosh, R.S. Chakraborty, Recycled and remarked counterfeit integrated circuit detection by image-processing-based package texture and indent analysis. IEEE Trans. Ind. Inf. **15**(4), 1966–1974 (2018)
138. B. Shakya, T. He, H. Salmani, D. Forte, S. Bhunia, M. Tehranipoor, Benchmarking of hardware trojans and maliciously affected circuits. J. Hardware Syst. Secur. **1**(1), 85–102 (2017)
139. H. Salmani, M. Tehranipoor, R. Karri, On design vulnerability analysis and trust benchmarks development, in *2013 IEEE 31st International Conference on Computer Design (ICCD)* (IEEE, Piscataway, 2013), pp. 471–474
140. X. Meng, S. Kundu, A.K. Kanuparthi, K. Basu, RTL-ConTest: concolic testing on RTL for detecting security vulnerabilities. IEEE Trans. Comput. Aided Des. Integr. Circuits Syst. **41**(3), 466–477 (2021)
141. A. Ardeshiricham, W. Hu, J. Marxen, R. Kastner, Register transfer level information flow tracking for provably secure hardware design, in *Design, Automation & Test in Europe Conference & Exhibition (DATE), 2017* (IEEE, Piscataway, 2017), pp. 1691–1696
142. N. Vashistha, M.T. Rahman, H. Shen, D.L. Woodard, N. Asadizanjani, M. Tehranipoor, Detecting hardware trojans inserted by untrusted foundry using physical inspection and advanced image processing. J. Hardware Syst. Secur. **2**(4), 333–344 (2018)
143. B.G. Marcot, J.D. Steventon, G.D. Sutherland, R.K. McCann, Guidelines for developing and updating Bayesian belief networks applied to ecological modeling and conservation. Can. J. For. Res. **36**(12), 3063–3074 (2006)
144. K. Beedkar, L.D. Corro, R. Gemulla, Fully parallel inference in markov logic networks, in *Datenbanksysteme für Business, Technologie und Web (BTW) 2026* (2013)
145. J. Shavlik, S. Natarajan, Speeding up inference in Markov logic networks by preprocessing to reduce the size of the resulting grounded network, in *Twenty-First International Joint Conference on Artificial Intelligence* (2009)

146. R. Daly, Q. Shen, Learning Bayesian network equivalence classes with ant colony optimization. J. Artif. Intell. Res. **35**, 391–447 (2009)
147. A.P.D. Nath, S. Ray, A. Basak, S. Bhunia, System-on-chip security architecture and CAD framework for hardware patch, in *2018 23rd Asia and South Pacific Design Automation Conference (ASP-DAC)* (IEEE, Piscataway, 2018), pp. 733–738
148. T. Beyrouthy, A. Razafindraibe, L. Fesquet, M. Renaudin, S. Chaudhuri, S. Guilley, P. Hoogvorst, J.-L. Danger, A novel asynchronous e-FPGA architecture for security applications, in *2007 International Conference on Field-Programmable Technology* (IEEE, Piscataway, 2007), pp. 369–372
149. S.K. Saha, C. Bobda, FPGA accelerated embedded system security through hardware isolation, in *2020 Asian Hardware Oriented Security and Trust Symposium (AsianHOST)* (IEEE, Piscataway, 2020), pp. 1–6
150. M.Md. Mashahedur Rahman, S. Tarek, K.Z. Azar, F. Farahmandi, EnSAFe: enabling sustainable SoC security auditing using eFPGA-based accelerators, in *2023 IEEE International Symposium on Defect and Fault Tolerance in VLSI and Nanotechnology Systems (DFT)* (IEEE, Piscataway, 2023), pp. 1–6
151. M.Md. Mashahedur Rahman, S. Tarek, K.Z. Azar, M. Tehranipoor, F. Farahmandi, Efficient SoC security monitoring: quality attributes and potential solutions. IEEE Design & Test, **early access**(1), pp. 1–6 (2023)
152. L. Delshadtehrani, S. Canakci, B. Zhou, S. Eldridge, A. Joshi, M. Egele, Phmon: a programmable hardware monitor and its security use cases, in *29th {USENIX} Security Symposium ({USENIX} Security 20)* (2020), pp. 807–824
153. K. Zhu, C. Wen, A. A. Aljarb, F. Xue, and X. Xu, V. Tung, X. A. Zhang, N. Husam, M. Lanza, The development of integrated circuits based on two-dimensional materials Nature Electronics, (Nature Publishing Group UK London, 2021) **4**(11), 775–785

Chapter 9
Secure Physical Design

9.1 Introduction

The worldwide semiconductor market has been growing due to the increase in demand for smartphones, 5G wireless devices, wearable devices, gaming, autonomous systems, servers, and artificial intelligence (AI)-based computing. Research firm International Data Corporation (IDC) expects the market to grow by 17.3% in 2021 versus 10.8% in 2020 [1]. To satisfy time-to-market in the semiconductor industry as well as the complex specification of an SoC, the reuse of intellectual property (IPs) and the global semiconductor supply chain are inevitable. Due to globalization, multiple parties are involved in the IC design. As a result, adversary can introduce malicious IPs in the design flow. This can lead to unintentionally serious hardware security breaches [2], such as asset leakage [3, 4], hardware trojan induced confidentiality and integrity violations [5–14], side-channel leakage [15–19], and fault injection vulnerability [20–25]. Apart from these, counterfeiting and recycling of integrated circuits have become a significant piracy threat in microelectronics supply chain [12, 26–35].

With the increase in complexity of the chip design and the involvement of multiple entities to create the design, it has become necessary to place security checks at all levels of the design cycle. Much in the literature has tried to address hardware security issues and countermeasures, starting from high-level synthesis to gate-level synthesis [36–41]. However, little work has been invested in security verification at the physical design stage to ensure the design is protected against vulnerabilities that take advantage at cell placement and wire routing. A secure gate-level netlist may become insecure after the generation of its physical layout. This is because the physical information (i.e., cell placement, power vias, power distribution network, placement of decaps, timing criticality of paths, obscurity of cells by wires in metal lines, etc.) is not available during the logic synthesis phase. Hence, making security sign-off during and after physical design[1] a necessity.

Over the past several years, hardware security community has concentrated mostly on securing logic designs at a higher level of abstraction. Suppose we have obtained a secure gate-level netlist after applying all security measures in a synthesized netlist. Subsequently, when the placement and routing (P&R) tool generates the layout of the chip, many physical information becomes available. The concern is: can an adversary utilize this physical information to extract vital information from the packaged chip, i.e., cause confidentiality violation? Or can an adversary manipulate logical values in the circuit to cause integrity violations? We have demonstrated a few such vulnerabilities in this chapter. That shows such questions are of major concerns that need to be carefully addressed. Therefore, we need to have a secure physical layout of the chip, free from any physical vulnerabilities, from which any information leakage is not possible. This is called *secure physical design* (SEPHYD). Overall, objective of this chapter is to demonstrate a perspective to identify physical characteristics that are related to security vulnerabilities and use those physical characteristics to guide the physical design process to ensure security sign-off.

This chapter will answer the following questions:

- What vulnerabilities are possible at the physical layout?
- How to identify vulnerabilities at the layout level?
- What countermeasure can be applied to establish security-aware physical design flow?
- What are the challenges concerning secure physical design?

9.2   Motivating Example

An integrated circuit can be made secure by applying countermeasures at different levels of abstractions, i.e., RTL, gate, and physical design levels, as shown in Fig. 9.1. The extent of security can be ensured if we have a security metric to evaluate the vulnerabilities of the final physical design, as shown in Case IV of Fig. 9.1. The usual practice of making secure integrated circuit possesses to apply the countermeasures at RTL or gate level. We already discussed in the previous section that applying only RTL level/gate-level countermeasure does not make the packaged chip secure, as many physical information is not available at the high-level functional design. Further, sometime situations may arise that make it difficult to apply RTL or gate-level countermeasures. It is also possible that logic designers have already transferred the design (after RTL and gate-level sign-off) to physical design team without applying the RTL/gate-level countermeasures. In such

---

[1] In this chapter, we have used the term "physical design" and "layout" interchangeably.

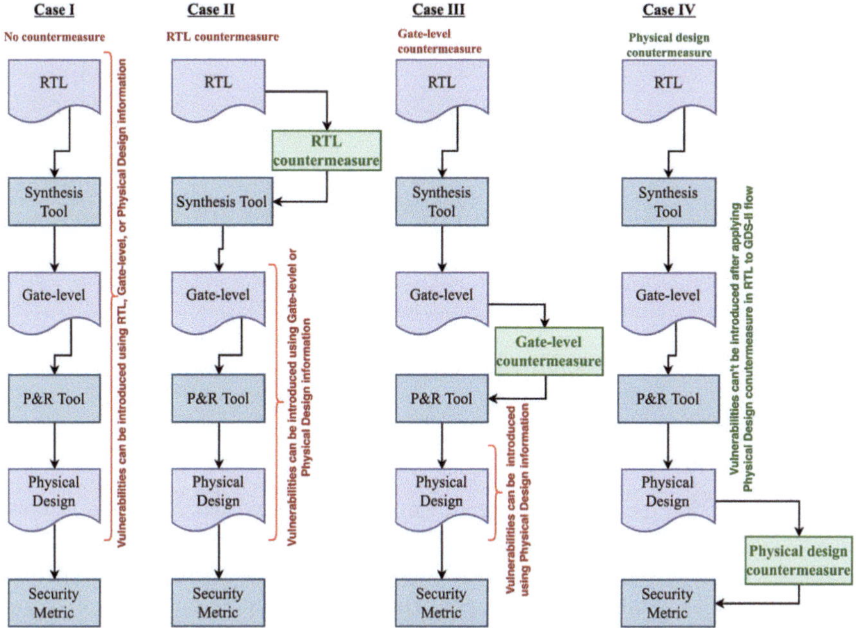

Fig. 9.1 Security metric evaluations in physical design level with various levels of countermeasures applied in RTL to GDS-II Flow

a situation, having a physical design level countermeasure becomes very essential. Therefore, irrespective of the countermeasures applied at RTL level/gate level or not, it is necessary to apply physical design countermeasures in order to obtain a secure chip package. Considering the above situation as our motivation, we discuss various aspects of realizing secure physical design in this chapter.

9.3 Preliminaries

9.3.1 Physical Design

The physical design phase starts with a synthesized RTL netlist as input, also known as a gate-level netlist. The physical design steps majorly consist of *Floorplanning, Placement,* and *Routing.* In the floorplanning step, estimated positions of macros and standard cells are determined. In this step, an on-chip power grid network is also generated. Once the initial positions of the blocks are determined, then it goes to the placement stage. Again, placement stage follows three sub-steps: *Global Placement, Legalization,* and *Detailed Placement.* In global placement, blocks are loosely placed. In the legalization stage, checks for overlapping blocks are performed.

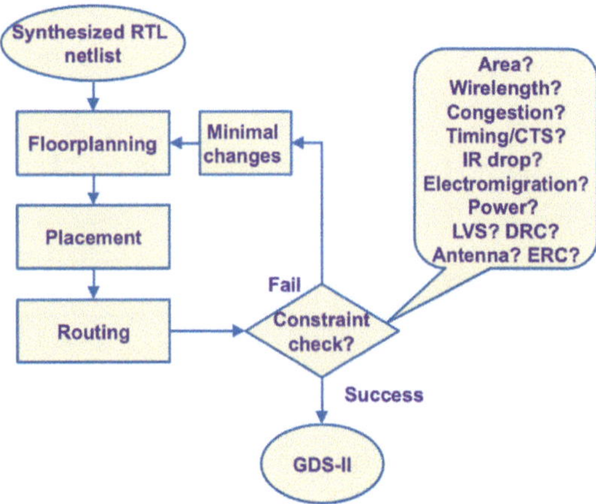

Fig. 9.2 Traditional physical design flow

Finally, we obtain the final placement of macros and standard cells in the detailed placement stage. The primary objective of the physical design stage is to build the geometrical and physical layout of the design with the exact specifications as defined in the RTL phase. Simultaneously, optimizing the area, power, and other objectives or constraints as shown in Fig. 9.2 are objectives of the physical design stage. There exist many other sub-steps of physical design stage, which include *Power Planning, Clock Tree Synthesis* (CTS), *Equivalence Check, Parasitic Extraction*, and several sign-off stages. These include *IR drop sign-off, timing sign-off, and physical sign-off.* In the power planning phase, the power grid is designed and initial IR/Electromigration sign-off is conducted. In CTS step, it is ensured that the clock reaches evenly in all the sequential elements of the design. The objective of the CTS step is to minimize clock skew and insertion delay. *Layout Versus Schematic* (LVS) is the process where equivalence logical connections and physical layout connections are checked. *Design Rule Checking* (DRC) is the process that verifies if the design is compatible with the foundry fabrication rules. In *Electrical Rule Checking* (ERC), a design is checked for substrate areas to verify proper contacts and spacing, which ensures correct power and ground connections. We have shown each of these sub-steps in Fig. 9.2 as a constraint check step. However, in practice, these checks are performed chronologically at a specific defined stage. Finally, when each check is successful, we receive the final layout in *Graphic Design System* (GDS)-II format. This GDS-II is sent to foundry for fabricating the layout into silicon.

9.3.2 Related Work

In the last two decades, there have been several works to devise efficient physical design algorithms that can produce cost-effective and fast time-to-market design solutions [42]. However, it has been observed that some vulnerabilities still exist even after the security sign-off in the RTL phase of the design, as many physical parameters are not present in the RTL design steps. Therefore, evaluating the security metrics in the physical layout phase has become more critical to produce trustworthy physical designs of an integrated circuit, free from any vulnerabilities. In this section, previous works related to vulnerability and security in the physical design are discussed. There are few works in literature that explores side-channel and fault injection vulnerabilities in the physical design stage. Firstly, we describe works related to side-channel vulnerability assessment and fault injection vulnerabilities at the physical design level. Later on, we also present a brief literature survey on secure physical design-related works.

9.3.2.1 Works Related to Side-Channel Vulnerability Assessment in Physical Design

Side-channel vulnerability assessment at the physical design level is crucial because there is more information in the layout then there is at the RTL and gate level. For example, the capacitance and resistance of the wires between blocks and standard cells are exploited as additional sources of side-channel leakage at the physical layout level. Crosstalk and IR drop, which are the potential of such security leakages, are not available at the RTL. This is why leakage analysis at the layout level offers the highest accuracy among the levels of abstraction in the pre-silicon stage. Very few works have been published regarding side-channel leakage at the layout level. Authors in [43] investigated the impact of the physical layout on side-channel security. They co-simulated the analog power delivery network (PDN) with a digital logic core. They quantified the impact of different layout parasitics such as parasitic resistors, inductors, capacitors, and power supply buffers. By examining the impact of these layout parasitics, they provided a deeper insight into potential layout sources. However, one major limitation of this work is the scalability of the approach. The authors applied their analysis to a minimal design (AND2, XOR2) because of the lack of computational ability to analyze the larger design. Cnudde et al. [44] claimed that placement and routing cause information leakage in FPGA. Some works [45, 46] used fast SPICE-based simulations to evaluate side-channel leakage analysis at back-end stages. However, these works also suffer from scalability problems for a larger design. Unlike other works, Lin et al. [47] developed fast simulation methodology for the layout-based power-noise side-channel leakage analysis. Their tool can improve the simulation time by 110 times for a SoC design compared to VCD-based analysis.

9.3.2.2 Works Related to Fault Injection Attack

Several fault injection attacks in security-critical applications are shown in recent research. These applications include an embedded system [48, 49], microprocessor-

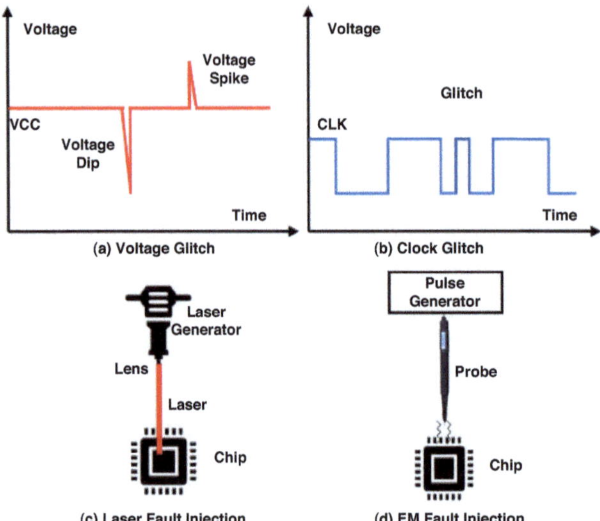

Fig. 9.3 Different fault injection techniques

based implementations [50], RFID tags [51], TRNG, SRAM, PLL, oscillators, and so on [52–55]. Among the global fault injection techniques, clock glitching, voltage glitching, and thermal glitching are widely used to inject exploitable faults in an Application-Specific Integrated Circuit (ASIC). Extensive research is performed to prove the feasibility and effectiveness of these global techniques [56–58]. In addition, local fault injection techniques are well researched, like laser/optical fault injection and EM fault injection [59–61]. Apart from noninvasive and semi-invasive attacks, faults can be injected using invasive techniques (e.g., FIB probing) proposed in literature [62, 63]. Clock glitching involves inserting glitches in the clock supply or disturbing the normal behavior to shorten the effective period of the clock, as shown in Fig. 9.3b. These clock glitches can cause setup and hold-time violations [64, 65], create metastability in the sequential elements, and inject timing faults. Since voltage starvation can also impact the clock and circuit timing, voltage and clock glitching attacks are combined to strengthen timing fault injection attacks. Voltage glitching involves a momentary power drop or spike in the supply voltage during the operation, as shown in Fig. 9.3a. These voltage spikes or drops increase the logic propagation delay, thus, causing timing violations [66, 67]. A voltage glitch can be caused by either disturbing the main power supply, creating a global effect, or by running a power-hungry circuit, like ring oscillators, or ROs, creating a localized voltage drop [68]. In recent years, remote fault injection attacks have been caused by disturbing the supply voltage remotely by using either software-based registers to control the supply voltage or by using ROs to cause a voltage drop in remote FPGAs [69–71]. Moreover, timing faults can be injected by inserting glitches in operating temperature. Overheating increases the scattering of charge carriers, decreasing mobility, and eventually affecting the logic propagation delay [66]. Optical/laser fault injection attacks use different wavelength laser/optics to inject transient faults from either the front or backside of the chip, as shown in Fig. 9.3c. To attack from the front-side, a higher wavelength laser ($1300 \, \mu$m) is used, whereas from the backside a near-infrared laser ($1064 \, \mu$m) is used (due to its lower absorption coefficient). The injected laser generates electron–hole pairs in the active region, which drift apart under the electric field's influence, resulting in transitory currents. These transient currents cause the transistors to conduct and thus cause the charging/discharging of the capacitive loads. Laser/optical fault injection generally requires depackaging of the chip to expose the die and thus qualifies as a semi-invasive attack. However, with the laser/optics, an attacker can target the precise locations and time of the laser injection, thus making it a powerful attack [59]. Similarly, EM fault injection attacks use electric or magnetic field flux to influence the normal functioning of the device, as shown in Fig. 9.3d. An electromagnetic field causes voltage and current fluctuations inside the device, leading to the faults [72].

9.3.2.3 Works Related to Security of Physical Design

Many works of literature address securing physical layout fabrication by split manufacturing [73–75], where front end of line (FEOL) interconnects are manufactured in untrusted foundries and back end of line (BEOL) interconnects are manufactured in trusted foundries to manufacture secure layouts and prevent counterfeit. However, split manufacturing does not ensure inherent layout security, as security is achieved by splitting the physical layout of an integrated circuit. There are few works which tried to obtain layout-level security assessment for hardware trojan-based and microprobing-based vulnerability [76–79]. There is a need to obtain security by design for physical layouts. There have been very few works that deal with this problem of obtaining secure physical layouts. We have listed these works in Table 9.1. In [80], authors have proposed intellectual property protection for VLSI physical design. In [81], authors have proposed a netlist scrambling-based technique to prevent reverse engineering. In [82–84], authors have proposed a physical design level hardware obfuscation technique to prevent reverse engineering. In [85], authors have proposed a differential fault analysis attack preventive physical design flow using floorplan heuristics. In [62, 86], authors have proposed a physical design flow considering anti-probing attacks. The flow utilizes internal shielding for protection. However, these approaches mentioned in Table 9.1 are not scalable due to the difficulty in implementing them separately, which makes the design flow costly. It is necessary to have a scalable IC design flow, where all security vulnerabilities in the physical design level are checked for a successful sign-off. Recent work [87] demonstrates a top-level overview of security closure of physical layouts. The work [87] mainly focuses on a case study of scanning and defending against Trojans and front-side probing attack. However, it does not

Table 9.1 Physical design security-related previous works

Work	Year	Physical vulnerabilities	Proposed countermeasure	Remarks
[80]	2007	IP protection	Robust IP scheme	Theoretically proved; MCNC benchmarks
[81]	2013	Reverse Engineering	Netlist Scrambling	w/ Random Scrambling; IWLS 2005 benchmarks
[74]	2015	Proximity Attack	Partitioning-based heuristic	Solution for 2.5D IC; ISCAS'85/ITC'99 benchmarks
[82]	2016	Reverse Engineering	Hardware Obfuscation	Device to Logic-level investigation
[75]	2017	Enhanced Security Split Manufacturing	Routing Perturbation	Split fabrication approach; ISCAS'85/ITC'99 benchmarks
[85]	2018	Differential Fault Attack	floorplan heuristic	Deals with localized faults; AES/Plantlet benchmarks
[88]	2019	Power Side-Channel	Gate reconfiguration	Divide & Conquer Approach; AES/SIMON/PRESENT benchmarks
[62]	2019	Probing Attack	Internal Shielding	Probing attack countermeasure; AES/DES benchmarks
[89, 90]	2021	Power/EM Side-Channel	Backside Power Grid	Distributed decap-based mitigation ; AES/ECC benchmarks
[91]	2021	EM Side-Channel	Modified Power Grid and Decap	Decap-based mitigation; AES/DES benchmarks
[87]	2021	Trojan and Front-side Probing	Scanning and Defending	DEFense countermeasure; AES, MIT-LL CEP benchmark

provide a comprehensive roadmap for obtaining a secure physical design. This work, however, presents a roadmap to obtain secure physical designs and considers various vulnerabilities at the physical design level.

9.4 Possible Vulnerabilities in Physical Design

Even after the security closure of gate-level synthesis, several vulnerabilities may arise utilizing physical design parameters at different steps. It has been observed that some vulnerabilities may arise due to the poor floorplanning stage. For example, while designing an SoC, an AES crypto module should be placed far away from the power supply pins in order to become less vulnerable to power side-channel attacks. Similarly, every step of physical design plays a crucial role in terms of security vulnerability. Any poor choices of floorplanning, placement, routing, or post-routing steps in physical design may increase the vulnerability level of the design. Based on the dependency of the vulnerabilities on various stages of physical design, we have broadly classified all the trust-hub physical vulnerabilities [92] into five categories which are floorplanning, placement (within a module), placement (SoC level), routing, and post-routing and listed it in Table 9.2. Further, the dependency of some of these vulnerabilities on the physical design stages are not well-defined. Most of the active invasive attacks can be prevented if we do security-aware floorplanning, placement, routing. Similarly, for other vulnerabilities we can establish security-aware physical design steps. Therefore, the problem of obtaining secure physical design lies in secure floorplanning, secure placement, secure routing, and secure post-routing measures. Out of these vulnerabilities mentioned in Table 9.2, physical designs are most vulnerable to the following categories of attacks, which are described in detail in subsequent subsections.

9.4.1 Power and EM Side-Channel Vulnerabilities

A side-channel attack is considered one of the most crucial threats and most studied security vulnerabilities. In 1996, Kocher drew the attention of the security community both from academia and industry through the first timing attack on different cryptography algorithms [93]. Later, through the invention of simple power analysis (SPA) and differential power analysis (DPA), he showed that there is a relationship between power consumption and the data being processed [94]. In the aftermath, researchers concentrated on the implementation, rather than the weakness in the algorithm, to recover the key. Since then, quality research has been conducted in the side-channel domain. Along with power consumption, other side-channel information, such as electromagnetic (EM) radiations [95], timing [93], sound [96], photonic [97], and microarchitectural elements [98] have been studied. It is found that there is a connection between these physical signals and the properties of the computational stack. In this work, we mainly discuss the power/EM side-channel attack.

The data dependency of power consumption [99] is prevalent at every level of abstraction. For example, different instructions might consume different amounts of power depending on the number of cycles and switching activity the instruction

Table 9.2 Classification of Passive Attacks of Trust-hub Physical Vulnerabilities-DB [92] based on its Dependency to Physical Design Steps

Passive attacks			Floorplanning	Placement (Within module)	Placement (SoC level)	Routing	Post-Routing
Non Invasive	Side Channel Observation	Acoustic	✓	✗	✓	✗	✗
		Photoemission	✓	✓	✓	✓	✓
		Voltage, Charge contrast	✓	✓	✓	✓	✓
		SEM Inspection	✓	✗	✓	✗	✗
		IREM Inspection	✓	✓	✓	✓	✓
		Temperature Imaging	✓	✗	✓	✗	✗
		E or M Fields	✓	✓	✓	✓	✓
		Current/Power Measure	✓	✓	✓	✓	✓
		Voltage Measure	✓	✓	✓	✓	✓
		Indirect Voltage Measure	✓	✓	✓	✓	✓
		Data Remanence	✓	✗	✓	✓	✓
		Black Box I / O	✓	✗	✓	✗	✗
	Logical Attacks	Brute Force Algorithm	✓	✗	✓	✗	✗
		Protocol Attacks	✓	✗	✓	✗	✗

requires [100, 101], which is defined at the system specification level and can be a source of side-channel leakage. Data-dependent switching activities and state transitions in the RTL cause side-channel leakage [18]. The data dependency can be eliminated by randomizing all intermediate results that occur during computations, called *masking* in the RTL [102]. However, the masked implementation can be vulnerable to power/EM side-channel attacks through *the effect of glitches* caused by timing properties of gates and interconnection delays at the gate level and the transistor level [103, 104]. Nikova et al. developed a side-channel resistant masking technique, called *threshold implementation*, even in the presence of glitches based on secret sharing, threshold cryptography, and multi-party computations [105]. Nevertheless, the threshold implementations also leak side-channel information to reveal secret keys caused by tightly placing logically independent shares, called *the effect of coupling capacitance*, and *IR drop* [106], and by *parasitic capacitance, resistance, and inductance* [43] as byproducts of the placement and routing in the layout level. Consequently, even though side-channel leakage cannot be detected at the higher level of abstraction, it can be determined at the advanced levels of abstraction.

To identify the side-channel leakage at various levels of abstraction, *post-silicon* and *pre-silicon side-channel leakage simulators* have been developed. The post-silicon simulator first generates a set of estimated power/EM traces and then performs a leakage detection test (e.g., TVLA [107]). The estimated power/EM traces are based on the stochastic modeling (e.g., a linear regression model) depending on consecutive instructions and operands in a target device, which requires a prior knowledge about the instruction-set architecture and power/EM measurements of the target device, such as ARM Cortex-M0 or RISC-V [100, 108–110]. These post-silicon simulators can detect side-channel vulnerable instructions and then rewrite codes to mitigate side-channel leakage [109, 110]. The post-silicon simulator can be used for developing side-channel resistant software in a typical device without side-channel measurements. On the other hand, the pre-silicon simulator can help hardware designers identify side-channel leakages in the design stages, which are the RTL, gate level, and layout level. For example, RTL-PSC [18] can detect vulnerable modules in a complete design at the RTL in such a way that it profiles power consumption by counting the number of transitions in each module during computations with randomly generated plaintext and fixed-key inputs in an AES design and then calculates the statistical distance between two different sets, which correspond to two different keys. SCRIPT [111] identifies target registers that leak side-channel information by utilizing information flow tracking at the gate level. It can estimate signal-to-noise (SNR) by dividing the power difference generated by a pair of specific patterns to make the maximum Hamming distance of target registers by the power consumption of the rest of the design based on vectorless power analysis. Karna [112] searches for vulnerable gates in the layout level and then changes the gate parameter through threshold voltage, supply voltage, or the size of gates to mitigate side-channel vulnerability without the cost of performance and area. Recent work [113] also demonstrates a side-channel evaluation platform for post-quantum algorithms.

Table 9.3 Comparison: of Pre- and Post-Silicon Side-Channel Leakage Simulator

	Pre-silicon Simulator			Post-silicon Simulator
	RTL	Gate level	Layout	
Time	Medium	High	Very High	Low
Accuracy	Low	Medium	High	Very High
Flexibility	High	Medium	Low	Not Feasible; Only Software
Tool	RTL-PSC [18], RTL-PAT [114], RTL-TG [116]	SCRIPT [111], EMSIM [115] Coco [117]	Karna [112]	ELMO [108], ROSITA [109] ROSITA++ [110]

Table 9.3 summarizes the evaluation time and accuracy of power/EM side-channel leakage estimation, the flexibility to make design changes at various pre-silicon design phases, and the post-fabricated device, and available tools. There is a trade-off between time/accuracy and flexibility at various design stages when assessing side-channel leakage. Although the post-silicon simulator offers the highest accuracy and fastest processing time, it does not allow design modifications to address potential vulnerabilities. Higher levels of pre-silicon abstraction, on the other hand, provide more flexibility at the expense of accuracy. Manufacturers prefer to discover side-channel vulnerabilities early in the product development process. Early detection of leakage is better because, according to the rule of 10, the cost associated with leakage detection, identification, and mitigation will increase by ten folds if detection is delayed by one stage.

9.4.2 Fault Injection Vulnerabilities

An attacker intentionally injects a fault in the system to change its runtime behavior. The attacker exploits this change to leak sensitive information or gain access to privileged functionality. Fault injection attacks are primarily physical attacks. However, recent software-assisted hardware attacks have shown that fault injection could also be performed in remote systems [69, 118]. Based on the means of fault injection [119], the attacks could be classified into voltage glitching [57], clock glitching [56], optical/laser injection [59, 120], and EM injection [121].

9.4.2.1 Optical/Laser Fault Injection

A laser is one efficient and precise method to inject faults into the ICs. A laser that interacts with layers that are silicon or metal creates photoelectric or thermoelectric effects, which can be exploited to inject faults.

From the backside of IC, when a laser beam with a high energy wavelength passes through silicon, it creates electron–hole pairs (EHPs) along the path. Most of these EHPs recombine together, having no impact on the IC. However, under the influence of a strong electric field of reverse-biased PN junction, these EHPs drift in opposite directions, causing a current pulse. These current pulses are transient and exist for a few nanoseconds after stopping laser injection. These transient currents can turn on the reverse-biased junctions, causing the charging/discharging of the capacitive load. The voltage change in the capacitive load causes faults, also known as *single-event transient* (SET). A laser can also induce faults directly in the memory elements (e.g., RAM and registers), leading to a *single-event upset* (SEU).

When the IC is exposed to the laser from the front-side, it can hit the metal layers or find gaps to reach the active region. If the laser penetrates the active region, the observed effect is equivalent to backside exposure. However, modern ICs have dense metal layers, therefore, the laser most exposed from the front-side interacts with the metal layers. A laser interacting with the metal layers is either reflected or absorbed and converted to heat. This heat can diffuse into the PN junction creating a thermoelectric effect. However, depending on the number of metal layers in an IC, an indirect heating effect from top layers toward bottom layers can be observed. Furthermore, if there are many metal layers, the heat generated may not be sufficient to diffuse in the PN junction and have any noticeable effect.

If the IC's backside is exposed, an attacker can target specific regions on the chip to expose to the laser [122, 123]. Thus, an attacker can inject faults in precise spatial locations and control the laser exposure time to have temporal controllability on the attack. These combinations make the LFI attacks lethal and can break crypto and secure systems to violate integrity and confidentiality. LFI can inject faults during regular crypto operations, causing faulty outputs. Faulty outputs can be used with differential fault analysis to guess the secret key used during the operations. Similarly, LFI can be used to inject fault to gain access to unauthorized functionality of the design or leak secrets about the machine learning/AI models. LFI attacks on crypto designs, such as AES and smart cards, have been demonstrated predominantly in the literature [124–127]. Sugawara et al. [128] and Vasselle et al. [129] demonstrated that LFI could be used to gain unauthorized voice-control access to the system and bypass secure-boot on smartphones. Similarly, fault assessments on neural networks have been presented to reverse engineer the model architecture or leak weights or biases [130].

9.4.2.2   Timing Fault Injection

Fault injection through glitch (e.g., clock, voltage, or temperature) is a global attack that is noninvasive and less expensive to perform. Inserting a glitch can cause timing violations at the target registers and can inject faults.

If a glitch is introduced at a specific clock cycle, it reduces the effective period of that particular clock cycle. Data propagating through a combinational logic must be stable inside the setup-time or hold-time margin to prevent a flop from going into a meta-stable state while latching the data at its output. Therefore, data must be latched without any timing violation, and the data transition must occur with

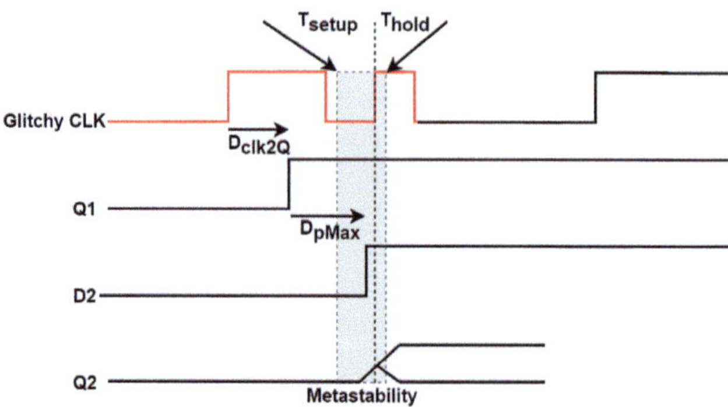

Fig. 9.4 Timing violation through clock glitch

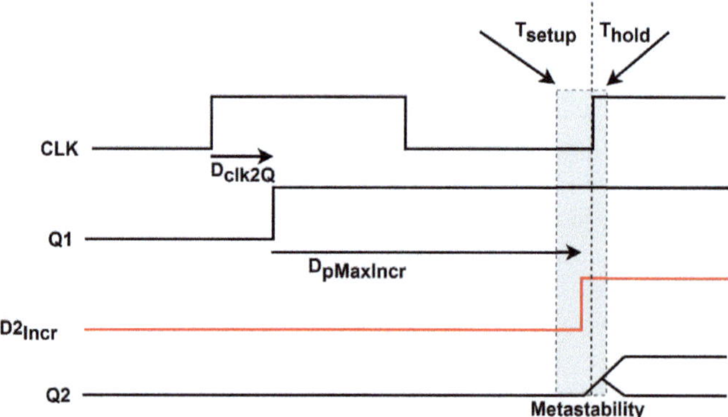

Fig. 9.5 Timing violation through voltage glitch/overheating

sufficient positive slack from both the setup-time and hold-time boundary. However, shortening the time period through glitch injection reduces the time for data to propagate through combinational logic and can violate the setup-time or hold-time constraint (shown in Fig. 9.4). Injecting spikes in power signals to reduce the supply voltage temporarily increases the propagation delay of a combinational logic. It causes the datapath delay to increase and can violate the setup-time constraints as well (shown in Fig. 9.5). Overheating can also increase the datapath delay and cause timing violations.These timing violations are responsible for causing bit flip at the output of the flop (single-event upset), which can be termed as a *timing fault injection* (TFI) [64, 65].

An attacker can tamper a clock port or power distribution network of a design to inject timing faults at some specific registers. As this is a global fault injection technique, the attacker cannot control the location of the fault injection. Nevertheless, the fault injection time can be controlled precisely. If a TFI can be successfully propagated to an observable output (exploitable fault injection), it can break crypto modules and secure designs by leaking secrets. The faulty outputs can be exploited along with fault analyses (e.g., differential fault analysis [131], fault sensitivity analysis [132], differential fault intensity analysis [133], etc.) to extract the secrets. Additionally, a TFI glitch can be used to violate the security properties related to the confidentiality, integrity, and availability of the assets of a design. Plenty of successful TFI attacks through glitches have been well demonstrated in various research works [49, 50, 52–55, 57, 58].

9.4.2.3 EM Fault Injection

In 1831, Michael Faraday found that a current-carrying coil can induce current in a nearby coil. Similarly, strong electric/magnetic fields can induce currents in loops within an IC chip. In modern chips, interconnect wires on one end are connected to the high-resistance CMOS gates, creating no impact due to EMFI [134]. On the other hand, routing power grids and rails form a large network of low resistive loops in the chip. Therefore, one can observe voltage drops and ground bounces on power rails in the presence of electric/magnetic fields. These power noises can impact the signal propagation to inject timing faults. The observed impact is similar to that of voltage glitching, however, due to the feasibility of generating targeted EM waves, EMFI can be local to achieve single-bit faults.

9.5 Secure Physical Design Flow for Possible Vulnerability Prevention

9.5.1 Security Rules for Physical Design

One of the important steps to establish a secure physical design is to have a database of rules that defines various vulnerabilities and its countermeasure. Based on the physical design level countermeasures and after performing experiments, several security rules are created. These security rules need to be implemented in P&R tools in order to ensure security, and checks need to be performed, similar to design rule checks (DRC) for physical layouts. Therefore, our proposed flow will have an extra level of security verification compared to the traditional physical design flow, as shown in Fig. 9.6. For example, to prevent the effect of coupling capacitance caused by tightly placing logically independent shares (mentioned in Sect. 9.4.1), which can violate an independent condition required in a side-channel resistant masking scheme, the shares should be placed separately not to influence each other. In this case, we can define a *Security Rule: the logically independent shares in a masking scheme should be placed far apart.* Suppose that the security checker finds the rule violation. In that case, the placement tool will change the location of the share logic cells without the violation and with little cost of performance and

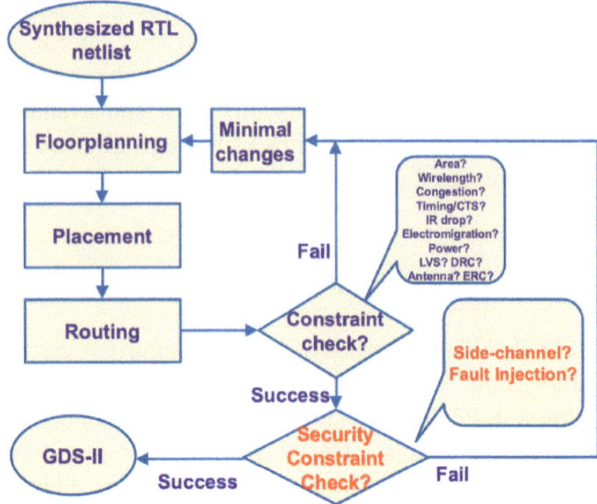

Fig. 9.6 Traditional physical design flow with security constraint check

area. Table 9.4 shows the summary of the security rules according to power/EM side-channel vulnerabilities and fault injection vulnerabilities at the physical design phases.

9.5.2 Implementation of Secure Physical Design Flow

In this section, we describe two ways of implementing secure physical design and formulate the problem of secure physical design. Security rules are discussed in the previous section. Traditional P&R-based CAD tools do not have those security rules. Security rules must be incorporated in P&R-based CAD tools to obtain secure physical design flow. Based on the source code availability of CAD tools, we can adapt the following approaches for realizing the secure physical design flow.

9.5.2.1 With Help of Open-Source CAD Tools

There are various open-source CAD tools and frameworks that deal with RTL to GDSII flows, like OpenROAD [136]. Also, some open-source tools perform the internal steps of physical design quite well, which include NTUPlace [137], RePlace [138], and DREAMPlace [139] for global placement, and NTUGr [140] for Global Routing. Since the source codes of the open-source tools are available to the public, the source code can be modified, and the security rules can be applied to establish a secure physical design flow using open-source tools.

Table 9.4 Security rules and countermeasures for secure physical design

Security vulnerability	Physical characteristic	Security rule	Implementation stage
Power side-channel	Minimize the IR drop of critical instances (s-box cells)	Place critical instances closer to decaps or power stripes	Placement
	Balance power consumption of shares	Place shares at equidistant from decaps and power stripes	Placement & Routing
	Interference from share placed to each other	Place shares far away from each other	Placement
EM-side-channel	EM signature related to the secret asset should not reach the topmost metal layer	Local low-level metal routing with a SAH	Routing
Laser Fault Injection (Front-side)	A laser should not be able to penetrate to active region	Shield cells with metal layers	Routing
Laser Fault Injection (Backside)	Critical cell placed nearby high switching & bulky cells are more susceptible	Metal shielding critical cells from nearby high switching & bulky cells	Routing
	Protecting active region from direct intensity of laser	Backside Buried Metal meander [135]	Placement & Routing
	Bulky cells are less susceptible to laser compared to small cells	Replace critical cells with bulky cells	Placement
Timing Fault Injection (clock/power/voltage glitch, voltage)	Faults injected and propagated by clock glitch, voltage glitch, power glitch can impact path delays	Placement & Routing of cells should be done in such a way to minimize path delays and clock skew	Placement & Routing
EM fault injection	Mitigate induced eddy currents	Connected loop structures should be avoided near critical cells	Placement & Routing

9.5.2.2   With Help of Commercial CAD Tools

The source codes of commercial CAD tools are not available. Therefore, we have to implement security rules as design constraints to restrict the layout designs to obtain desired security closure in commercial CAD tools.

Based on the discussion above, we can formally define our problem statement of a secure physical design as the following:

Problem 9.1 Suppose there is a regular layout L generated by the P&R tool. The regular layout may have some vulnerabilities. In that case, a security verification engine needs to be run to find the vulnerabilities. If a vulnerability exists, then changes need to make to this layout to minimize the security constraint violations and obtain a new secure layout L_{secure}.

We believe that AI/ML will help us to solve this problem and to achieve a secure physical layout, which can be best described as shown in Fig. 9.9. We have presented a detailed AI/ML roadmap for a secure physical design in Sect. 9.6. Before going to the AI/ML roadmap, let us discuss our proposed flow for secure physical design and secure physical verification sign-off.

9.5.3   Secure Physical Design and Its Verification Sign-off

Our proposed flow for secure physical design and secure physical design verification sign-off is shown in Fig. 9.7. The main components of our proposed flow are listed below:

- **Input**, the input to our flow can be either gate-level netlist or a physical design.
- **P&R tool**, if the input is in gate-level netlist, generate corresponding regular layout from the netlist. We use commercial P & R Tool such as Cadence Innovus [141] or Synopsys ICC2 [142] for generating layout. Or we can also employ OpenROAD [136] for generating the regular layout from synthesized netlist.
- **Secure Physical Design Verification (SPDV) tool**, which evaluates several physical vulnerabilities based on the various security metrics described in Sect. 9.5.4.
- **Security Metrics**, these are same security metrics mentioned in Sect. 9.5.4.
- **Security Rule Check database**, which basically contains several secure P&R recommendation as mentioned in Sect. 9.5.1. Some of the preliminary Security Rules are listed in Table 9.4.
- **SEPHYD tool** takes a gate-level netlist as its input (for the first time) and generates a secure physical design based on several security rules. These security rules are developed from several physical design level countermeasures (mentioned in Sects. 9.5.5 and 9.5.6), which in turn are constructed to ensure security from any physical vulnerabilities. For applying countermeasures in terms of security rules

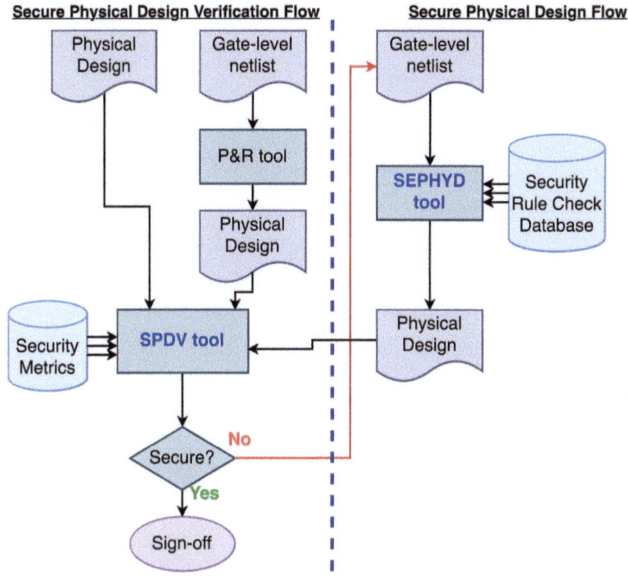

Fig. 9.7 Proposed Flow for Secure Physical Design (SEPHYD) and Secure Physical Design Verification (SPDV) Sign-off

and to bring automation, it employs AI/ML approaches. A detailed roadmap of the AI/ML approach for realizing SEPHYD tool is described in Sect. 9.6.

- **Output**, a secure physical design free from any physical vulnerabilities.

In the proposed flow, we consider the input gate-level netlist is secure where all high-level security measures are implemented. However, when the physical design is created from this secure gate-level netlist using the P&R tool, some security vulnerabilities are unintentionally injected, which some adversaries can exploit to compromise the secure design. Such threats can occur only at the physical design level, as much information is not available in the logic synthesis stage. For example, power grid network and clock tree synthesis information can only be obtained in the physical design stage.

Once the layout is generated using the P&R tool, we employ our security verification engine (SPDV tool) in order to verify the security of the regular layout. The security verification is performed by evaluating several security metrics depending on the vulnerabilities mentioned in Sect. 9.5.4. If the security metrics report that there is any vulnerability in the layout, then it goes through the *SEPHYD* tool, which performs minimal change in the layout and reconstructs the layout considering the security rules discussed in Sect. 9.5.1. In this way, we obtain a secure physical design. To ensure that generated physical design from the SEPHYD tool is actually secure, we again verify our generated physical design using our SPDV tool. If the verification engine reports that no physical vulnerability is present in the layout, we sign off our secure physical design and verification flow. Therefore, our proposed flow goes into a loop of several iterations between the security verification engine (SPDV tool) and SEPHYD tool, until the security metrics suggest that the desired level of security has been achieved. Once the security metrics suggest an acceptable level of security of physical layouts, we say that we have obtained the secure physical design.

For our initial experiments, SEPHYD tool is being implemented using a heuristic-based AI/ML approach. More about this is discussed in Sect. 9.6. We are also exploring data-driven AI/ML techniques to implement SPDV tool and SEPHYD tool.

9.5.4 Security Metrics for Evaluating Physical Vulnerabilities

The metrics used for evaluating the security vulnerability is described here. At present, we have defined preliminary metrics for evaluating Side-channel vulnerability and Timing Fault injection metrics.

9.5.4.1 Side-Channel Vulnerability Metrics

SEPHYD framework performs the side-channel vulnerability analysis based on the Kullback–Leibler (KL) divergence metric. Details on the KL divergence metric and its relation with the attacker's success rate (SR) based on maximum likelihood estimation are described as follows.

Let $f_x(z)$ and $f_y(z)$ be the probability density functions of random variables X and Y. KL divergence is defined as the following equation:

$$D_{KL}(X||Y) = \int f_x(z) log \frac{f_x(z)}{f_y(z)} dz \qquad (9.1)$$

If X and Y are of normal distributions with means (μ_x, μ_y) and variances (σ_x^2, σ_y^2), then the KL divergence equation can be simplified as follows:

$$D_{KL}(X||Y) = \frac{(\mu_x - \mu_y)^2 + \sigma_x^2 - \sigma_y^2}{2\sigma_y^2} + ln\left(\frac{\sigma_y}{\sigma_x}\right) \qquad (9.2)$$

The KL divergence is expected to be high if power leakage probability distributions for two different keys are distinguishable, meaning an adversary can easily correlate the power consumption between the keys. Hence, the maximum KL divergence for allowable failure probability (Pr) can be obtained, where failure probability is the adversary's probability of an incorrect inference based on PSC attacks. The higher the failure probability is, the more secure the design is against PSC attacks. For example, if we want the failure probability of more than 0.9, the

Table 9.5 KL Divergence Threshold for Different Failure Probabilities (**Pr**)

Pr	KL	Pr	KL
>0.96	<0.01	>0.53	<0.78
>0.90	<0.03	>0.45	<1.12
>0.80	<0.12	>0.38	<1.53
>0.71	<0.28	>0.32	<2.00
>0.61	<0.50	>0.26	<2.53

KL divergence should be less than 0.03 [143]. Table 9.5 provides the required KL divergence threshold for different failure probabilities (Pr).

Also, to assert with a high confidence level of $100(1-\alpha)\%$ that the two Gaussian distributions X and Y differ, it is necessary to account for the number of traces N in the KL divergence metric. The number of traces N contributes significantly to quantifying a lower constraint on side-channel attack complexity in terms of the required number of power traces. The smallest number of traces to satisfy that $\Pr[|\bar{X} - \bar{Y} - (\mu_X - \mu_Y)| < \epsilon] = (1 - \alpha)$ is

$$N \geq \frac{(\sigma_X + \sigma_Y)^2}{\epsilon^2(\mu_X - \mu_Y)^2} z_{1-\alpha/2}^2 \tag{9.3}$$

where the quantile $z_{1-\alpha/2}$ of the standard normal distribution has the property that $\Pr[Z \leq z_{1-\alpha/2}] = 1 - \alpha/2$.

One can correlate the KL divergence analysis with the probability of an attacker's success in leaking the key. Assuming for a given key K, the probability density function of the switching activity T follows a Gaussian distribution, then SR could be derived as follows:

$$f_{T|K}(t) = \frac{1}{\sqrt{2\pi}\sigma_k} e^{-\frac{(t-\mu_k)^2}{2\sigma_k^2}} \tag{9.4}$$

where μ_k and σ_k^2 are the mean and variance of T, respectively. The likelihood function is defined as $\mathcal{L}(k; t) = \frac{1}{n}\sum_{i=1}^{n} \ln f_{T|K}(t_i)$. Based on the maximum likelihood estimation, an adversary typically selects a guess key $\hat{k}$ as follows:

$$\hat{k} = \arg\max_{k \in K} \mathcal{L}(k; t) = \arg\max_{k \in K} \frac{1}{n} \sum_{i=1}^{n} \ln f_{T|K}(t_i) \tag{9.5}$$

If the guess key ($\hat{k}$) is equal to the correct key (k^*), the side-channel attack is successful. Thus, the attacker's success rate can be defined as follows:

$$SR = \Pr[k_g = k^*] = \Pr[\mathcal{L}(k^*; t) - \mathcal{L}(\langle \bar{k}^* \rangle; t) > 0] \tag{9.6}$$

where $\langle \bar{k^*} \rangle$ denotes all wrong keys, i.e., the correct key k^* is excluded from $\mathcal{K} = \{k_0, k_1, \ldots, k_{n_k-1}\}$, where n_k is the number of all possible keys.

The mathematical expectation of $\mathcal{L}(k^*; t) - \mathcal{L}(k_i; t)$ in Eq. (9.6) is equal to KL divergence between $T|k^*$ and $T|k_i$ [144]. It shows that both SR and KL divergence are closely related and follow a similar trend. Therefore, leakage obtained through KL analysis can be associated with an attacker's success in leaking the key.

9.5.4.2 Timing Fault Injection Security Metric

We define a security metric, V_{TFI}, that quantifies the susceptibility of a design against TFI attacks [145]. It is proportional to the probability of vulnerability to DFA attacks introduced by timing faults, P_{sp} for all the security properties of a crypto-design against DFA. It is also inversely proportional to the number of security properties defined in the database, $N_{DB_{sp}}$ since the attacker has to deal with more options. Here P_{sp} depends on the probability of finding feasible fault locations and the probability of the dispersion of the security-critical paths within the timing margin. The following equation represents the mathematical expression of the security metric.

$$V_{TFI} = \frac{1}{N_{DB_{sp}}} \sum_{p=1}^{p=N_{DB_{sp}}} P_{sp} \tag{9.7}$$

A larger value of P_{sp} of Eq. 9.7 means a higher probability of security property violations due to DFA attacks, and ultimately a higher susceptibility to DFA attacks.

9.5.5 Possible Countermeasures for Side-Channel Vulnerabilities

Based on the violation of the defined security rules (shown in Table 9.4), suitable countermeasures can be applied at the physical design phases as follows:

- *Minimizing IR drop*: IR drop or power supply noise is caused by the finite resistivity of each metal layer in the PDN. Instantaneous power consumption of each share in a masking scheme can be affected by adjacent shares significantly if not considering the IR drop effect. The IR drop should be minimized by inserting a sufficient number of decap cells or reconstructing the PDN to remove power dependency between logically independent shares. Interconnect width optimization-based approaches can also be implemented to minimize IR drop [146–148].
- *Separation of independent shares*: Due to capacitive coupling between two adjacent wires, when a wire switches a value, another wire can be influenced by

the inter-wire capacitance, called crosstalk. Logically independent shares should be placed with enough distance to remove the capacitance coupling between them [106].

- *Local low-level metal routing with a signature attenuating hardware (SAH)*: Das et al. [149] proposed a signature attenuation method with local low-level metal routing and a signature attenuating hardware (SAH), called *STELLAR*. The SAH significantly suppresses the EM/power signature with low overhead before it reaches the top metal layer of the chip that has the most contribution to measurable EM side-channel leakage [150]. This method protects the AES-128 encryption against EM/power SCAs and achieves *Measurements to disclosure (MTD)* $> 1M$ with a low-overhead physical countermeasure ($1.5\times$ power and $1.23\times$ area overhead).

- *Supply isolation*: The critical power signature of cryptographic modules can be isolated from the external supplies, which eliminates power side-channel information measured by adversaries. For example, a switched capacitor current equalizer isolates critical activities by equalizing currents into the cryptographic module [151]. The capacitor current equalizer consists of an array of capacitor circuits, and each circuit has three different switching states: S1-charge the capacitor from the supply, S2-provide charge to the cryptographic module, and S3-discharge the capacitor to a pre-programmed value. The charged capacitor serves as a voltage source for the cryptographic module. Since the capacitor is disconnected from the external supply during cryptographic operations, power signatures of the cryptographic module can not be measured. Three independent capacitor modules do not overlap switching states for uninterrupted operations of the cryptographic module.

The first and second countermeasures require a masking technique at the gate level, consisting of multiple logically independent shares. In contrast, the third and fourth approaches do not require any approaches at the higher level of design abstract. However, there is a need for a specific attenuating circuit with a routing method and a current equalizer at the physical level. Figure 9.8 shows the possible countermeasures against side-channel vulnerabilities at the physical design phase.

9.5.6   Possible Countermeasures for Fault Injection Vulnerabilities

For protecting against fault injection vulnerabilities, various countermeasures could be applied based on logical or physical parameters. Some of these countermeasures or parameters are discussed below.

- *Spatial or temporal redundancy:* Most prominent FI countermeasures include creating redundancies in the design to exploit the fact that it is difficult to

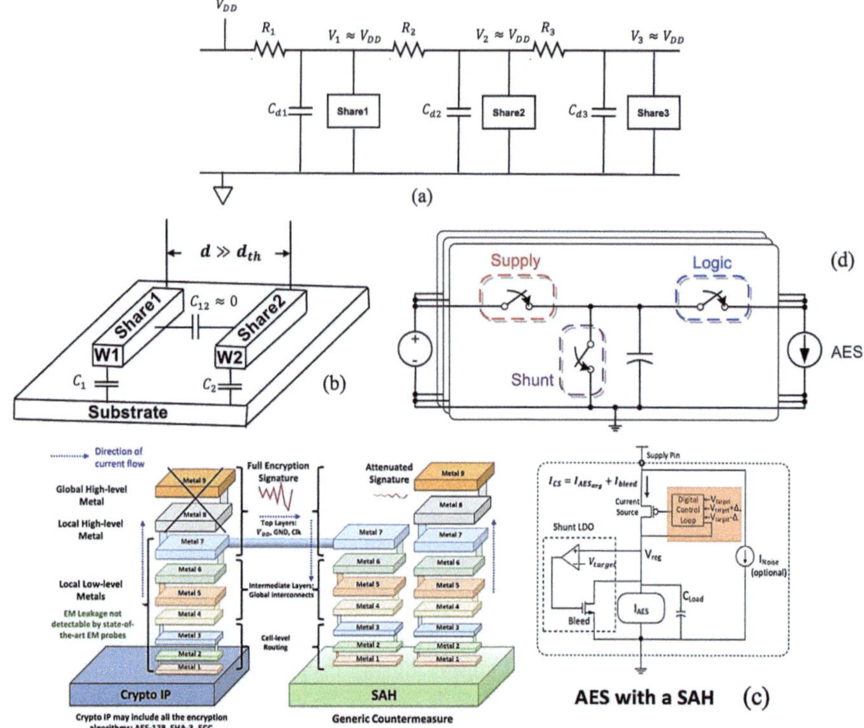

Fig. 9.8 SCA Countermeasures at the Physical Level: (**a**) minimizing IR drop using decap cells; (**b**) separation of independent shares; (**c**) local low-level metal routing with a SAH [149]; (**d**) supply isolation [151]

precisely inject faults in multiple redundant circuits at the same location and time.

- *FI standard cell library:* Different cell parameters, such as size, threshold, cell layout, etc., can play a significant role in determining the design's resiliency against FI vulnerabilities. For example, changing the gate size impacts the path delays and the cell's driving strength. A high threshold of the cells may require higher laser energy to flip the output. Similarly, the layout of a cell (sharing of p/n-well regions, poly widths, etc.) can impact how a cell behaves under different fault injection methods.

- *FI-aware layout design:* Standard cells in addition to different physical parameters can also impact the design's resiliency against fault injection methods. For example, power distribution networks, clock tree synthesis, placement, routing, and other factors can be customized to account for fault injection security.

9.6   AI/ML Roadmap for secure physical design and verification

Recently, AI/ML has attracted the attention of chip design community [152]. Google has also claimed successful deployment of reinforcement learning (RL) for macro placement [153] in its AI processor design. Although RL-based solution of Google has not provided good results for standard-cell placement yet [154]. There are several other works on machine learning (ML) in applications of CAD domain [155–160]. We also believe ML can be implemented in order to obtain a secure physical design. Here, we present the AI/ML roadmap for obtaining secure physical design. Initially, we describe AI/ML approach for developing our SEPHYD tool. Later on, we discuss AI/ML approach for developing our SPDV tool. Overall, the final goal is to implement the proposed flow of Fig. 9.7 using AI/ML automation (Fig. 9.9).

9.6.1   AI/ML Roadmap for SEPHYD Tool

Here, we describe AI/ML roadmap to implement the proposed SEPHYD tool. When we say SEPHYD, it basically means secure floorplanning, secure placement and secure routing. Here, our discussion is only limited to floorplannning and placement. However, similar approach can be adapted for secure routing. We have already observed one such vulnerability in floorplanning and placement stage using the timing fault injection parameters. The delay distributions of critical datapath vary widely in the gate and physical layout levels. Adversaries can utilize these vulnerabilities to inject fault and extract secret information from the chip. Initially, we propose a gate sizing-based solution to mitigate delay variation-based

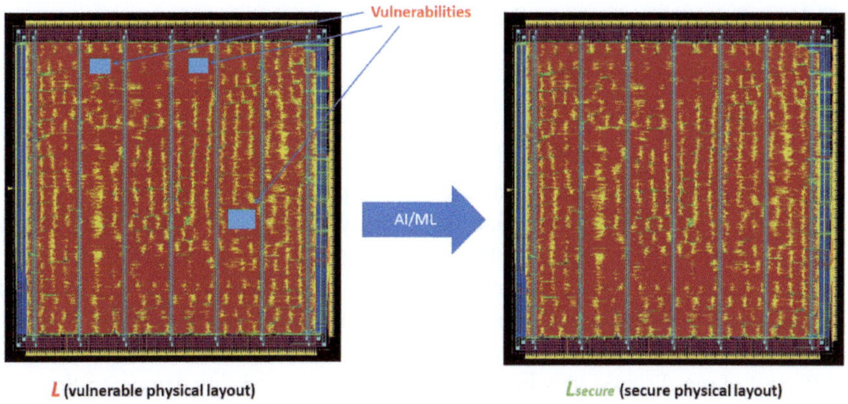

Fig. 9.9 Objective of AI/ML roadmap for secure physical layout generation

vulnerabilities [145]. However, we observe that this vulnerability can be mitigated using the AI/ML approach described below.

Suppose, we have n critical paths. Let the path delays of critical paths in the gate level be t_{g_i} $\forall i \in \{1, 2, \cdots, n\}$ and path delays of critical paths in physical design level be t_{pd_i} $\forall i \in \{1, 2, \cdots, n\}$. The absolute path delay difference between gate level and physical design level can be represented by

$$\Delta t_i = \left| t_{g_i} - t_{pd_i} \right| \ \forall i \in \{1, 2, \cdots, n\} \tag{9.8}$$

We have the values of t_{g_i} during the gate-level simulations, which cannot be changed during the physical design stage. Therefore, the objective is to place the standard cells connecting critical paths in such a way during the physical design stage that Δt_i becomes zero or minimum. However, if consider Δt_i as the only objective, and perform minimization, then the total wirelength and congestion may increase. Therefore, along with Δt_i, we can also consider the traditional placement metrics, such as total wirelength and congestion in our cost function. By considering all of these objectives, our cost function for Timing Fault Injection-Aware secure placement can be written as

$$\min \left(\sum_{e \in E} WL(e; x, y) \right) + \lambda D(x, y) + \Delta t_i \tag{9.9}$$

where $\left(\sum_{e \in E} WL(e; x, y) \right)$ represents total wirelength, $D(x, y)$ represents density [153], and Δt_i represents delay differences between gate-level and physical design level $\forall i \in \{1, 2, \cdots, n\}$ critical paths. The challenge is to obtain the locations of the standard cells, where Δt_i becomes zero or minimum, which is a NP-hard problem, as solution space of this problem is of exponential order. That is where we feel several AI/ML techniques can be implemented in order to obtain a near-optimal solutions for our TFI-Aware secure placement problem. Similarly, for other kinds of physical design level vulnerabilities, the cost function becomes as

$$\min \left(\sum_{e \in E} WL(e; x, y) \right) + \lambda D(x, y) + \Phi_i \tag{9.10}$$

where Φ_i represents the security metrics for the physical vulnerabilities in concern. Therefore, it is necessary to define the security metrics to evaluate the vulnerabilities. Subsequently, optimize the cost function that also includes the security metrics. To find solutions of such cost functions, we can apply two classes of AI/ML approaches. The first class is a heuristic-based approach and the second class is a data-driven approach. Since we have already mentioned about TFI vulnerability in (9.9), rest of our discussion is limited only for TFI vulnerability. However, the same approaches can be extended to any other kinds of physical vulnerabilities.

9.6.1.1   Heuristic-Based Approach

In heuristic-based approach, based on some predefined cost functions, agents perform search in the solution space to find near-optimal solutions, satisfying the cost functions and associated constraints. For our problem, we change the floorplanning and placement of standard cells to reduce the vulnerability by employing heuristics. We employ (9.9) as a cost function and subsequently apply well-known heuristic approaches, (e.g., simulated annealing [161], genetic algorithm [162], or gradient-descent search [163]). These heuristics interact with each other to produce near-optimal solutions. The heuristics start from random solutions and gradually improve the quality of solutions to converge in a desired level of tolerance for the cost function. These heuristics are slow, as it takes a considerable time to converge. However, we anticipate to obtain TFI-aware secure macro and standard cell locations using these heuristics.

9.6.1.2   Data-Driven Approach

When we talk about data-driven approaches it basically means traditional machine learning approaches, recent deep learning approaches, and deep reinforcement learning approaches. Each of these techniques employs datasets to find near-optimal solutions. One important aspect of data-driven approaches is to have a proper dataset. However, unlike the computer vision domain, there is no public open-source dataset available as of now, for developing machine learning models for electronic design automation (EDA). This leads us to generate our own dataset and perform proper feature engineering to find suitable features to develop machine learning models. Currently, we are creating datasets following the format mentioned in Tables 9.6 and 9.7. Basically, using the dataset of Tables 9.6 and 9.7, it is possible to define a floorplan with connections among macros and standard cells. Therefore, we engineer these datasets to obtain a suitable machine learning model. Now we will describe various data-driven approaches, which can be viable options for solving the problem of security-aware floorplanning and placement.

Supervised Learning

For supervised learning, relations between dependent and independent variables need to be established. For the delay variance as a timing fault injection parameter, we can label the locations of macros and standard cells to distances/datapath delays to create a dataset. After that, we can train a neural network and generate a model to deal with vulnerabilities. This is the supervised regression problem, and the prediction accuracy needs to be measured using Mean-Squared Error (MSE). However, from the understanding of the TFI-aware secure placement, MSE can be very high, as this approach may not produce the refined level of placement desired.

Table 9.6 Position dataset format for the sample Fig. 9.10

Item name	Location	Length	Height
P0	(xP0,yP0)	–	–
S0	(xS0,yS0)	lS0	hS0
M0	(xM0,yM0)	lM0	hM0
S1	(xS1,yS1)	lS1	hS1
M1	(xM1,yM1)	lM1	hM1
P1	(xP1,yP1)	–	–
P0_M0	(xP0_M0, yP0_M0)	–	–
P1_M0	(xP1_M0, yP1_M0)	–	–
P0_M1	(xP0_M1, yP0_M1)	–	–
P1_M1	(xP1_M1, yP1_M1)	–	–

Table 9.7 Empty table entries representing the manhattan distances between items at row and column

Item name	P0	S0	M0	S1	M1	P1	P0_M0	P1_M0	P0_M1	P1_M1
P0	0									
S0		0				$md(i, j)$				
M0			0							
S1				0						
M1					0					
P1						0				
P0_M0							0			
P1_M0								0		
P0_M1									0	
P1_M1										0

Sequential Supervised Learning

The task of TFI-aware secure placement is sequential in nature. Because the position of second macro depends on the first macro's position, in order to reduce the vulnerability and optimize other placement objectives. Therefore, we can employ several sequential machine learning techniques, such as Sliding Window, Recurring Sliding Window, Hidden Markov Model, Maximum Entropy Markov Model, Input–Output Markov Model, Conditional Random Field, and Graph Transformer Networks [164]. Accordingly, we perform engineering using the datasets mentioned in Tables 9.6 and 9.7.

Unsupervised Learning

Unsupervised learning works by finding patterns in an unlabeled dataset. This ML approach may not be a good candidate for this particular task, as TFI-aware secure macro or standard cell placement task does not rely on finding any pattern. However,

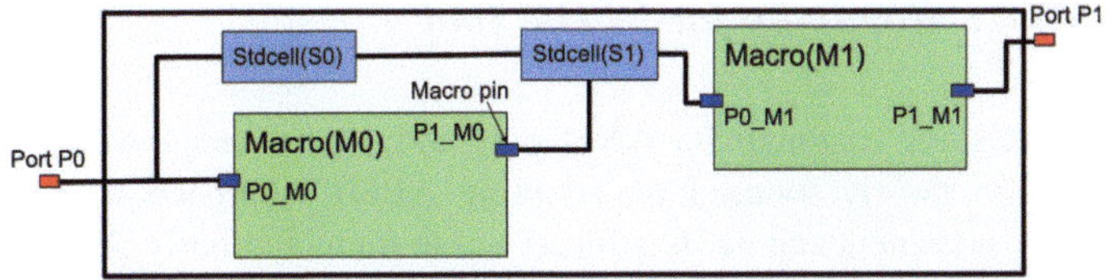

Fig. 9.10 Example of a chip floorplanning with macros and standard cells

more research and development is required before giving any final verdict about this approach.

Semi-Supervised Learning

In this case, we must have a mix of labeled and unlabeled data, with which predictions are performed. Subsequently, this technique may present viable options to predict secure placement or to solve the cost function of concern. However, more research and development is required before giving any final judgment about this approach.

Reinforcement Learning Approach

The reinforcement learning approach would be a suitable choice for the practical implementation of the TFI-aware secure macro and cell placement task, as this approach depends solely on the reward of agents. The reward is high for more secure positions and the reward is less for less secure positions. Already, Google has received success in their macros placement task [153]. However, Google's work does not consider the security aspect while performing the placement. The cost function of (9.9) needs to be addressed using deep reinforcement learning with appropriate reward, state definition, and policy function to have TFI-aware secure placement. Again, we need the datasets mentioned in Tables 9.6 and 9.7 if we want to develop deep reinforcement learning-based solutions.

Overall, the goal is to develop a single security metric (Φ) which can evaluate the security measures of all the vulnerabilities using a single cost function, instead of having separate cost functions for each vulnerability. Subsequently, we can solve the cost functions using the AI/ML approaches mentioned above. Likewise, we can also employ these AI/ML approaches to generate secure routing in the physical design. We discuss the AI/ML roadmap for side-channel assessment as part of our SPDV tool in the next section.

9.6.2 AI/ML Roadmap for SPDV Tool

Here, we describe a roadmap for developing SPDV tool using AI/ML approach. Our description mostly focused on adopting AI/ML approach for side-channel vulnerability assessment and verification as a case study. Similarly, AI/ML approach can be implemented for other types of physical vulnerabilities.

Side-channel analysis has seen remarkable growth in the last 5 years because of the inclusion of machine learning-based approaches for side-channel vulnerability assessment. The practice has been noticed only in post-silicon side-channel analysis, though the pre-silicon analysis is essential, especially at the physical design level. As a roadmap for side-channel analysis at the physical layout level, we propose two approaches: estimation and measurement. To elaborate, we define the estimation approach as a quick way to assess the side-channel vulnerability at the physical layout. The estimation approach will completely avoid any simulation, i.e., no EDA tool will be used in this approach for calculation. Tools can only give physical information. All these constraints are applied to make the process fast for deciding whether a physical design is vulnerable or not.

The measurement approach will be based on simulation. However, the measurement approach will give a more robust decision about side-channel vulnerability. No constraint is imposed upon this method like estimation. Since EDA tools will be used for power calculation, this approach will be time-consuming. The measurement approach starts with collecting power side-channel traces at the layout level of a full-blown SoC design through dynamic power analysis using EDA tools. A corresponding power trace is collected for a random plaintext and random key in each simulation. The vulnerability of the physical design can be assessed by performing a side-channel attack on these collected traces. A portion of the collected traces is used for training the neural network, and the attack is performed on the rest. The quality of the attack is evaluated through performance metrics like guessing entropy, success rate, etc. If the required number of traces to recover a key is less than a predefined threshold value, the design is vulnerable.

In this work, we provide a roadmap for AI-based side-channel assessment or measurement approach. The roadmap is shown in Fig. 9.11. At first, the collected power trace is processed through the data pre-processing step. The most common practice of data pre-processing is to apply the typical normalization or standardization technique. Additionally, signal decomposition techniques such as empirical mode decomposition (EMD) [165], Hilbert vibration decomposition (HVD) [166], variational mode decomposition (VMD) [167], and other techniques can be applied. Such decomposition techniques depict the intrinsic features unseen in raw signals. Notably, EMD can be used as a denoising step to remove unnecessary information from the power traces. In the case of a hiding countermeasure, dynamic time warping (DTW) is an effective way to bring alignment into the misaligned traces. Data augmentation is another data processing technique that can improve the quality of an attack by including newer observations of data in the training set and preventing overfitting.

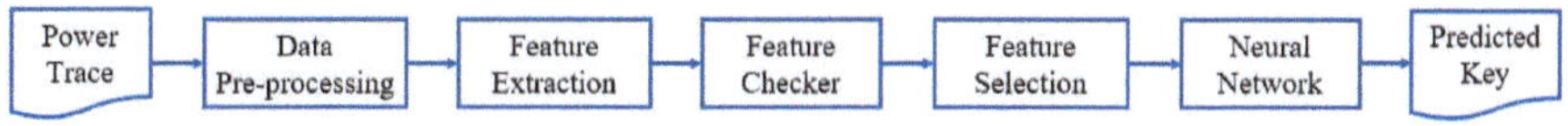

Fig. 9.11 Roadmap for AI-based side-channel assessment

The roadmap suggests several ways to perform appropriate feature selection with the help of a feature checker. The purpose of a feature checker is to ensure the choice of selected feature set is correct. The feature checker can be formed in different ways. One way is to calculate feature importance using a Random Forest classifier. In this case, the relative feature importance of each type of feature can be calculated in terms of Gini impurity. The features also show that higher feature importance can be kept, and other features can be discarded. Correlation analysis can be also a viable way to find out efficacy of a set of features. After selecting the appropriate features, one must choose a neural network. Convolutional neural networks (CNN) and multilayer perceptron networks are the most widely used neural architectures for side-channel analysis. Zaid et al. [168] discovered efficient but shallow convolutional neural network (CNN) architectures through weight visualization, gradient visualization, and heatmaps. However, with such techniques, achieving appropriate hyperparameter tuning can be challenging. Wu et al. [169] showed how automated hyperparameter tuning using Bayesian optimization assists in performing superior side-channel attacks. Perin et al. [170] preferred ensemble models to the single best performance for improving the generalization of the attack. Integration of automated hyperparameter tuning and model ensembling to the framework can boast the performance of the neural network. Using proper evaluation metrics to assess the performance of the neural network architecture is another important step. Guessing entropy and ranking loss are the top evaluation metrics for such task.

9.7 Open Challenges

In the previous sections, we have proposed several possible countermeasures to mitigate physical design level vulnerabilities, including an AI/ML roadmap. However, many challenges exist and it may be difficult to create countermeasures efficiently. These challenges are described in the following sections.

9.7.1 Challenges in Developing Physical Design Security Metrics

At present, there are no properly defined metrics that can evaluate the security of a physical design for any specific kind of vulnerability. To progress in secure physical design research, it is necessary to define secure physical design metrics so that we can evaluate how secure a physical design is. Once we have well-defined security metric, the objective of the physical design steps would be to optimize the layout in order to minimize the security metric and optimize other physical design constraints at the same time.

9.7.2 Challenges in Vulnerability Realization

9.7.2.1 Challenges in Side-Channel Vulnerability Assessment in Physical Level

For side-channel assessment, we have already proposed an AI/ML roadmap in Sect. 9.6.2. Although AI/ML-based approaches have a high potential for the successful assessment of side-channel vulnerability at the physical layout level, there are multiple critical challenges for performing AI/ML approaches, as explained below:

- Lack of data is one of the main challenges for the ML-based solution for side-channel vulnerability assessment at the physical level. The growth of deep learning-based approaches for side-channel attacks at post-silicon traces has been around since 2018 when the open-source large dataset ASCAD [171] was introduced. However, no AI/ML-based approach is performed for the pre-silicon side-channel analysis at layout because datasets of side-channel traces are not collected at the physical design level. The main reason for such unavailability of data at the physical layout is the high simulation time required for data collection. Trace collection at the layout level is a time-consuming and tedious task, as shown in Table 9.3. Lin et al. [47] shows that for a chip with a node count of 3.465M, the traditional VCD-based analysis takes around 671 hours runtime to generate 10000 power traces. ML-based approaches are expected to require millions of traces to train the neural network, which indicates that it may take many months to collect adequate data for an ML-based solution of side-channel assessment at the layout level. The challenge remains to introduce a well-defined and complete dataset of power traces (collected from the physical layout at the platform level) which may pave the way to new research direction for side-channel analysis at the physical design level (similar to what happened with the ASCAD dataset at the post-silicon stage).
- As mentioned before, the profiled DL-based side-channel attack requires the following steps: data pre-processing, feature engineering, algorithm selection, and attack evaluation. There is still no proper guideline for a post-silicon or pre-silicon side-channel attack for any of these steps. It is unclear what type of pre-processing and feature extraction approaches would work efficiently. For AI-based side-channel assessment, security metrics used for performance evaluation are different from typically used ML metrics. Still, many recent studies suggest

new security metrics for side-channel attack evaluation mentioning possible drawbacks of existing metrics.

- Lack of explainability is another open challenge for the ML-based approach to side-channel vulnerability assessment. Neural networks have shown great success, even in robust countermeasures. However, how these networks deal with masking countermeasures is unclear. If an unsuccessful attack, it is tough to say whether it happens because of a weak countermeasure or an ineffective AI approach. It is unclear why the attack did not succeed if it was not successful. Similarly, in case of a successful attack, no one has been able to point out the exact weakness of the design. Such a lack of explainability makes it challenging to propose effective countermeasures.

9.7.2.2 Challenges in Laser Fault Injection in Physical Level

Pre-silicon assessment for LFI poses many challenges at the physical level, as discussed below.

- Precise modeling of laser effects during pre-silicon conditions is challenging. Various SPICE models have been proposed to emulate the laser's transient current effect, yet fail to account for physical design parameters (power distribution network, nearby elements, etc.).
- Most models are presented on small components (like an inverter or a D-flip flop). Modeling the laser impact on a large SoC is still a challenge.
- With the reduction in technology nodes, the smallest laser spot size can impact multiple cells in the region, causing multiple faults. Laser modeling for the multi-faults scenario in a large SoC is also challenging.

9.7.2.3 Challenges in Clock Glitch Fault Injection Assessment in Physical Level

- Timing margin analysis: Defining the exact time range within which an attacker can inject a feasible and controllable fault is challenging. The upper and lower margin of this time range is also dependent of the input stimuli. An exhaustive simulation is required with random inputs to figure out the exact margin of feasible fault injection.
- Feasibility analysis: Timing fault feasibility analysis requires several factors to be considered simultaneously. A feasible fault location depends on input stimuli, RC-delay corner of static timing analysis, pulse width of glitch signal and delay distribution of the datapath delays of critical registers. Finding a feasible location where an attacker can inject a feasible fault through clock glitching is a very difficult task.

9.7.2.4   Challenges in AI Model Realization for Secure Physical Design

As discussed in the Sect. 9.6, designing AI/ML-based models for secure physical designs is challenging, which are a must to ensure trust in every stage of the hardware flow. Many challenges need to be addressed in order to move ahead. A few of the challenges are listed below:

- Data Requirement: To create a model that trains and produces accurate results, a model needs data. Acquiring circuit and layout data and labeling them is a critical stage. The most significant drawbacks of acquiring such data are specialized equipment and the time required to collect and label them. It will also be a challenge to collect data that is scalable from one design to another.
- Feature Extraction: This would be another challenge based on the data that is collected (1D and/or 2D) and how each feature or set of features can be utilized to make the physical design secure.
- Developing Models: Training supervised models has a huge bottleneck since the data required for training such models is very limited. Some research has shown progress when utilizing heuristic-based approaches. However, such approaches need to analyze the vulnerabilities. It also depends on the formulation of cost functions based on our observed vulnerability and how it scales.

9.7.3   Challenges in Reaching Optimum Design Point

Suppose we have the physical design security metrics, with which we can obtain the level of vulnerability of any physical design. Now, when we optimize the physical design to mitigate the physical vulnerabilities, what will be the Power-Performance-Area (PPA) overhead of the physical design. With the advancement of technology nodes, it is already very challenging to obtain a PPA trade-off. With security as an added objective, it will be more difficult for designers to obtain an acceptable PPA trade-off.

9.7.4   Challenges in Achieving Competitive Time-to-Market

The authors believe that defining physical design security metrics involves lots of dynamic data and statistical computations at all levels of metal stacks and throughout the design. For example, power side-channel assessment may require dynamic power traces at all levels of metal stacks and obtaining some kind of correlation from these dynamic power traces, which is a time-consuming and costly affair. Thus, several additional assessments need to be done. Subsequently, mitigating those security metrics (or vulnerabilities) will increase the time-to-market of such secure physical design-based chips. The need is a smart way of assessing the vulnerability which reduces time and resources needed in the process.

9.8 Concluding Remarks

To ensure trust at every level of the hardware design flow, having a secure physical design is crucial. In this chapter, we have introduced the concept of obtaining a *secure physical design* that is free from any physical vulnerability. We demonstrated how vulnerabilities can occur in the physical layout and utilize physical vulnerabilities despite having a secure synthesized gate-level netlist. Also, we have listed a brief overview of possible countermeasures to combat physical vulnerabilities. Moreover, we described our proposed framework for secure physical design and verification. We also presented an AI/ML roadmap to obtain a secure physical layout. Several challenges for obtaining a secure physical layout are also listed in the chapter. Overall, this chapter presents various aspects of obtaining a secure physical design.

References

1. IDC, *Semiconductor Market to Grow By 17.3 % in 2021 and Reach Potential Overcapacity by 2023, IDC Reports.* https://www.idc.com/getdoc.jsp?containerId$=$prAP48247621. Accessed: 2021-12-9.
2. B. Ahmed, M.K. Bepary, N. Pundir, M. Borza, O. Raikhman, A. Garg, D. Donchin, A. Cron, M.A. Abdel-moneum, F. Farahmandi, et al., Quantifiable Assurance: From IPs to Platforms. arXiv preprint arXiv:2204.07909 (2022)
3. N. Farzana, A. Ayalasomayajula, F. Rahman, F. Farahmandi, M. Tehranipoor, SAIF: automated asset identification for security verification at the register transfer level, in *2021 IEEE 39th VLSI Test Symposium (VTS)* (IEEE, New York, 2021), pp. 1–7
4. G.K. Contreras, A. Nahiyan, S. Bhunia, D. Forte, M. Tehranipoor, Security vulnerability analysis of design-for-test exploits for asset protection in SoCs, in *2017 22nd Asia and South Pacific Design Automation Conference (ASP-DAC)* (IEEE, New Year, 2017), pp. 617–622
5. M. Tehranipoor, F. Koushanfar, A survey of hardware trojan taxonomy and detection. IEEE Des. Test Comput. **27**(1), 10–25 (2010)
6. H. Salmani, M. Tehranipoor, J. Plusquellic, A novel technique for improving hardware trojan detection and reducing trojan activation time. IEEE Trans. Very Large Scale Integr. VLSI Syst. **20**(1), 112–125 (2011)
7. H. Salmani, M. Tehranipoor, J. Plusquellic, New design strategy for improving hardware Trojan detection and reducing Trojan activation time, in *2009 IEEE International Workshop on Hardware-Oriented Security and Trust* (IEEE, New York, 2009), pp. 66–73
8. S. Bhunia, M. Tehranipoor. *Hardware Security: A Hands-on Learning Approach* (Morgan Kaufmann, Burlington, 2018)
9. X. Wang, H. Salmani, M. Tehranipoor, J. Plusquellic, Hardware Trojan detection and isolation using current integration and localized current analysis, in *2008 IEEE International Symposium on Defect and Fault Tolerance of VLSI Systems* (IEEE, New York, 2008), pp. 87–95

10. M. Tehranipoor, H. Salmani, X. Zhang, M. Wang, R. Karri, J. Rajendran, K. Rosenfeld, Trustworthy hardware: Trojan detection and design-for-trust challenges. Computer **44**(7), 66–74 (2010)

11. H. Salmani, M. Tehranipoor, Analyzing circuit vulnerability to hardware Trojan insertion at the behavioral level, in *2013 IEEE International Symposium on Defect and Fault Tolerance in VLSI and Nanotechnology Systems (DFTS)* (IEEE, New York, 2013), pp. 190–195

12. M. Tehranipoor, H. Salmani, X. Zhang, *Integrated Circuit Authentication: Hardware Trojans and Counterfeit Detection* (2016)

13. M. Li, A. Davoodi, M. Tehranipoor, A sensor-assisted self-authentication framework for hardware Trojan detection, in *2012 Design, Automation & Test in Europe Conference & Exhibition (DATE)* (IEEE, New York, 2012), pp. 1331–1336

14. K.Z. Azar, M.M. Hossain, A. Vafaei, H. Al Shaikh, N.N. Mondol, F. Rahman, M. Tehranipoor, F. Farahmandi, Fuzz, penetration, and ai testing for soc security verification: challenges and solutions, in *Cryptology ePrint Archive* (2022)

15. G.T. Becker, F. Regazzoni, C. Paar, W.P. Burleson, Stealthy Dopant-Level Hardware Trojans, in *CHES* (Springer, Berlin, 2013), pp. 197–214. https://doi.org/10.1007/978-3-642-40349-1_12. https://www.iacr.org/archive/ches2013/80860203/80860203.pdf.

16. T. Zhang, J. Park, M. Tehranipoor, F. Farahmandi, PSC-TG: RTL power side-channel leakage assessment with test pattern generation, in *2021 58th ACM/IEEE Design Automation Conference (DAC)* (IEEE, New York, 2021), pp. 709–714

17. A. Nahiyan, J. Park, M. He, Y. Iskander, F. Farahmandi, D. Forte, M. Tehranipoor, Script: a cad framework for power side-channel vulnerability assessment using information flow tracking and pattern generation. ACM Trans. Des. Autom. Electron. Syst. (TODAES) **25**(3), 1–27 (2020)

18. M. He, J. Park, A. Nahiyan, A. Vassilev, Y. Jin, M. Tehranipoor, RTL-PSC: automated power side-channel leakage assessment at register-transfer level, in *2019 IEEE 37th VLSI Test Symposium (VTS)* (IEEE, New York, 2019), pp. 1–6

19. U. Das, M.S. Rahman, N.N. Anandakumar, K.Z. Azar, F. Rahman, M. Tehranipoor, F. Farahmandi, PSC-Watermark: power side channel based IP watermarking using clock gates, in *2023 IEEE European Test Symposium (ETS)* (IEEE, New York, 2023), pp. 1–6

20. C. Momin, O. Bronchain, F.-X. Standaert, A stealthy Hardware Trojan based on a Statistical Fault Attack. Cryptogr. Commun. **13**(4), 587–600 (2021). https://doi.org/10.1007/s12095-021-00480-4

21. N. Ahmed, M. Tehranipoor, V. Jayaram, Transition delay fault test pattern generation considering supply voltage noise in a SOC design, in *Proceedings of the 44th annual Design Automation Conference* (2007), pp. 533–538

22. H. Wang, H. Li, F. Rahman, M.M. Tehranipoor, F. Farahmandi, Sofi: Security property-driven vulnerability assessments of ics against fault-injection attacks, in *IEEE Transactions on Computer-Aided Design of Integrated Circuits and Systems* (2021)

23. A. Nahiyan, F. Farahmandi, P. Mishra, D. Forte, M. Tehranipoor, Security-aware FSM design flow for identifying and mitigating vulnerabilities to fault attacks. IEEE Trans. Comput. Aided Des. Integr. Circuits Syst. **38**(6), 1003–1016 (2018)

24. F. Bao, K. Peng, M. Yilmaz, K. Chakrabarty, L. Winemberg, M. Tehranipoor, Efficient pattern generation for small-delay defects using selection of critical faults. J. Electron. Test. **29**(1), 35–48 (2013)

25. M. Tehranipoor, K. Peng, K. Chakrabarty. *Test and Diagnosis for Small-Delay Defects* (Springer, Berlin, 2011)

26. U. Guin, X. Zhang, D. Forte, M. Tehranipoor, Low-cost on-chip structures for combating die and IC recycling, in *2014 51st ACM/EDAC/IEEE Design Automation Conference (DAC)* (IEEE, New York, 2014), pp. 1–6

27. H. Dogan, D. Forte, M.M. Tehranipoor, Aging analysis for recycled FPGA detection, in *2014 IEEE International Symposium on Defect and Fault Tolerance in VLSI and Nanotechnology Systems (DFT)* (IEEE, New York, 2014), pp. 171–176

28. U. Guin, D. DiMase, M. Tehranipoor, A comprehensive framework for counterfeit defect coverage analysis and detection assessment. J. Electron. Test. **30**(1), 25–40 (2014)

29. M.M. Alam, M. Tehranipoor, D. Forte, Recycled FPGA detection using exhaustive LUT path delay characterization and voltage scaling. IEEE Trans. Very Large Scale Integr. VLSI Syst. **27**(12), 2897–2910 (2019)

30. M. Alam, S. Chowdhury, M.M. Tehranipoor, U. Guin, Robust, low-cost, and accurate detection of recycled ICs using digital signatures, in *2018 IEEE International Symposium on Hardware Oriented Security and Trust (HOST)* (IEEE, New York, 2018), pp. 209–214

31. Z. Guo, M.T. Rahman, M.M. Tehranipoor, D. Forte, A zero-cost approach to detect recycled SoC chips using embedded SRAM, in *2016 IEEE International Symposium on Hardware Oriented Security and Trust (HOST)* (IEEE, New York, 2016), pp. 191–196

32. J. Wurm, Y. Jin, Y. Liu, S. Hu, K. Heffner, F. Rahman, M. Tehranipoor, Introduction to cyber-physical system security: a cross-layer perspective. IEEE Trans. Multi-Scale Comput. Syst. **3**(3), 215–227 (2016)

33. K. Ahi, N. Asadizanjani, S. Shahbazmohamadi, M. Tehranipoor, M. Anwar, Terahertz characterization of electronic components and comparison of terahertz imaging with x-ray imaging techniques, in *Terahertz Physics, Devices, and Systems IX: Advanced Applications in Industry and Defense*, vol. 9483 (International Society for Optics and Photonics, Bellingham, 2015), p. 94830K

34. U. Guin, K. Huang, D. DiMase, J.M. Carulli, M. Tehranipoor, Y. Makris, Counterfeit integrated circuits: a rising threat in the global semiconductor supply chain. Proc. IEEE **102**(8), 1207–1228 (2014)

35. K.Z. Azar, H.M. Kamali, F. Farahmandi, M. Tehranipoor, *Understanding Logic Locking* (2023)

36. W. Hu, C.-H. Chang, A. Sengupta, S. Bhunia, R. Kastner, H. Li, An overview of hardware security and trust: threats, countermeasures, and design tools. IEEE Trans. Comput. Aided Des. Integr. Circuits Syst. **40**(6), 1010–1038 (2020)

37. N. Pundir, F. Farahmandi, M. Tehranipoor, Secure high-level synthesis: challenges and solutions, in *2021 22nd International Symposium on Quality Electronic Design (ISQED)* (IEEE, New York, 2021), pp. 164–171

38. H.M. Kamali, K.Z. Azar, K. Gaj, H. Homayoun, A. Sasan, LUT-Lock: A Novel LUT-based Logic Obfuscation for FPGA-bitstream and ASIC-hardware Protection, in *IEEE Computer Society Annual Symposium on VLSI (ISVLSI)* (2018), pp. 405–410

39. H.M. Kamali, K.Z. Azar, H. Homayoun, A. Sasan, Full-Lock: hard distributions of SAT instances for obfuscating circuits using fully configurable logic and routing blocks, in *Proceedings of Design Automation Conference (DAC)* (2019), p. 89

40. M.R. Muttaki, R. Mohammadivojdan, H.M. Kamali, M. Tehranipoor, F. Farahmandi, Hlock+: a robust and low-overhead logic locking at the high-level language, in *IEEE Transactions on Computer-Aided Design of Integrated Circuits and Systems* (2022)

41. H.M. Kamali, K.Z. Azar, F. Farahmandi, M. Tehranipoor, Advances in logic locking: past, present, and prospects, in *Cryptology ePrint Archive* (2022)

42. A.B. Kahng, J. Lienig, I.L. Markov, J. Hu, *VLSI Physical Design: From Graph Partitioning to Timing Closure* (Springer Science & Business Media, Berlin, 2011)

43. D. Šijačić, J. Balasch, I. Verbauwhede, Sweeping for leakage in masked circuit layouts, in *2020 Design, Automation & Test in Europe Conference & Exhibition (DATE)* (IEEE, New York, 2020), pp. 915–920

44. T. De Cnudde, B. Bilgin, B. Gierlichs, V. Nikov, S. Nikova, V. Rijmen, Does coupling affect the security of masked implementations?, in *International Workshop on Constructive Side-Channel Analysis and Secure Design* (Springer, Berlin, 2017), pp. 1–18

45. S. Bhasin, J.-L. Danger, T. Graba, Y. Mathieu, D. Fujimoto, M. Nagata, Physical security evaluation at an early design-phase: a side-channel aware simulation methodology, in *Proceedings of International Workshop on Engineering Simulations for Cyber-Physical Systems* (2013), pp. 13–20

46. F. Regazzoni, S. Badel, T. Eisenbarth, J. Grobschadl, A. Poschmann, Z. Toprak, M. Macchetti, L. Pozzi, C. Paar, Y. Leblebici, et al., A simulation-based methodology for evaluating the DPA-resistance of cryptographic functional units with application to CMOS and MCML technologies, in *2007 International Conference on Embedded Computer Systems: Architectures, Modeling and Simulation* (IEEE, New York, 2007), pp. 209–214

47. L. Lin, D. Selvakumaran, D. Zhu, N. Chang, C. Chow, M. Nagata, K. Monta, Fast and comprehensive simulation methodology for layout-based power-noise side-channel leakage analysis, in *2020 IEEE International Symposium on Smart Electronic Systems (iSES)(Formerly iNiS)* (IEEE, New York, 2020), pp. 133–138

48. C. O'Flynn, Fault injection using crowbars on embedded systems, in *IACR Cryptol. ePrint Arch.*, vol. 2016 (2016), p. 810

49. J. Balasch, B. Gierlichs, I. Verbauwhede, An in-depth and black-box characterization of the effects of clock glitches on 8-bit MCUs, in *2011 Workshop on Fault Diagnosis and Tolerance in Cryptography* (IEEE, New York, 2011), pp. 105–114

50. A. Barenghi, G.M. Bertoni, L. Breveglieri, M. Pellicioli, G. Pelosi, Injection technologies for fault attacks on microprocessors, in *Fault Analysis in Cryptography* (2012), pp. 275–293

51. M. Hutter, J.-M. Schmidt, T. Plos, Contact-based fault injections and power analysis on RFID tags, in *2009 European Conference on Circuit Theory and Design* (IEEE, New York, 2009), pp. 409–412

52. H. Martin, T. Korak, E.S. Millán, M. Hutter, Fault attacks on STRNGs: impact of glitches, temperature, and underpowering on randomness. IEEE Trans. Inf. Forensics Secur. **10**(2), 266–277 (2014)

53. G. Canivet, P. Maistri, R. Leveugle, J. Clédière, F. Valette, M. Renaudin, Glitch and laser fault attacks onto a secure AES implementation on a SRAM-based FPGA. J. Cryptol. **24**(2), 247–268 (2011)

54. B. Selmke, F. Hauschild, J. Obermaier, Peak clock: fault injection into PLL-based systems via clock manipulation, in *Proceedings of the 3rd ACM Workshop on Attacks and Solutions in Hardware Security Workshop* (2019), pp. 85–94

55. T. Bonny, Q. Nasir, Clock glitch fault injection attack on an FPGA-based non-autonomous chaotic oscillator. Nonlinear Dyn. **96**3, 2087–2101 (2019)

56. B. Ning, Q. Liu, Modeling and efficiency analysis of clock glitch fault injection attack, in *2018 Asian Hardware Oriented Security and Trust Symposium (AsianHOST)* (IEEE, New York, 2018), pp. 13–18

57. N. Timmers, C. Mune, Escalating privileges in linux using voltage fault injection, in *2017 Workshop on Fault Diagnosis and Tolerance in Cryptography (FDTC)* (IEEE, New York, 2017), pp. 1–8

58. D. Ha, K. Woo, S. Meninger, T. Xanthopoulos, E. Crain, D. Ham, Time-domain CMOS temperature sensors with dual delay-locked loops for microprocessor thermal monitoring. IEEE Trans. Very Large Scale Integr. VLSI Syst. **20**(9), 1590–1601 (2011)

59. J.G.J. Van Woudenberg, M.F. Witteman, F. Menarini, Practical optical fault injection on secure microcontrollers, in *2011 Workshop on Fault Diagnosis and Tolerance in Cryptography* (IEEE, New York, 2011), pp. 91–99

60. M. Yilmaz, K. Chakrabarty, M. Tehranipoor, Interconnect-aware and layout-oriented test-pattern selection for small-delay defects, in *2008 IEEE International Test Conference* (IEEE, New York, 2008), pp. 1–10

61. H. Ziade, R.A. Ayoubi, R. Velazco, et al., A survey on fault injection techniques. Int. Arab J. Inf. Technol. **1**(2), 171–186 (2004)

62. H. Wang, Q. Shi, A. Nahiyan, D. Forte, M.M. Tehranipoor, A physical design flow against front-side probing attacks by internal shielding. IEEE Trans. Comput. Aided Des. Integr. Circuits Syst. **39**(10), 2152–2165 (2019)

63. H. Wang, Q. Shi, D. Forte, M.M. Tehranipoor, Probing assessment framework and evaluation of antiprobing solutions. IEEE Trans. Very Large Scale Integr. VLSI Syst. **27**(6), 1239–1252 (2019)

64. M. Agoyan, J.-M. Dutertre, D. Naccache, B. Robisson, A. Tria, When clocks fail: on critical paths and clock faults, in *International Conference on Smart Card Research and Advanced Applications* (Springer, Berlin, 2010), pp. 182–193
65. L. Zussa, J.-M. Dutertre, J. Clédiere, B. Robisson, A. Tria, et al., Investigation of timing constraints violation as a fault injection means, in *27th Conference on Design of Circuits and Integrated Systems (DCIS), Avignon, France* (Citeseer, New York, 2012), pp. 1–6
66. B. Razavi, *Fundamentals of Microelectronics* (Wiley, New York, 2021)
67. K.Z. Azar, H.M. Kamali, H. Homayoun, A. Sasan, SMT attack: next generation attack on obfuscated circuits with capabilities and performance beyond the SAT attacks, in *IACR Transactions on Cryptographic Hardware and Embedded Systems (TCHES)* (2019), pp. 97–122
68. K.M. Zick, M. Srivastav, W. Zhang, M. French, Sensing nanosecond-scale voltage attacks and natural transients in FPGAs, in *Proceedings of the ACM/SIGDA International Symposium on Field Programmable Gate Arrays* (2013), pp. 101–104
69. K. Murdock, D. Oswald, F.D. Garcia, J. Van Bulck, D. Gruss, F. Piessens, Plundervolt: software-based fault injection attacks against Intel SGX, in *2020 IEEE Symposium on Security and Privacy (SP)* (IEEE, New York, 2020), pp. 1466–1482
70. J. Krautter, D.R.E. Gnad, M.B. Tahoori, FPGAhammer: remote voltage fault attacks on shared FPGAs, suitable for DFA on AES, in *IACR Transactions on Cryptographic Hardware and Embedded Systems* (2018), pp. 44–68
71. N.N. Anandakumar, S.K. Sanadhya, M.S. Hashmi, FPGA-based true random number generation using programmable delays in oscillator-rings. IEEE Trans. Circuits Syst. II Express Briefs **67**(3), 570–574 (2019)
72. M. Dumont, M. Lisart, P. Maurine, Electromagnetic fault injection: how faults occur, in *2019 Workshop on Fault Diagnosis and Tolerance in Cryptography (FDTC)* (IEEE, New York, 2019), pp. 9–16
73. T.D. Perez, S. Pagliarini, A survey on split manufacturing: attacks, defenses, and challenges. IEEE Access **8**, 184013–184035 (2020)
74. Y. Xie, C. Bao, A. Srivastava, Security-aware design flow for 2.5 D IC technology, in *Proceedings of the 5th International Workshop on Trustworthy Embedded Devices* (2015), pp. 31–38
75. Y. Wang, P. Chen, J. Hu, J. Rajendran, Routing perturbation for enhanced security in split manufacturing, in *Asia and South Pacific Design Automation Conference (ASP-DAC)* (2017), pp. 605–510
76. H. Salmani, M.M. Tehranipoor, Vulnerability analysis of a circuit layout to hardware trojan insertion. IEEE Trans. Inf. Forensics Secur. **11**(6), 1214–1225 (2016)
77. Q. Shi, N. Asadizanjani, D. Forte, M.M. Tehranipoor, A layout-driven framework to assess vulnerability of ICs to microprobing attacks, in *2016 IEEE International Symposium on Hardware Oriented Security and Trust (HOST)* (IEEE, New York, 2016), pp. 155–160
78. H. Salmani, M. Tehranipoor, Layout-aware switching activity localization to enhance hardware Trojan detection. IEEE Trans. Inf. Forensics Secur. **7**(1), 76–87 (2011)
79. H. Salmani, M. Tehranipoor, J. Plusquellic, A layout-aware approach for improving localized switching to detect hardware Trojans in integrated circuits, in *2010 IEEE International Workshop on Information Forensics and Security* (IEEE, New York, 2010), pp. 1–6
80. D. Saha, S. Sur-Kolay, Fast robust intellectual property protection for VLSI physical design, in *10th International Conference on Information Technology (ICIT 2007)* (IEEE, New York, 2007), pp. 1–6
81. S. Zamanzadeh, A. Jahanian, Automatic netlist scrambling methodology in ASIC design flow to hinder the reverse engineering, in *2013 IFIP/IEEE 21st International Conference on Very Large Scale Integration (VLSI-SoC)* (IEEE, New York, 2013), pp. 52–53
82. A. Vijayakumar, V.C. Patil, D.E. Holcomb, C. Paar, S. Kundu, Physical design obfuscation of hardware: a comprehensive investigation of device and logic-level techniques. IEEE Trans. Inf. Forensics Secur. **12**(1), 64–77 (2016)

83. H.M. Kamali, K.Z. Azar, H. Homayoun, A. Sasan, ChaoLock: yet another SAT-hard logic locking using chaos computing, in *International Symposium on Quality Electronic Design (ISQED)* (2021), pp. 387–394

84. S. Rahman, N. Varshney, F. Farahmandi, N.A. Zanjani, M. Tehranipoor, LLE: mitigating IC piracy and reverse engineering by last level edit, in *ISTFA 2023* (ASM International, New York, 2023), pp. 360–369

85. M. Khairallah, R. Sadhukhan, R. Samanta, J. Breier, S. Bhasin, R.S. Chakraborty, A. Chattopadhyay, D. Mukhopadhyay, DFARPA: differential fault attack resistant physical design automation, in *2018 Design, Automation & Test in Europe Conference & Exhibition (DATE)* (IEEE, New York, 2018), pp. 1171–1174

86. M. Gao, M.S. Rahman, N. Varshney, M. Tehranipoor, D. Forte, iPROBE: internal shielding approach for protecting against front-side and back-side probing attacks, in *IEEE Transactions on Computer-Aided Design of Integrated Circuits and Systems* (2023)

87. J. Knechtel, et al., Security closure of physical layouts, in *2021 IEEE/ACM International Conference on Computer-Aided Design (ICCAD)* (IEEE, New York, 2021)

88. P. Slpsk, P.K. Vairam, C. Rebeiro, V. Kamakoti, Karna: a gate-sizing based security aware EDA flow for improved power side-channel attack protection, in *2019 IEEE/ACM International Conference on Computer-Aided Design (ICCAD)* (IEEE, New York, 2019), pp. 1–8

89. H. Sonoda, K. Monta, T. Okidono, Y. Araga, N. Watanabe, H. Shimamoto, K. Kikuchi, N. Miura, T. Miki, M. Nagata, Secure 3D CMOS chip stacks with backside buried metal power delivery networks for distributed decoupling capacitance, in *2020 IEEE International Electron Devices Meeting (IEDM)* (IEEE, New York, 2020), pp. 31–5

90. K. Monta, H. Sonoda, T. Okidono, Y. Araga, N. Watanabe, H. Shimamoto, K. Kikuchi, N. Miura, T. Miki, M. Nagata, 3-D CMOS chip stacking for security ICs featuring backside buried metal power delivery networks with distributed capacitance. IEEE Trans. Electron Devices **68**(4), 2077–2082 (2021)

91. M. Wang, V.V. Iyer, S. Xie, G. Li, S.K. Mathew, R. Kumar, M. Orshansky, A.E. Yilmaz, J.P. Kulkarni, Physical design strategies for mitigating fine-grained electromagnetic side-channel attacks, in *2021 IEEE Custom Integrated Circuits Conference (CICC)* (IEEE, New York, 2021), pp. 1–2

92. S. Brown, S. Aftabjahani, M. Tehranipoor, *Trust-hub Physical Vulnerabilities-DB*. https://trust-hub.org/#/vulnerability-db/physical-vulnerabilities

93. P.C. Kocher, Timing attacks on implementations of Diffie-Hellman, RSA, DSS, and other systems, in *Annual International Cryptology Conference* (Springer, Berlin, 1996), pp. 104–113

94. P.C. Kocher, J. Jaffe, B. Jun, Differential power analysis, in *Proceedings of the 19th Annual International Cryptology Conference on Advances in Cryptology (CRYPTO '99)* (Springer, Berlin, 1999), pp. 388–397. isbn: 3540663479

95. D. Agrawal, B. Archambeault, J.R. Rao, P. Rohatgi, The EM Side-Channel(s), in *Revised Papers from the 4th International Workshop on Cryptographic Hardware and Embedded Systems (CHES '02)* (Springer, Berlin, 2002), pp. 29–45. ISBN: 3540004092

96. D. Genkin, A. Shamir, E. Tromer, RSA key extraction via low-bandwidth acoustic cryptanalysis, in *Advances in Cryptology—CRYPTO 2014*, ed. by J.A. Garay, R. Gennaro (Springer, Berlin, 2014), pp. 444–461. ISBN: 978-3-662-44371-2

97. A. Schlösser, D. Nedospasov, J. Kämer, S. Orlic, J.-P. Seifert, Simple photonic emission analysis of AES, in *Cryptographic Hardware and Embedded Systems—CHES 2012*, ed. by E. Prouff, P. Schaumont (Springer, Berlin, 2012), pp. 41–57. ISBN: 978-3-642-33027-8

98. Y. Yarom, K. Falkner, FLUSH+RELOAD: a high resolution, low noise, L3 cache side-channel attack, in *Proceedings of the 23rd USENIX Conference on Security Symposium. SEC'14* (USENIX Association, San Diego, CA, 2014), pp. 719–732. ISBN: 9781931971157

99. N. Ahmed, M.H. Tehranipour, M. Nourani, Low power pattern generation for BIST architecture, in *2004 IEEE International Symposium on Circuits and Systems (IEEE Cat. No. 04CH37512)*, vol. 2 (IEEE, New York, 2004), pp. II–689

100. D. McCann, E. Oswald, C. Whitnall, Towards practical tools for side channel aware software engineering: 'Grey Box' modelling for instruction leakages, in *26th USENIX Security Symposium (USENIX Security 17)* (USENIX Association, Vancouver, BC, 2017), pp. 199–216. ISBN: 978-1-931971-40-9. https://www.usenix.org/conference/usenixsecurity17/technical-sessions/presentation/mccann

101. J. Park, X. Xu, Y. Jin, D. Forte, M. Tehranipoor, Power-based side-channel instruction-level disassembler, in *2018 55th ACM/ESDA/IEEE Design Automation Conference (DAC)* (2018), pp. 1–6. https://doi.org/10.1109/DAC.2018.8465848

102. M.-L. Akkar, C. Giraud, An implementation of DES and AES, secure against some attacks, in *Cryptographic Hardware and Embedded Systems—CHES 2001*, ed. by Ç.K. Koç, D. Naccache, C. Paar (Springer, Berlin, 2001), pp. 309–318. ISBN: 978-3-540-44709-2

103. S. Mangard, T. Popp, B.M. Gammel, Side-channel leakage of masked CMOS gates, in *Topics in Cryptology—CT-RSA 2005*, ed. by A. Menezes (Springer, Berlin, 2005), pp. 351–365. ISBN: 978-3-540-30574-3

104. M.H. Tehranipour, N. Ahmed, M. Nourani, Testing soc interconnects for signal integrity using boundary scan, in *Proceedings. 21st VLSI Test Symposium, 2003* (IEEE, New York 2003), pp. 158–163

105. S. Nikova, C. Rechberger, V. Rijmen, Threshold implementations against side-channel attacks and glitches, in *Information and Communications Security*, ed. by P. Ning, S. Qing, N. Li (Springer, Berlin, 2006), pp. 529–545. ISBN: 978-3-540-49497-3

106. T. De Cnudde, B. Bilgin, B. Gierlichs, V. Nikov, S. Nikova, V. Rijmen, *Does Coupling Affect the Security of Masked Implementations?* Cryptology ePrint Archive, Report 2016/1080 (2016). https://ia.cr/2016/1080

107. G. Goodwill, B. Jun, J. Jaffe, P. Rohatgi, *A Testing Methodology for Side Channel Resistance* (2011)

108. N. Sehatbakhsh, B.B. Yilmaz, A. Zajic, M. Prvulovic, EMSim: A microarchitecture-level simulation tool for modeling electromagnetic side-channel signals, in *2020 IEEE International Symposium on High Performance Computer Architecture (HPCA)* (2020), pp. 71–85. https://doi.org/10.1109/HPCA47549.2020.00016

109. M.A. Shelton, N. Samwel, L. Batina, F. Regazzoni, M. Wagner, Y. Yarom, Rosita: towards automatic elimination of power-analysis leakage in ciphers, in *28th Annual Network and Distributed System Security Symposium, NDSS 2021, virtually, February 21–25, 2021* (The Internet Society, Reston, 2021)

110. M.A. Shelton, L. Chmielewski, N. Samwel, M. Wagner, L. Batina, Y. Yarom, Rosita++: automatic higher-order leakage elimination from cryptographic code, in *Proceedings of the 2021 ACM SIGSAC Conference on Computer and Communications Security.* CCS '21. Virtual Event, Republic of Korea: Association for Computing Machinery (2021), pp. 685–699. ISBN: 9781450384544. https://doi.org/10.1145/3460120.3485380

111. A. Nahiyan, J. Park, M. He, Y. Iskander, F. Farahmandi, D. Forte, M. Tehranipoor, SCRIPT: A CAD framework for power side-channel vulnerability assessment using information flow tracking and pattern generation. ACM Trans. Des. Autom. Electron. Syst. **25**(3), 1–27 (2020). ISSN: 1084-4309. https://doi.org/10.1145/3383445

112. P. SLPSK, P.K. Vairam, C. Rebeiro, V. Kamakoti, Karna: a gate-sizing based security aware EDA flow for improved power side-channel attack protection, in *Proceedings of the International Conference on Computer-Aided Design, ICCAD 2019, Westminster, CO, USA, November 4–7, 2019*, ed. by D.Z. Pan (ACM, New York, 2019), pp. 1–8. https://doi.org/10.1109/ICCAD45719.2019.8942173.

113. J. Park, N.N. Anandakumar, D. Saha, D. Mehta, N. Pundir, F. Rahman, F. Farahmandi, M.M. Tehranipoor, PQC-SEP: power side-channel evaluation platform for post-quantum cryptography algorithms, in *IACR Cryptol. ePrint Arch.*, vol. 2022 (2022), p. 527

114. N. Pundir, J. Park, F. Farahmandi, M. Tehranipoor, Power side-channel leakage assessment framework at register-transfer level. IEEE Trans. Very Large Scale Integr. VLSI Syst. **30**(9), 1207–1218 (2022)

115. N. Sehatbakhsh, B.B. Yilmaz, A.G. Zajic, M. Prvulovic, EMSim: a microarchitecture-level simulation tool for modeling electromagnetic side-channel signals, in *IEEE International Symposium on High Performance Computer Architecture, HPCA 2020, San Diego, CA, USA, February 22–26, 2020* (IEEE, New York, 2020), pp. 71–85. https://doi.org/10.1109/HPCA47549.2020.00016

116. T. Zhang, J. Park, M.M. Tehranipoor, F. Farahmandi, PSC-TG: RTL power side-channel leakage assessment with test pattern generation, in *58th ACM/IEEE Design Automation Conference, DAC 2021, San Francisco, CA, USA, December 5–9, 2021* (IEEE, New York, 2021), pp. 709–714. https://doi.org/10.1109/DAC18074.2021.9586210

117. B. Gigerl, V. Hadzic, R. Primas, S. Mangard, R. Bloem, Coco: Co-Design and co-verification of masked software implementations on CPUs, in *30th USENIX Security Symposium (USENIX Security 21)* (USENIX Association, New York, 2021), pp. 1469–1468. ISBN: 978-1-939133-24-3. https://www.usenix.org/conference/usenixsecurity21/presentation/gigerl

118. M.M. Alam, S. Tajik, F. Ganji, M. Tehranipoor, D. Forte, Ram-jam: remote temperature and voltage fault attack on fpgas using memory collisions, in *2019 Workshop on Fault Diagnosis and Tolerance in Cryptography (FDTC)* (IEEE, New York, 2019), pp. 48–55

119. M.R. Muttaki, M. Tehranipoor, F. Farahmandi, FTC: a universal fault injection attack detection sensor, in *IEEE International Symposium on Hardware-Oriented Security and Trust (HOST)* (IEEE, New York, 2022)

120. A. Tria, Frontside laser fault injection on cryptosystems-Application to the AES'last round, in: *2013 IEEE International Symposium on Hardware-Oriented Security and Trust (HOST)* (IEEE Computer Society, New York, 2013), pp. 119–124

121. P. Maurine, Techniques for em fault injection: equipments and experimental results, in *2012 Workshop on Fault Diagnosis and Tolerance in Cryptography* (IEEE, New York, 2012), pp. 3–4

122. N. Pundir, H. Li, L. Lin, N. Chang, F. Farahmandi, M. Tehranipoor, Security properties driven pre-silicon laser fault injection assessment, in *2022 IEEE International Symposium on Hardware-Oriented Security and Trust (HOST)* (2022)

123. N. Pundir, H. Li, L. Lin, N. Chang, F. Farahmandi, M. Tehranipoor, SPILI: security properties and machine learning assisted pre-silicon laser fault injection assessment, in *International Symposium for Testing and Failure Analysis (ISTFA)* (2022)

124. C. Roscian, J.-M. Dutertre, A. Tria, Frontside laser fault injection on cryptosystems-Application to the AES'last round, in *2013 IEEE International Symposium on Hardware-Oriented Security and Trust (HOST)* (IEEE, New York, 2013), pp. 119–124

125. M. Agoyan, J.-M. Dutertre, A.-P. Mirbaha, D. Naccache, A.-L. Ribotta, A. Tria, Single-bit DFA using multiple-byte laser fault injection, in *2010 IEEE International Conference on Technologies for Homeland Security (HST)* (IEEE, New York, 2010), pp. 113–119

126. B. Selmke, J. Heyszl, G. Sigl, Attack on a DFA protected AES by simultaneous laser fault injections, in *2016 Workshop on Fault Diagnosis and Tolerance in Cryptography (FDTC)* (IEEE, New York, 2016), pp. 36–46

127. F. Cai, G. Bai, H. Liu, X. Hu, Optical fault injection attacks for flash memory of smartcards, in *2016 6th International Conference on Electronics Information and Emergency Communication (ICEIEC)* (IEEE, New York, 2016), pp. 46–50

128. T. Sugawara, B. Cyr, S. Rampazzi, D. Genkin, K. Fu, Light commands: laser-based audio injection attacks on voice-controllable systems, in *29th {USENIX} Security Symposium ({USENIX}Security 20)* (2020), pp. 2631–2648

129. A. Vasselle, H. Thiebeauld, Q. Maouhoub, A. Morisset, S. Ermeneux, Laser-induced fault injection on smartphone bypassing the secure boot, in *2017 Workshop on Fault Diagnosis and Tolerance in Cryptography (FDTC)* (IEEE, New York, 2017), pp. 41–48

130. J. Breier, D. Jap, X. Hou, S. Bhasin, Y. Liu, SNIFF: reverse engineering of neural networks with fault attacks. IEEE Trans. Reliab. **71**(4), 1527–1539 (2021)

131. E. Biham, A. Shamir, Differential fault analysis of secret key cryptosystems, in *Annual International Cryptology Conference* (Springer, Berlin, 1997), pp. 513–525

132. Y. Li, K. Sakiyama, S. Gomisawa, T. Fukunaga, J. Takahashi, K. Ohta, Fault sensitivity analysis, in *International Workshop on Cryptographic Hardware and Embedded Systems* (Springer, Berlin, 2010), pp. 320–334

133. N.F. Ghalaty, B. Yuce, M. Taha, P. Schaumont, Differential fault intensity analysis, in *2014 Workshop on Fault Diagnosis and Tolerance in Cryptography* (IEEE, New York, 2014), pp. 49–58

134. S. Ordas, L. Guillaume-Sage, K. Tobich, J.-M. Dutertre, P. Maurine, Evidence of a larger EM-induced fault model, in *Inter-national Conference on Smart Card Research and Advanced Applications* (Springer, Berlin, 2014), pp. 245–259

135. T. Miki, M. Nagata, H. Sonoda, N. Miura, T. Okidono, Y. Araga, N. Watanabe, H. Shimamoto, K. Kikuchi, A Si-backside protection circuits against physical security attacks on flip-chip devices, in *2019 IEEE Asian Solid-State Circuits Conference (A-SSCC)* (IEEE, New York, 2019), pp. 25–28

136. T. Ajayi, V.A. Chhabria, M. Fogaça, S. Hashemi, A. Hosny, A.B. Kahng, M. Kim, J. Lee, U. Mallappa, M. Neseem, et al., Toward an open-source digital flow: First learnings from the openroad project, in *Proceedings of the 56th Annual Design Automation Conference 2019* (2019), pp. 1–4

137. T.-C. Chen, Z.-W. Jiang, T.-C. Hsu, H.-C. Chen, Y.-W. Chang, NTUplace3: an analytical placer for large-scale mixed-size designs with preplaced blocks and density constraints, in IEEE Trans. Comput. Aided Des. Integr. Circuits Syst. **27**(7), 1228–1240 (2008)

138. C.-K. Cheng, A.B. Kahng, I. Kang, L. Wang, Replace: advancing solution quality and routability validation in global placement. IEEE Trans. Comput. Aided Des. Integr. Circuits Syst. **38**(9), 1717–1730 (2018)

139. Y. Lin, Z. Jiang, J. Gu, W. Li, S. Dhar, H. Ren, B. Khailany, D.Z. Pan, Dreamplace: deep learning toolkit-enabled gpu acceleration for modern vlsi placement, in IEEE Trans. Comput. Aided Des. Integr. Circuits Syst. **40**(4), 748–761 (2020)

140. C.-H. Hsu, H.-Y. Chen, Y.-W. Chang, Multi-layer global routing considering via and wire capacities, in *2008 IEEE/ACM International Conference on Computer-Aided Design* (IEEE, New York, 2008), pp. 350–355

141. *Innovus Implementation System* (Cadence, San Jose, 2022). https://www.cadence.com/en_US/home/tools/digital-design-and-signoff/soc-implementation-and-floorplanning/innovus-implementation-system.html

142. *IC Compiler II* (Synopsys, Sunnyvale, 2022). https://www.synopsys.com/implementation-and-signoff/physical-implementation/ic-compiler.html

143. J. Park, A. Tyagi, Security metrics for power based SCA resistant hardware implementation, in *2016 29th International Conference on VLSI Design and 2016 15th International Conference on Embedded Systems (VLSID)* (2016), pp. 541–546. https://doi.org/10.1109/VLSID.2016.43

144. Y. Fei, A.A. Ding, J. Lao, L. Zhang, *A Statistics-based Fundamental Model for Side-channel Attack Analysis.* Cryptology ePrint Archive, Report 2014/152. https://eprint.iacr.org/2014/152.2014

145. A.M. Shuvo, N. Pundir, J. Park, F. Farahmandi, M. Tehranipoor, LDTFI: layout-aware timing fault-injection attack assessment against differential fault analysis, in *2022 IEEE Computer Society Annual Symposium on VLSI* (2022)

146. S. Dey, S. Dash, S. Nandi, G. Trivedi, PGIREM: reliability-constrained IR drop minimization and electromigration assessment of VLSI power grid networks using cooperative coevolution, in *2018 IEEE Computer Society Annual Symposium on VLSI (ISVLSI)* (IEEE, New York, 2018), pp. 40–45

147. S. Dey, S. Nandi, G. Trivedi, PGOpt: multi-objective design space exploration framework for large-Scale on-chip power grid design in VLSI SoC using evolutionary computing technique. Microprocess. Microsyst. **81**, 103440 (2021)

148. S. Dey, S. Nandi, G. Trivedi, PGRDP: reliability, delay, and power-aware area minimization of large-scale VLSI power grid network using cooperative coevolution, in *Intelligent Computing Paradigm: Recent Trends* (Springer, Berlin, 2020), pp. 69–84

149. D. Das, M. Nath, B. Chatterjee, S. Ghosh, S. Sen, *STELLAR: A Generic EM Side-Channel Attack Protection through Ground-Up Root-cause Analysis*. Cryptology ePrint Archive, Report 2018/620 (2018). https://ia.cr/2018/620

150. P.P. Sarker, U. Das, M.B. Monjil, H.M. Kamali, F. Farahmandi, M. Tehranipoor, GEM-Water: generation of EM-based watermark for SoC IP validation with hidden FSMs, in *ISTFA 2023* (ASM International, New York, 2023), pp. 271–278

151. C. Tokunaga, D. Blaauw, Securing encryption systems with a switched capacitor current equalizer. IEEE J. Solid State Circuits **45**(1), 23–31 (2010). https://doi.org/10.1109/JSSC.2009.2034081

152. *4th ACM/IEEE Workshop on Machine Learning for CAD* (2022). https://mlcad-workshop.org/

153. A. Mirhoseini, A. Goldie, M. Yazgan, J.W. Jiang, E. Songhori, S. Wang, Y.-J. Lee, E. Johnson, O. Pathak, A. Nazi, et al., A graph placement methodology for fast chip design. Nature **594**(7862), 207–212 (2021)

154. Y.-J. Lee, et al.. *Learning to Play the Game of Macro Placement with the help of Deep Reinforcement Learning* (2021). https://youtu.be/EKjlr2k%5C_wBM

155. G. Huang, J. Hu, Y. He, J. Liu, M. Ma, Z. Shen, J. Wu, Y. Xu, H. Zhang, K. Zhong, et al., Machine learning for electronic design automation: a survey. ACM Trans. Des. Autom. Electron. Syst. (TODAES) **26**(5), 1–46 (2021)

156. M. Rapp, H. Amrouch, Y. Lin, B. Yu, D.Z. Pan, M. Wolf, J. Henkel, MLCAD: a survey of research in machine learning for CAD keynote paper, in *IEEE Transactions on Computer-Aided Design of Integrated Circuits and Systems* (2021)

157. S. Dey, S. Nandi, G. Trivedi, Machine learning for VLSI CAD: a case study in on-chip power grid design, in *2021 IEEE Computer Society Annual Symposium on VLSI (ISVLSI)* (IEEE, New York, 2021), pp. 378–383

158. S. Dey, S. Nandi, G. Trivedi, Machine learning approach for fast electromigration aware aging prediction in incremental design of large scale on-chip power grid network. ACM Trans. Des. Autom. Electron. Syst. (TODAES) **25**(5), 1–29 (2020)

159. S. Dey, S. Nandi, G. Trivedi, PowerPlanningDL: reliability-aware framework for on-chip power grid design using deep learning, in *2020 Design, Automation & Test in Europe Conference & Exhibition (DATE)* (IEEE, New York, 2020), pp. 1520–1525

160. S. Dey, *Design Methodology for On-Chip Power Grid Interconnect: AI/ML Perspective* (IIT, Guwahati, 2021)

161. P.J.M. Van Laarhoven, E.H.L. Aarts, Simulated annealing, in *Simulated annealing: Theory and applications* (Springer, Berlin, 1987), pp. 7–15

162. J.H. Holland, Genetic algorithms. Sci. Am. **267**(1), 66–73 (1992)

163. H.B. Curry, The method of steepest descent for non-linear minimization problems. Q. Appl. Math. **2**(3), 258–261 (1944)

164. R. Maclin, *Machine Learning for Sequential Data* (University of Minnesota, Minnesota, 2002)

165. N.E. Huang, Z. Shen, S.R. Long, M.C. Wu, H.H. Shih, Q. Zheng, N.-C. Yen, C.C. Tung, H.H. Liu, The empirical mode decomposition and the Hilbert spectrum for nonlinear and non-stationary time series analysis. Proc. R. Soc. London, Ser. A Math. Phys. Eng. Sci. **454**(1971), 903–995 (1998)

166. M. Feldman, Time-varying vibration decomposition and analysis based on the Hilbert transform. J. Sound Vib. **295**(3–5), 518–530 (2006)

167. K. Dragomiretskiy, D. Zosso, Variational mode decomposition. IEEE Trans. Signal Proc. **62**(3), 531–544 (2013)

168. G. Zaid, L. Bossuet, A. Habrard, A. Venelli, Methodology for efficient CNN architectures in profiling attacks. IACR Trans. Cryptographic Hardware and Embedded Systems **2020**(1), 1–36 (2020)

169. L. Wu, G. Perin, S. Picek, I choose you: automated hyperparameter tuning for deep learning-based side-channel analysis, in *IACR Cryptol. ePrint Arch.*, vol. 2020 (2020), p. 1293

170. G. Perin, L. Chmielewski, S. Picek, Strength in numbers: improving generalization with ensembles in machine learning-based profiled side-channel analysis, in *IACR Transactions on Cryptographic Hardware and Embedded Systems* (2020), pp. 337–364
171. E. Prouff, R. Strullu, R. Benadjila, E. Cagli, C. Dumas, Study of deep learning techniques for side-channel analysis and introduction to ASCAD database, in *Cryptology ePrint Archive* (2018)

Chapter 10
Secure Heterogeneous Integration

10.1 Introduction

Electronic devices are becoming deeply ingrained in our lifestyle by transforming the way we live and work. We live in a digital economy with the widespread connectivity of high-speed devices generating big data. At the same time, some systems need to capture, store, and analyze this big data to process further transactions (autonomous cars, data centers, and AI-based systems). This data evolution is supported at the ground level by state-of-the-art semiconductor ICs that provide multiple processing cores, high bandwidth memory, and high-speed I/O ports. These advanced ICs are available because of Moore's law, pushing the semiconductor industry to supply faster, smaller, and cheaper semiconductor ICs. However, this law is ending due to the fabrication cost, power dissipation, and yield issues at advanced technology nodes. The ITRS 2015 has set a long-term vision to sustain the historical scaling of CMOS technology to keep Moore's law alive by using Heterogeneous Integration (HI) [1]. The HI refers to integrating individually designed, and fabricated components that can be assembled on a substrate layer called an interposer to perform a function like an SoC [2].

Heterogeneous integration combines (see Fig. 10.1) separately manufactured components of different technology nodes and functionality to form a higher-level assembly called System-in-Package (SiP)[1] or Multichip Module (MCM). A SiP provides greater functionality and achieves better-operating characteristics which are challenging to achieve on a single-die systems-on-chip (SoCs). In a SiP, components such as chiplets, MEMS devices, and active/passive parts are integrated into a single package. A chiplet is an individually fabricated silicon die

---

[1] In a real sense, SiP is a couple of decade-old technology developed to shrink PCBs and achieve higher bandwidth and lower power. However, these days, the SiP term is reused for marketing purposes as a synonym for heterogeneous integration (HI). In this chapter, the term SiP stands for packages made through HI technology.

© The Author(s), under exclusive license to Springer Nature Switzerland AG 2024
M. Tehranipoor et al., *Hardware Security*,
https://doi.org/10.1007/978-3-031-58687-3_10

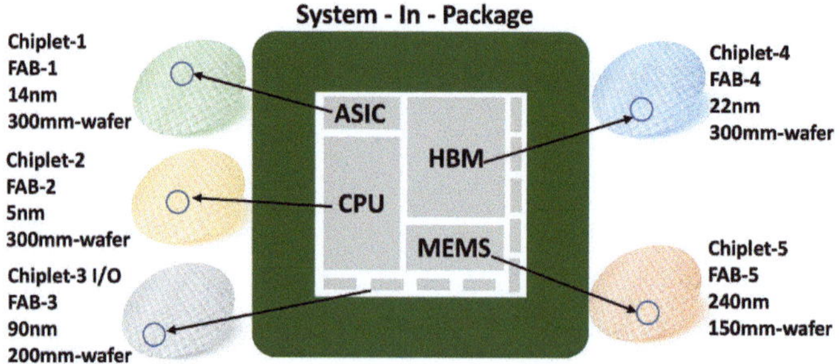

Fig. 10.1 Heterogeneous integration for building System-in-Package

(also known as hardened IP) for a targeted function such as memory, analog-mixed signal, RF, or processor. The System-in-Package can be a vertical stacking (3D) or adjacent placement (2.5D) of chiplets on the substrate layer called an interposer. Several integrated device manufacturers (Intel, Micron, and Samsung), fabless design houses (AMD and IBM), foundries (TSMC), and outsourced semiconductor assembly and testing companies (TSMC and Amkor) are working on developing heterogeneous integration solutions. For example, Intel Agilex and the AMD EPYC are the commercially available heterogeneous 3D System-in-Package (SiP). DARPA has a similar vision for the US government's DoD application designs and technologies through the Common Heterogeneous Integration and IP Reuse Strategies (CHIPS) program for trusted microelectronics.

Despite the many lucrative advantages of HI, it requires further research and development, including packaging technology, standardization of interconnecting interfaces, communication protocols, and secure design. For example, the packaging methods should consume less space to support small form factors. In addition, the design of the interconnecting interface of chiplets must conform with the speed, power requirements, and crosstalk issues. Some organizations (ODSA and CHIPS Alliance) are independently developing Die-to-Die interface protocols such as Advanced Interface Bus (AIB), Bunch of Wires (BoW), and open High Bandwidth Interconnect (openHBI). Recently, leaders in semiconductors design, packaging, IP suppliers, foundries, and cloud service providers have formed a consortium to develop an open standard for interfacing chiplets, i.e., Universal Chiplet Interconnect Express (UCIe) [3]. However, this interface standard's security vulnerabilities and associated risks are yet to be evaluated.

Like SoCs, the SiPs can be vulnerable to hardware security attacks. Hence, there is a need for trust validation of chiplets, substrate layer/interposer security, and security assessment of SiPs. In the SiP supply chain, a SiP OEM or designer obtains necessary chiplets from different chiplet equipment/component manufacturers (OEMs/OCMs) with no information about the IPs design of these chiplets. Further, the horizontal business model renders offshore chiplet foundries to control the fabrication and testing of the chiplets. This business model further

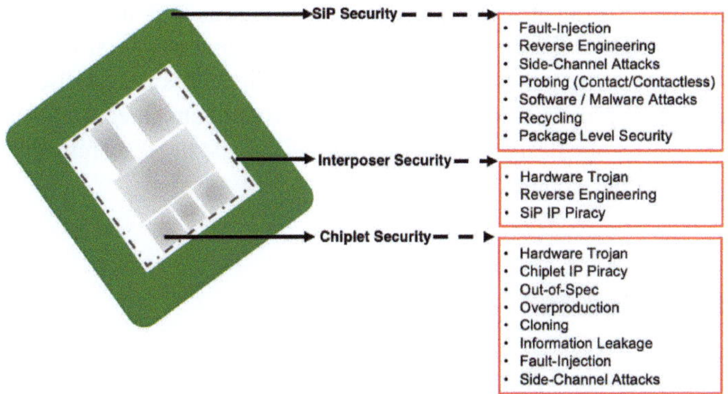

Fig. 10.2 Possible identified security risk and threats at various levels of heterogeneous integration

raises significant security concerns about the SiP's confidentiality, integrity, and availability (CIA) by making it vulnerable to various threats (e.g., hardware Trojans, out-of-spec, cloned, and overproduction). Therefore, a trust validation mechanism must be developed to validate the authenticity of chiplets acquired from different OCMs. In addition to trusted chiplets, the interposer layer should be free from any malicious changes (Interposer level Trojans) and immune to reverse engineering attacks.

Further, before deploying the SiP package in a system, a security assessment is required to ensure it is resistant to any security attacks. For example, similar to SoCs, a SiP package may be physically or remotely attacked (fault injection, contact-less probing, or side-channel analysis). In addition, a SiP designer needs to ensure secure communication between chiplets (anti-bus snooping or preventing unauthorized memory access) for a secure SiP. Hence, these security concerns should be addressed, from procurement of chiplets to interposer design and fabrication to the integration stage for a trusted & secure SiP (Fig. 10.2).

To summarize, we propose a comprehensive three-tier (chiplet-to-interposer-to-SiP) and end-to-end hardware security approach for secure heterogeneous integration by identifying risk and threats at every level as depicted in Fig. 10.4. Therefore following will be covered in this chapter:

- We have analyzed the chiplet security problem by identifying risks, threats, and vulnerabilities by determining attack vectors and surface. Furthermore, we identified adversarial entities that can compromise chiplet security and suggested respective trust validation methods.
- We have investigated security vulnerabilities at the interconnect (interposer) level of heterogeneous integration with a comprehensive set of possible threats in the supply chain, followed by specified the associated attack vectors & surface and recommended possible countermeasures.

- We analyzed SiP level security for secure heterogeneous integration, including the attack surface, threat model, vulnerabilities, and potential countermeasure. The role of security policies in protecting on-chip assets is also discussed.
- A road map is proposed toward a secure heterogeneous integration for a trusted System-in-Package design. This road map will assist the research community by opening multiple avenues for future research.

10.2 Preliminaries

10.2.1 Motivation for HI

Heterogeneous integration is expected to keep the pace of More-than-Moore (MtM) progress. It is moving forward with higher performance, yield, reduced latency, compact size, lighter weight, low power, and cost of semiconductor. Here are the prime motivations for the semiconductor industry to pace toward HI:

10.2.1.1 Features of Heterogeneous Integration

There are three primary drivers for innovation in heterogeneous integration, which can be seen as its critical features, presented by DARPA [4]:

- **Technological Diversity**—Heterogeneous integration involves using a variety of chiplets that can differ by technology node and foundry when integrated on the common interposer. For example, a 14 nm transistor SRAM memory chiplet fabricated by foundry A can be integrated with a 22 nm-based processor chiplet fabricated by foundry B on the same interposer. This approach allows newer technology node chiplets to be integrated with older but high-yielding technology node chiplets, adding to this technological diversity of the system. This way, chiplets with their respective matured process nodes can be integrated into the same package.
- **Functional Diversity**—Another feature of heterogeneous integration is that chiplets with diverse functions can be integrated on the same package. For example, memory, logic chiplets, analog I/O, and MEMS sensor chiplets can be integrated on the interposer to design a SiP for an end-user application. These diverse chiplets perform specific and unique roles in the integrated SiP, allowing the modular and custom design of the SiP package.
- **Materials Diversity**—Heterogeneous integration also allows for diversity in the materials used to create these chiplets—the chiplet acts as a black box in the overall system. As long as the chiplet's materials do not affect the functionality of the integrated system in an unintended way, then the materials of each independent chiplet can differ. Certain chiplets may be optimized for a specific function and have enhanced capabilities with newer materials.

10.2.1.2 The Continuation of Moore's Law

Over the past half-century, Moore's Law has been the guiding framework for predicting the direction of innovation in the semiconductor industry. This law pushes researchers to scale CMOS devices to double the density of transistors in an IC every two years. However, there is contention in the community as to whether this principle is becoming less apparent as further decreases in transistor size can lead quantum factors to become more relevant and increase manufacturing costs for these classical processes [5]. Revolutionary innovations, especially in packaging and design like heterogeneous integration, have enabled the viewing of Moore's law with a new lens, using the functional density rather than scaling of transistor density as a predictor for performance.

10.2.1.3 Higher Yields with Lower Development Cost

Heterogeneous integration can lead to an overall increase in the yield of SiP due to the incorporation of known good dies (KGD) or chiplets, which can, in turn, be manufactured with a higher yield. Further technological advances have increased integration and stacking yields while decreasing manufacturing and research costs. Collective die-to-wafer bonding has been proposed even further to increase transfer bonding and electrical die yields [6]. Also, because chiplets are used from a matured process node, the development cost of SiP is reduced as post-silicon validation is rarely required. Previously researchers have demonstrated high yield and superior reliability manufacturing capability in high-performance 3D-ICs [7].

10.2.1.4 SiP for Better Performance

As performance gains by increasing transistor count per area on a die might be flat-lining, heterogeneous integration by incorporating various dies from different OEMs can enable higher performance by increased memory access speeds. For example, with 3D packaging technology, CPU and memory dies can be stacked, allowing increased memory bandwidth and decreased transmission latency as the dies have much shorter interconnects [8]. Furthermore, 2.5D, 3D, and 5.5D packaging increase functional density as the dies are integrated into the same package, communicating through a silicon interposer and through silicon vias (TSV). The interposers are currently under research to increase their communication quality and rate with a reduction in overall thickness and complexity of the redistribution layer in the package. Active interposers are even being proposed where transistor-based logic circuits are embedded in the interposer, further increasing the functional density of the SiP.

10.2.1.5   Form Factor (Space/Size) Reduction

The 2.D and 3D packaging paradigms have led to smaller area and size requirements. The smaller size compared to traditional packaging can be attributed to integrating these various dies in one package rather than multiple separate dies connected using traces on a printed circuit board (PCB). Interconnects are, as a result, smaller in these integrated technologies, further increasing speed and decreasing power usage. 3D packaging gives the best functional density benefits as these dies are stacked vertically and horizontally. However, it introduces many thermal challenges, as these dies emit heat within the stack. 2.5D has also increased functional density compared to traditional packaging but has less functional density than 3D since chiplets are not vertically stacked but horizontally integrated.

Hence, HI can achieve a compact form factor by 2.5D and 3D packaging, increase performance, high manufacturer yields, and reduce overall area [9].

10.2.2   Challenges Toward RoadMap for HI

Although heterogeneous integration offers numerous advantages and sounds promising in the More-Than-Moore (MTM) approach, incorporating functionally diverse dies adds value to the SiP package but may not necessarily scale according to "Moore's Law." In addition, various architectural design and security-related challenges need to be overcome while developing a System-in-Package.

10.2.2.1   Interface Standards

One big challenge in SiP design is the interfacing of chiplets due to the variety of chiplets and I/O interfaces. Therefore, attempts have been made by mainstream SiP integrators to develop fast and simple chiplet-to-chiplet interfaces so that various types of chiplets can be connected. These efforts will eventually result in shorter SiP design and assembly time.

- **Serial Interfaces**—Based on transmission distance, the serial interface can be divided into the following categories long reach (LR), medium reach (MR), very short reach (VSR), extremely short reach (XSR), and ultra short-reach (USR) SerDes. LM/MR/VSR SerDes are used for inter-chip and chip-to-module communication in PCB boards. They are also used for PCIe, Ethernet, and Rapid I/O communicating interfaces. However, advantages like reliable transmission of these SerDes do not apply to heterogeneous integration areas, performance, and power. XSR SerDes accommodates the SerDes standard of Die-to-Die and Die-to-Optical engines. As the bandwidth increases, the power consumption and delay also increase. Compared to XSR, USR is suitable for high-speed interconnection in Die-to-Die communication via 2.5D and 3D packaging technologies.

Ultimately, the transmission distance of USR hinders the large scale integration of chiplets [8].

- **Parallel Interfaces** —The common chiplet Die-to-Die communication interface for intel's AIB, EMIB TSMC's Low-voltage-In-Package-INterCONnect (LIP-INCON), etc. Intel's AIB sends and receives data via microbumps from one chiplet to another. The benefit of a parallel interface, such as AIB, is that it has extremely low latency, power, and area requirements. The main disadvantage is that it necessitates using a silicon interposer or similar packaging technology, which adds significant cost [8].

- **Universal Chiplet Interconnect Express (UCIe)**—It is an open industry standard interconnect that provides chiplets with high-bandwidth, low latency, power-efficient, and cost-effective on-package connectivity. It covers computation, memory, storage, and connectivity demands throughout the computing spectrum, including cloud, edge, corporate, 5G, automotive, high-performance computing, and hand-held segments. UCIe can package dies from various sources, including diverse fabs, designs, and packaging technologies. It is the first package-level integration that provides energy-efficient and cost-effective results. UCIe can be used in two different ways. First, as a memory, graphic accelerators, networking devices, modems, and other board-level components can be integrated at the package level, with applications ranging from hand-held devices to high-end servers, with dies from numerous sources coupled through different packaging methods even on the same package. The second application is to provide off-package connectivity using various types of media (e.g., optical, electrical cable, mmWave) and UCIe Retimers to transport the underlying protocols (e.g., PCIe, CXL) at the rack or even pod level. These protocols support resource pooling, resource sharing, and even message passing using load-store semantics to derive better power-efficient and cost-effective performance at data centers [3]. The latest UCIe 1.0 specification maps PCIe and CXL protocols natively as those are widely deployed at the board level across all computing segments. Therefore, it relies on the security solutions already deployed for the previously developed ones PCIe and CXL protocols. Furthermore, it is "silent" about the security policies and methods a SiP designer can use for a secure System-in-Package. Therefore, it may need further research to investigate vulnerabilities and incorporate security solutions in this standard.

10.2.2.2 Less Power

The SoCs' Performance-Per-Watt (PPW) design metric also holds for SiPs that use diverse functional chiplets. The SiP's ultimate objective is to provide the highest possible processing bandwidth at the cost of the lowest power consumption. The low PPW objective can save battery power in handled battery power devices. It can also reduce heating issues in constantly powered (car and data centers) and battery-powered devices.

10.2.2.3 Thermal Management

Heterogeneous integration can increase the overall power density of the SiP, which can, in turn, increase total package power dissipation. However, power is generally dissipated as heat, and it can increase thermal crosstalk, and temperature-sensitive components need further thermal isolation [10].

10.2.2.4 Secure Heterogeneous Integration

During the last few decades, hardware security and trust assurance (free from any counterfeiting issues) has emerged as a vital parameter during circuit design and system development. After the emergence of various threats and vulnerabilities at the system level, the integrated circuits, PCB, and systems are now designed for security during their design phase. However, not many security assessments have been done on heterogeneous integration technology. For a secure heterogeneous integration, a bottom-up security approach is required from the root level (chiplets) to the packaging (interposer) to the final deployment (system) level.

- **Chiplet Security: Trust Validation of Chiplets**—A SiP designer must source trusted chiplets from various chiplet OEMs for heterogeneous integration. Like fabless packaged IC design companies, chiplet fabless OEMs will rely on pure-play overseas foundries for fabrication. These foundries have full access to GDSII, test patterns, and a confidential fabrication process beyond design houses' control. An adversarial foundry can insert malicious changes or hardware Trojan during the manufacturing of a chiplet. A chiplet with a Trojan inside a heterogeneous integration can expose that System-in-Package to various attacks. For a secure heterogeneous integration, all chiplets need trust validation by the SiP designer.

  The trust validation of these chiplets can be very challenging for a SiP designer. They do not access proprietary information such as chiplet GDSII design and test patterns. Besides foundry, a chiplet design house cannot share the above-mentioned proprietary information with anyone in the supply chain. Finally, the chiplets are sold through various distributors, which cannot be trusted due to the involvement of an overseas foundry. In this trust-less and IP confidentiality scenario, only the design house can validate the trustworthiness of the chiplets. Such trust validation steps require much effort in terms of time and money. Unfortunately, no universal standards or independent trust validation entities exist for chiplet security assurance.

- **Interposer Level Security: Secure Packaging**—With the insurgence of hardware-based attack reports and added attack vectors from various entities associated with heterogeneous integration, security assessment of packaging is crucial for hardware assurance of SIPs used in the military, space, and automobiles. The threat model exploiting the vulnerabilities pertinent to material and fabrication in IC packaging can be extended to SiPs. An adversary can alter the package material composition to cause chip failure during deployment. Unfortunately, current research trends for semiconductor packaging in industry and academia mainly focus on packaging reliability while its security is barely addressed. An adversarial integration facility can maliciously insert Trojans in an active interposer and alter packaging materials to create vulnerabilities in SiP that may be very difficult to detect in the postmanufacturing stages.

The current process to assess packaging integrity primarily focuses on testing the chip's reliability during failure analysis and in-process testing. The physical inspection methods such as X-ray Photoelectron Spectroscopy (XPS), X-ray Fluorescence, Scanning Electron Microscopy, and Tera-Hertz Imaging can be effective for material composition analysis, interface anomalies, and malicious change detection [11, 12]. However, the effectiveness of these methods greatly depends on the complexity and material composition. For example, the sub-micron micro-bumps that connect the die and an interposer cannot be detected using a Scan Acoustic Microscopy (SAM). In addition, it may be challenging to detect without decapsulating the assembled SiP. Therefore, destructive detection techniques are required to provide robust integrity checks at the cost of time.

For this reason, these checks are applied to random samples and cannot offer extensive hardware assurance. Existing methods may suffice to address the inherent process-induced reliability issues. However, new detection techniques are needed to ensure packaging integrity and provide hardware assurance.

10.2.3 Current Advancements in Heterogeneous Integration

This section will discuss the heterogeneous integration supply chain, applications, and secure design initiatives.

- **SiP Supply Chain**—As compared with SiP, SoC supply chains are straightforward due to fewer manufacturing steps and entities involved. However, the focus of SiPs has shifted from monolithic systems to the consumer-focused realm, where computing has become pervasive and increasingly heterogeneous. As a result, supply chain dynamics have inevitably become far more complex (see Fig. 10.3).

  Recent trends and research illustrate the improvement of chiplet integration technology in heterogeneous systems, which necessitates the generation of a new business model and modification of the conventional supply chain of an SoC. In the SiP supply chain, an entity similar to the SoC integrator called SiP designer or OEM sources various chiplets for heterogeneous integration to develop custom silicon "chips" or System-in-Package. As the chiplet ecosystem evolves continuously to provide better system-level scaling, higher bandwidth communication, higher performance, higher functional and integration complexity, less power consumption with reduced development costs, and reduced time-to-market brackets, new entities are participating in the supply chain of

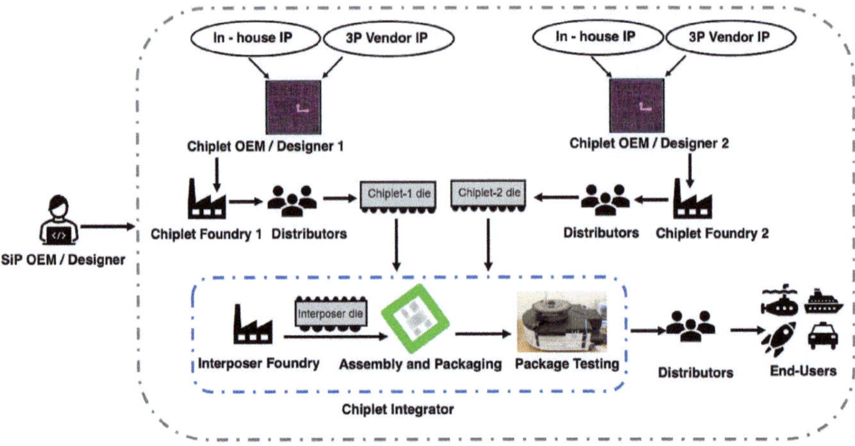

Fig. 10.3 SiP supply chain for heterogeneous integration

the heterogeneous system. Chiplet OEMs (Fabless or IDMs), IP vendors, CAD tool developers, and pure-play foundries are the primary entities who control the availability of the chiplet in the supply chain to a great extent [8]. Chiplet OEMs with advanced research facilities and state-of-the-art design technology (e.g., Intel, AMD, Micron, Apple) have their chiplets in the market and high-volume production.

To achieve high computing performance at a reduced cost, AMD provides chiplets to the supply chain that includes multi-core processor dies, IO dies, and high bandwidth memory dies in different technology nodes. Intel has demonstrated its chiplet integration feasibility by designing high-performance, high bandwidth heterogeneous systems by integrating an FPGA die and IO die with an AIB interface. AIB connections can be made using wires on an interposer and bridge technologies such as Intel's EMIB bridge [13]. Some start-up fabless semiconductor companies such as zGlue are trying to establish a set of basic EDA toolchains for chiplets. Some advanced packaging technologies (Substrate-based, Silicon Interposer-based, Silicon Bridge-based, and Redistribution Layer-based) have been exploited to achieve higher IO density with reduced transmission delay and power consumption in the heterogeneous integration of chiplets. TSMC has implemented substrate-based fanout packaging based on RDL in Apple's A10 processor to make it more cost-effective than silicon interposer-based packaging.

Although the roadmap of the current chiplet supply chain demonstrates much advancement in its development and implementation, it raises some unique challenges that need to be resolved. The standardization of interface and communication protocols in the design stage of a chiplet must consider the new fabrication process, packaging technology, and integration technology to achieve flexibility and scalability [8]. The current chiplet design and integration technology requires comprehensive EDA tool support to reduce the failure in post-silicon analysis and improve the quality of the manufacturing process. Moreover, the security vulnerabilities of the chiplet supply chain have emerged from the entities that can be potentially untrusted such as 3PIP, untrusted chiplet OEMs (fabless design house or IDM), chiplet pure-play foundry (for fabless chiplet OEMs), chiplet integrator. A chiplet integrator can be a single entity providing complete integration (interposer design, packaging, assembly, and testing) or segregated into multiple separate entities, depending upon their expertise or business model. Finally, an adversarial end-user is the entity that procures a SiP package to perform unethical attacks on the in-field SiP package to exploit its vulnerabilities to violate the CIA triad or produce counterfeit packages.

- **Secure Design and Packaging Programs**—Concerns about hardware security threats and vulnerabilities have presented an opportunity for the Department of Defense (DoD) to reduce barriers by utilizing mainstream electronics technology while protecting critical defense technologies and manufacturing. As a result, the DoD has reevaluated trusted and assured access to advanced node foundry production over the last several years. The goal is to incorporate commercial industry ASIC and SoC (including SiP) capabilities while maintaining the integrity and security of defense systems [14]. Furthermore, in order to achieve more rapid modernization while reducing the size and increasing the performance of DoD systems, the department has launched various trusted and assured microelectronics programs as follows:

1. **SHIP Prototype Project**—This program is sponsored by the Office of the Under Secretary of Defense for Research and Engineering and funded by the Trusted and Assured Microelectronics program. They established the State of the Art (SOTA) Heterogeneous Integrated Packaging (SHIP) program to create a sustainable industry and functioning standard for addressing government needs in Microelectronics (ME) packaging. SHIP will use commercial industry expertise to design a novel standard for ensuring DoD access to secure advanced packaging and testing. The standard will provide the DoD, and the Defense Industrial Base (DIB) with continuous access to a catalog of proven IP and chiplets that can be used to design and build customized multichip modules using commercial off-the-shelf state-of-the-art devices [15].

   The program's second phase will create multichip package prototypes and speed up the advancement of interface standards, protocols, and security for heterogeneous systems. SHIP prototypes will combine special-purpose government chips with advanced commercial silicon products from Intel, such as FPGAs, ASICs, and CPUs. This combination of technologies opens new avenues for the US government's industry partners to develop and modernize mission-critical systems while leveraging Intel's US fabrication facility [16].

2. **DAHI Program**—The Diverse Accessible Heterogeneous Integration (DAHI) program was established to create transistor-scale heterogeneous integration processes integrating cutting-edge compound semiconductor (CS) devices and other emerging materials and devices with high-density silicon

CMOS technology. DAHI's ultimate goal was to create a manufacturable, accessible foundry technology for the monolithic heterogeneous co-integration of various devices and complex silicon-enabled architectures on a common substrate platform. This kind of integration would boost the capabilities of high-performance microsystems for the US military. This program sought to address the following critical technical issues: (a) development of heterogeneous integration processes, (b) establishment of high-yield manufacturing and foundries, and (c) circuit design and architecture innovation [17].

3. **CHIPS Program**—The demand for faster, more compact, and cheaper electronic devices has pushed the semiconductor industry to integrate various circuit blocks such as digital, analog, and analog-mixed signal blocks into an SoC. This integration has been enabled by advanced CMOS technology but has also increased design and processing costs. IP reuse has emerged as a tool to reduce overall design costs associated with advanced SoCs, owing to aggressive digital CMOS scaling for high-volume products. However, due to factors such as high initial prototype costs and requirements for alternative material sets, the monolithic nature of cutting-edge SoCs is not always acceptable for DoD or other low-volume applications. The Common Heterogeneous Integration and Intellectual Property (IP) Reuse Strategies (CHIPS) program seeks to create a new paradigm in IP reuse to improve overall system flexibility and reduce design time for next-generation products. CHIPS envision an ecosystem of discrete modular, reusable IP blocks that can be assembled into a system with existing and emerging integration technologies. Modularity and reusability of IP blocks will necessitate widespread adoption of electrical and physical interface standards by the CHIPS ecosystem community. As a result, the CHIPS program will create the design tools and integration standards needed to demonstrate modular IC designs that combine the best of DoD and commercial design and technology [4].

- **Heterogeneous Integration Roadmap for Electronics Applications**—The Heterogeneous Integration Roadmap spans the complete semiconductor and electronics technology ecosystem in detail. It functions as a knowledge-based blueprint for future electronic technology. The roadmap is market- and application-driven, beginning with six market segments: high-performance computing and data centers, IoT, 5G communications and beyond, smart mobile, automotive, wearable and health, and aerospace and defense.

  The demand for next-generation systems has increased significantly, owing to emerging trends driven by numerous applications such as smart devices (House, TV, mobile, automotive), data centers, high-speed wireless communication, and medical and health care devices. As a result, technologies are evolving rapidly. However, they face complexity in keeping pace with the increased demand. For example, the number of globally connected IoT devices/endpoints reached 12.3 billion by 2021, and the projection is to be 27 billion by 2025 [18]. Manufacturing such a massive number of products requires a lower time-to-market and lower production costs, making the challenge surrounding conventional SoCs even more difficult.

Similarly, newer medical and healthcare products are primarily driven by miniaturization, implantability, and portability [19]. The recent trend toward various medical and healthcare equipment (e.g., pacemakers, neurostimulators, insulin pumps) focuses on smaller form factors and increased performance. In addition, modern medical wearables utilize different sensors to detect heart rate, blood pressure, oximetry, respiratory monitoring, hearing aid, and body temperature and incorporate IoT-based technology and cloud computing to provide better healthcare facilities.

Furthermore, there has been a momentous technological shift in the field of the automobile toward autonomous driving, and the industry emphasizes various security and safety features. Therefore, modern autonomous cars require high-performance embedded machine learning chips to collect and process thousands of images from various camera sensors embedded at numerous locations of the automobile. In addition, various safety and performance sensors (blind-spot detection, pedestrian detection, park assistance, collision detection, emergency braking system, cruise control, lane assistance) are installed, which require low power and high bandwidth data communication among internal components and processors, as well as intra/inter-vehicular communication [20].

To summarize the above discussion, next-generation electronic systems must ensure (1) high bandwidth, (2) low power, (3) smaller form factor, (4) increased functionality, and (5) flexibility [21]. Packaging different discrete IP blocks/components on a PCB may achieve the increased functionality requirement; however, the form factor is a large and less flexible design, requiring high chip-to-chip bandwidth, and the overall system power may increase. In addition, monolithic integration of the IP blocks inside an SoC design may solve some challenges with PCB; however, time-to-market is extensive and can incur IP maturity problems [21]. For example, a designer purchased an IP block from a vendor with a 40 nm technology node, but the SoC is designed with a 14 nm technology node. In this case, plug-n-play for the IP block is impossible as the design is less flexible, has increased time-to-market, and incurred a lower yield.

Heterogeneous integration of SIPs is the most appropriate solution to achieve the goal. It allows flexibility to choose known good dies from various vendors fabricated with different process nodes, integrates over active/passive interposer substrate, and allows higher bandwidth. For example, Intel used Stratix 10 FPGAs, and SoC utilized EMIB and AIB technology to integrate different IP blocks with shorter interconnects, thus creating lesser delay with support up to 58 Gbps of data [21]. In addition, the known good dies are available from various vendors and are easy to integrate, thus reducing time-to-market significantly. Although heterogeneous integration is more suitable for designing next-generation systems, it introduces newer attack surfaces that can be exploited to access various security assets. Therefore, appropriate security measures are needed to utilize the overall benefits provided by heterogeneous integration.

The roadmap to HI has opened an avenue for research and development to resolve HI design and security issues.

10.3   Chiplet Security: Risks, Threats, Vulnerabilities, and Assurance

Chiplets are the main component of a System-in-Package created through heterogeneous integration. Therefore, chiplet security is critical as a building component of the SiP because a system created with vulnerable chiplets can eventually result in a SiP package vulnerable to security attacks.

10.3.1   Threat Model for Chiplet Security

The chiplet design and fabrication process involves various entities such as a design house, third-party IP vendors, and a foundry. Like IC IDM, a chiplet OEM can do their chiplet design and fabrication, but a fabless chiplet design house relies on a pure-play foundry for fabrication. Therefore, chiplet security involves security from the design phase (RTL) to the SiP OEMs who procure chiplets for heterogeneous integration. So, ensuring the chiplet security will be a step toward the security of the SiP package. However, due to the severity and multitude of security threats associated with the untrusted pure-play foundries, we will primarily focus on the supply chain involving fabless design houses [22].

The fabless design house follows a horizontal business model in which fabrication, assembly, and testing of integrated circuits (ICs) are outsourced to offshore foundries and OSATs to reduce cost and time-to-market. Similarly, for chiplets, the fabless design houses are expected to follow the same business model (see Fig. 10.4). The supply entities related to the chiplet security are presented in the threat model (see Fig. 10.4). The SiP designer, also known as the original equipment manufacturer, is trusted because it is responsible for the security of SiP. A chiplet design house may use in-house and third-party IPs to develop a chiplet design. Then, it sends the design for fabrication to an offshore foundry. Depending on the geo-location or market reputation of a chiplet design house, it may be either trusted or untrusted. However, providing a chiplet IP to an offshore foundry can make the chiplet vulnerable to insertion of hardware Trojans or overproduction. Furthermore, a competitor design house can steal IP by reverse engineering, or a rival foundry can clone chiplets and sell them as authentic. Also, recycled chiplets can be a threat if carefully extracted from the SiP package by an untrusted distributor or an end-user. Typically, recycled chiplets may not be perceived as a considerable threat due to the significant effort required to remove a chiplet from a SiP package. The chiplet attack vectors can impact the security of mission-critical applications, such as defense

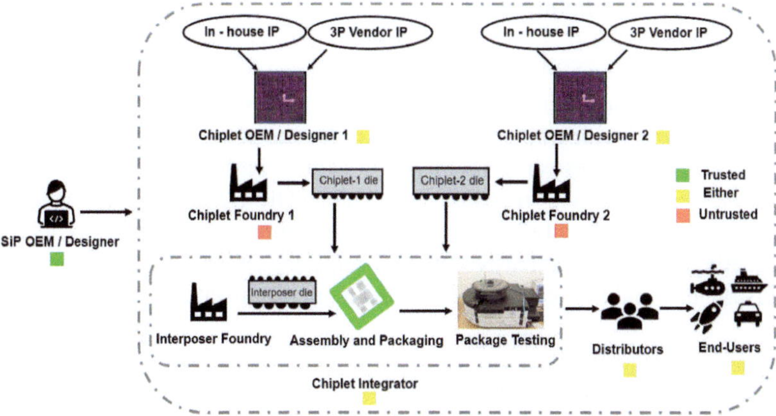

Fig. 10.4 Threat model of "chiplet security" for secure heterogeneous integration at different phases of the lifecycle

systems, airplanes, and health care, by causing early failure, data breaches, and reliability problems. So, in our threat model, the SiP designer/OEM is considered trusted. All other entities in the supply chain are considered untrusted. The design tools used during the generation of the chiplet IP are also considered trusted.

10.3.2   Hardware Attack Vectors

10.3.2.1   Hardware Trojans

A hardware Trojan is a malicious change in chiplets to sabotage the security of the SiP package or an electronic system in which the SiP package will be deployed. These Trojans can cause significant security concerns regarding the SiP's confidentiality, integrity, and availability (CIA- triad) [23]. For example, it can cause a denial of service to cause reliability issues, Man-in-the-middle attacks to get unauthorized access to confidential data such as encryption keys, and the bias of neural networks [23]. Hence these SiP packages cannot be trusted for critical government data-center and national security applications such as defense, space, and energy sectors [24]. In a SiP supply chain, a hardware Trojan can be inserted by either an adversarial IP designer or an untrusted foundry [25]. Based on the adversary, there can be different attack models based on the trust assumption with any of these entities [26]. Among them, the threat model of the untrusted foundry has been widely discussed in the hardware security community [26, 27].

10.3.2.2 Reverse Engineering

Concerns about reverse engineering (RE), a process of extracting an RTL of design by de-processing and imaging various device layers from fabricated integrated circuits, still exist with chiplets [28–30]. Competitor semiconductor design houses or adversarial foundries can perform RE to gain a competitive and financial edge. However, it can cause a revenue loss to the chiplet OEM, and the reverse-engineered chiplet may have reliability and trust issues. Furthermore, like chips, chiplet reverse engineering can be an extensive process requiring much effort and time, so an adversary may find SiP reverse engineering more beneficial than chiplet.

10.3.2.3 Counterfeit Chiplets

A counterfeit chiplet can be defined as (a) an unlicensed chiplet, (b) does not meet the specification and performance of the OEM, (c) is not manufactured by the authorized contractors, (d) is a non-conforming, faulty, or used OEM product offered as "new" (e) has inaccurate or erroneous markings, or documentation [31]. Counterfeit is a billion-dollar business and is easy money for the adversary. Nevertheless, on the other hand, the OEM faces revenue loss and tarnished brand reputation. For chiplets counterfeit chiplets can be classified in the following categories [32]:

- **Recycled**—Recycled chiplets like recycled ICs [33, 34] can be one of counterfeit types. Recycled ICs are a big industry; chiplets can be recycled like recycled chips. However, as chiplet recycling is at the silicon die level counterfeiting, the adversary must put more effort and use sophisticated methods for this purpose. The chiplets may be taken from a used SiP, repackaged, and then sold to the market as new. During this process, a chiplet may break or lose its functionality. A successfully recycled chiplet may perform poorly and cause reliability issues in the System-in-Package. An adversary may find the recycling of SiP packages more motivating.
- **Remarked**—The counterfeiters remove the old marking on the die package of the chiplet and mark the package with a new identification number and generate forged documentation to sell it to the market as higher grade ICs [33]. In the remarking process, the chiplets old markings can be removed via chemical processes and new marks applied by the adversary. For example, an adversary can remark a chiplet to change the grade of the chiplets, such as commercial-grade to military grade chiplets.
- **Overproduced**—A offshore foundry can fabricate more chiplets than the number of chiplet a design house has ordered. These overproduced chiplets can enter the SiP supply chain through unauthorized channels [34].
- **Out-of-Specification**—After fabrication of chiplets, they are tested for the designed electrical parameters. A chiplet may still work, even if it fails to meet the design specification during post-silicon validation. However, it may

be unreliable, slow, or vulnerable to security attacks. In addition, an adversarial distributor or reseller may try to sell these chiplets to a SiP designer as a high-quality chiplet. When the out-of-specification chiplets are integrated into the supply chain, it can be challenging to detect them during SiP system-level testing (SLT) [34, 35].

- **Cloned**—Chiplet can be cloned by adversarial foundries like competitors/counterfeiters to reduce the significant development costs of a component. A cloned chiplet is a replica of the original chiplet and sold as an authentic chiplet fabricated by or for the chiplet OEM [35].

10.3.3 Trust Validation of Chiplets

Trust validation of chiplets involves identifying tampering, counterfeiting, or malicious changes using various suitable electrical testing or physical inspection methods.

- **Testing—Logical Testing and Side-Channel Analysis:** Logical Testing is a proposed method for hardware Trojan detection in SoC by using test vectors and observing a deviation in the output. It has its pros and cons, such as it is a non-destructive method, but it needs to trigger a Trojan to detect the presence of malicious change. For chiplets, a fabrication facility performs a wafer-level testing, and it can be a challenge at the SiP designer end due to:

  - SiP designer needs to test every kind of chiplet, which requires time and monetary efforts.
  - SiP designer needs to acquire or develop a testing infrastructure for wafer-level testing of chiplet.
  - SiP designer needs test patterns from chiplet OEMs, which is subject to chiplet OEM's discretion.

  Side-channel analysis can also be challenging due to the limited or absence of test infrastructure for chiplets at the SiP designer end, and besides this, it needs a golden model or signatures for trust validation, which is highly debatable when it comes to an untrusted foundry threat model.

- **Physical Inspection:** Physical inspection-based techniques typically capture nano-imaging (SEM-scanning electron microscopy) data from a polished thinned die [36]. There are two popular techniques, the first one, reverse engineering (RE), and the second *Trojan Scanner* [37]. RE required SEM imaging data from all integrated circuit layers to reconstruct the netlist to compare with the original netlist. RE process is time-consuming, error-prone, requires highly skilled engineers, and many die samples are wasted during sample preparation. However, using *Trojan Scanner* [27, 38, 39], an SoC design house can perform trust validation of the die by using only active (or diffusion) layer SEM images and comparing them with a golden layout (trusted layout) to detect any malicious change. Therefore, it requires lesser time and fewer samples as compared to RE. In the SiP domain, there can be multiple chiplet OEMs involved during the SiP design and development. RE can be a suitable method for counterfeit such as IP piracy detection, but it is a poor approach for Trojan detection due to the above-mentioned reasons. Chiplet trust validation using *Trojan Scanner* method can be challenging, and it cannot be directly applied for chiplets due to the following reasons:

1. It can be challenging for a chiplet OEM to entertain multiple requests from several SiP designers.
2. A chiplet OEM cannot share its golden layout to SiP OEM for trust validation to protect IP confidentiality.
3. Moreover, in various scenarios, a chiplet OEM cannot be trusted by a SiP designer due to various reasons, including geo-location of the chiplet OEM and foundry, the duration of presence in the supply chain, and brand image.

Hence there is a "void" for an entity or service or a validation mechanism in the SiP ecosystem for trust validation of chiplets using a physical inspection approach. It needs further research to develop a framework for SiP designers to perform trust validation of chiplets.

10.3.4 Attack Mitigation or Countermeasures

Chiplets are hardened IP which means their logic circuit is hardwired. Therefore, it can prevent an SiP designer from deploying a security feature if the chiplet security is ignored during the chiplet design phase. For example, to protect an SiP from an optical probing attack, the chiplet must have a security mechanism to avert an unauthorized optical probing attempt. Hence for SiP attack mitigation, a chiplet designer needs to consider various security vulnerabilities during the design phase. Later on, when the chiplet is integrated into a System-in-Package, the package already has the necessary sensors or security features, which can be used to detect and trigger a necessary response against the SiP level attacks.

- **Design Obfuscation**—Hardware obfuscation is done with the aim of hiding design details and implementation details against reverse engineering and using design as a black box against IP cloning [40, 41]. The obfuscation can be done at pre-synthesis, postsynthesis, and physical layout levels [42–44]. In pre-synthesis, IPs are encrypted with IEEEP1735 [45, 46]. Postsynthesis hides the actual functionality of the circuit using structural modifications of the design. Finally, at the physical level, the connections between the cells are obfuscated, such as doping-based methods and dummy contact insertion, so that the adversary cannot understand the layout design to perform reverse engineering and, in some cases, insert malicious circuits such as hardware Trojans.

- **Side-Channel Resistant Designs**—The chiplet circuit designers need to consider the threat of side-channel attacks in the chiplets storing encryption keys and confidential data. In order to design a side-channel attack resistant chiplet running a cryptographic circuit or having confidential data, the circuit's power consumption needs to be made independent of the performed operation and processed data. Hiding and masking are two methods of designing side-channel resistant cryptographic circuits [47]. However, these countermeasure cost additional circuit area and degrade performance. Furthermore, recent research shows that these countermeasures can be further attacked. So, higher-order hiding and masking techniques are recommended for designing side-channel resistant integrated circuits [48]. Finally, chiplet designers need to consider the trade-off between the level of security and the chiplet's power, performance, and area.

- **Sensors for Temper Detection and Prevention**—Like on-chip sensors are used for detecting probing attacks. Similarly, a chiplet sensor (a hardened or designed as programmable logic similar to FPGAs) can detect probing and prevent further tampering. An optically active layer with angular-dependent reflectivity on an IC's backside protects it from semi-invasive physical attacks and optical fault injection attacks [49]. Light emitted from a transistor's drain is detected in another transistor's drain after reflecting on the active layer that ensures no backside tampering. In addition, this layer is opaque to IR illumination, thus preventing photon-induced fault injection. Moreover, any damage to this protection from backside tampering can be detected by IC electronics [50, 51]. Another advanced solution can be CMOS compatible structures called nanopyramids that can mitigate electro-optical probing (EOP) and electro-optical frequency mapping (EOFM) attacks by scrambling the light signal reflected by these structures. Furthermore, the Nanopyramid structure can be applied to designated chiplets that require protection against EOP and EOFM attacks [52].

- **Security Primitives**—Security primitives are secure circuits that can be physically embedded into chiplets for mitigation against supply chain threats such as cloning, recycling, or overproduction. For example, physical unclonable functions (PUF) can be used to fingerprint a chiplet for authentication purposes to detect cloning. In addition, a silicon odometer can detect recycled components.

- **IP Secure Zones and Trust Modules (CHSM): Tamper Resistant Design**—The main objective of the chiplet hardware security module (CHSM) is to detect any security threat (e.g., Trojan, physical tampering) and provide a countermeasure and attack mitigation against hardware security attacks on individual components (chiplets) or complete System-in-Package [53]. CHSM considers both the chiplet-level and system-level tampering scenarios to make a heterogeneous integration tamper-resistant. To prevent system-level tampering, monitoring the electrical parameters of the interconnecting and interfacing components of a SiP is controlled by the CHSM module. In addition, monitoring the output signals from the backside (FEOL) and front-side (BEOL) IC tampering is processed by the CHSM as well. When a tampering or malicious modification has been detected, CHSM can stop communication between chiplets by blocking the communication channel and altering the data in confidential information-carrying signal nets to prevent confidentiality, integrity, and availability violation. Furthermore, the aforementioned detection methods of tampering through noninvasive and semi-invasive fault injection attacks (e.g., clock glitch, voltage glitch, EM, laser) at the chiplet-level are incorporated in CHSM to take preventive security measures during the run time of an SiP. The preventive approaches include:

- Blocking secure communication between IPs in a chiplet.
- Stopping the propagation of any malicious signal outside a chiplet.
- Masking the injected faults.
- Blocking the means of fault injection upon detection of the chiplet-level tampering.

10.4   Interposer/Substrate Security: Threats and Assurance

The chiplet integrator uses redistribution layers (RDLs) to connect different components on a SiP. These components can be integrated using various advanced assembly methods such as build-up substrate, PoP (package-on-package), FOW/PLP (fan-out wafer/panel-level packaging), WLCSP (wafer-level chip-scale package), silicon interposer, Foveros, and EMIB (embedded multi-die interconnect bridge), etc. [54]. In the following discussion, interposer-based packaging technology will be used as an example to explain the vulnerabilities added to the existing system interconnect technology and the origin of the new assurance problems in heterogeneous integration.

10.4.1   Threat Model for Interposer Security

Let us look into the supply chain for heterogeneous integration after the chiplets are fabricated by the foundry (see Fig. 10.5). There is an added complexity in the SiP design and manufacturing steps compared to the conventional SoC manufacturing process as chiplets need to be connected using advanced packaging technologies such as an interposer. A SiP designer procures the relevant chiplets for heterogeneous integration and sends the chiplets to a chiplet integrator. The chiplet integrator can perform multiple tasks based on their business model. While this integration process improves time-to-market and production cost, the added process steps and requirement for separate layers to establish interconnection can create a new dimension of threats such as Trojan insertion, SiP piracy, and reverse engineering. Furthermore, a chiplet integrator can play an adversarial role due to access to interposer design, chiplet types, and specifications. Hence it may or may not be trusted.

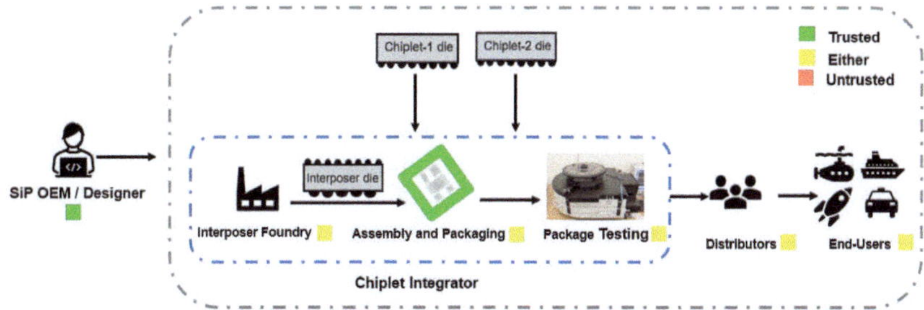

Fig. 10.5 Threat model for interposer level security

10.4.2 Hardware Attack Vectors

In the supply chain, the SiP designer creates the GDSII files for the interposer layer to be fabricated in the "Interposer Foundry." Once the interposer is fabricated, the interposer layer and the chiplets are sent to the assembly and packaging facilities, where all the components are integrated to create SiP. Unfortunately, the involvement of untrusted entities in the supply chain renders the heterogeneous integration vulnerable to attacks.

We consider three scenarios with combinations of trusted and untrusted entities in the supply chain to get a comprehensive attack surface for the interposer layer in heterogeneous integration.

- **Scenario-1:** The SiP OEM sends the interposer GDSII to an untrusted offshore foundry to fabricate the interposer layer. Once the interposer is fabricated, it is returned to a trusted facility for assembly and packaging. An untrusted foundry may perform malicious modification/alterations to the GDSII of the interposer layer and change the parameters of the RDL or TSVs to cause reliability issues or incite leakage. For active interposers, the foundry may insert trojans [55] in the interposer layer. Because of the complexity of heterogeneous integration, these Trojans may be harder to detect during testing and verification [56, 57]. Furthermore, the foundry, having access to GDSII, has the potential to be complicit in IP piracy by giving away critical information about the interconnect network of the chiplet. The untrusted foundry can extract meaningful information from the interposer GDSII about interfaces and possible functionalities of chiplets based on the interconnect and TSV locations.
- **Scenario-2:** The SiP OEM owner is a fabless design house that depends on offshore facilities to realize the SiP design. In this case, the Interposer layer fabrication, along with assembly and packaging, is done in an offshore untrusted environment like the service provided by many advanced packaging companies [58]. Therefore, the vulnerabilities discussed in Scenario-1 apply here. On top of that, since the integration is also done offshore, adversaries in any part of these supply chain entities can better understand the whole SiP design. This scenario

leaves the SiP vulnerable to stealthier Trojan-based attacks, counterfeiting such as overproduction, and IP piracy.

- **Scenario-3:** The whole SiP manufacturing process is done in a trusted environment, but the end-user (adversarial nature) or SiP distributor is untrusted. Because of the availability of high resolution failure analysis tools such as X-rays, photon emission microscopy, and SEM, the packaged SiP remains susceptible to hardware attacks such as reverse engineering [30]. An adversary with little to no involvement in the manufacturing process can easily reveal the interconnections between the dies by reverse-engineering the integrating interposer layer. The separate interposer layer makes it easier than the monolithic SoCs to trace all the active and passive components integrated into the interposer. As the chiplets on a SiP become readily available in the market, adversaries may reverse engineer the SiP and fabricate their interposer layer to produce cloned SiPs.

All three scenarios are summarized in Table 10.1.

10.4.3 Trust Validation by Substrate Verification

As mentioned earlier, the present microelectronics assurance techniques in the advanced packaging industry are focused only on reliability. Security assurance can barely be achieved with in-comprehensive tests that check for defects, reliability, and durability in specific conditions. As the fabricated active interposer layer goes through a similar manufacturing process as chiplets, many of the security threats and assurance techniques already discussed for chiplets can also be extended to the interposing silicon layer. The material and structure of the interposer layer can be inspected to establish trust and assurance in the post-fabrication or post-assembly stage in the horizontal supply chain. While both destructive and non-destructive detection techniques discussed earlier can be used to evaluate the interposer layer's integrity during in-process testing and failure analysis, the effectiveness may vary depending on the technology node and structural complexity of the interposer layer [59].

10.4.4 Attack Mitigation or Countermeasures

Rigorous testing and physical inspection of the structural integrity can effectively detect malicious modification or alteration of the interposer layer in an untrusted foundry. However, this comes at the cost of time. This method becomes even more difficult for scenario 2, where the system-level integration is outsourced. A possible way to circumvent this is to adopt innovative design-based solutions as a countermeasure. An active interposer-based root of trust was proposed by Nabeel

Table 10.1 Attack vectors concerning advanced packaging

Scenario	Entity						Attack vector
	SiP designer	Chiplets	Interposer/Si-bridge foundry	Assembly and packaging	Package testing	Distributors and end-users	
Scenario-1	Trusted	Trusted/ untrusted	Untrusted	Trusted	Trusted	Trusted	• Trojan insertion • Reliability • IP piracy
Scenario-2	Trusted	Trusted/ untrusted	Untrusted	Untrusted	Untrusted	Trusted	• Trojan insertion • IP theft • Overproduction
Scenario-3	Trusted	Trusted/ untrusted	Trusted	Trusted	Trusted	Untrusted	• Reverse engineering

et al. [60] to monitor data transactions between processor and memory modules. This active monitoring can be leveraged to detect malicious events between concerned chiplets. A sensor module may be devised to check for anomalies in run time power consumption, electromagnetic radiation, and temperature that may serve as a trigger signal. Nevertheless, the integrity of these active schemes may become compromised if they are fabricated or assembled in an untrusted environment. Another way to ensure the integrity of this security module is to use reconfigurable FPGA. The FPGA can be embedded inside the interposer layer as an embedded FPGA (eFPGA) [61] or in the case of passive interposer-based SiP as a separate die which will serve as the root of trust. Even if the SiP is fabricated in an untrusted environment, the FPGA can be configured in a trusted facility. This FPGA-based solution can be devised to create a key-based permutation block [62] to obfuscate the design/ interconnection of interposed dies, making it harder to perform reverse engineering attack on the interposer layer [63]. Furthermore, different layout-based obfuscation techniques in conjunction with logic locking to lock the dies or the interposer layer can be adopted to prevent IP piracy and SiP overproduction threats discussed in the next section.

10.5 SiP Security: Risks, Threats, Vulnerabilities, and Assurance

Even if an individual chiplet and interposer go through trust validation and assurance checks, the final System-in-Package (SiP) can also face similar security threats as an SoC from an adversarial end-user. These security attacks can exploit unidentified chiplet, interposer, and system-level vulnerabilities; hence, a SiP package can still be vulnerable to further attacks. This section discusses the security threats faced by SiP and their possible countermeasures.

10.5.1 Threat Model for Heterogeneous Integration for SiP

A SiP package is vulnerable to attacks by an adversarial user to compromise its security, intellectual property, and counterfeiting purposes (such as recycling or cloning). The risks associated with untrusted chiplet and untrusted foundry have been discussed in the previous sections. After procuring chiplets and interposer fabrication, they are sent to the chiplet integrator for heterogeneous integration. After packaging and testing, the SiP package is available through various channels (direct selling or distributors). Figure 10.6 shows the security risks involved at the package level, and the following subsection discusses various attack vectors in detail.

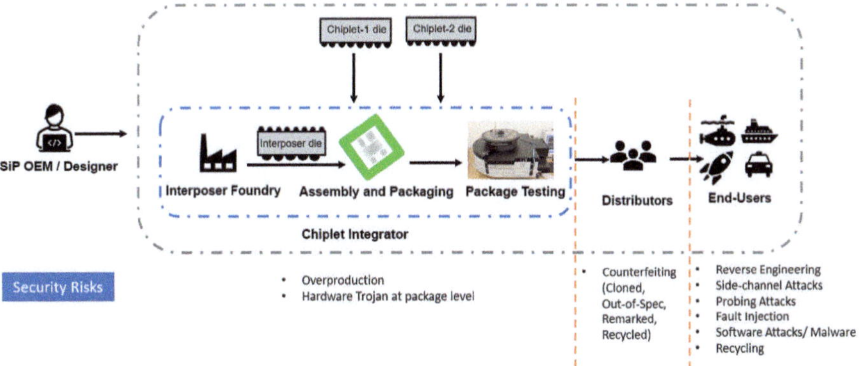

Fig. 10.6 Threat model for SiP security

10.5.2 *Hardware Attack Vectors and Surface*

Even a SiP package is securely assembled using trusted chiplets and interposer. However, the SiP package can be vulnerable to further attacks for various reasons. First, in some scenarios, it is not possible to perform trust validation, or the only option is to use a potentially untrusted supply chiplet to build a SiP [64]. Another reason can be an unknown vulnerability at the chiplet or interposer level. Possible attacks by malicious chiplets can be described as snooping on data intended for other chiplets, modifying data transferred between other chiplets, or an untrusted chiplet masquerading as a trusted chiplet. Also, in a heterogeneous SiP, individual chiplets are physically placed significantly closer than a PCB. Therefore, it increases the communication bandwidth and vulnerability to side-channel attacks by a malicious chiplet or an adversarial end-user. Furthermore, various SiP level attack vectors are as follows:

- **Supply Chain Problems (Counterfeiting)**—The supply chain for heterogeneously integrated (HI) chips, like for monolithic chips, can experience threats of counterfeiting at chiplets and the package level. However, due to the structure and the involvement of multiple entities in the supply chain of HI chips, the threat of a chiplet integrator overproducing the entire SiP is unlikely, as the Assembly and Packaging entity needs to procure extra chiplets for the overproduction of the entire SiP. Moreover, that malpractice will fall under cloning. In addition, other package-level threats from the SoC supply chain still exist in the HI supply chain, such as recycling, remarking, and out-of-specification. These counterfeiting problems can occur post-assembly and packaging in the supply chain and follow a similar threat model to the SoC domain [22].

  There may be potential threats in the HI supply chain which are not found in the SoC domain because of the introduction of chiplets. For example, a SiP package may be integrated with low-grade chiplets instead of the higher grade, which can affect the overall grade of a SiP package. Hence it can impact the SiP package's functionality and reliability. For example, a chiplet distributor may sell commercial-grade chiplets to the SiP designer who expects military grade chiplets. Although this can be seen as a threat at the chiplet level, it affects the authenticity of the entire SiP package chip, which is intended to be military grade.

- **Side-Channel Attacks**—In HI, chiplets are placed in much closer physical proximity than an IC on a PCB board. It is especially true for 2.5D, 3D, and 5.5D designs. Unfortunately, this placement strategy increases the ability of a malicious chip to perform various power, timing, and EM-based side-channel attacks. For example, a malicious chiplet introduced into the System-in-Package improves the sensitivity and spatial resolution of collecting EM, thermal, and power signatures from the other chiplets immediately above-below and adjacent to it by orders of magnitude [65, 66].

- **Fault Injection Attacks**—The adversaries can exploit several fault injection attacks to leak secret information from SiP like the conventional SoC [67]. Heterogeneous integration of chiplets creates multiple attack points inside the SiP package, such as memory, processor (registers, or functional logic area), the vertical interconnects (TSVs), and the logic circuit in the active interposer. Besides this, the communication bus between different chiplets, the distribution network of the clock, and power can be the possible surfaces of fault injection attacks. A fault can be injected into the system using laser illumination, EM radiation, clock, or voltage glitching, propagating to an observable node. In the fault analysis phase, an attacker may perform a differential or sensitivity analysis to steal the assets, such as the encryption key. Although the 3D die stacking technology and heterogeneous integration provide the defense against some fault injection attacks naturally [67], assessment of some attack surfaces is still required since design and integration complexity increases in a SiP. In addition, new attack surfaces have evolved beyond the traditional 2D packaging because the SiP package is built after integrating chiplets from various process nodes; their speed and supply voltage can be different. Furthermore, the fault injection techniques mentioned earlier make the SiP package vulnerable to timing attacks. In addition, the CAD tools for heterogeneous system design and verification may need a security assessment feature to perform security validation against the security threat posed by fault injection attacks.

- **Tampering**—The possibilities of tampering with a System-in-Package (SiP) can be classified into two areas, chiplet-level tampering, and SIP level tampering. Untrusted supply chain entities associated with a heterogeneous integration can be linked to these possibilities of tampering. For example, an adversarial IP owner or chiplet vendor can tamper with a chiplet during its design and manufacturing process (e.g., chiplet-level tampering). On the other hand, tampering is also possible during the heterogeneous integration of SiP by a chiplet integrator (e.g., system-level tampering). In addition, an adversarial end-user or black-hat attackers can tamper with physical attacks on chiplets or the integrating components of a heterogeneous system (e.g., vertical interconnects, communication interfaces, or interposer) by exploiting unidentified vulnerabilities. These tampering approaches may pose security threats and create hardware assurance issues that must be addressed during a heterogeneous system's design and integration step. The ways of tampering and their potential threats are briefly discussed as follows:

(1) Chiplet-level Tampering: There are various ways of SoC-level tampering have been reported by black-hat (adversarial users), and white-hat (security researchers) attackers, which can be further extrapolated to chiplet [23]. The ultimate goal of chiplet-level tampering is to leak confidential information (e.g., secret key or any asset). A chiplet can be maliciously modified during design time to insert hardware Trojans which facilitates confidentiality, integrity, and availability violations. An adversary can also perform a noninvasive attack without any physical alteration of a target device, and a semi-invasive attack requires a slight hardware change. For example, photon-induced current and EM radiations can be exploited to inject faults by flipping the bits of a memory chiplet, retrieving the stored bitstream of an FPGA chiplet, or registers of a chiplet having crypto hardware (e.g., AES, RSA, or SHA). A clock port, clock network, and power distribution network can tamper with advanced probing techniques (e.g., electro-optical, nano-probing) to assist clock glitching or voltage glitching attacks. Moreover, semi-invasive techniques such as focused-ion beam (FIB) help an attacker edit circuits to bypass security features or reroute a net carrying asset to attack the security of the system [68–70]. Tampering with probing can also enable an adversary to modify or read a signal directly inside the logic circuit of a chiplet.

(2) System-level Tampering: The main objective of system-level tampering is to alter the interaction between various functional chiplets, attack inter-chiplet communication, and reduce the strength of system-level security incorporated in heterogeneous integration. Malicious modification can degrade the reliability and signal integrity of the vertical interconnects that transfer signal and power from one chiplet to another. The change of design parameters of TSVs (through silicon vias) may affect reliable performance during runtime and cause failure in operation [71]. During the design and fabrication of interconnects, a hardware Trojan (e.g., modification of logic in active interposer or silicon bridges of passive interposer) can be inserted in the interposer to bypass any security feature or perform DoS (Denial of Service) attack.

In the case of HI, a monolithic SoC is partitioned into different chiplets based on their functionalities. As a result, many communications within a single die now happen as inter-die communication. Generally, probing inter-chiplet interconnects is easier than probing a bus buried in a die between metal layers. Therefore, these interconnect between chiplet make the SiP vulnerable to probing attacks. Advanced probing techniques leverage the direct access to the interconnects and interfaces to circumvent the security module (e.g., remove and change connections between chiplets) and intercept secure communication among chiplets (e.g., modification or read-out of any signal nets).

- **Malware and Ransomware Attacks**—Malware is malicious software code that can harm an electronic system by leaking, destroying data, or in extreme cases, causing a denial of service (DoS) attack [72]. The government, non-profit agencies, and private companies worldwide have been attacked by malware and cyber-attacks, which cost them billions of dollars [73, 74]. Malware and ransomware must be handled at the SiP level since electronic systems use SiPs assembled through heterogeneous integration. An electronic system cannot be considered secure until all its parts are secure. To have a safe system, any system that employs SiP, such as an IoT, high-performance computing, or cloud server, must perform the security assurance of the SiPs used in that system. The system-level operation of the devices eventually comes down to hardware, and malware detection in hardware is very challenging unless the designers of the hardware/chiplet integrators are aware of such attacks. So, by adding proper detection methods to the SiP package, malware and ransomware can be prevented to a great extent. Such methods are discussed in Sect. 10.5.4.

10.5.3   *Trust Validation by System Verification*

Heterogeneous Integration (HI) delivers flexibility advantages over monolithic SoC development due to its ability to integrate separately manufactured components into a higher-level assembly (SiP). However, since chiplets can be sourced from various supply chain entities, Die-to-Die communication can be vulnerable and lead to multiple attack vectors and surfaces. Further, a SiP package can be a counterfeit one that can cause reliability issues in the end-user applications and a significant revenue loss to the SiP designer. Package level potential attacks and vulnerabilities can be detected and avoided by enforcing security policies to monitor the transaction between chiplets. Additionally, an end-user can use provenance-based methods to verify the authenticity of a chiplet.

- **Security Policies**—In the context of security verification, security policies specify the confidentiality, integrity, and availability requirements for specific security-critical assets and provide mitigation strategies that need to be implemented [75, 76]. For example, an encryption module can only respond to encryption requests by other IPs and take encryption keys as input if the system is not in debug mode. Otherwise, an untrusted debugger can violate confidentiality and integrity requirements. Therefore, a security policy can be enforced which monitors the debug signal and deny all encryption requests when a debug request is accepted by raising a flag. This policy ensures that a secured communication standard between the encryption module and other IPs can be established. Similarly, a set of security policies can be derived from the design specification for secured communication between chiplets in HI. A chiplet-based hardware security module (CHSM) can be utilized to enforce these policies during chip operation. The security policies can be placed in the form of sensors or monitors in the silicon to operate in the run time, and the CHSM can raise a flag if any violation of the security policy is detected during the communication between chiplets. Hence, a policy-driven post-silicon security verification methodology can be established for secured communication between chiplets. We will discuss more security policies in Sect. 10.5.5.

- **Provenance Detection**—With the globalization of the semiconductor industry, the supply chain entities are spread across the globe. Due to these outsourcing practices of fabrication, assembly, and packaging, many package-level threats can impact SiPs, such as counterfeiting. In the SoC domain, counterfeiting includes recycled, cloned, remarked, or overproduction of SiPs. These nefarious and lucrative threats can create devastating threats to mission control applications such as space, military, energy, and healthcare which may be using chiplets integrated on a SiP. Various strides have been taken to identify these threats in the supply chain, and methods are created to prevent and detect some or all of these counterfeit scenarios. However, in reality, methods rarely protect or prevent all threats but are specialized in preventing a few. For example, a framework called Secure Split-Test (SST) was introduced in 2013 to instantiate a more robust and secure interface between the IP owner and foundry/assembly facilities by mandating specific interactions between these entities like sending keys and validating results [77]. Furthermore, on-chip security primitives such as physically unclonable functions (PUF) and silicon odometer readings can detect cloned ICs [78]. However, these methods cannot verify all previously mentioned counterfeit instances. Furthermore, not everyone in the supply chain is familiar with the technology for performing these time-consuming verification processes.

  Another recent innovative strategy proposed involves incorporating provenance methods using a blockchain framework called *eChain* to create an ecosystem of trusted microelectronics to thwart the recycled, cloned, remarked, and overproduced ICs [22, 79]. Blockchain technology, introduced by Satoshi Nakamoto in 2008, provides a framework for a distributed database with an increasing ledger of blocks ensuring decentralization and consensus by all nodes as to the validity of a transaction [80]. The *eChain* framework is similar to that of the Bitcoin blockchain network, with changes (such as a consortium of already vetted members) geared explicitly toward ensuring electronics supply chain integrity.

- **Limited testability**—Heterogeneous Integration introduces new challenges for chiplet and complete SiP package testability. For example, in the case of 3D ICs, one can access DFT pins only through the base die. Currently, many types of interface protocols and packaging technology exist for chiplets. However, there is no interoperable testing and debug standard for HI. Furthermore, because chiplets can come from a diverse supply chain, the lack of standard testing infrastructure creates a challenge for effectively testing the final SiP. Additionally, a testing infrastructure can pose a major security risk if proper design-for-test or design-for-debug algorithms are not used. For example, scan chains are frequently utilized as one of the most effective attack surfaces. To provide a standard test infrastructure, IEEE has proposed IEEE 1838 standard [81] for 3D IC testing. This standard can also be applied to 2.5D ICs. Also, IEEE EPS has published the best-known methods for HI testability which manufacturers and designers can use to ensure proper testing of a HI product [82].

10.5.4 Attack Mitigation or Possible Countermeasures

As discussed earlier, a SiP package can face in-field threats by an adversarial end-user. Therefore, the SiP level vulnerabilities need to be addressed by deploying relevant countermeasures against a particular attack as follows:

- **Fault Injection Detection:** Different types of fault injection techniques, including fault injection through clock glitch, voltage glitch, laser illumination, and EM radiation, can impact the timing delay in the victim device. A TDC (Time-to-digital converter) sensor capable of sensing the fluctuation in the delays of several components (logic gates, interconnects) of a heterogeneous system can be used to monitor fault injector attempts. This kind of sensor is proposed as FTC (Fault-to-time converter) sensor and demonstrates success in detecting attempts of different fault injections [49, 83, 84].
- **Timing Fault injection Mitigation:** A heterogeneous system has a set of security policies that define the security of the secret information of the system. An adversary can leak assets or reduce the strength of the security of the system by injecting controlled timing faults at the security-critical locations and ultimately by violating the defined security properties. Pre-silicon analysis paired with modifying the physical design parameters (e.g., gate size, interconnect length and width, power pads) can tune the delay of the security-critical paths and make the delay distribution uniform. These modifications will ultimately make a controlled timing fault injection uncontrollable and hide the impact of controlled faults. In summary, a preliminary security property-driven assessment with a set of physical design rules can reduce a design's susceptibility to timing fault injection attacks [85].
- **On-chip Cryptography:** It can ensure the CIA triad for messages, signals, and confidential information exchanged between chiplets. For example, hash-based message authentication code techniques can be used to send-receive signals between two chiplets which requires message integrity. In addition, the authenticated encryption protocol can be used if the signals require both confidentiality and integrity.
- **3D Trojan Detection and Mitigation:** The complementary characteristics of DCVSL (Differential Cascade Voltage Switch Logic) can detect malicious modification (parametric hardware Trojans on TSV) or external voltage glitches. Any tampering or malicious modification on the power or ground lines prevents the output of a DCVSL from being complementary, thus asserting a warning signal [71].
- **Clock Tampering Detection:** A buffer-based delay chain can detect a clock glitch or tampering. The upper boundary of the clock frequency is predefined by the propagation delay (T_d) of the delay chain. Timing violations may occur if the tampered or glitched clock runs faster than the upper bound $(1/T_d)$. Therefore, the upper bound is set to equal to the critical path delay to ensure the best performance. The comparison result of the delay chain detects the possible glitch injection or clock tampering [86].

- **Sensors and Protective Shields:** Active internal shielding techniques can be exploited to prevent front-side probing attacks in a heterogeneous system [69, 87]. These shields can be integrated into the SiP after the assembly and testing. An active protection technique can detect system-level tampering that senses any change in electrical parameters (e.g., resistance, capacitance, or inductance) of a heterogeneous system due to tampering [88]. Anti-tampering sensors such as light, pressure, and electrostatic or electromagnetic sensors can be integrated into the chiplets to notify the security controller unit whether the SiP has physically tampered or attacked. The security module can also periodically monitor the resistance of different components (e.g., TSVs, horizontal traces through the interposer, interfaces, I/O pads) that compares the measured values with pre-stored initial values and gives a decision on whether tampering is performed or not. One of the challenges with these approaches is that the chiplet is a hardened IP with almost nil scope for modification. In this scenario, a chiplet designer has to incorporate these security sensors during design to ensure they are fabricated on the chiplet, or a chiplet designer leaves some programmable logic space similar to FPGA. Then, SiP design security engineers can configure this programmable logic to design a circuit that can work as a sensor.
- **Side-Channel Attack Mitigation:** This section will focus on 2.5 D HI packages because 3D packaging provides various attack mitigation due to its inherent stacking architecture that creates a natural countermeasure against fault and cache-based timing side-channel attacks [67]. At the 2.5 D SiP level, side-channel attack simulations can assist security researchers in uncovering system-level vulnerabilities. Commercial CAD tools provide fast and efficient transistor-level simulations for EM and power side channels. However, to make a system resilient against EM, thermal, timing, and power side-channel attacks, it needs countermeasures at different levels, such as low-leakage transistors, randomized operation at the system level, and side-channel attack resistant algorithms.
- **Malware and Ransomware Detection:** Malware and ransomware can frequently mutate (employ polymorphism) to evade signature-based detection techniques such as anti-viruses. Therefore, advanced detection schemes based on hardware performance counters and machine learning have been proposed in the literature to identify zero-day exploits by classifying microarchitectural hardware events associated with the program's execution [89–91]. Another way to detect malware is by putting power sensors [92] within the SiPs where a power side channel based disassembler [93, 94] can be utilized in real-time to detect malware. Despite the space overhead caused by the addition of power monitoring sensors, the approach can still be used depending on how crucial the application is. Another possibility is to use machine learning techniques to improve the systems' ability to detect malware [95, 96].
- **On-Chip Security Module:** Like SoC, a SiP level security controller can monitor chiplet and their interfaces for any potential security property violation, run time detection of hardware trojans, and tampering. Furthermore, it can use a machine learning-based approach to detect new Trojans. Its firmware can be

updated to provide the controller with new Trojan signatures and updated security policies. A chiplet hardware security module (CHSM) can be one of the futuristic approaches as a security controller that be implemented on SiP package, which can co-ordinate possibly with chiplets to detect and neuter above mention threats. Concept and working of CHSM is discussed in details in Sect. 10.6.3.

10.5.5   Security Policies

Security policies refer to a set of rules or requirements to protect the assets and IPs. For example, security policies are widely used in an SoC device to prevent unauthorized access, or transaction inside the device [97]. Furthermore, for secure heterogeneous integration of chiplets, security policies can help designers to develop design constraints and forbidden user actions to prevent any CIA triad violation.

There are two types of assets, primary and secondary. Primary assets may include the cryptographic keys and configuration registers which are the main target of attackers. Secondary assets mean the infrastructures that require protection to ensure the security of the primary assets. Security policies can enforce the confidentiality, integrity, and availability requirements of SiP packages on the system level as follows:

- **Confidentiality:** Only authorized entities will have access to an asset. Temporal validation will be required for those permitted entities. For example, Chiplet A may only receive hash data from the crypto core if it is the entity that sent out the data to be hashed in the first place. It will not be able to obtain hash output from the crypto core at any other time.
- **Integrity:** An asset cannot be modified by an unauthorized entity. For example, an unauthorized transaction (by a user, chiplet, or firmware) cannot change the data transferred from chiplet A to chiplet B while it is in transit.
- **Availability:** An authorized entity should be able to reliably access an asset when needed. For example, if a logic locked entity requires an unlocking key to function, this key should be available to it whenever required.

Based on these requirements, many security policies can be formulated. Below, we enlist some of the standard policies. Of course, this list is not exhaustive but can provide a good understanding of various policies that can be implemented.

10.5.5.1   Access Control Policies

Access control policies specify how one chiplet can access an asset during different execution points for the SiP. These are:

- An unauthorized chiplet cannot access memory in the protected address range.

- An unauthorized chiplet cannot write out data to a restricted memory region (information leakage).

10.5.5.2 Information Flow Policies

Information Flow Policies restrict leakage or modification of information related to secure assets. Examples of such policies are:

- An unauthorized chiplet cannot access data intended for other chiplets during transit.
- An unauthorized chiplet cannot modify data intended for other chiplets.
- Chiplet A cannot pose as chiplet B to receive data intended for chiplet B.
- Data intended for a chiplet cannot be blocked by an unauthorized chiplet.

10.5.5.3 Liveness Policies

Liveness policies ensure that SiP can execute normal tasks without interruption. Examples include:

- A chiplet cannot flood communication fabric with messages to disrupt normal behavior (DDoS).
- During operation, the number of messages sent by an untrusted chiplet should not exceed the threshold of the maximum limit.
- The limit on the number of packets generated by an untrusted chiplet can only be assigned and updated at secure boot time.

10.5.5.4 Active Monitoring Policies

Active monitoring policies ensure the secure operation of the system during runtime. Examples of such policies are given below:

- The frequency of the clock signal cannot vary out of range to prevent the clock glitching.
- The voltage supplied to individual dies cannot vary out of range to prevent power glitching.

10.5.5.5 Security Policy Enforcement

CHSM is responsible for enforcing security policies. For this purpose, the high-level policies mentioned earlier need to be converted to formal constraints [98]. First, the assets in the system that need to be protected should be listed. Then, security policies involving that asset will be identified. These security policies may need to be modified or expanded while moving from one abstraction layer to another [99]. After that, possible attack surfaces will be identified, and necessary protections can be developed. Finally, these policies can be implemented in CHSM firmware, using a secure boot mechanism so that an adversary cannot modify the policies. Also, if the authorized owner needs to update the security policies, he/she can do it by upgrading the firmware.

10.6 Roadmap Toward Future Research

In previous sections, based on our previous research and development in the SoC domain, we have drawn parallelism between SoC and SiP level threats and vulnerabilities and their respective probable trust validation attack mitigation. However, these SoC-level pre-existing solutions need further research and development (including enhancement and modification) before a security researcher can directly import these solutions for chiplet or interposer level security and, ultimately, secure heterogeneous integration of SiPs. Here are the following a few areas that open up a roadmap for future research work in the area of secure heterogeneous integration:

10.6.1 Trust Validation of Chiplets for Chiplet Security

As discussed earlier in Sect. 10.3.3 about a strong requirement for trust validation of chiplets in SiP ecosystem. In the future, a better trust validation approach will be available on demand. This approach will not require a chiplet OEM to share its IP for trust validation of chiplets, and still, it can bring trust between chiplet OEMs and SiP OEMs. One possible approach could be an independent entity that can handle trust validation requests from SiP OEMs and perform trust validation on behalf of chiplet OEMs to provide validation results in a couple of hours to a day. This independent entity can work like *Interactive proof system*, in which a *prover* has access to computational resources, and *verifier* has limited computation power but seeks an answer to the problem, which is here trust validation. This independent entity can certify a chiplet as authentic or flag it as malicious based on the validation outcome; henceforth, we call this entity by Certificate Authority (CA) [100]. The notion of CA is based on an incentivized approach over an enforcement approach, which encourages most chiplet OEMs to join this certification authority, as they can be benefited by becoming trusted chiplet OEMs in the supply chain, which can boost their revenue and brand reputation. Also, penalties can be imposed if malpractice is detected to keep a sanity check on chiplet OEMs.

The trust validation process can start with a voluntary enrollment of a chiplet OEM with the Certificate Authority (CA) by sharing the active region footprint of logic cells; henceforth, we call it as *Minimal Layout* (see Fig. 10.7). In this way, chiplet OEM can protect its IP by sharing not all but necessary information to the

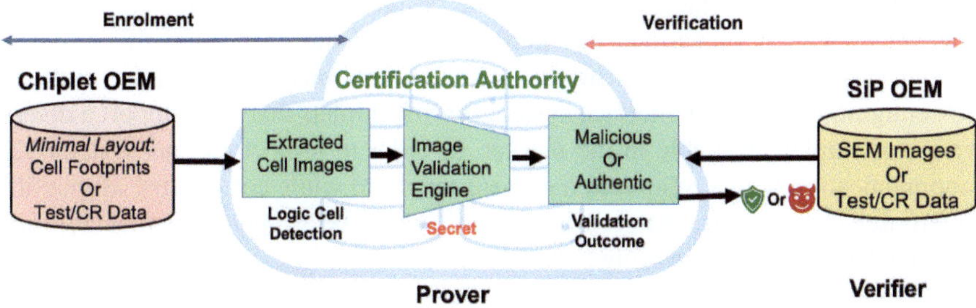

Fig. 10.7 Certification authority for trust validation of chiplets

CA. After this, CA can develop an SEM imaging specification and a validation model. Once a SiP OEM approach CA for chiplet validation of chiplet(s), CA shares SEM imaging requirements and classifies the chiplet as malicious or authentic based on the outcome of the validation model developed earlier. Various steps of CA design and development and validation process are described through Algorithm 15.

Algorithm 15 Certification authority (CA) for trust validation of chiplets

---

 1: **procedure** ENROLMENT
 2:     Chiplet OEM joins CA
 3:     Agrees to terms & conditions of CA
 4:     Share Minimal Layout to CA
 5:     SiP OEM sign up as verifier

 6: **procedure** CA DESIGN AND DEVELOPMENT
 7:     SEM imaging specifications
 8:     Trust validation model
 9: **procedure** TRUST VALIDATION PROCESS
10:     SiP OEM request for chiplet validation
11:     CA shares SEM imaging requirements
12:     SiP OEM shares chiplet's SEM images
13:     **if** Validation = No Change Detected **then**
14:         Chiplet Certified as Authentic
15:     **else**
16:         Chiplet Flagged as Malicious

10.6.2   Secure Die-to-Die Interface for Interposer Security

Protecting the Die-to-Die (D2D) communication system in heterogeneously integrated systems is critical in preventing eavesdropping and the growing number of hardware hacking attacks. A hardware solution is required to ensure a secure chiplet-to-chiplet communication interface via advanced interface protocol. In order to meet adaptability, low-cost terms, and varying bus performance requirements, a scalable wrapper/interface IP-core must be designed and implemented. It enables the authentication of various bus participants and the encryption of chiplet-to-chiplet buses using a single primitive. The solution must be transparent and easy to configure in any D2D interface system for SiP chiplets that will support a diverse range of chiplets available in the market. In order to protect the bus system/D2D communication with a minimum hardware overhead while considering all possible interposer level threats, the security feature should be simply implementable into the standardized interface bridge.

10.6.3   Chiplet Hardware Security Module (CHSM) for Secure SiP

It is clear from the earlier discussions that the heterogeneous integration has instigated various vulnerabilities in the SIP from its multi-party supply chain and manufacturing flow which can cause various counterfeit, hardware assurance, and trust problems. As heterogeneous integration follows a different manufacturing flow compared to the SoC, existing solutions developed for typical SoC design may not be enough to solve the challenges mentioned above. For example, prior research proposed various trusted computing architectures to prevent attacks on SoCs. AEGIS [101] is one of the earliest secure architectures which protects memory tampering from possible software/hardware-based attacks. It protects the integrity of the software program by calculating the hash of the secure kernel at the initial boot.

Moreover, the secure kernel is responsible for ensuring AEGIS secure functionalities. Here, the processor is also trusted, and the on-chip cache is assumed secure against physical attacks. Furthermore, the trusted platform module (TPM) [102] also protects the software by measuring the integrity metric of the successive programs starting from root of trust for measurement (RTM) and stores the hash of programs inside the platform configuration register (PCR), thus creating a chain of trust. In addition, TPM uses endorsement keys (EK) unique to each chip serving as a master derivation key, attestation identity key (AIK) for creating digital signatures, and storage keys to protect the memory by storing encrypted programs and data. However, none of the above architectures are applicable in HI as the processor because memory, processor, and other components are integrated as separate chiplets and prone to attacks on the insecure communication channel.

In addition, the ARM TrustZone [103] is a well-known industry-developed architecture for ensuring SoC security. TrustZone protects the trusted hardware and software resources by isolating the Trusted Execution Environment (TEE) or the secure world and Rich Execution Environment (REE) or the normal world using a Non-Secure (NS) bit at the bus architecture. The hardware and software resources running in the normal world do not have access to the components in the secure world, although the opposite is allowed. Therefore, this hardware-based security

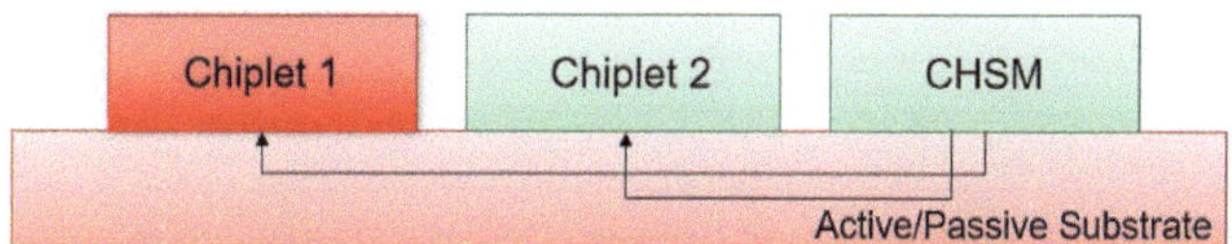

Fig. 10.8 High-level block diagram of CHSM inside SiP

solution is well suited for SoC as the communication within the secure world happens in plain text, which is prone to physical attacks in the heterogeneous integration. Similarly, Intel Software Guard Extensions (SGX) [104] create "enclaves" where the trusted execution of programs happens, and unauthorized access to the enclaves is not allowed. There are other security architectures (e.g., Bastion [105], Sancus [106], and TyTan [107]) which are developed to provide security for an SoC design, however, significant changes are needed in the designs considering the threat model of heterogeneous integration.

Therefore, one of the possible solutions for the challenges mentioned above is to design a centralized chiplet that will act as a hardware security module to prevent all possible attacks based on the HI threat model. We refer to the chiplet as CHSM, which is assumed to be trusted (see Fig. 10.8). As stated earlier, the heterogeneous SIP designer has no control over many of the chiplets procured from third-party vendors and can not trust those chiplets. Thus, the trusted CHSM plays a vital role in ensuring the overall system's security. In [108], the authors proposed an end-to-end secure SoC lifecycle flow where the SoC contains a root of trust security engine (SE). Similarly, the heterogeneous SiP requires a secure manufacturing flow to make it free from all possible vulnerabilities mentioned earlier, and the CHSM acts as a root of trust. The CHSM contains various static security assets (e.g., a device-specific identity (ID), private keys, keys stored in effuse.) to perform various cryptographic operations and encrypted communication between CHSM and chiplets. Therefore, these security assets must be provisioned inside the CHSM securely on the manufacturing floor or in a trusted facility. POCA [109] provisions security assets securely inside the chip at the zero trust stage of the manufacturing floor. Thus, CHSM must integrate POCA infrastructure inside it to provide security assets securely on the manufacturing floor.

CHSM can be implemented in the embedded FPGA platform, which may contain a fully FSM-based hardware controller or a processor with firmware support. During in-field operation, the CHSM authenticates the chiplets and generates a shared secret key to encrypt the communication between the two chiplets. Encrypting every communication between chiplets can be hazardous in terms of performance. In addition, the System-in-Package may not need all chiplets to be trusted and all encrypted communications. However, the channel must be encrypted between the CHSM and the chiplets when the chiplet contains security assets and runs any mission/security-critical applications. Thus, the performance and security of the SiP and security applications can be ensured efficiently. The CHSM monitors the data communication among chiplets to determine any malicious activity by Trojan

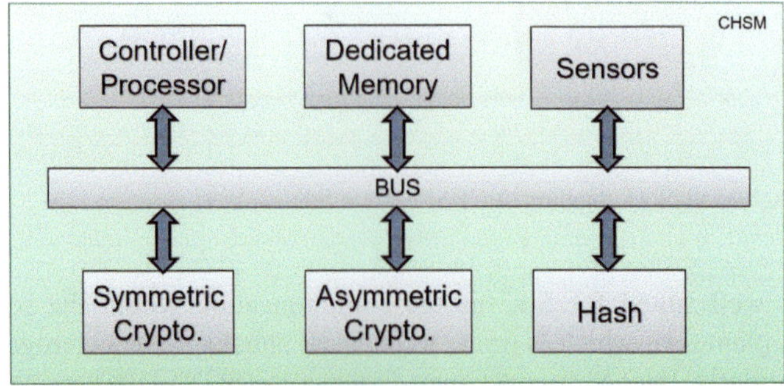

Fig. 10.9 High-level block diagram of CHSM IP

implanted inside SiP package or malicious software. The CHSM contains security policies to detect the anomalies of the data communication or illegal access by untrusted software/hardware entities to the protected region. Once an anomaly is detected, the CHSM protects the security assets, prevents malicious activity based on the security policies, and creates a protected boundary where the malicious implant/software cannot access the trusted resources. The CHSM contains various sensors to detect physical attacks (e.g., laser, X-ray, voltage/clock glitching), and it protects the SIP based on the security policies. In a nutshell, CHSM acts as a root of trust inside the SIP and protects it from all possible vulnerabilities from the manufacturing floor to the end of life. Figure 10.9 shows the high-level block diagram of the CHSM IP and its components.

10.6.3.1 Provenance Method to Identity Counterfeit SiP

For SiP hardware assurance, the complete package and its components must be traced back to its source of origin using provenance methods. Assuming all supply chain entities are known and vetted, it is very challenging for an adversary to inject a counterfeit SiP package into the supply chain. However, if an adversary successfully infiltrates a counterfeit SiP in the supply chain, that entity can be identified using provenance techniques such as blockchain [110, 111].

The question arises: Can preventative measures like these align with the newer heterogeneous integration processes? Assuming a Chiplet Hardware Security Module (CHSM), introduced in the previous section, can be utilized in the HI, many pre-integration threats could be alleviated. However, this leaves provenance post-integration as a vital goal; once the HI chip is integrated and assembled, the authenticity of the chip needs to be asserted in the remaining phases of its lifespan: as an element assembled in a PCB, in the field, and at the end of its life.

- *SiP Manufacturing Phase:* This phase involved procuring individual chiplets, assembly on the interposer layer, and producing the final SiP package. Tracing and tracking individual components of a SiP can thwart counterfeit chiplets and further impede a counterfeit SiP package from the supply chain.

- *Assembly on PCB Phase:* As a PCB component, the integrated SiP performs its complete operations as one part in an extensive system of components. The component is assembled onto the PCB along with the other components by a PCB design and assembly house.
- *Field Deployment:* As the PCB is deployed to the field (typically inside a system) and operating at its intended application, the SiP is an active member functioning in the system.
- *End of Life:* Once a system is not in service or an individual component has been stripped from its PCB due to a failure or preventive maintenance. In this scenario, that system or component is no longer utilized in its intended functioning system, and it has reached the end of its life.

The goal of provenance methods is to verify the authenticity of heterogeneously integrated packages by tracking and tracing a package to its origin, recording ownership records, and its life events, such as the end of the life.

References

1. B. Hoefflinger, ITRS: the international technology roadmap for semi-conductors, in *Chips 2020: A Guide to the Future of Nanoelectronics* (Springer, Berlin, 2011), pp. 161–174
2. Md.S. Ul Islam Sami, H.M. Kamali, F. Farahmandi, F. Rahman, M. Tehranipoor, Enabling security of heterogeneous integration: from supply chain to in-field operations. IEEE Des. Test **40**(5), 86–95 (2023)
3. D.D. Sharma, Universal chiplet interconnect express (UCIe): building an open chiplet ecosystem, in *White Paper published by UCIe Consortium* (2022)
4. G. Keeler, S. Shumarayev, M. Wade, B. Hamilton, ERI summit, in *Co-packaging Heterogeneous Microelectronics and Optical Input/Output*
5. E. Sperling, Quantum effects at 7/5nm and beyond. Semicond. Eng. **10585**(1), 81–124 (2018)
6. T. Uhrmann, J. Burggraf, M. Eibelhuber, Heterogeneous integration by collective die-to-wafer bonding, in *2018 International Wafer Level Packaging Conference (IWLPC)* (IEEE, Piscataway, 2018), pp. 1–7
7. K.F. Yang, T.J. Wu, W.C. Chiou, M.F. Chen, Y.-C. Lin, F.W. Tsai, C.C. Hsieh, C.H. Chang, W.J. Wu, Y.H. Chen, et al., Yield and reliability of 3DIC technology for advanced 28nm node and beyond, in *2011 Symposium on VLSI Technology-Digest of Technical Papers* (IEEE, Piscataway, 2011), pp. 140–141
8. T. Li, J. Hou, J. Yan, R. Liu, H. Yang, Z. Sun, Chiplet heterogeneous integration technology—status and challenges. Electronics **9**(4), 670 (2020)
9. R. Chau, Process and packaging innovations for Moore's law continuation and beyond, in *2019 IEEE International Electron Devices Meeting (IEDM)* (IEEE, Piscataway, 2019), pp. 1–1
10. IEEE, Heterogeneous integration roadmap - chapter 20: thermal, in *Heterogeneous Integration Roadmap* (IEEE, Piscataway, 2021), pp. 1–24
11. K. Ahi, N. Asadizanjani, S. Shahbazmohamadi, M. Tehranipoor, M. Anwar, Terahertz characterization of electronic components and comparison of terahertz imaging with x-ray imaging techniques, in *Terahertz Physics, Devices, and Systems IX: Advanced Applications in Industry and Defense*, vol. 9483 (International Society for Optics and Photonics, Bellingham, 2015), 94830K

12. J. True, C. Xi, N. Jessurun, K. Ahi, M. Tehranipoor, N. Asadizanjani, Terahertz based machine learning approach to integrated circuit assurance, in *2021 IEEE 71st Electronic Components and Technology Conference (ECTC)* (IEEE, Piscataway, 2021), pp. 2235–2245

13. https://www.intel.com/content/dam/www/public/us/en/documents/white-papers/accelerating-innovation-through-aib-whitepaper.pdf

14. Trusted and assured microelectronics. https://rt.cto.mil/ddre-rt/dd-rtl/tam/

15. State of The Art (SOTA) Heterogeneous Integrated Packaging (SHIP) program. https://www.ndia.org/events/2021/12/14/287m-ship-day

16. Intel-Newsroom. Intel wins US Government Advanced Packaging Project. https://www.intel.com/content/www/us/en/newsroom/news/us-government-advanced-packaging-project.html

17. Diverse Accessible Heterogeneous Integration. https://www.darpa.mil/program/diverse-accessible-heterogeneous-integration

18. S. Sinha, State of IOT 2021: number of connected IOT Devices (2021). https://iot-analytics.com/number-connected-iot-devices/

19. Chapter 4: Medical, Health and Wearables (2021). https://eps.ieee.org/images/files/HIR_2020/ch04_health.pdf

20. Chapter 5: Automotive - IEEE Electronics Packaging Society (2019). https://eps.ieee.org/images/files/HIR_2019/HIR1_ch05_automotive.pdf

21. M. Deo, Enabling next-generation platforms using Intel's 3d System-in-Package technology, in *White Paper* (2017), pp. 1–7

22. N. Vashistha, M.M. Hossain, Md.R. Shahriar, F. Farahmandi, F. Rahman, M. Tehranipoor, eChain: a blockchain-enabled ecosystem for electronic device authenticity verification. IEEE Trans. Consum. Electron. **68**(1), 23–37 (2021)

23. S. Bhunia, M. Tehranipoor, *Hardware Security: A Hands-On Learning Approach* (Morgan Kaufmann, Burlington, 2018)

24. J. Wurm, Y. Jin, Y. Liu, S. Hu, K. Heffner, F. Rahman, M. Tehranipoor, Introduction to cyber-physical system security: a cross-layer perspective. IEEE Trans. Multi-Scale Comput. Syst. **3**(3), 215–227 (2016)

25. N. Vashistha, M.T. Rahman, H. Shen, D.L. Woodard, N. Asadizanjani, M. Tehranipoor, Detecting hardware trojans inserted by untrusted foundry using physical inspection and advanced image processing. J. Hardware Syst. Secur. **2**(4), 333–344 (2018)

26. K. Xiao, D. Forte, Y. Jin, R. Karri, S. Bhunia, M. Tehranipoor, Hardware trojans: lessons learned after one decade of research. ACM Trans. Des. Autom. Electron. Syst. **22**(1), 1–23 (2016)

27. N. Vashistha, H. Lu, Q. Shi, M.T. Rahman, H. Shen, D.L. Woodard, N. Asadizanjani, M. Tehranipoor, Trojan scanner: detecting hardware trojans with rapid sem imaging combined with image processing and machine learning, in *ISTFA 2018: Proceedings from the 44th International Symposium for Testing and Failure Analysis* (ASM International, Geauga County, 2018), p. 256

28. T. Meade, Y. Jin, M. Tehranipoor, S. Zhang, Gate-level netlist reverse engineering for hardware security: control logic register identification, in *2016 IEEE International Symposium on Circuits and Systems (ISCAS)* (IEEE, Piscataway, 2016), pp. 1334–1337

29. S. Chen, J. Chen, D. Forte, J. Di, M. Tehranipoor, L. Wang, Chip-level anti-reverse engineering using transformable interconnects, in *2015 IEEE International Symposium on Defect and Fault Tolerance in VLSI and Nanotechnology Systems (DFTS)* (IEEE, Piscataway, 2015), pp. 109–114

30. S.E. Quadir, J. Chen, D. Forte, N. Asadizanjani, S. Shahbazmohamadi, L. Wang, J. Chandy, M. Tehranipoor, A survey on chip to system reverse engineering. ACM J. Emerging Technol. Comput. Syst. **13**(1), 1–34 (2016)

31. UDO Commerce, Defense industrial base assessment: counterfeit electronics, in *Bureau of Industry and Security, Office of Technology Evaluation*. Technical Report (2010)

32. M. Tehranipoor, H. Salmani, X. Zhang, *Integrated Circuit Authentication*, vol. 10 (Springer, Cham, 2014), pp. 978–3

33. M.M. Tehranipoor, U. Guin, D. Forte, Counterfeit test coverage: an assessment of current counterfeit detection methods, in *Counterfeit Integrated Circuits* (Springer, Berlin, 2015), pp. 109–131

34. U. Guin, K. Huang, D. DiMase, J.M. Carulli, M. Tehranipoor, Y. Makris, Counterfeit integrated circuits: a rising threat in the global semiconductor supply chain. Proc. IEEE **102**(8), 1207–1228 (2014)

35. U. Guin, D. DiMase, M. Tehranipoor, A comprehensive framework for counterfeit defect coverage analysis and detection assessment. J. Electron. Test. **30**(1), 25–40 (2014)

36. N. Asadizanjani, M.T. Rahman, M. Tehranipoor, M.H. Tehranipoor, *Physical Assurance: For Electronic Devices and Systems* (Springer, Berlin, 2021)

37. M.M. Tehranipoor, H. Shen, N. Vashistha, N. Asadizanjani, M.T. Rahman, D. Woodard, Hardware trojan scanner. US Patent **11**, 030–737 (2021)

38. Q. Shi, N. Vashistha, H. Lu, H. Shen, B. Tehranipoor, D.L. Woodard, N. Asadizanjani, Golden gates: a new hybrid approach for rapid hardware Trojan detection using testing and imaging, in *2019 IEEE International Symposium on Hardware Oriented Security and Trust (HOST)* (IEEE, Piscataway, 2019), pp. 61–71

39. N. Vashistha, H. Lu, Q. Shi, D.L. Woodard, N. Asadizanjani, M. Tehranipoor, Detecting hardware trojans using combined self testing and imaging. *IEEE Transactions on Computer-Aided Design of Integrated Circuits and Systems* **41**(6), 1730–1743 (2021)

40. H.M. Kamali, K.Z. Azar, F. Farahmandi, M. Tehranipoor, Advances in logic locking: past, present, and prospects. Cryptology ePrint Archive **2022**(260), 1–39 (2022)

41. K.Z. Azar, H.M. Kamali, F. Farahmandi, M.H. Tehranipoor, *Understanding Logic Locking* (Springer, 2024)

42. H.M. Kamali, K.Z. Azar, K. Gaj, H. Homayoun, A. Sasan, LUT-Lock: a Novel LUT-based logic obfuscation for FPGA-bitstream and ASIC-hardware protection, in *IEEE Computer Society Annual Symposium on VLSI (ISVLSI)* (2018), pp. 405–410

43. H.M. Kamali, K.Z. Azar, H. Homayoun, A. Sasan, Full-lock: hard distributions of SAT instances for obfuscating circuits using fully configurable logic and routing blocks, in *Proceedings of Design Automation Conference (DAC)* (2019), p. 89

44. M.S. Rahman, R. Guo, H.M. Kamali, F. Rahman, F. Farahmandi, M. Tehranipoor, ReTrustFSM: toward RTL hardware obfuscation-a hybrid FSM approach. IEEE Access **11**, 19741–19761 (2023)

45. IEEE recommended practice for encryption and management of electronic design intellectual property (IP), in *IEEE Std 1735-2014 (Incorporates IEEE Std 1735-2014/Cor 1-2015)* (2015), pp. 1–90. https://doi.org/10.1109/IEEESTD.2015.7274481

46. Md.R. Muttaki, S. Saha, H.M. Kamali, F. Rahman, M. Tehranipoor, F. Farahmandi, RTLock: IP protection using scan-aware logic locking at RTL, in *2023 Design, Automation & Test in Europe Conference & Exhibition (DATE)* (IEEE, Piscataway, 2023), pp. 1–6

47. S. Mangard, E. Oswald, T. Popp, *Power Analysis Attacks: Revealing the Secrets of Smart Cards*, vol. 31 (Springer Science & Business Media, Berlin, 2008)

48. S. Chari, C.S. Jutla, J.R. Rao, P. Rohatgi, Towards sound approaches to counteract power-analysis attacks, in *Proceedings 19 of the Advances in Cryptology—CRYPTO'99: 19th Annual International Cryptology Conference Santa Barbara, California*, August 15–19, 1999 (Springer, Berlin, 1999), pp. 398–412

49. Md.R. Muttaki, M. Tehranipoor, F. Farahmandi, FTC: a universal fault injection attack detection sensor, in *IEEE International Symposium on Hardware-Oriented Security and Trust (HOST)* (IEEE, Piscataway, 2022)

50. E. Amini, A. Beyreuther, N. Herfurth, A. Steigert, B. Szyszka, C. Boit, Assessment of a chip backside protection. J. Hardware Syst. Secur. **2**(4), 345–352 (2018)

51. N. Vashistha, M.T. Rahman, O.P. Paradis, N. Asadizanjani, Is backside the new backdoor in modern socs? in *2019 IEEE International Test Conference (ITC)* (IEEE, Piscataway, 2019), pp. 1–10

52. H. Shen, N. Asadizanjani, M. Tehranipoor, D. Forte, Nanopyramid: an optical scrambler against backside probing attacks, in *International Symposium for Testing and Failure Analysis (ISTFA)* (2018), p. 280

53. P.E. Calzada, Md.S. Ul Islam Sami, K.Z. Azar, F. Rahman, F. Farahmandi, M. Tehranipoor, Heterogeneous integration supply chain integrity through blockchain and CHSM. ACM Trans. Des. Autom. Electron. Syst. **29**(1), 1–25 (2023)
54. J.H. Lau, *Heterogeneous Integrations* (Springer, Berlin, 2019)
55. M. Tehranipoor, F. Koushanfar, A survey of hardware trojan taxonomy and detection. IEEE Des. Test Comput. **27**(1), 10–25 (2010)
56. M.M. Hossain, A. Vafaei, K.Z. Azar, F. Rahman, F. Farahmandi, M. Tehranipoor, SoCFuzzer: SoC vulnerability detection using cost function enabled fuzz testing, in *2023 Design, Automation & Test in Europe Conference & Exhibition (DATE)* (IEEE, Piscataway, 2023), pp. 1–6
57. H. Al-Shaikh, A. Vafaei, M.Md. Mashahedur Rahman, K.Z. Azar, F. Rahman, F. Farahmandi, M. Tehranipoor, Sharpen: SoC security verification by hardware penetration test, in *Proceedings of the 28th Asia and South Pacific Design Automation Conference* (2023), pp. 579–584
58. J.H. Lau, Advanced packaging, in *Semiconductor Advanced Packaging* (Springer, Berlin, 2021), pp. 1–25
59. M. Shafkat, M. Khan, C. Xi, A.A. Khan, M.T. Rahman, M.M. Tehranipoor, N. Asadizanjani, Secure interposer-based heterogeneous integration. IEEE Des. Test **39**(6), 156–164 (2022)
60. M. Nabeel, M. Ashraf, S. Patnaik, V. Soteriou, O. Sinanoglu, J. Knechtel, An interposer-based root of trust: seize the opportunity for secure system-level integration of untrusted chiplets. Preprint. arXiv:1906.02044 (2019)
61. P.D. Schiavone, D. Rossi, A. Di Mauro, F.K. Gürkaynak, T. Saxe, M. Wang, K.C. Yap, L. Benini, Arnold: an eFPGA-augmented RISC-V SoC for flexible and low-power IoT end nodes. IEEE Trans. Very Large Scale Integr. VLSI Syst. **29**(4), 677–690 (2021)
62. Z. Guo, M. Tehranipoor, D. Forte, J. Di, Investigation of obfuscation-based anti-reverse engineering for printed circuit boards, in *Proceedings of the 52nd Annual Design Automation Conference* (2015), pp. 1–6
63. M. Shafkat, M. Khan, C. Xi, Md.S. Ul Haque, M.M. Tehranipoor, N. Asadizanjani, Exploring advanced packaging technologies for reverse engineering a System-in-Package (SiP). IEEE Trans. Compon. Packag. Manuf. Technol. **13**(9), 1360–1370 (2023)
64. M. Nabeel, M. Ashraf, S. Patnaik, V. Soteriou, O. Sinanoglu, J. Knechtel, 2.5 D root of trust: secure system-level integration of untrusted chiplets. IEEE Trans. Comput. **69**(11), 1611–1625 (2020)
65. J. Park, N.N. Anandakumar, D. Saha, D. Mehta, N. Pundir, F. Rahman, F. Farahmandi, M.M. Tehranipoor, PQC-SEP: power side-channel evaluation platform for post-quantum cryptography algorithms. IACR Cryptol. ePrint Arch. **2022**, 527 (2022)
66. B. Ahmed, Md.K. Bepary, N. Pundir, M. Borza, O. Raikhman, A. Garg, D. Donchin, A. Cron, M.A. Abdel-moneum, F. Farahmandi, et al., Quantifiable assurance: from IPs to platforms. Preprint. arXiv:2204.07909 (2022)
67. Y. Xie, C. Bao, C. Serafy, T. Lu, A. Srivastava, M. Tehranipoor, Security and vulnerability implications of 3D ICs. IEEE Trans. Multi-Scale Comput. Syst. **2**(2), 108–122 (2016)
68. H. Wang, D. Forte, M. Tehranipoor, Q. Shi. Probing attacks on integrated circuits: challenges and research opportunities. IEEE Des. Test **34**(5), 63–71 (2017)
69. H. Wang, Q. Shi, A. Nahiyan, D. Forte, M.M. Tehranipoor, A physical design flow against front-side probing attacks by internal shielding. IEEE Trans. Comput. Aided Des. Integr. Circuits Syst. **39**(10), 2152–2165 (2019)
70. H. Wang, Q. Shi, D. Forte, M.M. Tehranipoor, Probing assessment framework and evaluation of antiprobing solutions. IEEE Trans. Very Large Scale Integr. VLSI Syst. **27**(6), 1239–1252 (2019)
71. J. Dofe, P. Gu, D. Stow, Q. Yu, E. Kursun, Y. Xie, Security threats and countermeasures in three-dimensional integrated circuits, in *Proceedings of the on Great Lakes Symposium on VLSI 2017* (2017), pp. 321–326
72. https://www.cisco.com/c/en/us/products/security/advanced-malware-protection/what-is-malware.html
73. https://www.sungardas.com/en-us/blog/ransomware-attacks-on-us-government-entities/

74. https://www.cybersecuritydive.com/news/ransomware-attacks-payouts-2021/622784/
75. N. Farzana, F. Rahman, M. Tehranipoor, F. Farahmandi, SoC security verification using property checking, in *2019 IEEE International Test Conference (ITC)* (IEEE, Piscataway, 2019), pp. 1–10
76. K.Z. Azar, M.M. Hossain, A. Vafaei, H. Al Shaikh, N.N. Mondol, F. Rahman, M. Tehranipoor, F. Farahmandi, Fuzz, penetration, and AI testing for SoC security verification: challenges and solutions. Cryptology ePrint Archive **2022**(394), 1–22 (2022)
77. G.K. Contreras, Md.T. Rahman, M. Tehranipoor, Secure split-test for preventing IC piracy by untrusted foundry and assembly, in *2013 IEEE International Symposium on Defect and Fault Tolerance in VLSI and Nanotechnology Systems (DFTS)* (IEEE, Piscataway, 2013), pp. 196–203
78. U. Guin, D. Forte, M. Tehranipoor, Design of accurate low-cost on-chip structures for protecting integrated circuits against recycling. IEEE Trans. Very Large Scale Integr. VLSI Syst. **24**(4), 1233–1246 (2015)
79. X. Xu, F. Rahman, B. Shakya, A. Vassilev, D. Forte, M. Tehranipoor, Electronics supply chain integrity enabled by blockchain. ACM Trans. Des. Autom. Electron. Syst. **24**(3), 1–25 (2019)
80. S. Nakamoto, Bitcoin: a peer-to-peer electronic cash system. Decentralized Bus. Rev., **2008/1**, 21260 (2008)
81. IEEE standard for test access architecture for three-dimensional stacked integrated circuits, in *IEEE Std 1838-2019* (2020), pp. 1–73. https://doi.org/10.1109/IEEESTD.2020.9036129
82. Heterogeneous Integrated Product Testability Best-Known Methods (BKM). https://cmte.ieee.org/eps-test/wp-content/uploads/sites/132/2022/01/IEEE_EPS_Test_Het_Int_Product_Testability_BKM_Final_v1_0-1-14-22-1.pdf
83. T. Zhang, Md.L. Rahman, H.M. Kamali, K.Z. Azar, M. Tehranipoor, F. Farahmandi, FISHI: fault injection detection in secure heterogeneous integration via power noise variation, in *2023 IEEE 73rd Electronic Components and Technology Conference (ECTC)* (IEEE, Piscataway, 2023), pp. 2188–2195
84. T. Zhang, Md.L. Rahman, H.M. Kamali, K.Z. Azar, F. Farahmandi, SiPGuard: run-time System-in-Package security monitoring via power noise variation. IEEE Trans. Very Large Scale Integr. VLSI Syst. **32**(2), 305–318 (2023)
85. A.M. Shuvo, N. Pundir, J. Park, F. Farahmandi, M. Tehranipoor, LDTFI: layout-aware timing fault-injection attack assessment against differential fault analysis, in *IEEE Computer Society Annual Symposium on VLSI (ISVLSI)* (IEEE, Piscataway, 2022)
86. H. Igarashi, Y. Shi, M. Yanagisawa, N. Togawa, Concurrent faulty clock detection for crypto circuits against clock glitch based DFA, in *2013 IEEE International Symposium on Circuits and Systems (ISCAS)* (IEEE, Piscataway, 2013), pp. 1432–1435
87. M. Gao, M.S. Rahman, N. Varshney, M. Tehranipoor, D. Forte, iPROBE: internal shielding approach for protecting against front-side and back-side probing attacks. IEEE Trans. Comput. Aided Des. Integr. Circuits Syst. **42**(12), 4541–4554 (2023)
88. S. Paley, T. Hoque, S. Bhunia, Active protection against PCB physical tampering, in *2016 17th International Symposium on Quality Electronic Design (ISQED)* (IEEE, Piscataway, 2016), pp. 356–361
89. N. Patel, A. Sasan, H. Homayoun, Analyzing hardware based malware detectors, in *Proceedings of the 54th Annual Design Automation Conference 2017* (2017), pp. 1–6
90. N. Pundir, M. Tehranipoor, F. Rahman, RanStop: a hardware-assisted runtime crypto-ransomware detection technique. Preprint. arXiv:2011.12248 (2020)
91. B. Zhou, A. Gupta, R. Jahanshahi, M. Egele, A. Joshi, Hardware performance counters can detect malware: myth or fact? in *Proceedings of the 2018 on Asia Conference on Computer and Communications Security* (2018), pp. 457–468
92. M. Zhao, G.E. Suh, FPGA-based remote power side-channel attacks, in *2018 IEEE Symposium on Security and Privacy (SP)* (IEEE, Piscataway, 2018), pp. 229–244
93. T. Eisenbarth, C. Paar, B. Weghenkel, Building a side channel based disassembler, in *Transactions on Computational Science X* (Springer, Berlin, 2010), pp. 78–99

94. J. Park, X. Xu, Y. Jin, D. Forte, M. Tehranipoor, Power-based side-channel instruction-level disassembler, in *Proceedings of the 55th Annual Design Automation Conference* (2018), pp. 1–6
95. Z. Pan, J. Sheldon, C. Sudusinghe, S. Charles, P. Mishra, Hardware-assisted malware detection using machine learning, in *2021 Design, Automation & Test in Europe Conference & Exhibition (DATE)* (IEEE, Piscataway, 2021), pp. 1775–1780
96. F. Kenarangi, I. Partin-Vaisband, Exploiting machine learning against on-chip power analysis attacks: tradeoffs and design considerations. IEEE Trans. Circuits Syst. I Regul. Pap. **66**(2), 769–781 (2018)
97. S. Ray, E. Peeters, M.M. Tehranipoor, S. Bhunia, System-on-chip platform security assurance: architecture and validation. Proc. IEEE **106**(1), 21–37 (2017)
98. S. Ray, Y. Jin, Security policy enforcement in modern SoC designs, in *2015 IEEE/ACM International Conference on Computer-Aided Design (ICCAD)* (IEEE, Piscataway, 2015), pp. 345–350
99. B. Ahmed, F. Rahman, N. Hooten, F. Farahmandi, M. Tehranipoor, AutoMap: automated mapping of security properties between different levels of abstraction in design flow, in *2021 IEEE/ACM International Conference On Computer Aided Design (ICCAD)* (IEEE, Piscataway, 2021), pp. 1–9
100. N. Vashistha, Md.M. Al Hasan, N. Asadizanjani, F. Rahman, M. Tehranipoor, Trust validation of chiplets using a physical inspection based certification authority, in *2022 IEEE 72nd Electronic Components and Technology Conference (ECTC)* (IEEE, Piscataway, 2022), pp. 2311–2320
101. G.E. Suh, D. Clarke, B. Gassend, M. Van Dijk, S. Devadas, AEGIS: architecture for tamper-evident and tamper-resistant processing, in *ACM International Conference on Supercomputing 25th Anniversary Volume* (2003), pp. 357–368
102. W. Arthur, D. Challener, K. Goldman, *A Practical Guide to TPM 2.0: Using the New Trusted Platform Module in the New Age of Security* (Springer Nature, Berlin, 2015)
103. A. Arm, Security technology-building a secure system using trustzone technology, in *ARM Technical White Paper* (2009)
104. F. McKeen, I. Alexandrovich, A. Berenzon, C.V. Rozas, H. Shafi, V. Shanbhogue, U.R. Savagaonkar, Innovative instructions and software model for isolated execution. Hasp@ isca **10**(1), 1–8 (2013)
105. D. Champagne, R.B. Lee, Scalable architectural support for trusted software, in *HPCA-16 2010 The Sixteenth International Symposium on High-Performance Computer Architecture* (IEEE, Piscataway, 2010), pp. 1–12
106. J. Noorman, P. Agten, W. Daniels, R. Strackx, A. Van Herrewege, C. Huygens, B. Preneel, I. Verbauwhede, F. Piessens, Sancus: low-cost trustworthy extensible networked devices with a zero-software trusted computing base, in *22nd {USENIX} Security Symposium ({USENIX} Security 13)* (2013), pp. 479–498
107. F. Brasser, B. El Mahjoub, A.-R. Sadeghi, C. Wachsmann, P. Koeberl, TyTAN: tiny trust anchor for tiny devices, in *Proceedings of the 52nd Annual Design Automation Conference* (2015), pp. 1–6
108. Md.S. Ul Islam Sami, F. Rahman, F. Farahmandi, A. Cron, M. Borza, M. Tehranipoor, End-to-end secure SoC lifecycle management, in *2021 58th ACM/IEEE Design Automation Conference (DAC)* (IEEE, Piscataway, 2021), pp. 1295–1298
109. Md.S. Ul Islam Sami, F. Rahman, A. Cron, D. Donchin, M. Borza, F. Farahmandi, M. Tehranipoor, Poca: first power-on chip authentication in untrusted foundry and assembly, in *2021 IEEE International Symposium on Hardware Oriented Security and Trust (HOST)* (IEEE, Piscataway, 2021), pp. 124–135
110. M.M. Hossain, N. Vashistha, J. Allen, M. Allen, F. Farahmandi, F. Rahman, M. Tehranipoor, *Thwarting Counterfeit Electronics by Blockchain* (2022)
111. T. Zhang, F. Rahman, M. Tehranipoor, F. Farahmandi, FPGA-chain: enabling holistic protection of FPGA supply chain with blockchain technology. IEEE Des. Test **40**(2), 127–136 (2022)

Chapter 11
Materials for Hardware Security

11.1 Introduction

Since the 1970s, integrated circuit (IC) packaging has been primarily used to isolate dies from the external environment and provide electrical connections. As shown in Fig. 11.1, the packaging of ICs has shifted from dual in-line packaging (DIP) to ball grid arrays (BGAs) in response to the rising demand for a large number of IO interfaces and small packaging sizes. Advanced approaches like 2.5D and 3D packaging have caused a shift from homogeneous to heterogeneous integration (HI) over the past two decades. Using modern connectivity technologies such as an interposer, rapid communications between chiplets enable them to function as if they were integrated on a single silicon substrate. These improvements in IC packaging may help address the demand for high-performance, low-power electronic devices and the rapid time-to-market of modern electronic gadgets. In accordance with Moore's Law, HI has emerged as one of the most preferred techniques for cramming more transistors into a confined space. HI, often referred to as chiplet integration, successfully combines many dies into a single device. Nevertheless, this module-based technology will create hardware security problems since its design involves the inclusion of many third-party chiplets, which allows malicious chips or hardware Trojans to be inserted by attackers [1]. This chapter will discuss the materials-based solutions as countermeasures for hardware security problems, especially for the ICs that use advanced packaging [2–5].

Previously, the researchers have discussed various electrical and physical inspection-based countermeasures to avoid hardware attacks on commercial off-the-shelf (COTS) semiconductor devices. Compared with the electrical-based approaches, the physical inspection methods attract some attention since they can provide straightforward results and show promising results in preventing hardware attacks such as counterfeit IC detection and hardware Trojans detection. Among these physical inspection methods, reverse engineering (RE) the circuit design of the targeted devices can prevent a number of the aforementioned hardware

M. Tehranipoor et al., *Hardware Security*,
https://doi.org/10.1007/978-3-031-58687-3_11

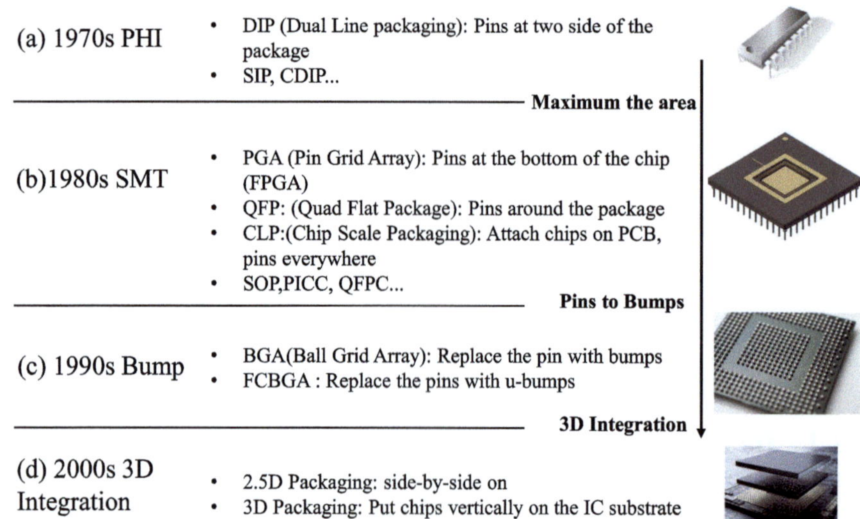

Fig. 11.1 IC packaging development: (**a**) pin through hole technology, (**b**) surface mount technology, (**c**) bump interconnection technology, and (**d**) 3D stack technology

security concerns [6, 7]. This technique requires original designs (known as "golden" samples) for comparison purposes, which are not always accessible to end-users. Advanced sample preparation, image capture of the highest quality, and data analysis skills are all required of using physical inspection for hardware assurance. Problematic is the end-user authentication of all small-scale components in HI, including through-silicon vias (TSVs), C-4 bumps, passive components, and transistors from different chiplets. In contrast, the non-destructive physical inspection can uncover harmful alterations without removing materials or causing structural damage [8]. In terms of 3D physical inspection tools for IC package hardware validation, materials in advanced packaging present both potential and challenges.

Several physical inspection methods have been utilized to authenticate individual components on an IC package in order to protect ICs from hardware attack. The current research focuses on enhancing existing and upcoming physical inspection techniques to identify IC internal information in a non-destructive and high-resolution manner. This chapter will review multi-domain characterization methods, including structural, material, and logic gate characterization. Various new methods for hardware assurance are made possible by the rapid development of HI technology. IR analysis and terahertz (THz) for semiconductors material characterization and Scan Acoustic Microscopy (SAM) for internal structure characterization are examined in depth in this chapter.

Pre-packaging techniques, such as incorporating micro-electro-mechanical systems (MEMS) or nano-electro-mechanical systems (NEMS) during the package design phase, can also be presented for hardware assurance. Active and passive protections based on MEMS/NEMS give security benefits while enabling low-cost, high-volume ASIC device production. MEMS primitives will be integrated at the system level, leading to significant advances in semiconductor technology and fundamental security goals. MEMS devices are mass-produced and packaged similarly to ICs, but they have distinct process differences that are essential for identification [9] or physical unclonable function (PUF) generation [10, 11]. Either actively or passively, PUFs can safeguard an electronic system against tampering. In a novel method, MEMS integration addresses a crucial aspect of electronic device security: system-level protections against counterfeiting and RE.

11.2 Packaging Structure Development History

As depicted in Fig. 11.1, semiconductor packaging for electronic devices has developed with the increasing technology required for high-density, dependable, and compact chips. DIP was used initially for pin insertion packages, in which lead pins are inserted into the through-hole of a printed wiring board (PWB) and welded [12, 13]. In response to the desire for reduced pin pitches, QFP (Quad Flat Package) with solder reflow on components mounted on printed wire bond (PWB) lands has become widespread. FC-BGA (Flip Chip-Ball Grid Array) technology was created to accommodate the increasing number of pins on logic semiconductor packages (PKG) utilized in microprocessors (MPUs). As the number of IOs and control signal terminals multiplied, certain requirements arose. This approach, which entails producing a solder ball at the back of a PKG for area array packing, also known as BGA, has been widely implemented in critical electronic equipment as a factor for pitch narrowing and PKG reduction [13].

Another trend in IC packaging is to embed more chips in one packaging to increase the transistor numbers and decrease the cost. Multichip modules (MCMs) debuted in the early 1990s, with many dies stacked on a single ceramic packing substrate, as depicted in Fig. 11.2. In terms of integration densities and performance, the System-in-Package (SiP) concept exceeded MCMs, which gained prominence around the year 2000 [14]. However, its scale has generated significant problems for the semiconductor industry [12]. Continuous scaling of integrated circuits lowered gate delays while increasing connection delays. The delay issue in advanced SiP packaging hastens the emergence of interconnection research within IC packaging. Using flip-chip technology, dies are frequently mounted to SiP substrates with 100 micrometer solder bump pitch. The enormous number of input/output ports thus restricts the number of dies that can be placed on the substrate. This discrepancy in SiP connection size has an influence on performance and power consumption. In

Fig. 11.2 Various structure
geometries of IC packaging:
(**a**) multi-chip module
(McM), (**b**) 2.5D system in
packaging (SiP), and (**c**) 3D
system in packaging

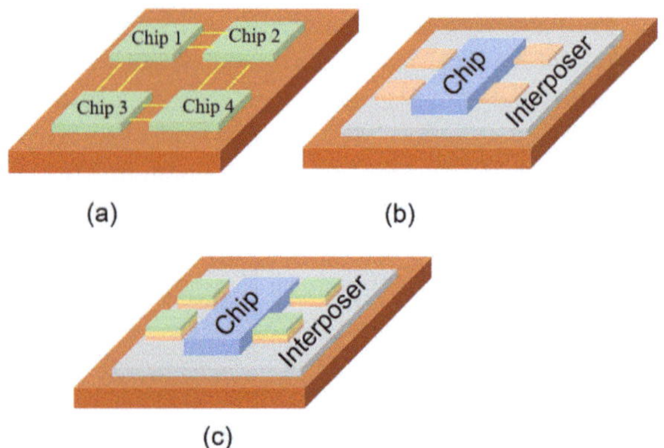

addition, the number of die-to-die connections is limited by the routing congestion
issue of interconnects on the SiP substrate [15]. Various types of SiP are detailed in
this section.

11.2.1   Homogeneous Integration

Despite advancements in on-chip integration, it has consistently failed to meet
the demands of the semiconductor system. This ushered in the first age of 2D
MCM for homogeneous integration in the 1980s, requiring up to 100 logic ICs
to make a single processor, as depicted in Fig. 11.2a. It was initially developed
using high-temperature co-fired ceramics (HTCC) with molybdenum conductors,
which led to the current industry standard: low-temperature co-fired ceramic
(LTCC) with 61 copper layers [16]. The earliest wave of integrated circuit vertical
stacking utilized non-TSV techniques. These techniques are considered low cost
because they employ low-cost wire bonding infrastructure with a high level of
flexibility for stacking and wiring stacked chips to substrates [16]. With ultra-thin
IC dimensions, however, stacked ICs result in increased inductance and resistance,
degraded electrical performance, and chip fractures during bonding procedures.
Hybrid approaches such as flip-chip and wire bonding have been developed to
overcome these crucial challenges within homogeneous packaging.

Beginning in the 1990s, the second trend in Homogeneous Integration packaging
was all about reduction [15]. This trend, researchers brought the "near die size"
packaging ideal, which aims to reduce package cost and area, was propelled by
portable devices such as laptops and mobile phones. Several packing technologies
were developed in response to these needs. Quad flat no-lead (QFN) packages were
launched as a type of CSP (Chip Scale Package) and have since become the most
popular package type due to their ease of use, high performance, and low cost. The
Wafer-Level Chip Scale Package (WLCSP) type was also developed, and it remains

the package with the smallest size (package size and die size are deemed to be comparable [17].

11.2.2 Heterogeneous Integration (HI)

In the 1990s, TSV-based 3D solutions were investigated to accommodate the rising bandwidth between logic and memory, which demanded more input and output density and improved electrical performance, as shown in Fig. 11.2b, c [18]. TSV fabrication, testability prior to stacking, stacking process yield, the need to mix incompatible chips, and the thermal dissipation of logic ICs in these stacks are some of the barriers to the development of this technology. This led to a variety of integration strategies: Package-on-Package (PoP). The PoP supported individual package testing and a scalable stacking system that could accommodate chips of varied sizes and functions. This strategy achieves heterogeneous integration by placing two distinct bare chips as close together as possible on a package substrate with adequate wire density for interconnections, hence improving bandwidth performance.

CPU or GPU ICs are typically positioned near to HBM stacks. Prior to 3D, 2.5D packaging provides HI with a short-term solution. During the past decade, several 2.5D MCMs have been developed, including 2.5D silicon interposers, which were initially developed by Xilinx and then by several others. Intel's Embedded Multi-die Interconnect Bridge (EMIB) architecture, which blends silicon interposer technology with low-cost organic panel packaging, is the performance benchmark for silicon-based 2.5D interposers. AMD has also developed 2.5D silicon interposer technology for connecting graphics to HBM with up to 256 GB/s of bandwidth, high power efficiency, and a compact form factor. The AMD Fiji GPU, which has four HBM stacks surrounding the GPU logic unit, exemplifies this characteristic. Another illustration is NVIDIA's 2.5D silicon interposers with two HBMs and a GPU chip, as well as their NV Link technology, which also employs a GPU chip [16].

Recent interest has been focused on HI packaging as a new type of advanced IC packaging, which refers to the packaging that integrates dies with varying functionalities. The packaging facilities can acquire dies, also known as chiplets, from many suppliers and integrate them according to the design specifications. This module-based integration can significantly boost chip yield and reduce development costs and time. Moreover, HI packaging has the potential to extend beyond Moore's Law, giving rise to a new expression: "greater than Moore's Law." Currently, most advanced packaging, such as 2.5D and 3D packaging, fall under this category. HI packaging is predominantly used in a restricted number of applications, such as integrating high bandwidth memory (HBM) with CPUs or GPUs to achieve better performance with greater bandwidth and shorter interconnects. In the next 5–10 years, more passive and active components, including MEMS, sensors, radar, and lidar, will be integrated, according to the IEEE HI roadmap [14].

11.2.2.1 2.5D Packaging

As shown in Fig. 11.2b, a typical 2.5D interposer package consists of one or more dies stacked on top of a thinned TSV interposer, which is then assembled on an organic flip-chip BGA (FCBGA) or ceramic substrate. A 2.5D interposer package is seen in Fig. 11.2b. As indicated in the schematics, the chips may be constructed next to one another on the interposer or replaced by one or more stacks of numerous dies (e.g., a memory cube). Using the 3D stacking of many chips increases the overall silicon area packed within a single package. This can also help to increase the device's performance by shorting the interconnection between different components.

11.2.2.2 3D Packaging

Researchers are also trying to reduce the total size of the packaging to meet the scale requirements of different devices. Three-dimensional (3D) packaging methods have been developed due to their high package density, excellent performance, and integration of several functional components; the structure is shown in Fig. 11.2c. For many SiP designs, 3D packaging technology is an ideal geometry due to its quicker signal routing and lower wiring density possibility. It has the benefits of a smaller size, lighter weight, increased silicon efficiency, decreased parasitic and noise, a shorter delay, and decreased power consumption [19, 20]. Wire bonding is a popular connecting method in 3D packaging because it is cost-effective and adaptable. Silicon carriers with through-silicon vias (TSV) interconnects have lately emerged as one of the most promising alternatives for stacking multifunctional devices in 3D SiP packaging due to their faster signal routing than conventional wire bond interconnects. Simultaneous thermal expansion of the silicon carrier and the silicon-based device enhanced dependability. Using TSV interconnects for microfluidic interconnects, as depicted in Fig. 11.2c, liquid cooling solutions for high-performance electronics may be incorporated into 3D packaging [19].

11.2.2.3 Interconnection

Diverse technologies have been developed in order to successfully and consistently incorporate chiplets. To efficiently integrate these chiplets into the HI packaging, new connectors such as interposer layers with multiple substrates, EMIB, and multilayer wire bond have been created. These new interconnections enable the use of HI for applications requiring high performance. Due to the high bandwidth and low latency, dies with varied functionality and technology nodes can function as if they were a single die, making SiP as powerful as System on Packaging (SoC). Also, the development of HI packaging necessitates the use of more complex materials to address reliability difficulties. The previous study has demonstrated a clear correlation between the material qualities of HI packaging and its reliability [21].

Interposers are substrates utilized in the packaging industry to connect diverse components, such as an integrated circuit (IC) to a package or an IC to a passive device. Originally comprised of silicon, they are now also composed of glass and organic elements. Interposers offer a number of other benefits in addition to efficient communication, including power routing with little latency and impedance. Passive devices, as previously stated, can be integrated into an interposer for advanced applications. Bell Labs and IBM invented the first interposers in the 1980s [22]. Large bandwidth, high performance, and low power consumption make silicon interposers a common choice. Silicon substrates are ideally suited for fine-pitch wiring, allowing for a high IO density. Because they are made of the same material, their CTE is identical to that of a silicon die [15, 23].

To enable the high-density information flow within HI, various interconnection architectures such as interposer-based through-silicon vias (TSV), Embedded Multi-die Interconnect Bridges (EMIB), and organic-based TSV have been created. The ultimate objective of interconnection architectures is to hold as many input–outputs (IOs) as feasible with safety and dependability. To meet the small-scale requirements for HI packaging, however, the wire pitch, bump pitch, and through pitch are reduced. Intel employs EMIB with a bump pitch of 40–50 microns, which is greater than 10 microns for Faveros.

11.3 Encapsulant Material of IC Packaging

Encapsulant of IC Package is also a vital and necessary part which enables the advanced packaging development. Encapsulant is typically composed of various organic and inorganic substances. It protects the semiconductor devices from contamination, moisture erosion, radiation, and associated environmental conditions [24]. However, pure epoxy is not a good thermal and mechanical material, which limits its application in high-performance devices with very high power consumption. Instead, high-performance devices such as GPUs or CPUs typically operate with bare dies or metal caps that serve as heat sinks. However, these devices still use encapsulant materials as underfilled material to provide extra mechanical support. The underfill serves a variety of purposes to enhance the reliability of packing, such as enhancing attachment, regulating the difference in heat conductivity between the various materials, and more [25]. In some of the recently developed systems-on-chip (SoCs) packaging and System in Packaging (SiP), the substance used to produce IC package encapsulants is also employed as a support material between the dies and the substrate or interposer, also known as underfill. Due to the increasing demand for high-performance microelectronic systems, high thermal transport, low interstress, and long reliability (up to 250°) are more important than ever.

Presently, the academic and industrial communities are placing a great deal of emphasis on enhancing encapsulant reliability through unique material compositions or manufacturing processes. Various encapsulant formulations have been created to suit the dependability requirements of various electronic device applications. In order to comprehend and prevent encapsulant-related package failures, the thermal and mechanical mechanisms of epoxy resin manufacturing have been examined. Typically, destructive mechanical property characterization is utilized to validate EMC material dependability. Accelerated cycle experiments, which are time-consuming and omit critical elements for various types of packaging, are nonetheless utilized by the scientific community and industry to validate the reliability of encapsulant samples [25]. However, these methods are difficult to be used by the end-user to verify whether manufactured encapsulants correspond to the original design. This allows malicious packaging facilities and employers to alter the design of the original material without being noticed. By creating unanticipated failure under specific working conditions, malicious entities that conduct these alterations on purpose offer a security risk. Due to the globalization and outsourcing of the semiconductor industry supply chain, attackers have a greater chance of implanting these types of hardware-level alterations.

In this section we will discuss a variety of encapsulant materials and production techniques, highlighting their effects on package dependability. The composition and fabrication process of these encapsulant materials relating to reliability are evaluated from a hardware security perspective, as each mentioned parameter is susceptible to modification by attackers. Considering these parameters, a taxonomy of encapsulant security vulnerabilities will subsequently be developed. The hardware security vulnerabilities and fabrication details can be used as a guide for the development of security assessments. Existing packaging characterization techniques will be evaluated as potential candidates for encapsulate material validation. There will be a discussion of well-known approaches for encapsulant dependability characterization, as well as its assurance limitations. The encapsulant material will then be characterized using a selection of non-destructive physical inspection techniques for the sake of hardware assurance, and various proof-of-concept demonstrations will be shown.

11.3.1 Encapsulant Material Composition

The packing encapsulant substance, commonly known as blacktopping, is comprised of distinct functional components. The EMC is often utilized as the main body of the encapsulant. As depicted in Fig. 11.3, the encapsulant is a complex material system comprised of more than ten distinct components, each with different functionalities. Epoxy resin and hardener are the most often utilized ingredients in plastic encapsulants, constituting just 5 wt% by weight (weight percentage). Typically, encapsulants based on epoxy are designed to have a glass transition temperature (Tg) above 200–250 °C. This design enables the encapsulant to have mechanical qualities up to the working temperature of the majority of microelectronic devices. Some microelectronic devices function in harsh environments,

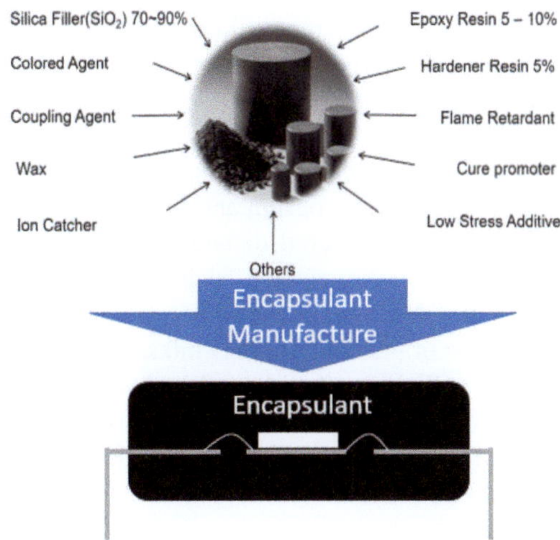

Fig. 11.3 Package encapsulant formation (upper images: different types of EMC)

including aerospace and automotive applications. Utilizing [26], the encapsulant material generated by diverse types of polymer matrices may function up to 350 °C for high-temperature applications. The majority of the encapsulant is composed of the inorganic filler (60–80 wt% weight percentage), which is often silica. Silica material can simultaneously decrease the thermal expansion coefficient (CTE) and boost the thermal transformation of the encapsulant [27]. The shape, composition, and percentage of the filler are directly tied to the mechanical qualities of the encapsulant and must be developed to suit the semiconductor specifications. Inappropriate proportions or shape of the silicon filler can result in a variety of reliability issues, including delamination produced by the CTE difference at the interface (as summarized in Fig. 11.3) and voids caused by uniform filling. A tiny amount of additional ingredients, such as curing acceleration, helps to manage the curing process, and coupling agents can assist in improving the surface's mechanical strength. Changes in material types, composition, and form will directly affect both long-term and short-term packing reliabilities.

11.3.2 Encapsulant Fabrication Process

The fabrication of the encapsulant is also critical to the reliability of the semiconductor devices. The four basic steps of encapsulant creation include EMC material storage, pre-curing, curing, and post-curing. Initial purchases of EMC raw materials must be kept at extremely low temperatures to prevent automatic curing prior to manufacturing. The raw EMC material will be pre-cured at room temperature (25 °C) prior to encapsulant manufacture, according to [28]. Then, the pre-cured material will be injected into the mold to generate the encapsulant around the packaging. Different techniques of transporting EMC pre-cured materials have been devised to improve the packing process yield. For instance, compression molding is utilized for large SoC devices or heterogeneous integration (HI) packed devices that contain several chiplets [29]. Compressing the EMC curing material eliminates material heterogeneity and undesirable warpage. Later, EMC will be poured into the mold and cured under high temperature and pressure for several minutes, according to the experimental design and the polymer types. The properties of polymers that can be thermally cured dictate the timing of specific operations. Note that there are more curing processes, such as air curing at room temperature and UV curing [30]. The EMC will be partially cured throughout the curing phase, resulting in the formation of the bulk of cross-links. In the last post-mold curing step, the packaged sample will be exposed to a high-temperature environment. This procedure is primarily intended to increase the curing percentage, also known as the cross-link percentage, which is essential for package dependability under high-temperature working conditions.

11.4 Security Threats in Advanced IC Packaging IC

11.4.1 Threat Models

It has long been recognized that the complexity and worldwide spread of the IC supply chain pose a threat to hardware security of semiconductor devices [31]. In accordance with Fig. 11.4, the supply chain model of the semiconductor devices, the GDSII files of various founders are first transmitted to the foundries. In addition, the interposer and other passive components' designs are also forwarded to foundries for manufacturing. The created chiplets, passive components, and interposer at the packaging facility will be integrated together as the final magnification process.

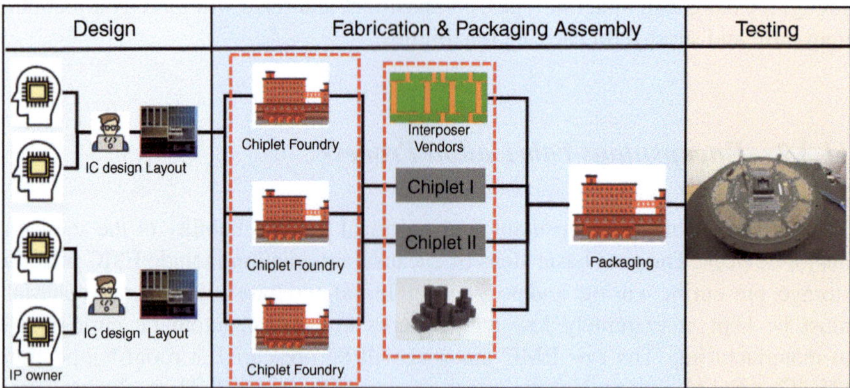

Fig. 11.4 Supply chain model and threat model of HI packaging (red dashed boxes: potential untrusted (adversarial) entities)

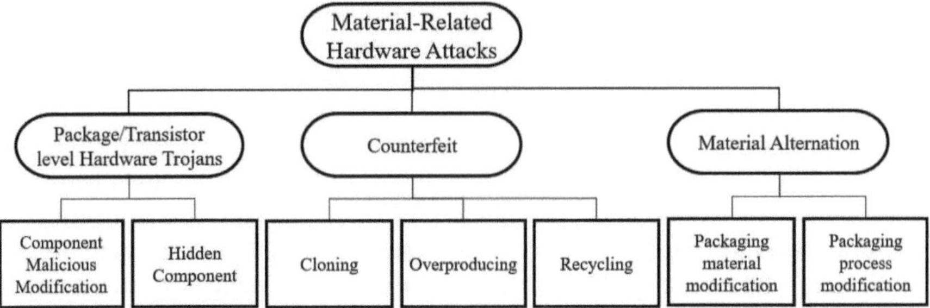

Fig. 11.5 IC packaging material-related hardware security vulnerabilities

The final test is conducted on the finished products in the packaging facilities. Due to the outsourcing of IC manufacture, adversaries are able to introduce malicious alterations that result in hardware vulnerabilities. Numerous fabrication processes contribute to the increased attack surface of HI packaging devices. In Fig. 11.4, the potential adversaries inside the supply chain are emphasized in red boxes.

- Chiplet foundries
- Passive components and interposer foundries
- Material vendors
- Packaging companies

These facilities have the capability to perform hardware-level changes and cause hardware security problems.

11.4.2 Potential Hardware Attacks

Due to the above-mentioned supply chain vulnerabilities of HI packaging, numerous attacks can be carried out by these entities by changing the original design. Figure 11.5 summarizes the various forms of advanced IC packaging hardware security issues, which are detailed in this section.

11.4.2.1 Hardware Trojans

A hardware Trojan is the purposeful removal or addition of logic gates from the original IP design. Typically, these modifications are performed for malicious reasons, such as compromising functioning or leaking data. Today's IC designs rely substantially on the intellectual property of third parties, whose owners or foundries can maliciously insert Trojans or modify the original design. These attacks are referred to as transistor-level hardware Trojans in this study. Prior study analyzed and categorized the numerous types of hardware Trojans at the transistor level [32].

Beyond this, developing HI packaging enables malicious actors to install destructive active and passive components, referred to as package-level hardware Trojans in this section. Since system complexity increases, both transistor- and package-level hardware Trojans will rise, as HI packaging will integrate more third-party passive and active chiplets.

Also, interconnections such as the interposer in HI packaging might be manipulated maliciously to become package-level hardware Trojans. It is difficult for an end-user to confirm the integrity of interconnects in advanced packaging because of their high IO density and compact technology nodes. Malicious actors, such as untrusted foundries, are able to install or remove interconnects, which they can then utilize to power malicious components or cause data leakage. In addition, malicious actors may alter the fabrication parameters of interconnects to introduce vulnerabilities like cross-talk and early failure, resulting in denial-of-service (DoS) problems. These modifications are difficult to detect at testing facilities and may go undetected until the failure occurs at a later stage.

Multiple countermeasures have been developed to protect ICs from the hardware Trojan type attack as listed above. By securing the design using runtime monitoring, obfuscation, camouflage, runtime monitoring, etc. [33], IP owners can avoid hardware Trojans. IP owners can also include design obfuscation or split production to prevent hostile foundries from having access to the whole original design [34, 35]. In addition, they are able to do pre-silicon verification to compare received IC designs to their original design [36]. Nevertheless, only post-silicon physical testing permits the end-user to validate the hardware design. Ideally, complete reverse engineering of the IC using scanning electronic microscope (SEM) imaging may provide this verification and eliminate hardware Trojans at the transistor/package level. Due to the small size of hardware Trojans, however, scanning Trojans becomes troublesome in practice. In 2008, Samuel et al. demonstrated this by constructing a little Trojan backdoor that provided the attacker full access at the highest level. Using the original architecture, the researchers added 1341 gates that operated as Trojan hardware [37]. To detect these Trojans effectively, humanless computer vision techniques based on machine learning mathematics are required [38, 39]. However, the limitations of physical inspection tools prevent end-users from identifying Trojans at the transistor and package level in a variety of low technology node devices.

11.4.3   Counterfeit ICs

Guin et al. classified counterfeits into seven distinct categories: recycled, noted, overproduced, out-of-specification/defective, cloned, fabricated paperwork, and manipulated [40]. When implanted in critical systems, these counterfeit samples will damage the reputation and income of the IP owners and pose significant confidentiality and dependability issues, such as data breaches and backdoors. Different physical inspection methods have been developed for chip package inspection in order to examine samples from multiple perspectives and make counterfeiting more challenging. Material characterization, texture analysis, and structure analysis are examples of these analyses [40]. However, finding counterfeit samples is still a challenging task because there is no distinguishing trait that can be relied upon. In addition, the rapid growth of electronic components has resulted in multiple generations of integrated circuits (ICs) on the market, making mass identification of counterfeits increasingly difficult. HI package containing many passive and active components is deemed partially counterfeit if, for example, only one chiplet or the passive component is fake. The opponents can also counterfeit a small section of the chips, making detection even more difficult. Due to these difficulties, HI package counterfeiting remains a significant hardware security concern.

11.4.4 Material-Based Hardware Attacks

Innovations in materials have an increasing importance in the development of high-performance devices. For instance, micro-bump material and production techniques are designed to minimize defects and long-term electrical migration. The fabrication process for TSVs with a high aspect ratio has been designed to achieve consistent, defect-free filling [41]. In general, innovation in HI packaging implies that the packaging material can be resilient in difficult operating conditions. Due to the importance of IC packaging material, it is feasible for malevolent entities to modify the formula of the material and cause undesirable reliability concerns. This is considered a material-level hardware attack. The work [42] specifies that the viscosity of epoxy molding compound (EMC) should be sufficiently low to fill the entire micrometer gap between the die and substrate. Different formulations of underfill materials with enhanced glass transition temperatures (Tg) are designed to function at elevated temperatures [26]. Any unauthorized alteration may result in nonuniform underfilling and threaten the dependability of the device. To change this molecule, however, only a modest amount of specialist knowledge is required. In order to get the desired effect, adversaries can arbitrarily vary the material composition or fabrication techniques. Additional material-related hardware security information will be presented in the coming section.

11.5 Security-Related Material Chemical and Mechanical Properties

According to the aforementioned threat models, attackers can deliberately induce reliability difficulties by modifying the package encapsulant's substance and fabrication process. Figure 11.6 lists six encapsulant settings that can be altered to produce reliability issues. This table also provides the threat level by examining the modifiable and detectable parameters of certain package types. Thermal con-

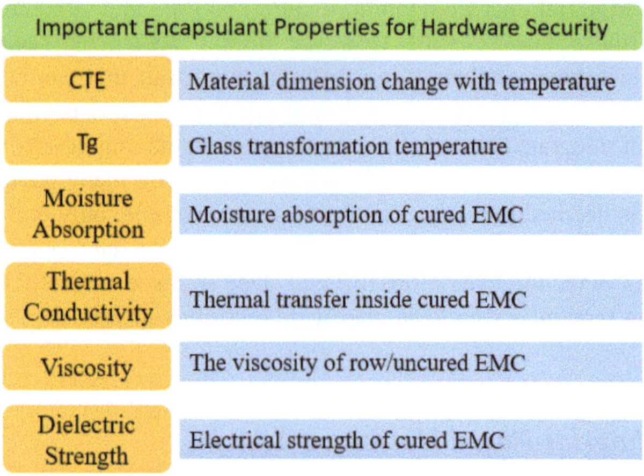

Important Encapsulant Properties for Hardware Security	
CTE	Material dimension change with temperature
Tg	Glass transformation temperature
Moisture Absorption	Moisture absorption of cured EMC
Thermal Conductivity	Thermal transfer inside cured EMC
Viscosity	The viscosity of row/uncured EMC
Dielectric Strength	Electrical strength of cured EMC

Fig. 11.6 The most important encapsulant properties for hardware security

ductivity, mechanical strength, glass transformation temperature (Tg), coefficient of thermal expansion (CTE), and other mechanical qualities are crucial to the reliability of encapsulants. Following consideration of the aforementioned threat model, CTE, Tg, stickiness, and moisture absorption can be altered undetected. These qualities are easily modified by imperceptible adjustments to the production method and the material that are modest and random. Certain variables are inextricably related. For instance, Tg values influence mechanical strength at a certain temperature. Multiple methods exist for attackers to alter the targeted parameters and so raise the difficulty of generating defenses. In this research, we construct a database that may be utilized to develop countermeasures for the most prevalent security vulnerabilities.

As shown in Fig. 11.6, modifications to encapsulate fabrication procedures and material composition will directly affect packaging thermal and mechanical properties. These impacts can easily translate into concerns about dependability of IC packaging. CTE defines the material dimension change with temperature and its variances across different packing materials. In the packaging encapsulant, the epoxy matrix has substantially higher CTE, which means it will expand more at the same temperature compared to the inorganic materials. This CTE difference between materials at the interface generates reliability concerns such as delamination or warping problems. To deal with the CTE discrepancy, silica is commonly employed as the filler ingredient in the encapsulant to lower the CTE of the epoxy molding compound (EMC). The silica filler will also affect other mechanical qualities like viscosity and mechanical strength. Different filler materials, sizes, forms, and homogeneity have been evaluated to enhance the amount of silica without impacting other mechanical properties [43]. In addition, the CTE of the encapsulant can be affected slightly by the epoxy, hardener type, ratio curing temperature, and manufacturing time. As the input and output density of new sophisticated packaging devices rises, the tolerance of CTE mismatch between different materials inside the package diminishes. This allows adversaries to modify the CTE to cause reliability problems.

Another security-related package material property is glass transformation temperature (Tg) which refers to the temperature of the encapsulant transitions from glass to rubber. The CTE of the encapsulant increases substantially above the Tg, and its thermal mechanical properties decrease exponentially. The Tg of the encapsulant material normally is designed to be higher than the maximum working temperature of the devices to prevent unforeseen reliability challenges in harsh operating situations. The Tg of a polymer is governed primarily by its cross-linking density, which can be influenced by the material's composition and fabrication method [44]. For example, the post-curing process will alter the cross-link density of the encapsulant. Inadequate post-curing time will reduce the density of cross-links, resulting in a decreased Tg for the encapsulant. Additionally, the cross-link density of different types of epoxy molding compounds and hardeners might alter the Tg.

The package's reliability will also be affected by the encapsulant material's moisture absorption. Throughout the lifetime of the encapsulant, the moisture absorption rate must be as low as possible. Inadvertently high moisture absorption is the result of an increase in working temperature and humidity levels. As a result, metal components such as the lead frame, wires, and bonding pad can rapidly corrode, compromising the connectivity reliability of the device. Additionally, the expansion caused by the encapsulant moisture absorption will result in delamination or disconnection. Several encapsulant characteristics, such as polymer kinds, influence the encapsulant's water content and influence the encapsulant's moisture absorption. In addition, the post-curing procedure will impact the moisture content by altering the proportion of cross-links [44].

Several more encapsulant factors are strongly correlated with the dependability and performance of electrical devices. As an electrical insulator, the encapsulant must be able to tolerate high electric stress conditions, particularly in power electronic applications. Epoxy kinds and filler content have a substantial effect on the dielectric strength of power electronic packages [45]. Moreover, thermal conductivity influences the performance of electrical devices at elevated temperatures. Primarily, the filler alters the encapsulant's heat conductivity. If the concentration of the filter is too high, the viscosity of the raw material will prevent it from being molded uniformly, resulting in fractures and voids. Prior to curing, the viscosity of EMC is mostly determined by filler concentration and epoxy type. Due to the relevance of these encapsulant characteristics to the dependability of the packaging, attackers can impair device quality by manipulating the original encapsulant design.

11.6 Post-packaging HI Assurance: Non-destructive Physical Inspection

In response to the threats mentioned in the preceding section, numerous defenses have been created. Due to the fact that end-users cannot simply incorporate pre-silicon assurance procedures into the design, alternate approaches such as electrical and physical inspection are developed as countermeasures. Even while electrical testing may typically find irregularities within detected samples, attackers can build components to circumvent this standard electrical examination [32]. To ensure full hardware assurance, as many IC samples as possible should be subjected to physical inspection ideally non-destructive testing (NDT) techniques. Physical inspection is used by both academics and industry to discover and prevent extra or missing components and functionality in packaged ICs, a process referred to in this chapter as post-packaging verification. In this section, a variety of non-destructive physical inspection approaches for structural and material characterization, as well as their limitations for HI packaging hardware assurance, will be discussed.

11.6.1 Structural Characterization

Typically, NDT imaging modalities are used to evaluate structural information such as surface, subsurface, and volumetric imaging of the package sample being evaluated. Using appropriate image processing and data analysis techniques, it is possible to identify suspicious components and features inside ICs for hardware assurance purpose. Figure 11.7 displays several examples of IC structural analysis results.

Optical imagings which can be collected by optical camera and microscope are generally utilized to identify the surface texture, surface color, surface labeling, and

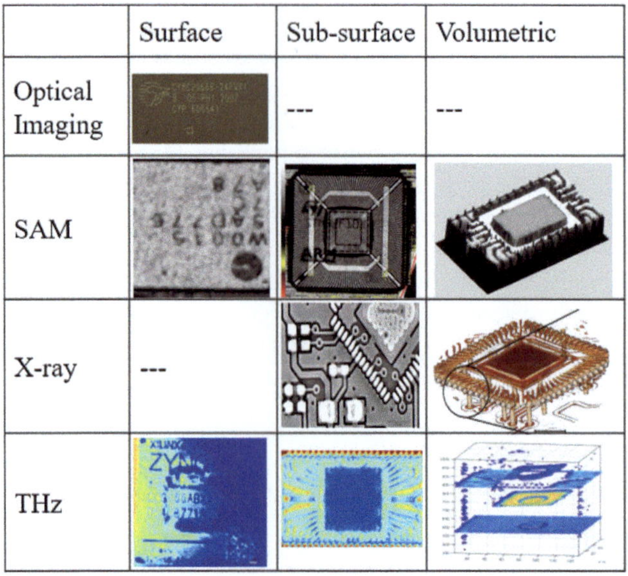

	Surface	Sub-surface	Volumetric
Optical Imaging		---	---
SAM			
X-ray	---		
THz			

Fig. 11.7 Non-destructive inspection for hardware security

size of IC samples [46]. It is straightforward, rapid, and economical to do this type of characterization. Nevertheless, optical imaging has some noticeable shortcomings. These modalities can only characterize information related with visible light, which offers relatively few descriptive features. In addition, optical imaging techniques are dependent on the surrounding light and variable illumination settings, such as hue and intensity. Different imaging collecting setups can offer diverse findings and mislead the inspection method. Normally, these concerns can be rectified with more thorough data preparation.

Scanning acoustic microscopy (SAM) is able to acquire both the surface and subsurface information of IC samples. By raster scanning the sample with ultrasonic beams, SAM is able to collect a high-resolution 2D image with a resolution of a few micrometers. By selecting the correct output signal waveform window, the images of a certain sample layer can be generated. SAM is commonly used in IC failure analysis to localize the failure spot by detecting the delamination and voids [47]. Moreover, SAM is quicker than comparable techniques; it can characterize an IC sample on a centimeter scale in minutes. However, SAM demands the sample be immersed in liquid such as water, which may cause damage. In addition, the high-frequency signals used to acquire high-resolution imaging have a restricted penetrating depth and can only reach an extremely shallow subsurface [48].

X-ray is able to accurately characterize the internal structure of integrated circuits and printed circuit boards [49]. For semiconductor device failure analysis, 2D X-ray inspection can discover internal faults by penetrating the packaging substrate. 2D X-rays may also deliver top-down imaging of typical IC samples, which is utilized for failure mechanisms analysis such as the lead frame architecture, bond wire quality, and connectivity interfaces. However, 2D X-rays cannot detect the HI packaging adequately, since HI packaging comprises multiple layers and demands 3D structural validation [50]. Using computed tomography (CT), 3D structure of HI packaging has the potential to be produced. Additionally, laminography and digital tomosynthesis are applied to improve IC characterization capabilities. However, X-rays can be detrimental to specific samples, such as storage devices, due to the fact that the high-energy beams can damage memory data [51].

11.6.2 Material Characterization

11.6.3 Packaging Material Verification Process

Packaging characterizations mainly focused on directly testing the reliability of the IC packaging without depending on inferred conclusions from static investigation of material attributes. This is performed by adding temperature and moisture stress to the sample during reliability evaluation, which examines the reliability of package design properties and the final products. When the test sample breaks down under the above-mentioned stress condition, physical inspection procedures will be utilized to localize the failure site to discover the causes and alter the design or material composition as appropriate. This procedure is repeated until satisfactory results are reached. The packaging reliability verification method can be separated into three basic phases as outlined below:

- Design: Prior to deciding on the material and creating the geometry of the IC packaging, the specifications will be compiled. This design specifications aid in determining the optimal material composition for the current requirements. On the designed system, material and system-level simulations are performed [52]. During this procedure, the packaging's thermal transfer, warpage, and electrical and mechanical properties are examined. A number of prototypes will be created and evaluated following the design phase, which is in-process testing.
- In-process Testing: During the in-process testing, the designed material and IC packaging structure will be confirmed. Several methods of material characterization will be utilized to confirm that the raw material supplied by the vendors has attributes equal to the parameters simulated. This characterization can also be used to ensure that the row materials were stored and post-cured properly [24]. Incorrect component storage by a vendor could result in micro-fractures that were not caused by the external stress loading during a subsequent failure study. In other words, this test ensures that any problems detected during failure analysis are the consequence of stress testing alone and not the manufacturing or design processes. In-process testing can also be performed on the manufactured characteristics before and after environmental stress loading. The in-process test employs a variety of physical inspection techniques to discover any package faults resulting from environmental stress or improper package manufacture.
- Failure Analysis: In order to simulate actual working conditions, the packed IC will be subjected to a variety of mechanical pressures during the failure analysis tests. Several variables, including temperature, humidity, and the length of time spent in challenging conditions, are changing during the test. Physical inspection procedures will detect potential faults such as delamination, fracture, and disconnection following the stress loading. A summary of typical failures caused by the mechanical qualities of IC packaging is provided in Fig. 11.6. After pinpointing the precise area of the failure, mechanical and material characterization techniques will be employed to discover what caused it and to identify solutions. Once the failure's root cause has been identified, previous procedures will be updated to improve the IC package's design.

11.6.4   Packaging Encapsulant for Reliability

As stated in the previous section, adversaries can exploit hardware weaknesses by modifying material recipes and IC package manufacture. The characterization of IC packaging material hence avoids this type of attack. Since the early 2000s, material characterization has been recognized as a possible approach for detecting counterfeits [40]. Since then, however, little research has been performed to specify IC packaging material for hardware assurance. The key causes of this lack of development are as follows: First, there are currently few non-destructive methods for material characterization, and the majority of existing approaches suffer from a low signal-to-noise ratio (SNR). In order to eliminate environmental and equipment noise, non-destructive material characterization requires numerous rounds of scanning, which can be exceedingly time-consuming and unsuitable for IC analysis. Second, the existing material characterization for hardware assurance requires a known authentic sample, sometimes known as a "golden sample," which is difficult or impossible to acquire. Even when in possession of a sample of gold, it is difficult to determine if observed material changes are due to magnification fluctuation, ambient noise, intentional manipulation, or counterfeit samples. Consequently, non-destructive material characterization with supplementary data analysis methods will be introduced as a golden-free method. Material characterization, in conjunction with data analysis and categorization, can give hardware assurance without the need for material understanding.

Thermal mechanical property characterization and physical inspection for encapsulant analysis are introduced in this section. These strategies assess the subject matter from several perspectives. The limitations and drawbacks of each strategy will also be explored.

11.6.4.1   Thermal and Mechanical Properties Characterization

For packaging encapsulant dependability and durability verification, numerous material and mechanical characterization methods have been developed. There are several destructive and non-destructive methods for characterizing the chemical and mechanical properties of an encapsulant material. Nonetheless, these procedures are not intended for hardware assurance. Therefore, they fail to differentiate between malicious encapsulant modification and defective samples, resulting in innovative hardware security evaluations. In the following section, a summary of the most prevalent approaches for characterizing encapsulants will be presented. Also covered will be the constraints of employing these methods for hardware assurance. Typically, the mechanical properties of the package encapsulant are measured using four distinct destructive analytical techniques: thermomechanical analyzer (TMA), differential scanning calorimetry (DSC), dynamic mechanical analysis (DMA), and thermogravimetric analysis (TGA) [45]. All of these characterizations require sample preparation and irreparably harm the sample.

- TMA: Thermomechanical analyzer (TMA) is able to capture the encapsulant sample material expansion at a certain temperature range with constant loading stress. TMA is also wildly used to measure the CTE of the test sample deformation below Tg in the unit of ppm/°C (parts per million/°C). To perform the TMA measurement, the tested sample is placed into a sample older with parallel sides and flat faces. TMA can also be used to quantify Tg, as Tg is defined as the temperature at which CTE rapidly changes.

- DSC: DSC is used to measure the Tg of encapsulant samples. DSC can determine the amount of heat required to raise a test sample's temperature to a specific level. The peak of the endothermic process determines the Tg of the encapsulant [26].
- DMA: DMA can also be used to determine the Tg of the encapsulant, as it can measure directly how the mechanical properties of the tested material vary with temperature. At Tg, the rate of mechanical strength loss will slow down.
- TGA: TGA measures the Tg of the encapsulant by monitoring mass changes with temperature. The temperature at which mass drops abruptly during the TGA measurement is known as Tg. As an example, the temperature of the test chamber rises from 25 to 800 °C at a pace of 40 °C every minute. The chamber will maintain the maximum temperature until the mass ceases to change. This marks the completion of the volatilization process. TGA is also used to aid in the analysis of the filler content, which is the residual substance.

Certain mechanical properties may only be evaluated following accelerated environmental loading, which was designed to simulate the long-term real-world loading of the IC package. For instance, the definition of moisture absorption is the change in weight after 24 hours at 85° 85% RH. The encapsulate moisture absorption can be tested using TMA in conjunction with a humidity chamber [53]. The majority of these thermal–mechanical encapsulant characterizations are relatively time-consuming and require sample preparation, limiting the volume of encapsulant characterization that can be performed for hardware assurance.

11.6.4.2 Characterization of Chemical Properties

Material component and composition characterization also play a vital part in the verification of package reliability. However, unlike thermal and mechanical approaches, material components and methods cannot directly predict encapsulant qualities. Instead, they are utilized mostly for failure analysis, which helps to identify the core cause of problems and prevent future occurrences. For qualitative and quantitative material characterization, infrared (IR) laser-based techniques, such as Fourier transform infrared spectroscopy [54] and Raman spectroscopy [55], are commonly utilized. These approaches identify chemical compounds based on the variable IR absorption of different materials. Due to the complexity of the encapsulant material and the condition of the test sample, their measurements are always plagued by noise signals. Nonetheless, these issues can be mitigated by employing the appropriate data preprocessing and machine learning methods. In one of the case studies in the coming section, further information on IR-based encapsulant material characterization for counterfeit IC detection is provided. Other methods have also been employed to characterize encapsulant materials:

- X-Ray fluorescence (XRF) measures the fluorescent light emitted from the examined samples.
- Energy-dispersive X-ray spectroscopy (EDS) is used to determine the surface materials [56].

11.7 Material Characterization for Hardware Security

11.7.1 Requirement Specification

As previously stated, adversaries must modify the original encapsulant or fabrication process in order to launch a hardware-level assault on the package encapsulant. These alterations may produce undetected reliability issues under harsh working conditions. In other words, these malicious alterations are capable of evading the accelerated cycling test, the most common benchmark for determining the reliability of finished items. Consequently, only chemical analysis can determine whether these material concentrations are aberrant and prevent these types of attacks on final goods.

Traditional methods of characterizing encapsulant materials, which focus primarily on qualitative and quantitative characterization, may not be appropriate or successful for hardware assurance. These methods need a human comparison of acquired information between test samples and the authentic sample in order to detect the substance-level anomaly. However, the authentic (or "golden" sample, as it is commonly known) is not always available to end customers. In addition, mechanical property characterization necessitates sample preparation, which cannot be conducted on shipping products. We describe the requirements as a starting point for investigating viable chemical material characterization methodologies for encapsulant assurance.

- The encapsulant assurance method must be non-destructive. According to the threat model, any sample could be manipulated maliciously. Therefore, destructive detection methods can only be employed as verification methods on a tiny subset of total manufacturing, rendering them inappropriate for comprehensive hardware assurance verification. Only non-destructive measurement provides the opportunity to validate each sample.
- The validation and anomaly detection procedure should be totally automated with little human intervention. First, the entire verification procedure drastically improves its efficiency, hence enhancing the potential of complete package verification. Due to the minute discrepancies between authentic and defective material properties, the manual comparison technique is susceptible to false positives and negatives. Involving human inspectors also introduces new attack vectors, such as social engineering.

- To avoid false alarms, the precision and repeatability of the material characterization methods are crucial. The measurement must produce the same precise result regardless of the testing conditions and surroundings.

Due to the significance of the encapsulant to both classic and sophisticated microelectronic device packages, it is necessary to develop effective inspection methods for checking the material properties of encapsulants for the purposes of hardware assurance verification.

11.7.2 Encapsulate Material Characterization Case Study

As a proof of concept for meeting the aforementioned characterization requirements for assurance reasons, selective encapsulant characterization will be done on the package encapsulant. Material information will be extracted using three non-destructive material characterization techniques designed to meet the hardware assurance violation criteria. The acquired material information from each test sample will serve as a hardware assurance fingerprint. The material fingerprint of the test sample can be determined fast and precisely using these techniques. This case study's test sample consists of two chips from reputable and untrusted vendors. Optical pictures reveal the distinction between the logos, as seen in Fig. 11.8a, b. The material difference between authentic and counterfeit IC samples, which could cause problems with hardware assurance, is successfully detected.

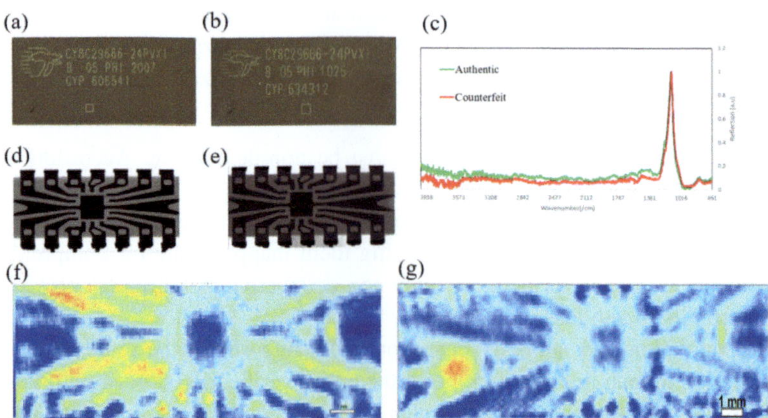

Fig. 11.8 (**a**) Optical imaging of authentic sample, (**b**) optical imaging of counterfeit sample, (**c**) IR spectrum of authentic and counterfeit ICs, (**d**) 2D X-ray images of counterfeit IC, (**e**) 2D X-ray images of authentic IC (X-ray different brightness may indicate material difference), (**f**) THz-TDS images of counterfeit IC, and (**g**) Hz-TDS images of authentic IC

11.7.2.1 IR-Based Material Characterization

Owing to the Infrared (IR) laser's low power, it has been widely accepted as a non-destructive method for material characterization [57]. In the mid-infrared spectrum, the 1300-910 cm^{-1} wave number region, sometimes known as the "fingerprint region," is recognized for its material information collection capabilities. Most molecular structures of polymers absorb infrared (IR) radiation at mid-IR range. Therefore, mid-IR spectroscopy may be used for qualitative and quantitative polymer material characterization, which is the material of the package encapsulant [54]. However, mid-infrared laser tests have weaker energy sources, resulting in a lower signal-to-noise ratio. It will be difficult to determine whether the mid-IR spectrum variations are the result of material changes of the encapsulant or environmental noise. To solve this issue, near-IR laser-based measurements with higher energy levels have been used to distinguish the minute material differences between real and counterfeit pharmaceutical samples [58]. However, the near-IR measurement will contain less relevant information than the mid-IR measurement. In addition, a variety of sample preparation methods are required to permit deep IR laser penetration and reduce impurity- or surface-roughness-related noise. As this is intrinsically hazardous, it does not meet the previously described encapsulant hardware assurance requirement.

Some non-destructive IR-based techniques, such as Attenuated total reflectance Fourier transform infrared (ATR-FTIR) and Diffuse Reflectance Infrared Fourier Spectroscopy (DRIFT), have been developed to eliminate the need for sample preparation. DRIFT is utilized to analyze the encapsulant substance of authentic and counterfeit samples to assess their noise impacts. During the non-destructive IR measurement, the overall absorption energy is low due to the large dispersion of the rough surface and the strong absorption of the epoxy molding material. Nevertheless, the distinction between real and counterfeit samples remains readily apparent, making these methods suitable for hardware assurance applications. As observed in Fig. 11.8c, the DRIFT spectrum difference indicates that the authentic and counterfeit EMC materials are distinct. In the region with large wave numbers, this distinction is obscured. Also, significant differences are visible in the "fingerprint" region of low bandwidth. When this method is used for a large number of these samples, there is no obvious difference between the authentic and counterfeit versions. This is mostly owing to the low signal-to-noise ratio of the material's mid-IR characterization. The issue can be resolved with suitable material data preparation and a machine learning categorization model.

11.7.2.2 X-ray for Encapsulant Characterization

X-rays can also be utilized to examine the material difference for hardware assurance. Typically, X-ray imaging is employed to determine the sample's internal structure [59]. Due to the variable X-ray absorption characteristics of various IC package materials, the detector will receive signals with differing energy levels. These changes in energy level are the source of contrast in X-ray 2D imaging.

This case study proves the X-ray capability for reviewing the material difference between authentic and counterfeit ICs. SKYSCAN Multiscale X-Ray Nano-CT System is utilized here to collect the 2D X-ray imaging. At the same voltage, current, and detecting distance, two legitimate and counterfeit sample readings are done. In the preceding section, the IR-based material measurement demonstrates that the material contained within these two samples is distinct. The figure's X-ray imaging in two dimensions reveals the same result Fig. 11.8d, e. The disparity in brightness between these two images may be attributable to the variation in encapsulant material between the two samples. X-rays can detect material differences more intuitively than IR-based material characterization technologies, which requires further material spectra analysis. However, other characteristics, such as the encapsulant's thickness and silica filler, might also alter the measured signal. To further describe the capabilities of X-rays for encapsulant material characterization, more tests must be conducted.

11.7.2.3 THz-TDS for Encapsulant Characterization

THz signals are positioned between microwaves and lasers on the electromagnetic spectrum, which is able to detect both material and structural information. Since the non-metallic packing material is transparent to the THz signal, non-destructive depth detection with THz-TDS is achievable. Due to this distinctive feature of THz imaging, THz-TDS has been developed as a tool for detecting encapsulant defects [60]. THz-TDS has also been applied to semiconductor device hardware assurance [61, 62].

In this case study, the THz-TDS reflection spectrum is utilized to evaluate the encapsulant substance of real and counterfeit samples. As illustrated in Fig. 11.8f, g, it is demonstrated that the THz-TDS signal strength distinguishes between authentic and counterfeit samples. The discrepancy between these two spectra is presumably due to a variation in substance. When raster scanning the entire sample with THz pulses, the obtained THZ-TDS can be used to determine the internal volumetric structure of the device. MAYBE ADD MORE.

11.7.3 *Other Physical Methods for Encapsulant*

In this section, three non-destructive physical inspection methods have been evaluated as material characterization techniques. X-rays are typically employed for structural investigation and THz-TDS for the detection of encapsulant defects. The material characterization can prevent unauthorized change at the material level and can also be used a key reference for hardware assurance. Each technique discovered the material distinction between the original and counterfeit samples.

In order to analyze packaging material for hardware assurance, additional non-destructive material characterization methods might be utilized. Several vibrational spectroscopic modalities and chemical imaging techniques have been developed to assess polymer materials non-destructively [57].

11.8 Post-packaging Assurance: Destructive Physical Inspection

Reverse engineering the entire IC design by the use of multiple imaging methods and sample preparation is the most dependable and comprehensive countermeasure for ensuring complete hardware integrity [63]. Reverse engineering (RE) can help detect any irregularity at the device or transistor level. Reconstructing a netlist requires uniformly deleting IC layers and obtaining photographs of each layer at regular intervals. This requires a milling machine with an accuracy of up to 2 micrometers to conduct the material removal. Other device layers, such as passivation, metal layers, vias, polysilicon, and active areas, can be removed by CNC polishing after each layer has been removed. This sample preparation procedure is very time-consuming due to the thickness of the silicon substrate.

Trojan Scanner (TS) is a method for detecting hardware Trojans utilizing SEM images of logic cells from backside-thinned silicon die. To detect Trojans, these images of logic cells can be compared to the known-good GDSII design. By removing the epoxy encapsulant from the silicon die, one can prepare a sample for TS. A bare die or flip-chip can be immediately thinned with mechanical milling to a remaining silicon thickness (RST) of 10 micrometer. The silicon dies can then be polished to a mirror-like 1 μm RST finish. After polishing, silicon can be further removed using a plasma FIB, and then high-resolution pictures can be captured using a SEM. TS requires an even polishing surface with a gradient of 100 nm or less. Evenly polished surfaces guarantee that SEM captures only logic cells and does not overlapping pictures from underlying device layers such as polysilicon, vias, and metal. Images that overlap can result in analysis errors and false positives.

11.9 Pre-packaging Assurance: MEMS and NEMS

The most of chips used in current electronic devices are made and packed at OSAT facilities, which are sometimes unstable and susceptible to assault. Due to the complexity of the global supply chain, individual trustworthy foundries or IP owners cannot oversee the fabrication and packaging process in its entirety. After going through multiple phases of the supply chain, the original IO design of the interposer created by IP owners for the production and packaging of the chiplets is still identifiable. It is consequently susceptible to rogue entities, as the design may be

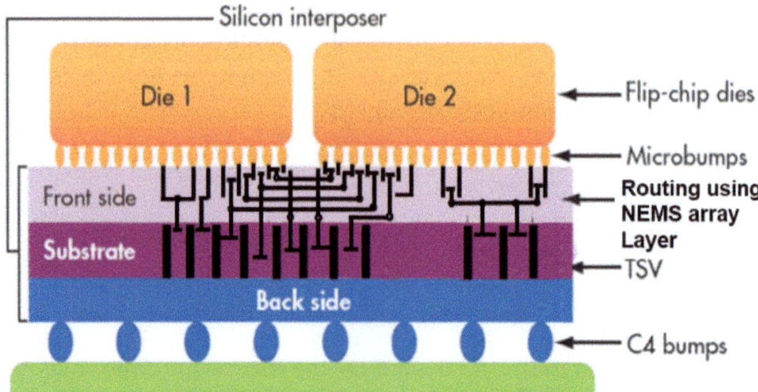

Fig. 11.9 Reconfigurable IC packaging routed with NEMS array

viewed by any entity in the supply chain using straightforward physical inspection techniques. A variety of attacks, including IP piracy, overproduction, reverse engineering, and counterfeiting, can be carried out by adversaries [64]. These adversaries may be end-users, reverse engineering organizations, systems-on-chip (SoCs) integrators, untrusted foundries, or interposer foundries. Numerous secure defensive measures have been developed in response to these threats, and they are usually regarded as effective. These countermeasures consist of camouflaging and logic locking [65–69]. However, the security of these sequential obfuscation techniques is uncertain because of the omission of genuine reverse engineering skills and the possibility of modification from security evaluations.

Combining current active and passive protection into a low-cost MEMS device, as depicted in Fig. 11.9, would enable system designers to select which IC connections should be protected. Due to the fact that the MEMS device conceals connections to several integrated circuits, it will be more challenging to develop a non-destructive X-ray system. If passive X-ray absorption materials are used in the packaging and active internal circuitry restructuring devices are implemented, the design of subsequent generations may become even more enigmatic.

11.10 Discussion and Conclusion

In this chapter, typical IC supply chain vulnerabilities are reevaluated for current HI packaging, and new IC hardware security issues are discussed. Due to globalization, IC design, implementation, and production include a multitude of unreliable suppliers and stakeholders. Consequently, a designer's intellectual property is visible to various parties, which further complicates the hardware verification process. Trojans, counterfeit ICs, and material alterations are introduced as three different forms of potential attacks on ICs (particularly in HI packaging) as a result of this discussion. On the basis of the present supply chain model and encapsulant characterization tools, adversaries can alter encapsulant formulations and compromise hardware security. Unlike other hardware security vulnerabilities, these attacks can be carried out without a comprehensive understanding of the package and are more difficult to identify using current reliability validation tools. In addition, the security analysis presented in this work suggests more dependable and exhaustive encapsulant verification procedures.

Traditional physical inspection techniques for material and volumetric characterization are described, along with their limitations in combating emergent threats. As a countermeasure for this type of hardware security issue, non-destructive material characterization techniques will mostly be employed. To decrease the effectiveness of untrusted entities, the amount of material or encapsulant parameters that can be altered will be evaluated. Focusing just on high-risk parameters during security verification can greatly enhance inspection efficiency and precision. DRIFT, X-ray, and THz-TDS have all successfully recognized the material variations between legitimate and counterfeit samples in this research. Since counterfeit detection is an important challenge in hardware assurance, these techniques can also be used for this purpose [40]. However, additional research is required to comprehend the relationship between encapsulant modification and the resulting security or dependability concerns. Future examination of additional samples will contribute to the development of a larger database of encapsulant properties, which is essential for this in-depth investigation. This data will also be collected and labeled in the event of dependability failures, allowing researchers to link the material difference to security concerns.

Despite the fact that numerous research have provided threat mitigation measures, the majority of assurance procedures are questionable due to the absence of comprehensive coverage. Material characterization and security evaluation are still disjointed in considerable measure. This study investigated the security and constraints of a variety of cutting-edge assurance methodologies for HI, focusing on the use of MEMS/NEMS for customization, functional obfuscation, and reconfigurability prior to packaging. With the provided information on the sample material, the suggested non-destructive processes can also aid in creating realistic and accurate simulations of advanced packaging. As packaging technology has evolved in accordance with Moore's Law, the complexity of packaging construction has increased. To optimize the design and provide a more sustainable life cycle, it is necessary to simulate the sample over its lifetime with a more precise model.

References

1. M.S.M. Khan, C. Xi, A.A. Khan, M.T. Rahman, M.M. Tehranipoor, N. Asadizanjani, Secure interposer-based heterogeneous integration. IEEE Des. Test **39**(6), 156–164 (2022)
2. P.E. Calzada, Md.S. Ul Islam Sami, K.Z. Azar, F. Rahman, F. Farahmandi, M. Tehranipoor, Heterogeneous integration supply chain integrity through blockchain and CHSM. ACM Trans. Des. Autom. Electron. Syst. **29**(1), 1–25 (2023)

3. T. Zhang, Md.L. Rahman, H.M. Kamali, K.Z. Azar, F. Farahmandi, SiPGuard: run-time System-in-Package security monitoring via power noise variation. IEEE Trans. Very Large Scale Integr. VLSI Syst. **32**(2), 305–318 (2023)
4. T. Zhang, Md.L. Rahman, H.M. Kamali, K.Z. Azar, M. Tehranipoor, F. Farahmandi, FISHI: fault injection detection in secure heterogeneous integration via power noise variation, in *2023 IEEE 73rd Electronic Components and Technology Conference (ECTC)* (IEEE, Piscataway, 2023), pp. 2188–2195
5. Md.S. Ul Islam Sami, H.M. Kamali, F. Farahmandi, F. Rahman, M. Tehranipoor, Enabling security of heterogeneous integration: from supply chain to in-field operations. IEEE Des. Test. **40**(5), 86–95 (2023)
6. N. Vashistha, H. Lu, Q. Shi, M.T. Rahman, H. Shen, D.L. Woodard, N. Asadizanjani, M. Tehranipoor, Trojan scanner: detecting hardware trojans with rapid SEM imaging combined with image processing and machine learning, in *ISTFA 2018: Proceedings from the 44th International Symposium for Testing and Failure Analysis* (ASM International, Geauga County, 2018), p. 256
7. M.S.M. Khan, C. Xi, Md.S. Ul Haque, M.M. Tehranipoor, N. Asadizanjani, Exploring advanced packaging technologies for reverse engineering a System-in-Package (SiP). IEEE Trans. Compon. Packag. Manuf. Technol. **13**(9), 1360–1370 (2023)
8. C. Xi, A.A. Khan, N. Jessurun, N. Vashisthan, M.M. Tehranipoor, N. Asadizanjani, Physical assurance for heterogeneous integration: challenges and opportunities, in *2022 IEEE International Symposium on the Physical and Failure Analysis of Integrated Circuits (IPFA)* (IEEE, Piscataway, 2022), pp. 1–6
9. G. Baldini, G. Steri, F. Dimc, R. Giuliani, R. Kamnik, Experimental identification of smartphones using fingerprints of built-in micro-electro mechanical systems (MEMS). Sensors **16**(6), 818 (2016)
10. M. Martin, J. Plusquellic, NotchPUF: printed circuit board PUF based on microstrip notch filter. Cryptography **4**(2), 11 (2020)
11. T. McGrath, I.E. Bagci, Z.M. Wang, U. Roedig, R.J. Young, A PUF taxonomy. Appl. Phys. Rev. **6**(1), 011303 (2019)
12. A. Chen, R.H.-Y. Lo, *Semiconductor Packaging: Materials Interaction And reliability* (Taylor & Francis, Milton Park, 2012)
13. Y. Nakamura, S. Katogi, Technology trends and future history of semiconductor packaging substrate material. Hitachi Chem. Rev. **6**(1), 1–6 (2013)
14. K.L. Tai, System-in-package (SiP) challenges and opportunities, in *Proceedings of the 2000 Asia and South Pacific Design Automation Conference* (2000), pp. 191–196
15. R. Wang, K. Chakrabarty, Testing of interposer-based 2.5 D integrated circuits, in *2016 IEEE International Test Conference (ITC)* (IEEE, Piscataway, 2016), pp. 1–10
16. R. Tummala, N. Nedumthakady, S. Ravichandran, B. DeProspo, V. Sundaram, Heterogeneous and homogeneous package integration technologies at device and system levels, in *2018 Pan Pacific Microelectronics Symposium (Pan Pacific)* (IEEE, Piscataway, 2018), pp. 1–5
17. X. Fan, Wafer level packaging (WLP): fan-in, fan-out and three-dimensional integration, in *2010 11th International Thermal, Mechanical & Multi-Physics Simulation, and Experiments in Microelectronics and Microsystems (EuroSimE)* (IEEE, Piscataway, 2010), pp. 1–7
18. X. Zhang, J.K. Lin, S. Wickramanayaka, S. Zhang, R. Weerasekera, R. Dutta, K.F. Chang, K.-J. Chui, H.Y. Li, D.S.W. Ho, et al., Heterogeneous 2.5 D integration on through silicon interposer. Appl. Phys. Rev. **2**(2), 021308 (2015)
19. V.S. Rao, H.S. Wee, L.W.S. Vincent, L.H. Yu, L. Ebin, R. Nagarajan, C.T. Chong, X. Zhang, P. Damaruganath, TSV interposer fabrication for 3D IC packaging, in *2009 11th Electronics Packaging Technology Conference* (IEEE, Piscataway, 2009), pp. 431–437
20. J. Keech, S. Chaparala, A. Shorey, G. Piech, S. Pollard, Fabrication of 3D-IC interposers, in *2013 IEEE 63rd Electronic Components and Technology Conference* (IEEE, Piscataway, 2013), pp. 1829–1833

21. C. Xi, N. Jessurun, N. Asadizanjani, A framework to assess the security of advanced integrated circuit (IC) packaging, in *2020 IEEE 8th Electronics System-Integration Technology Conference (ESTC)* (IEEE, Piscataway, 2020), pp. 1–7

22. A. Usman, E. Shah, N.B. Satishprasad, J. Chen, S.A. Bohlemann, S.H. Shami, A.A. Eftekhar, A. Adibi, Interposer technologies for high-performance applications. IEEE Trans. Compon. Packag. Manuf. Technol. **7**(6), 819–828 (2017)

23. Z. Wu, C. Dou, Y. Wu, H. Yang, Design, fabrication of through silicon vias based on suspended photoresist thin film and its application in silicon interposer, in *2011 12th International Conference on Electronic Packaging Technology and High Density Packaging* (IEEE, Piscataway, 2011), pp. 1–4

24. H. Ardebili, J. Zhang, M.G. Pecht, *Encapsulation Technologies for Electronic Applications* (William Andrew, Norwich, 2018)

25. Y. He, B.E. Moreira, A. Overson, S.H. Nakamura, C. Bider, J.F. Briscoe, Thermal characterization of an epoxy-based underfill material for flip chip packaging. Thermochim. Acta **357**, 1–8 (2000)

26. Y. Yao, G.-Q. Lu, D. Boroyevich, K.D.T. Ngo, Survey of high-temperature polymeric encapsulants for power electronics packaging. IEEE Trans. Compon. Packag. Manuf. Technol. **5**(2), 168–181 (2015)

27. L. Lantz, S. Hwang, M. Pecht, Characterization of plastic encapsulant materials as a baseline for quality assessment and reliability testing. Microelectron. Reliab. **42**(8), 1163–1170 (2002)

28. Y. Yu, H. Su, W. Gan, Effects of storage aging on the properties of epoxy prepregs. Ind. Eng. Chem. Res. **48**(9), 4340–4345 (2009)

29. T. Braun, S. Raatz, S. Voges, R. Kahle, V. Bader, J. Bauer, K.-F. Becker, T. Thomas, R. Aschenbrenner, K.-D. Lang, Large area compression molding for fan-out panel level packing, in *2015 IEEE 65th Electronic Components and Technology Conference (ECTC)* (IEEE, Piscataway, 2015), pp. 1077–1083

30. S.C. Yang, S.-Y. Kwak, J.H. Jin, J.-S. Kim, Y. Choi, K.-W. Paik, B.-S. Bae, Thermally resistant UV-curable epoxy–siloxane hybrid materials for light emitting diode (LED) encapsulation. J. Mater. Chem. **22**(18), 8874–8880 (2012)

31. K.Z. Azar, H.M. Kamali, H. Homayoun, A. Sasan, Threats on logic locking: a decade later, in *Proceedings of the 2019 on Great Lakes Symposium on VLSI* (2019), pp. 471–476

32. R.S. Chakraborty, S. Narasimhan, S. Bhunia, Hardware Trojan: threats and emerging solutions, in *2009 IEEE International High Level Design Validation and Test Workshop* (IEEE, Piscataway, 2009), pp. 166–171

33. H.M. Kamali, K.Z. Azar, F. Farahmandi, M. Tehranipoor, Advances in logic locking: past, present, and prospects. Cryptology ePrint Archive **2022**(260), 1–39 (2022)

34. J. Rajendran, O. Sinanoglu, R. Karri, Is split manufacturing secure? in *Design, Automation & Test in Europe Conference & Exhibition (DATE)* (2013), pp. 1259–1264

35. H.M. Kamali, K.Z. Azar, H. Homayoun, A. Sasan, InterLock: an intercorrelated logic and routing locking, in *IEEE/ACM International Conference on Computer-Aided Design (ICCAD)* (2020), pp. 1–9

36. K.Z. Azar, M.M. Hossain, A. Vafaei, H. Al Shaikh, N.N. Mondol, F. Rahman, M. Tehranipoor, F. Farahmandi, Fuzz, penetration, and AI testing for SoC security verification: challenges and solutions. Cryptology ePrint Archive **2022**(394), 1–22 (2022)

37. S.T. King, J. Tucek, A. Cozzie, C. Grier, W. Jiang, Y. Zhou, Designing and implementing malicious hardware. Leet **8**, 1–8 (2008)

38. M. Tehranipoor, F. Koushanfar, A survey of hardware trojan taxonomy and detection. IEEE Des. Test Comput. **27**(1), 10–25 (2010)

39. N.N. Mondol, A. Vafaei, K.Z. Azar, F. Farahmandi, M. Tehranipoor, RL-TPG: automated pre-silicon security verification through reinforcement learning-based test pattern generation, in *Design, Automation and Test in Europe (DATE)* (IEEE, Piscataway, 2024), pp. 1–6

40. U. Guin, K. Huang, D. DiMase, J.M. Carulli, M. Tehranipoor, Y. Makris, Counterfeit integrated circuits: a rising threat in the global semiconductor supply chain. Proc. IEEE **102**(8), 1207–1228 (2014)

41. B. Banijamali, S. Ramalingam, H. Liu, M. Kim, Outstanding and innovative reliability study of 3D TSV interposer and fine pitch solder micro-bumps, in *2012 IEEE 62nd Electronic Components and Technology Conference* (IEEE, Piscataway, 2012), pp. 309–314
42. J.H. Lau, M. Li, D. Tian, N. Fan, E. Kuah, W. Kai, M. Li, J. Hao, Y.M. Cheung, Z. Li, et al., Warpage and thermal characterization of fan-out wafer-level packaging. IEEE Trans. Compon. Packag. Manuf. Technol. **7**(10), 1729–1738 (2017)
43. S.-Y. Fu, X.-Q. Feng, B. Lauke, Y.-W. Mai, Effects of particle size, particle/matrix interface adhesion and particle loading on mechanical properties of particulate–polymer composites. Composites, Part B **39**(6), 933–961 (2008)
44. M. Mengel, J. Mahler, W. Schober, Effect of post-mold curing on package reliability. J. Reinf. Plast. Compos. **23**(16), 1755–1765 (2004)
45. Y. Yao, Z. Chen, G.-Q. Lu, D. Boroyevich, K.D.T. Ngo, Characterization of encapsulants for high-voltage high-temperature power electronic packaging. IEEE Trans. Compon. Packag. Manuf. Technol. **2**(4), 539–547 (2012)
46. P. Ghosh, R.S. Chakraborty, Recycled and remarked counterfeit integrated circuit detection by image-processing-based package texture and indent analysis. IEEE Trans. Ind. Inf. **15**(4), 1966–1974 (2018)
47. D. Johnson, P.-W. Hsu, C. Xi, N. Asadizanjani, Scanning acoustic microscopy package finger-print extraction for integrated circuit hardware assurance, in *ISTFA 2021* (ASM International, Geauga County, 2021), pp. 59–64
48. G.-M. Zhang, D.M. Harvey, D.R. Braden, Microelectronic package characterisation using scanning acoustic microscopy. NDT E Int. **40**(8), 609–617 (2007)
49. A. Roberts, J. True, N.T. Jessurun, N. Asadizanjani, et al., An overview of 3D X-ray reconstruction algorithms for PCB inspection, in *ISTFA 2020* (2020), pp. 188–197
50. E.L. Principe, N. Asadizanjani, D. Forte, M. Tehranipoor, R. Chivas, M. DiBattista, S. Silverman, M. Marsh, N. Piche, J. Mastovich, Steps toward automated deprocessing of integrated circuits, in *ISTFA 2017: Proceedings from the 43rd International Symposium for Testing and Failure Analysis* (ASM International, Geauga County, 2017), p. 285
51. A. Ditali, M.K. Ma, M. Johnston, X-ray radiation effect in DRAM retention time. IEEE Trans. Device Mater. Reliab. **7**(1), 105–111 (2007)
52. C.Y. Khor, M.Z. Abdullah, C.-S. Lau, I.A. Azid, Recent fluid–structure interaction modeling challenges in IC encapsulation–a review. Microelectron. Reliab. **54**(8), 1511–1526 (2014)
53. Y. He, Z. Alam, Moisture absorption and diffusion in an underfill encapsulant at $T > T$ g and $T < T$ g. J. Therm. Anal. Calorim. **113**, 461–466 (2013)
54. Y.-S. Ng, W.-K. Hooi, FTIR method to distinguish mold compound types and identify the presence of post-mold-cure (PMC) treatment on electronic packaging, in *2007 9th Electronics Packaging Technology Conference* (IEEE, Piscataway, 2007), pp. 652–656
55. I.D. Wolf, C. Jian, W.M. van Spengen, The investigation of microsystems using Raman spectroscopy. Opt. Lasers Eng. **36**(2), 213–223 (2001)
56. T. Moore, *Characterization of Integrated Circuit Packaging Materials* (Elsevier, Amsterdam, 2013)
57. S. Mukherjee, A. Gowen, A review of recent trends in polymer characterization using non-destructive vibrational spectroscopic modalities and chemical imaging. Anal. Chim. Acta **895**, 12–34 (2015)
58. F. Mustata, N. Tudorachi, D. Rosu, Curing and thermal behavior of resin matrix for composites based on epoxidized soybean oil/diglycidyl ether of bisphenol A. Composites, Part B **42**(7), 1803–1812 (2011)
59. J. True, N. Jessurun, D. Mehta, N. Asadi, QUAINTPEAX QUantifying algorithmically INTrinsic properties of electronic assemblies via X-ray CT. Microsc. Microanal. **27**(S1), 1222–1225 (2021)
60. K. Ahi, S. Shahbazmohamadi, N. Asadizanjani, Quality control and authentication of packaged integrated circuits using enhanced-spatial-resolution terahertz time-domain spectroscopy and imaging. Opt. Lasers Eng. **104**, 274–284 (2018)

61. K. Ahi, M. Anwar, Advanced terahertz techniques for quality control and counterfeit detection, in *Terahertz Physics, Devices, and Systems X: Advanced Applications in Industry and Defense*, vol. 9856 (SPIE, Bellingham, 2016), pp. 31–44

62. J. True, C. Xi, N. Jessurun, K. Ahi, N. Asadizanjani, Review of THz-based semiconductor assurance. Opt. Eng. **60**(6), 060901–060901 (2021)

63. R. Wilson, H. Lu, M. Zhu, D. Forte, D.L. Woodard, Refics: a step towards linking vision with hardware assurance, in *Proceedings of the IEEE/CVF Winter Conference on Applications of Computer Vision* (2022), pp. 4031–4040

64. M. Nabeel, M. Ashraf, S. Patnaik, V. Soteriou, O. Sinanoglu, J. Knechtel, An interposer-based root of trust: seize the opportunity for secure system-level integration of untrusted chiplets. Preprint. arXiv:1906.02044 (2019)

65. H.M. Kamali, K.Z. Azar, H. Homayoun, A. Sasan, Full-lock: hard distributions of SAT instances for obfuscating circuits using fully configurable logic and routing blocks, in *Proceedings of Design Automation Conference (DAC)* (2019), p. 89

66. H.M. Kamali, K.Z. Azar, K. Gaj, H. Homayoun, A. Sasan, LUT-lock: a novel LUT-based logic obfuscation for FPGA-bitstream and ASIC-hardware protection, in *IEEE Computer Society Annual Symposium on VLSI (ISVLSI)* (2018), pp. 405–410

67. M.S. Rahman, R. Guo, H.M. Kamali, F. Rahman, F. Farahmandi, M. Tehranipoor, ReTrustFSM: toward RTL hardware obfuscation-a hybrid FSM approach. IEEE Access **11**, 19741–19761 (2023)

68. M.S. Rahman, R. Guo, H.M. Kamali, F. Rahman, F. Farahmandi, M. Abdel-Moneum, M. Tehranipoor, O'Clock: lock the clock via clock-gating for SoC IP protection, in *Design Automation Conference (DAC)* (2022), pp. 1–6

69. H.M. Kamali, K.Z. Azar, F. Farahmandi, M. Tehranipoor, SheLL: shrinking eFPGA fabrics for logic locking, in *2023 Design, Automation & Test in Europe Conference & Exhibition (DATE)* (IEEE, Piscataway, 2023), pp. 1–6

Index

© The Editor(s) (if applicable) and The Author(s), under exclusive license to
Springer Nature Switzerland AG 2024
M. Tehranipoor et al., *Hardware Security*,
https://doi.org/10.1007/978-3-031-58687-3

MIX
Papier aus verantwortungsvollen Quellen
Paper from responsible sources
FSC® C105338

If you have any concerns about our products,
you can contact us on
ProductSafety@springernature.com

In case Publisher is established outside the EU,
the EU authorized representative is:
Springer Nature Customer Service Center GmbH
Europaplatz 3, 69115 Heidelberg, Germany

Printed by Libri Plureos GmbH
in Hamburg, Germany